Reagents for Organic Synthesis

Fieser and Fieser's

Reagents for Organic Synthesis

VOLUME EIGHT

Mary Fieser

A WILEY-INTERSCIENCE PUBLICATION
JOHN WILEY & SONS
NEW YORK • CHICHESTER • BRISBANE • TORONTO

Library of Congress Cataloging in Publication Data:

Library of Congress Catalog Card Number: 66-27894

ISBN 0-471-04834-8

Printed in the United States of America

10 9 8 7 6 5 4 3 2 1

PREFACE

This volume of reagents includes references to papers published in 1977 and the first half of 1978. I am very grateful for the continuing advice and help of my colleagues. Professor Daniel Sternbach and Gary Amstutz have read portions of the manuscript. Professor John Cooper has given advice on nomenclature of organometallic compounds, particularly of metal carbonyls. The following chemists have been especially helpful in the proofreading—Professor William Roush, Professor Rick Danheiser, Dr. Mark A. Wuonola, Dr. William Moberg, Dr. James V. Heck, Professor Bruce Lipshutz, Dr. Ving Lee, Dr. Mary Lee, Dr. John Secrist, Professor Dale L. Boger, Stephen Kamin, Anthony Feliu, Jeffrey C. Hayes, Don Landry, Marcus A. Tius, John Maher, Jay W. Ponder, Paul Hopkins, John Munroe, Donald Wolanin, Charles Manley, Howard Simmons, III, Stuart Schreiber, Craig Shaefer, Stephen Hall, Stephen Pesceckis, John Morin, David Carini, Howard Sard, Stephen Brenner, Katharine Brightly, Chad G. Miller, Robert Wolf, Stephen Freilich, William Roberts, Jeffrey Howbert, and William McWhorter.

The picture of the Reagents group was taken by Professor Janice Smith.

MARY FIESER

Cambridge, Massachusetts
March 1980

CONTENTS

Reagents for Organic Synthesis

A

Acetic acid, **2**, 5–7; **5**, 3; **7**, 1.

2-Alkyl-5-hydroxycyclopentene-2-ones (**3**). Epoxides (**1**) of 2-alkylcyclopen-tene-2-ones when heated in acetic acid rearrange to acetoxyenones (**2**), which can be hydrolyzed by sodium carbonate at room temperature to the hydroxy-enones (**3**). These can be rearranged by strong base or acid to **4**.[1]

Cyclization. Treatment of the urea **1** with acetic acid at 50° results in cycliza-tion to **2**. Treatment with a stronger acid, such as trifluoroacetic acid, results in cyclization to a mixture of **3** and **2**. Since **2** and **3** are not interconvertible two different cyclizations are possible.[2] The ring system of **2** is present in the neuro-toxins related to saxitoxin, which are present in certain clams and mussels.

[1] M. P. L. Caton, G. Darnbrough, and T. Parker, *Syn. Comm.*, **8**, 155 (1978).
[2] H. Taguchi, H. Yazawa, J. F. Arnett, and Y. Kishi, *Tetrahedron Letters*, 627 (1977).

Acetic anhydride, **1**, 3; **2**, 7–10; **5**, 3–4; **6**, 1–2; **7**, 1.

Enol lactonization. Acetic anhydride[1] is superior to acetyl chloride[2] for enol lactonization of **1** to give **2**. Prolonged treatment, particularly in the presence of sodium acetate, leads to **3**. The enol lactone **2** is a useful precursor to oxygenated γ-butyrolactones.

[1] T. A. Eggelte, J. J. J. de Boer, H. de Koning, and H. O. Huisman, *Syn. Comm.*, **8**, 353 (1978).
[2] K. W. Rosenmund, H. Herzberg, and H. Schütt, *Ber.*, **87**, 1258 (1954).

Acetic anhydride–Acetic acid, 6, 1.

Pummerer reaction. The reaction of the 3-*exo*-methylenecepham sulfoxide **1** with Ac_2O–HOAc (2:1) at reflux gives a mixture of **2** and **3**, probably by way of

a 1,4-Pummerer type intermediate (**a**). Treatment of the mixture with *m*-chloroperbenzoic acid at 0° gives the Δ^3-3′-acetoxycephem sulfoxide **4** in 84% overall yield.[1]

[1] G. A. Koppel and L. J. McShane, *Am. Soc.*, **100**, 288 (1978).

2-Acetoxyisobutyryl chloride, $(CH_3)_2\overset{|}{\underset{OAc}{C}}COCl$ **(1)**. Mol. wt. 164.59, b.p. 55–56°/
6 mm. The material can be prepared on a kilogram scale by acetylation of 2-hy-droxyisobutyric acid (AcCl, 2 hours reflux) followed by reaction with thionyl chloride (100°, 2 hours); yield 84.6%.[1]

Reactions with diols. A few years ago Mattocks[2] reported a curious reaction of 2-acetoxy-2-methylbutyryl chloride (**2**) with the 1,4-diol **3** to give **4** and **5**. The same paper reported other abnormal reactions of **2** with 1,2- and 1,3-diols.

3	**2**	**4 (54%)**	**5 (20%)**

Greenberg and Moffatt[1] have confirmed and extended these results, but they used α-acetoxyisobutyryl chloride, which is not chiral, to simplify the reactions. These chemists first examined the reaction of **1** with a primary alcohol, *p*-nitro-benzyl alcohol (**6**). The reaction in acetonitrile and with added triethylamine (3 equiv.) gives the 1,3-dioxolane-4-one **7** in 84% yield. The reaction in pure triethylamine gives the crystalline ester **8** in 47% yield.

The reagent (**1**) reacts with *trans*-cycloalkane-1,2-diols to give a multitude of products. Reaction with *cis*-cycloalkane-1,2-diols gives *trans*-2-chlorocycloalkyl acetates in high yield. An example is formulated in equation (I). This reaction involves a cyclic acetoxonium ion, which can form only from a *cis*-1,2-diol.

Greenberg and Moffatt also examined the reaction of **1** with ribonucleosides, in particular, a protected form of uridine (**9**). In this case, chlorine is introduced

exclusively into the 2′-position with overall retention of configuration, presumably because of participation of the carbonyl group in the base. The major product (71% yield) of reaction of **1** with uridine itself (no solvent) is **11**. This reaction, however, is solvent dependent.

The reaction of **1** with pseudouridine (**12**), a C-nucleoside, differs from the reaction with uridine in that a mixture of 2′- and 3′-chlorodeoxynucleosides are formed. Thus the reaction of **1** with **12**, either neat or in acetonitrile, followed by hydrogenolysis with tri-*n*-butyltin hydride and deblocking gives about equal amounts of 2′- and 3′-deoxypseudouridine, (**13** and **14**).[3]

[1] S. Greenberg and J. G. Moffatt, *Am. Soc.*, **95**, 4016 (1973).
[2] A. R. Mattocks, *J. Chem. Soc.*, 1918, 4840 (1964).
[3] M. J. Robins and W. H. Muhs, *J.C.S. Chem. Comm.*, 677 (1978).

Acetoxymethyl methyl selenide, $CH_3SeCH_2OCOCH_3$. Mol. wt. 151.06, b.p. 51–52°/10 mm. The selenide is prepared by the Pummerer reaction of dimethyl selenoxide (**6**, 224) with acetic acid (29.2% yield).

Oxyselenation. In the presence of hydrogen peroxide the selenide reacts with alkenes to form acetoxyselenated products with loss of formaldehyde. The reaction apparently involves formation of the selenoxide, and indeed no reaction occurs in the absence of H_2O_2. Presumably the selenoxide then loses formaldehyde to form methaneselenenyl acetate, CH_3SeOAc, which is the actual reagent.[1]

Examples:

$$C_4H_9CH{=}CH_2 \xrightarrow[60\%]{} \underset{\overset{|}{OAc}}{C_4H_9CH}{-}CH_2SeCH_3 + \underset{\overset{|}{SeCH_3}}{C_4H_9CH}{-}CH_2OAc$$

$$92:8$$

$$C_6H_5CH{=}CH_2 \xrightarrow[74\%]{} \underset{\overset{|}{OAc}}{C_6H_5CH}{-}CH_2SeCH_3$$

[1] N. Miyoshi, S. Murai, and N. Sonoda, *Tetrahedron Letters*, 851 (1977).

Acetyl hexachloroantimonate, $CH_3CO^+SbCl_6^-$. Mol. wt. 374.51, colorless crystals sensitive to moisture. The salt is prepared from acetyl chloride and antimony(V) chloride.[1]

Acylation of alkenes. In the presence of a nonnucleophilic base (*e.g.*, ethyldiisopropylamine), the reagent reacts with alkenes in CH_2Cl_2 at -50 to $-25°$ under kinetic control to form β,γ-unsaturated ketones in moderate to high yield.

Examples:

$$(CH_3)_2C{=}CH_2 \xrightarrow[>90\%]{} CH_3COCH_2C{\overset{\displaystyle CH_2}{\underset{\displaystyle CH_3}{\diagdown}}}$$

This acylation is considered to be an ene reaction, in which the acetyl cation abstracts an allylic hydrogen from the alkene in concerted fashion (equation I).[2]

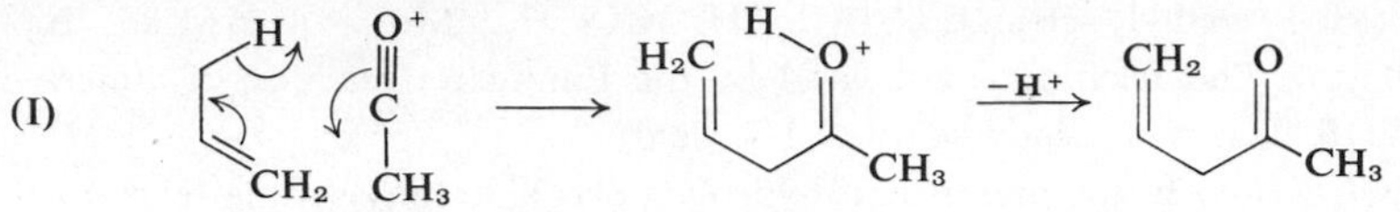

(I)

[1] G. A. Olah, S. J. Kuhn, W. S. Tolgyesi, and E. B. Baker, *Am. Soc.*, **84**, 2733 (1962).
[2] H. M. R. Hoffmann and T. Tsushima, *ibid.*, **99**, 6008 (1977).

B-1-Alkenyl-9-borabicyclo[3.3.1]nonanes (1). These substances are prepared by hydroboration of alkynes with 9-BBN (**2**, 31; **3**, 24–29; **4**, 41; **5**, 46–47; **6**, 62–64).[1]

Allylic alcohols.[2] Usually reaction of organoboranes with carbonyl groups results in reduction. However, B-1-alkenyl-9-borabicyclo[3.3.1]nonanes (**1**) add to simple aldehydes to give, after oxidation, allylic alcohols, with retention of configuration (equation I). The reaction is a Grignard-like synthesis and should be a useful route to allylic alcohols, even to those containing functional groups.

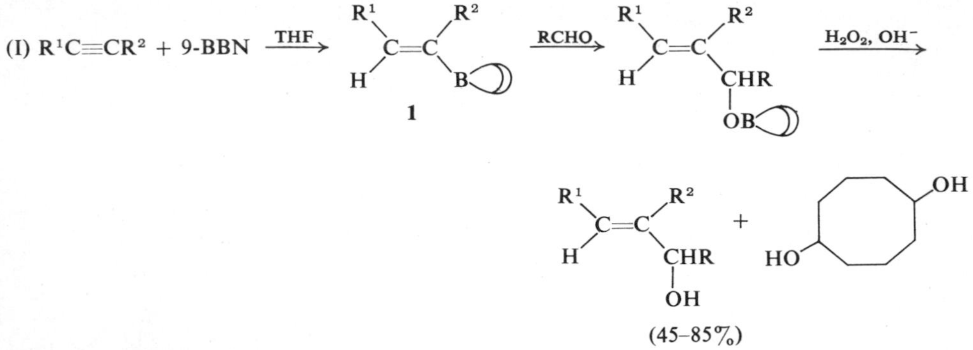

(I) $R^1C{\equiv}CR^2$ + 9-BBN $\xrightarrow{\text{THF}}$

(45–85%)

[1] H. C. Brown, E. F. Knights, and C. G. Scouten, *Am. Soc.*, **96**, 7765 (1974).
[2] P. Jacob III and H. C. Brown, *J. Org.*, **42**, 579 (1977).

B-1-Alkynyl-9-borabicyclo[3.3.1]nonanes (1).
   Preparation[1]:

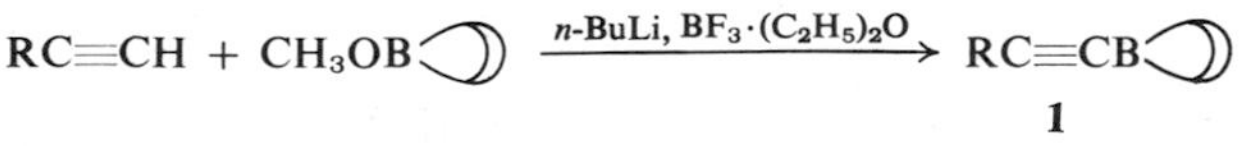

$$RC{\equiv}CH + CH_3OB \xrightarrow{\text{n-BuLi, BF}_3\cdot\text{(C}_2\text{H}_5\text{)}_2\text{O}} RC{\equiv}CB$$

1

Conjugate additions. The reagent undergoes conjugate addition to α,β-unsaturated ketones to form γ,δ-acetylenic ketones.[2]
   Examples:

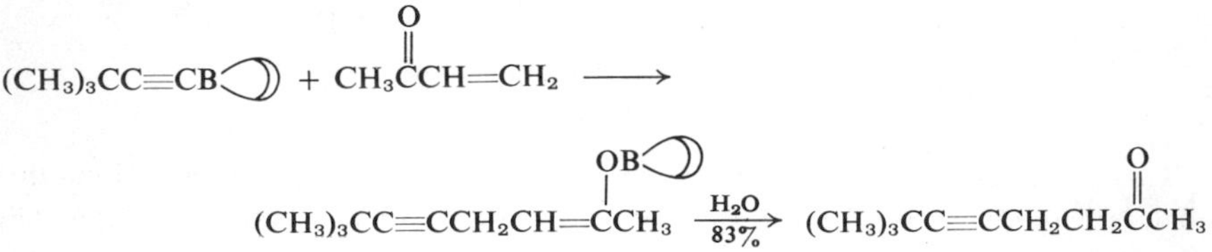

$$CH_3(CH_2)_3C\equiv CB\!\!\bigcirc\!\!\bigcirc + (CH_3)_2C\!=\!CHCOCH_3 \xrightarrow[70\%]{} CH_3(CH_2)_3C\equiv C\overset{\overset{\displaystyle CH_3}{|}}{\underset{\underset{\displaystyle CH_3}{|}}{C}}CH_2COCH_3$$

4-Methoxy-3-butene-2-one also undergoes rapid conjugate addition with B-1-alkynyl-9-borabicyclo[3.3.1]nonanes with elimination of B-methoxy-9-BBN to give a 4-*trans*-alkynyl-3-butene-2-one in almost quantitative yield.[3]

$$RC\equiv CB\!\!\bigcirc\!\!\bigcirc \underset{THF}{} + \underset{CH_3O}{\overset{H}{\diagdown}}\!\!C\!=\!C\!\!\underset{H}{\overset{\overset{\textstyle O}{\overset{\|}{C}}CH_3}{\diagup}} \xrightarrow[98\%]{} \underset{RC\equiv C}{\overset{H}{\diagdown}}\!\!C\!=\!C\!\!\underset{H}{\overset{\overset{\textstyle O}{\overset{\|}{C}}CH_3}{\diagup}} + \bigcirc\!\!BOCH_3$$

The reaction is general for other β-methoxy-α,β-unsaturated ketones that can assume the cisoid conformation and provides a useful synthesis of conjugated enynones.

[1] J. A. Sinclair and H. C. Brown, *J. Org.*, **41**, 1078 (1976).
[2] J. A. Sinclair, G. A. Molander, and H. C. Brown, *Am. Soc.*, **99**, 954 (1977).
[3] G. A. Molander and H. C. Brown, *J. Org.*, **42**, 3106 (1977).

Allylacetophenone, $C_6H_5COCH_2CH_2CH\!=\!CH_2$ **(1).** Mol. wt. 160.21, b.p. 125–127°/16 mm. This compound can be prepared from acetophenone and allyl bromide in the presence of sodium *t*-amyloxide.[1]

The dianion **(b)** of this unsaturated ketone **(1)** can be obtained by deprotonation first with KH in THF at 0° and then with *sec*-butyllithium (TMEDA). The dianion reacts with electrophiles exclusively at the terminal position; in

$$1 \xrightarrow{KH,\ THF} \underset{\textbf{a}}{C_6H_5\overset{\overset{\textstyle O^-K^+}{|}}{C}\!=\!CHCH_2CH\!=\!CH_2} \xrightarrow{sec\text{-BuLi, TMEDA}} \underset{\textbf{b}}{C_6H_5\overset{\overset{\textstyle O}{\|}}{C}\!\cdots\!\overset{-}{C}H\!\cdots\!CH\!\cdots\!\overset{=}{C}H\!\cdots\!CH_2}\ \text{Li}^+K^+$$

$$\textbf{b} + CH_3I \xrightarrow[60\text{–}72\%]{} C_6H_5\!\!\overset{\textstyle O}{\diagup}\!\!\diagdown\!\!\diagup\!\!\diagdown\!\!CH_2CH_3$$

$$\textbf{b} + CH_3CH_2CHO \xrightarrow[20\text{–}35\%]{} C_6H_5\!\!\overset{\textstyle O}{\diagup}\!\!\diagdown\!\!\diagup\!\!\diagdown\!\!\underset{\underset{\textstyle OH}{|}}{CH_2CHCH_2CH_3}$$

$$\textbf{b} + C_6H_5CHO \xrightarrow[38\text{–}53\%]{} C_6H_5\!\!\overset{\textstyle O}{\diagup}\!\!\diagdown\!\!\diagup\!\!\diagdown\!\!\underset{\underset{\textstyle OH}{|}}{CH_2CHC_6H_5}$$

the reaction with alkyl halides, epoxides, and aliphatic aldehydes or ketones only the *cis*-compound is formed, whereas only *trans*-adducts are formed with aromatic carbonyl compounds.[2]

[1] G. Vavon and J. Conia, *Compt. rend.*, **223**, 245 (1946).
[2] M. Pohmakotr and D. Seebach, *Angew. Chem., Int. Ed.*, **16**, 320 (1977).

Allyl triflate, $CH_2=CHCH_2OSO_2CF_3$ **(1).** Mol. wt. 191.16, colorless liquid, must be stored at $-78°$ in a vented flask. This triflate can be prepared in pure form in 69% yield by the reaction of allyl alcohol with trifluoromethanesulfonic anhydride and pyridine in dichloromethane.[1]

Caution: Avoid contact with or inhalation of this volatile, very reactive alkylating reagent.

Ring expansion. Vedejs[2] and Schmid[3] have devised a repeatable ring expansion process involving successive [2,3]sigmatropic shifts as illustrated in scheme (I). An α-vinyl heterocycle such as 2-vinylthiacyclopentane (**2**) is alkylated with allyl triflate to give the salt **3**. On treatment with a base (DBU) in acetonitrile,

Scheme (I)

3 is converted into the endocyclic (**a**) and exocyclic (**b**) ylides in a 1:1 ratio. These ylides rearrange under the conditions of generation into **4** and **5**, respectively. The desired ring expanded cyclic sulfide **5** can be converted in the same way into cyclic sulfides with 11-, 14-, and 17-membered rings. In these later ring expansions the undesired shifts to products such as **4** are not observed.

The same sequence has been applied to 2-vinylthiane (**6**). In this case formation of the undesired ylide was minimized by use of LDA in place of DBU (scheme II).

Scheme (II)

Conformational factors involved in these [2,3] sigmatropic shifts have been discussed.[4]

[1] E. Vedejs, D. A. Engler, and M. J. Mullins, *J. Org.*, **42**, 3109 (1977).
[2] E. Vedejs and J. P. Hagen, *Am. Soc.*, **97**, 6878 (1975); E. Vedejs, M. J. Mullins, J. M. Renga, and C. P. Singer, *Tetrahedron Letters*, 519 (1978).
[3] R. Schmid and H. Schmid, *Helv.*, **60**, 1361 (1977).
[4] E. Vedejs, M. J. Arco, and J. M. Renga, *Tetrahedron Letters*, 523 (1978).

Alumina, 1, 19–20; **2**, 17; **3**, 6; **4**, 8; **6**, 16–17; **7**, 5–7.

Vinyl halides. These compounds can be prepared from vinylsilanes by halogenation followed by elimination of $ClSi(CH_3)_3$ with sodium methoxide or neutral alumina.

Examples:

The stereoselectivity is high with *trans*-vinylsilanes but somewhat lower with *cis*-vinylsilanes; the same selectivity is observed with both $NaOCH_3$ and Al_2O_3.[1]

2-Alkyl(aryl)-5-hydroxy-3-ketocyclopentenes.[2] These substances can be obtained in 85–95% yield by column chromatography of 4-alkyl(aryl)-5-hydroxy-3-ketocyclopentenes.[3]

Doped aluminas. Posner[4] has reviewed organic reactions on alumina surfaces (93 references). Although the exact role of alumina is not understood, alumina may activate both the substrate and the reagent by adsorption in the proper orientation for a chemical reaction. Some recent examples are discussed.

Complete details of nucleophilic opening of epoxides (**6**, 16–17) catalyzed by commercially available alumina (W-200-N) have been published.[5] Some generalizations can be drawn from the extensive investigations. The reactivity of symmetrical cycloalkene oxides decreases in the order of ring size: 6 > 8 > 12; and reactivity of the nucleophiles decreases in the order: ROH ∼ HOAc > RNH_2. One great advantage of this method is that usually only one product is formed.

This reaction has also been applied to arene oxides. These oxides can be opened by alumina doped with alcohols and amines, but this reaction competes with 1,2-hydride shifts to give phenols.[6]

Examples:

A complete report on use of isopropanol- and diisopropylcarbinol-doped alumina for reduction of carbonyl compounds is available. The paper concludes that the method is not particularly useful for standard reductions, but has a place for selective reductions, as in the two examples cited.[7]

Examples:

$$CH_3CO(CH_2)_8CHO \xrightarrow[68\%]{Al_2O_3, (CH_3)_2CHOH} CH_3CO(CH_2)_8CH_2OH$$

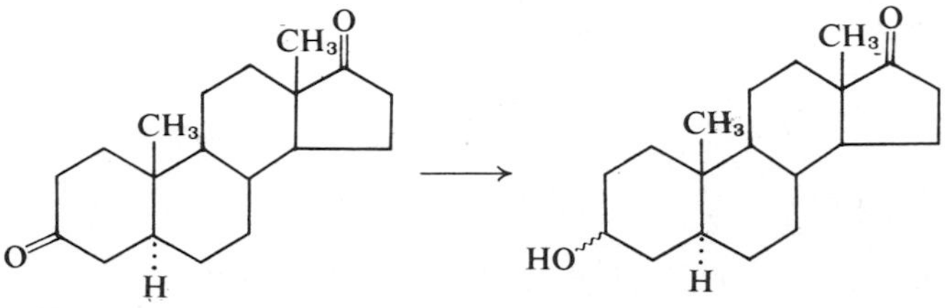

(3β, 95%; 3α, 2%)

Some sulfonate esters are converted into alkenes (**7**, 6) conveniently by treatment with W-200-N alumina. In some cases the commercial material can be used directly; alumina dehydrated at 400° under vacuum is sometimes more efficient. The reaction can be carried out by stirring solutions of the sulfonate esters over the alumina at room temperature. This dehydrosulfonation by alumina is particularly useful for some secondary sulfonate esters when elimination in one direction is favored by stereoelectronic factors or by the absence of a β-hydrogen atom. The method is generally less useful with primary systems. It also is useful for dehydrofluorination of *gem*-difluorides.[8]

Examples:

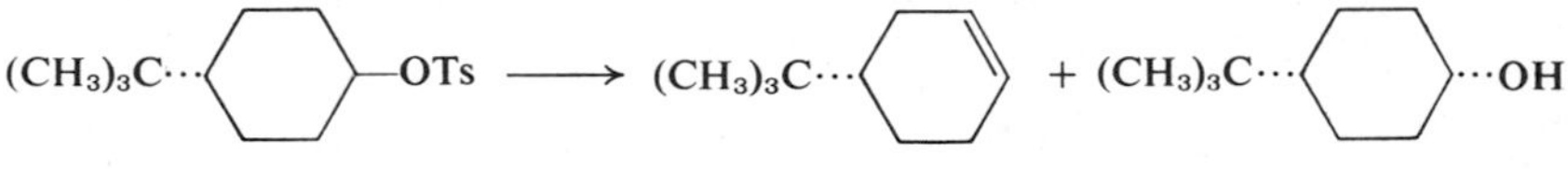

(80–90%)   (trace)

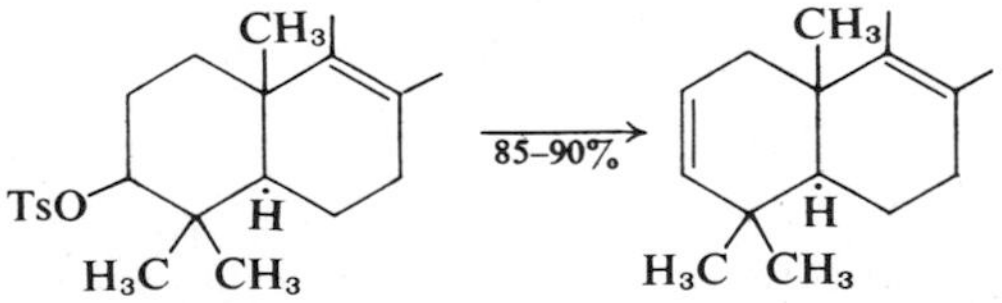

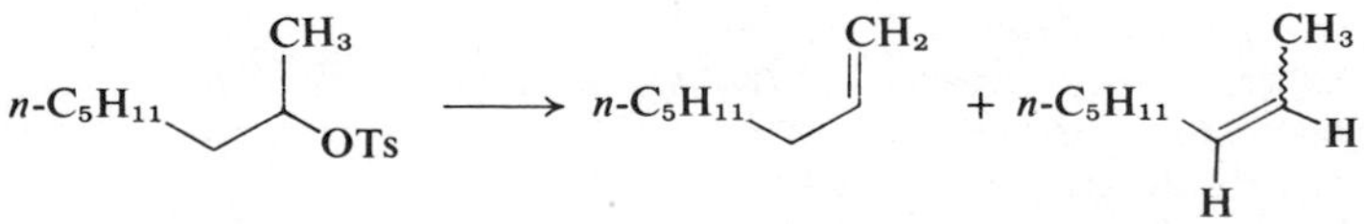

(22%)   (*cis/trans* = 33:12)

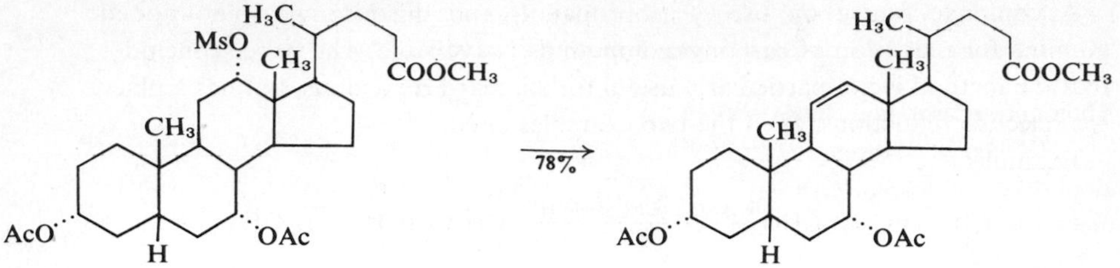

(60%)   (trace)

β-Hydroxy sulfides and selenides can be prepared by reaction of epoxides with thiols or selenols promoted by activated alumina (**6**, 16–17). These products can be oxidized to β-keto sulfides and selenides by chloral on activated Woelm alumina without oxidation of the S or Se atoms or of a C=C bond.[9]

Examples:

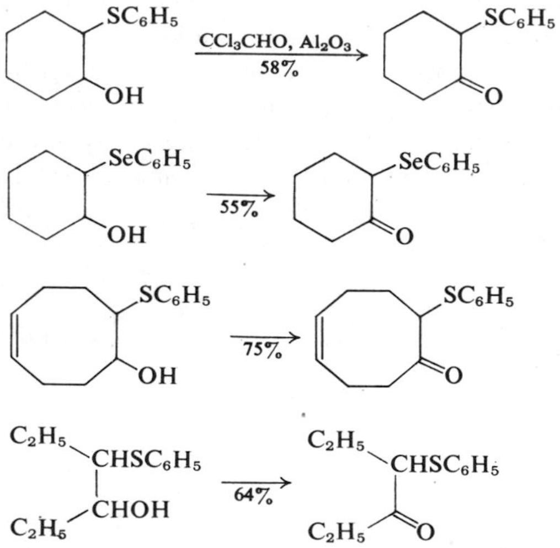

[1] R. B. Miller and T. Reichenbach, *Tetrahedron Letters*, 543 (1974); R. B. Miller and G. McGarvey, *Syn. Comm.*, **7**, 475 (1977).

[2] G. Piancatelli and A. Scettri, *Synthesis*, 116 (1977).

[3] G. Piancatelli, A. Scettri, and S. Barbadoro, *Tetrahedron Letters*, 3555 (1976).

[4] G. H. Posner, *Angew. Chem., Int. Ed.*, **17**, 487 (1978).

[5] G. H. Posner and D. Z. Rogers, *Am. Soc.*, **99**, 8208 (1977).

[6] *Idem, ibid.*, **99**, 8214 (1977).

[7] G. H. Posner, A. W. Runquist, and M. J. Chapdelaine, *J. Org.*, **42**, 1202 (1977).

[8] G. H. Posner, G. M. Gurria, and K. A. Babiak, *ibid.*, **42**, 3173 (1977).
[9] G. H. Posner and M. J. Chapdelaine, *Tetrahedron Letters*, 3227 (1977).

Aluminum t-butoxide–Raney nickel.

Alkylation of amines. Primary and secondary amines can be alkylated by an alcohol in the presence of catalytic amounts of aluminum *t*-butoxide and Raney nickel W-2.[1]

Examples:

$$C_6H_5NH_2 + n\text{-}C_3H_7OH \xrightarrow[98\%]{(t\text{-}C_4H_9O)_3Al,\ Raney\ Ni,\ \Delta} C_6H_5NHC_3H_7\text{-}n$$

[1] M. Botta, F. De Angelis, and R. Nicoletti, *Synthesis*, 722 (1977).

Aluminum chloride, 1, 24–34; 2, 21–23; 3, 7–9; 4, 10–15; 5, 10–13; 6, 17–19; 7, 7–9.

Ene reaction. Aluminum chloride catalyzes the ene reaction of methyl propiolate with citronellyl acetate to give **1** as a mixture of diastereomers in 55% yield. The conjugated double bond in **1** can be selectively reduced with $Fe(CO)_5$ and NaOH in aqueous methanol (**4**, 269) to give **2**, which was converted by known methods into **3**, a component of the female sex pheromone of the California red scale.[1]

Diels-Alder catalyst. Hoffmann-LaRoche chemists[2] have reported an asymmetric total synthesis of a D-homosteroid (**4**) in which one step is a diastereo-selective Diels-Alder reaction mediated by $AlCl_3$ (*cf.*, **6**, 65–66). Thus addition of

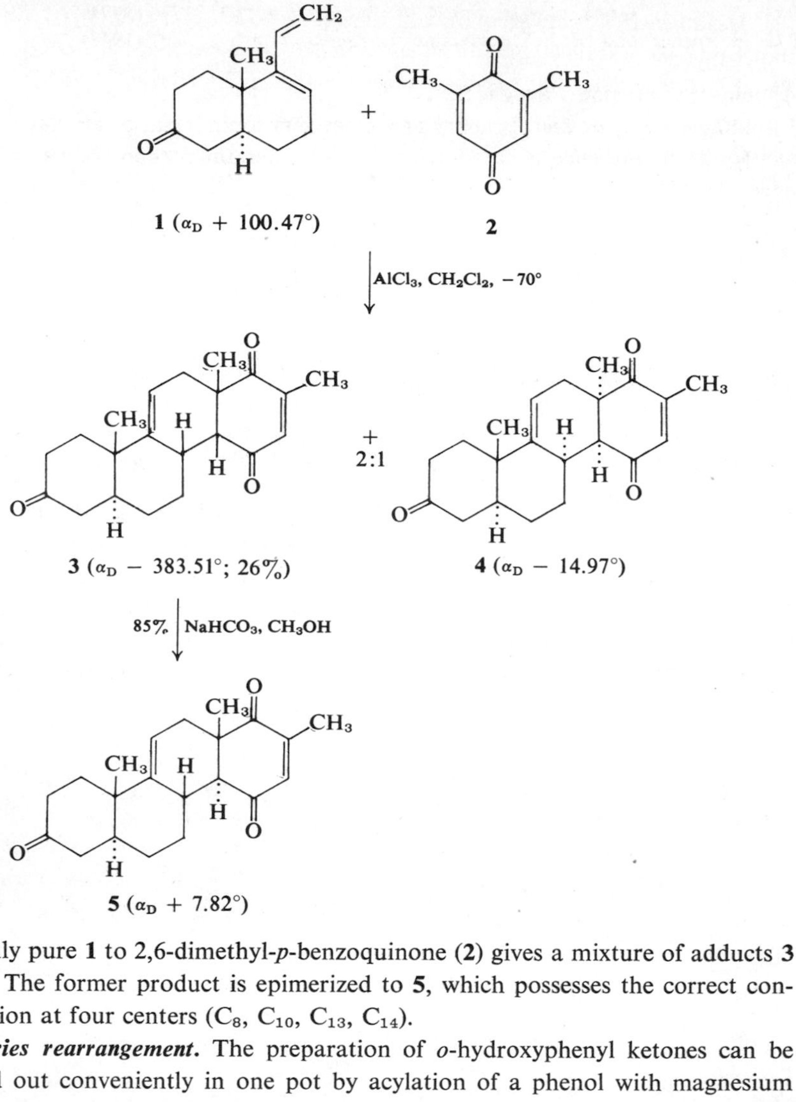

1 (α_D + 100.47°)   **2**

AlCl₃, CH₂Cl₂, − 70°

3 (α_D − 383.51°; 26%)   **4** (α_D − 14.97°)

+
2:1

85% NaHCO₃, CH₃OH

5 (α_D + 7.82°)

optically pure **1** to 2,6-dimethyl-*p*-benzoquinone (**2**) gives a mixture of adducts **3** and **4**. The former product is epimerized to **5**, which possesses the correct configuration at four centers (C_8, C_{10}, C_{13}, C_{14}).

Fries rearrangement. The preparation of *o*-hydroxyphenyl ketones can be carried out conveniently in one pot by acylation of a phenol with magnesium perchlorate catalysis followed by addition of AlCl₃ to the crude phenyl ester. Overall yields are 80–90%. Perchloric acid can replace AlCl₃, but yields are then lower.[3]

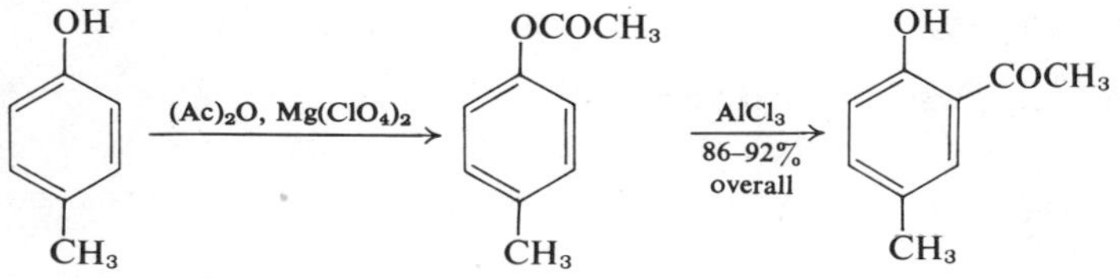

Hexamethyl(Dewar) benzene (**2**, 23). A detailed preparation of this substance on a large scale by bicyclotrimerization of 2-butyne is available. The yield is 30–50%, lower than that reported in the original publication (60–70%).[4]

[1] B. B. Snider and D. Rodini, *Tetrahedron Letters*, 1399 (1978).
[2] N. Cohen, B. L. Banner, and W. F. Eichel, *Syn. Comm.*, **8**, 427 (1978).
[3] J. Gray, *Org. Syn.*, submitted (1977).
[4] S. A. Shama and C. C. Wamser, *Org. Syn.* submitted (1978).

Aluminum chloride–Ethanol.

—CHO → —CONHR. This transformation can be accomplished by conversion of an aldehyde by known reactions into an α-cyanoenamine (**1**). Amides (**2**) are obtained on reaction of **1** with anhydrous aluminum chloride in absolute alcohol (reflux). The result is replacement of the CN group of **1** by OH.[1]

[1] N. De Kimpe, R. Verhé, L. De Buyck, J. Chys, and N. Schamp, *Org. Prep. Proc. Int.*, **10**, 149 (1978).

Aluminum isopropoxide, 1, 35–37.

Fragmentation of an epoxy mesylate. The key step in a synthesis of the sesquiterpene lactone saussurea lactone (**4**) involved fragmentation of the epoxy mesylate (**1**), obtained from α-santonin by several steps. When treated with aluminum isopropoxide in boiling toluene (N₂, 72 hours), **1** is converted mainly into **3**. The minor product (**2**) is the only product when the fragmentation is quenched after 12 hours. Other bases such as potassium *t*-butoxide, LDA, and lithium diethylamide, cannot be used. Aluminum isopropoxide is effective probably because aluminum has marked affinity for oxygen and effects cleavage of the epoxide ring. Meerwein-Ponndorf reduction is probably involved in one step.[1]

[1] M. Ando, K. Tajima, and K. Takase, *Chem. Letters*, 617 (1978).

(S)-1-Amino-2-methoxymethyl-1-pyrrolidine (1). Mol. wt. 101.16, b.p. 56–57°/ 3 mm., stable on refrigeration.

Preparation from (S)-prolinol (55% overall yield)[1]:

Asymmetric synthesis of α-alkylated ketones[1] *and aldehydes.*[2] Chiral hydrazones of ketones prepared from 1 on metalation (LDA) and alkylation are converted into α-alkyl hydrazones of the ketones. The α-alkyl ketones are obtained on ozonolysis or hydrolysis. Optical yields in this conversion are 30–85% and depend on the structures of the ketone and alkylating agents. α-Chiral aldehydes can be obtained in the same way and with similar optical yields.

Enantioselective aldol reactions. Enders *et al.*[3] have reported a regiospecific and enantioselective aldol synthesis. The method involves conversion of a methyl ketone into the chiral hydrazone 2, α-metalation, reaction with a carbonyl compound, and silylation to form a doubly protected ketol 3. The chiral ketols 4 are obtained by oxidative hydrolysis with H_2O_2 (30%) at pH 7 or by sensitized photooxidation in THF followed by reduction with dimethyl sulfide. Optical yields are 30–60%. The absolute configuration of the ketols is not known at present.

[1] D. Enders and H. Eichenauer, *Angew. Chem., Int. Ed.*, **15**, 549 (1976).
[2] *Idem, Tetrahedron Letters*, 191 (1977).
[3] H. Eichenauer, E. Friedrich, W. Lutz, and D. Enders, *Angew. Chem., Int. Ed.*, **17**, 206 (1978).

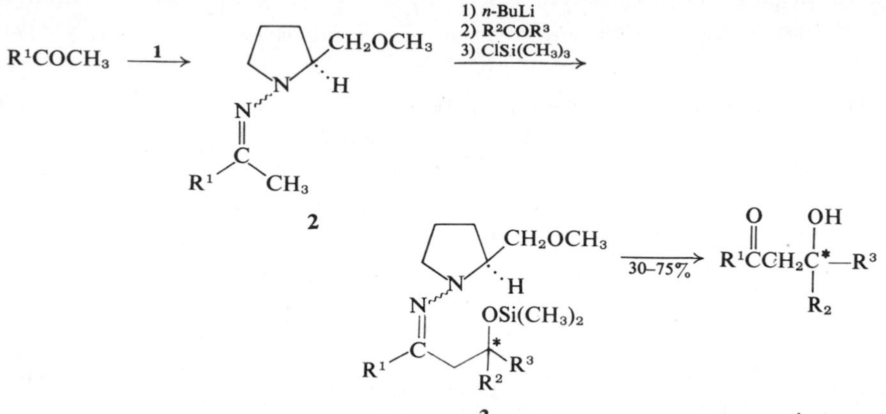

R^1COCH_3 $\xrightarrow{\textbf{1}}$

[structure 2: pyrrolidine ring with CH₂OCH₃ and H, N, N= C with R¹ and CH₃]

1) n-BuLi
2) R²COR³
3) ClSi(CH₃)₃ →

2

[structure 3: pyrrolidine ring with CH₂OCH₃ and H, N, N=C with R¹ and CH₂ C*R³, R², OSi(CH₃)₂]

$\xrightarrow[30-75\%]{}$

$R^1CCH_2C^*\!\!-R^3$ with O, OH, R₂

3   **4**

(+)-3-Aminomethylpinane,

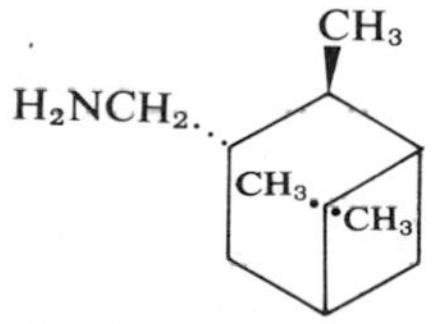

[structure 1: bicyclic pinane with CH₃, H₂NCH₂, CH₃, CH₃]

(1). Mol. wt. 165.28, b.p.

110°/20 mm., α_D +41.5°, configurationally stable.

Preparation from (−)-α-pinene[1]:

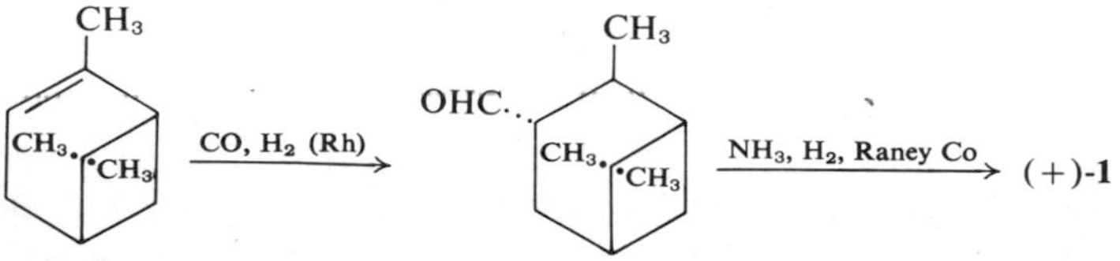

[structure of (−)-α-pinene with CH₃, CH₃, CH₃]

$\xrightarrow{\text{CO, H}_2\text{ (Rh)}}$

[structure with OHC, CH₃, CH₃, CH₃]

$\xrightarrow{\text{NH}_3,\ \text{H}_2,\ \text{Raney Co}}$ (+)-**1**

(−)

Resolution of pantoic acid (2). This precursor of (R)-pantothenic acid **(4)** has been resolved with (+)-**1**. The salts of the (R)-acid and the (S)-acid with this amine show a marked difference in solubility in water at 60°, which makes separation relatively simple. In fact, (−)-(R)-pantolactone **(3)** can be obtained in 90% yield with a purity of 94%.[2]

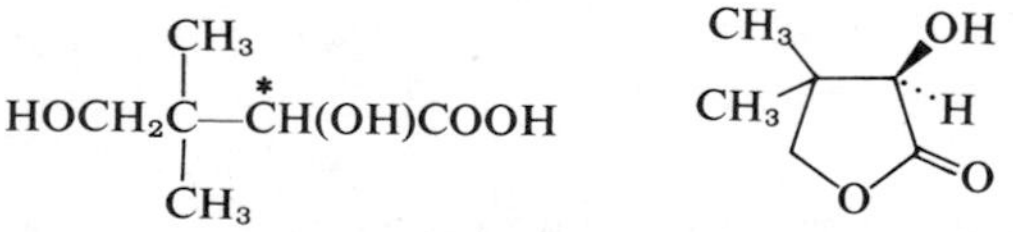

$HOCH_2C\!\!-\!\!CH(OH)COOH$ with CH₃, *, CH₃

[structure (−)-(R)-3: lactone ring with CH₃, CH₃, OH, H, O, O]

(R, S)-**2**   (−)-(R)-**3**

[structure (+)-(R)-4: HO, H, H, N, HO, CH₃, CH₃, O, COOH]

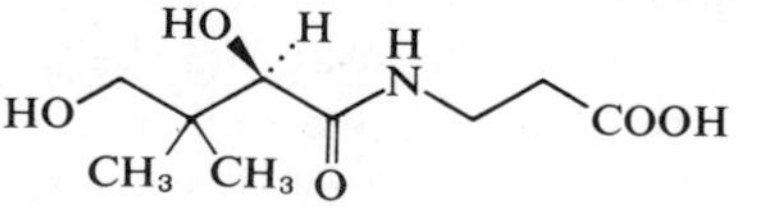

(+)-(R)-**4**

[1] W. Himmele and H. Siegel, *Tetrahedron Letters*, 907, 911 (1976).
[2] J. Paust, S. Pfohl, W. Reif, and W. Schmidt, *Ann.*, 1024 (1978).

Ammonium persulfate–Silver nitrate, **5**, 16–17.

Alkylation of quinones (**5**, 17). Several examples of the free-radical alkylation of 1,4-benzoquinones and 1,4-naphthoquinones have been published. Conditions can usually be found to cause precipitation of the monoalkylation product, thereby suppressing dialkylation. The method is not useful if the radical formed on decarboxylation is subject to oxidation. Yields vary from 40 to 85%.[1]

[1] N. Jacobsen, *Org. Syn.*, **56**, 68 (1977).

9-Anthrylmethyl *p*-nitrophenyl carbonate (**1**). Mol. wt. 373.36, m.p. 108–109°. Preparation:

Protection of amines (*cf.* **5**, 626, 627). Amines (RNH_2) can be protected as the 9-anthrylmethylcarbamates (**2**), formed by reaction with **1** at room temperature (75–95% yield). The carbamates are resistant to various acids and bases, but are readily deblocked by reaction with the sodium salt of methyl mercaptan (equation I). The amines are also liberated by TFA in CH_2Cl_2 at 0° (90–95%).[1]

[1] N. Kornblum and A. Scott, *J. Org.*, **42**, 399 (1977).

$$(I) \quad \underset{2}{\text{[anthracene with } CH_2OC(O)NHR]} \quad \xrightarrow{CH_3SNa, \ DMF, \ 20°} \quad [RNHCOONa] + \quad [anthracene with } CH_2SCH_3]$$

$$\downarrow 80\text{–}95\%$$

$$RNH_2$$

Antimony(III) chloride, $SbCl_3$. Mol. wt. 228.11.

Scholl reaction (**6**, 524–525). This reaction is commonly conducted in molten aluminum chloride or sodium aluminum chloride. Molten antimony(III) chloride (m.p. 73.4°) has some advantages over aluminum chloride because it is more easily handled and is also a milder Lewis acid. Naphthalene is stable in molten $SbCl_3$ at 125°, but anthracene under these conditions is converted mainly into 1,9'-bianthracene (**1**) and anthra[2,1-*a*]-aceanthrylene (**2**); the latter is a known hydrocarbon, but this reaction represents the first synthesis of the unsymmetrical bianthracene (**1**).[1] It has been generally assumed that the hydrogen formed in the coupling reactions is evolved as gaseous H_2. Actually the hydrogen converts anthracene into 9,10-dihydroanthracene and, to a lesser extent, 1,2,3,4-tetra-hydroanthracene.

1   **2**

[1] M. L. Poutsma, A. S. Dworkin, J. Brynestad, L. L. Brown, B. M. Benjamin, and G. P. Smith, *Tetrahedron Letters*, 873 (1978).

1,1'-(Azodicarbonyl)dipiperidine, $\quad [\text{piperidine}]N-\overset{O}{\underset{}{C}}-N=N-\overset{O}{\underset{}{C}}-N[\text{piperidine}] \quad$ (**1**). Mol. wt. 252.32, m.p. 134–135°. The reagent is prepared by the reaction of diethyl azodicarboxylate with piperidine.[1]

Oxidation of alcohols.[2] Primary and secondary alcohols are oxidized to carbonyl compounds in 90–99% yield by reaction with a slight excess of *n*-pro-pylmagnesium bromide in THF followed by treatment with **1** (1.2 equiv.) at room temperature.

Examples:

$$C_6H_5CH_2CH_2CH\!\!=\!\!CHCH_2CH_2CH_2OH \xrightarrow{\textit{n}\text{-PrMgBr}}$$

$$[C_6H_5(CH_2)_2CH\!\!=\!\!CH(CH_2)_3OMgBr] \xrightarrow[89\%]{1,\ THF,\ 20°} C_6H_5(CH_2)_2CH\!\!=\!\!CH(CH_2)_2CHO$$

$$(C_6H_5)_2CHOH \xrightarrow[99\%]{} (C_6H_5)_2C\!\!=\!\!O$$

For oxidation of hydroxy ketones, *t*-butoxymagnesium bromide, which does not add to ketones, is used as the Grignard reagent. Under these conditions testosterone (**2**) is oxidized to Δ^4-androstene-3,17-dione (**3**) in 96% yield.

Yields of carbonyl compounds are lower when **1** is replaced by *m*-chloroperbenzoic acid, NCS, $C_6H_5I(OAc)_2$, or other azo compounds.

[1] E. E. Şmissman and A. Makriyannis, *J. Org.*, **38**, 1652 (1973).
[2] K. Narasaka, A. Morikawa, K. Saigo, and T. Mukaiyama, *Bull. Chem. Soc. Japan*, **50**, 2773 (1977).

B

Barium manganate, $BaMnO_4$. Mol. wt. 256.28, dark green crystals. Supplier: Aldrich. The reagent is prepared by the reaction of aqueous solutions of $KMnO_4$ (1 equiv.), $BaCl_2$ (1 equiv.), NaOH (1 equiv.), and KI (0.12 equiv.). The reagent separates as dark green crystals and is dried by azeotropic removal of water with benzene.

Oxidation of alcohols. The reagent was first prepared by Stamm,[1] but its use as an oxidant for organic substrates has only been reported recently.[2] It oxidizes primary and secondary alcohols in CH_2Cl_2 at room temperature (4–16 hours) to aldehydes and ketones in high yields in the five examples reported.

[1] H. Stamm, *Z. Angew. Chem.*, **47**, 791 (1930).
[2] H. Firouzabadi and E. Ghaderi, *Tetrahedron Letters*, 839 (1978).

Benzenechromium tricarbonyl, 6, 27–28.

Other arenechromium tricarbonyls. Semmelhack[1] has extended his study of the reactions of carbanions with benzenechromium tricarbonyl to anisole- and toluenechromium tricarbonyl. The distribution of products depends on the particular anion used, but almost complete absence of *para*-substitution is consistently observed. A preference for *meta*-substitution over *ortho*-substitution is general.

Examples:

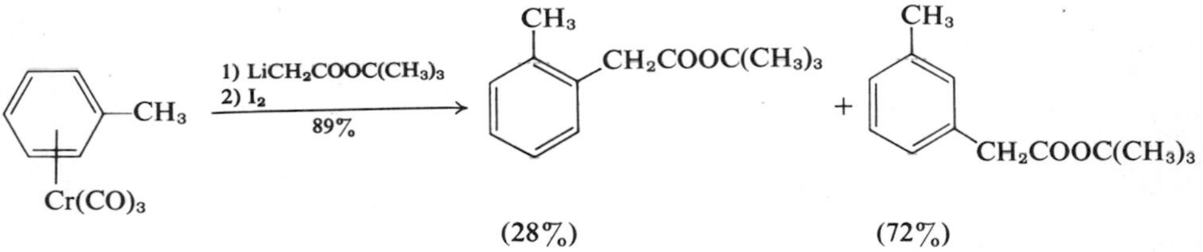

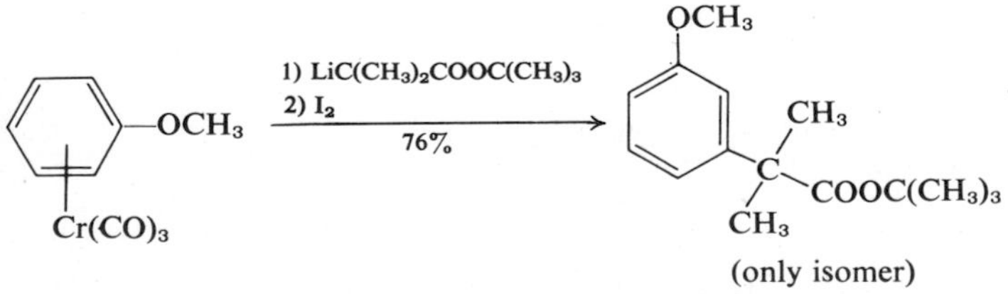

(28%)   (72%)

(only isomer)

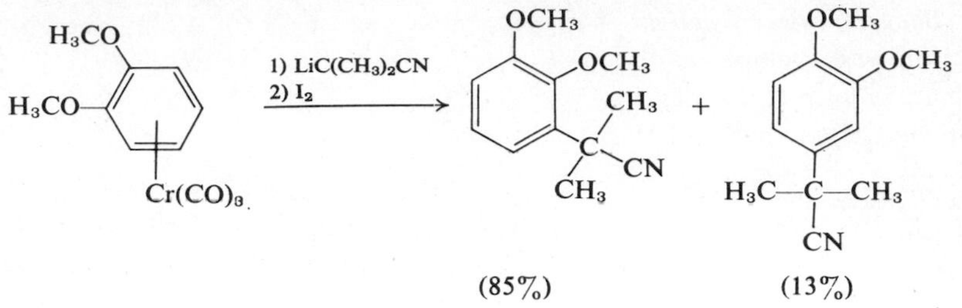

(85%)   (13%)

[1] M. F. Semmelhack and G. Clark, *Am. Soc.*, **99**, 1675 (1977).

Benzenediazonium tetrafluoroborate, 1, 43–44; **2**, 25; **3**, 16.

Japp-Klingemann reaction.[1] A variant of this reaction involves the reaction of a cyclic or acyclic β-keto ester with benzenediazonium tetrafluoroborate in aqueous pyridine followed by cleavage of the azo compound with sodium borohydride.[2]

Examples:

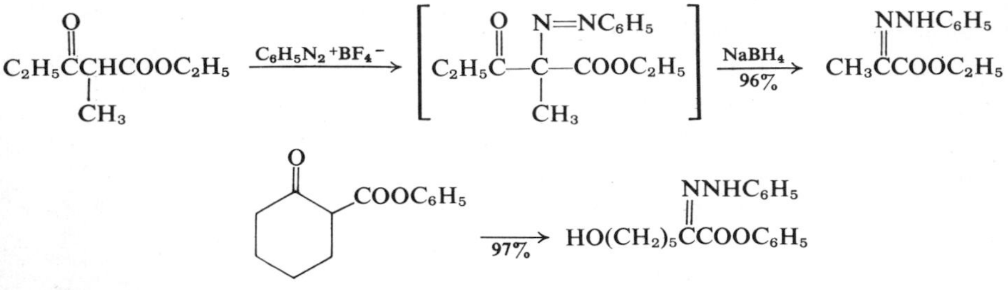

The reaction was developed as a route to the C-β-ribofuranosyl derivative **2** from **1**.[2]

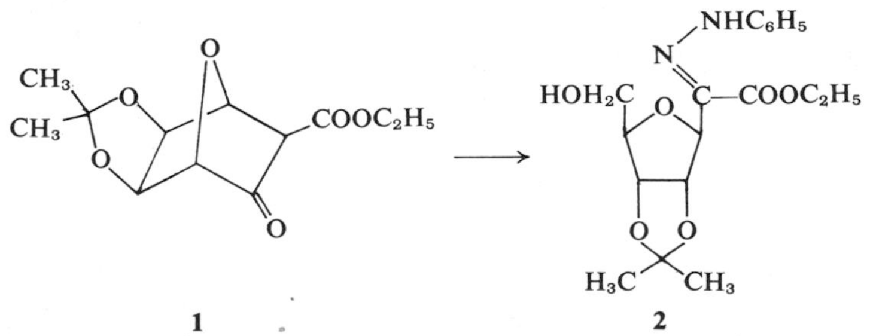

1   **2**

[1] R. R. Phillips, *Org. React.*, **10**, 143 (1959).
[2] A. P. Kozikowski and W. C. Floyd, *Tetrahedron Letters*, 19 (1978).

Benzeneperoxyseleninic acid, C_6H_5SeOOH. Mol. wt. 205.07. This peracid is prepared *in situ* from benzeneseleninic acid, $C_6H_5Se(O)OH$,[1] and 30% aqueous H_2O_2.

Baeyer-Villiger reaction.[2] This peracid has been used for the Baeyer-Villiger reaction with cyclopentanones and cyclohexanones. The reaction is carried out with 1.25 molar equiv. of benzeneseleninic acid and 10 molar equiv. of H_2O_2 in THF or CH_2Cl_2 buffered to pH 7. Yields are comparable to those with other peracids. One example, **1** → **2**, was reported where oxidation was achieved only with this peracid.

This peracid has been used to convert damsin (**3**) directly into psilostachyin (**4**).[3]

Epoxidation.[4,5] The peracid is useful for epoxidation of olefinic double bonds. This epoxidation differs from epoxidation with *t*-butyl hydroperoxide–transition metals in that a double bond furthest removed from the hydroxyl group is more readily attacked than a double bond in the allylic position.

Examples:

[1] J. D. McCullough and E. S. Gould, *Am. Soc.*, **71**, 674 (1949).

[2] P. A. Grieco, Y. Yokoyama, S. Gilman, and Y. Ohfune, *J.C.S. Chem. Comm.*, 870 (1977).

[3] P. A. Grieco, Y. Ohfune, and G. Majetich, *Am. Soc.*, **99**, 7393 (1977).

[4] P. A. Grieco, Y. Yokoyama, S. Gilman, and M. Nishizawa, *J. Org.*, **42**, 2034 (1977).

[5] H. J. Reich, F. Chow, and S. L. Peake, *Synthesis*, 299 (1978); T. Hori and K. B. Sharpless, *J. Org.*, **43**, 1689 (1978).

Benzeneselenenic acid, C_6H_5SeOH. Mol. wt. 173.07. The reagent is generated *in situ* from benzeneseleninic acid and diphenyl diselenide:

$$C_6H_5SeO_2H + C_6H_5SeSeC_6H_5 + H_2O \rightleftharpoons 3C_6H_5SeOH$$

Allylic alcohols. The reagent adds to trisubstituted double bonds with high regioselectivity (Markownikoff addition), particularly in the presence of $MgSO_4$ to sequester excess water. Selenoxide elimination gives an allylic alcohol.[1] Hydrogen peroxide is the usual reagent for effecting oxidation of alkyl aryl selenides, but is not satisfactory in this case because of epoxidation of the double bond in the product.[2] *t*-Butyl hydroperoxide is a much more satisfactory oxidant. Excess is used to prevent addition of the eliminated C_6H_5SeOH to the double bond.

Examples:

[1] T. Hori and K. B. Sharpless, *J. Org.*, **43**, 1689 (1978).

[2] *See also* H. J. Reich, S. Wollowitz, J. E. Trend, F. Chow, and D. F. Wendelborn, *ibid.*, **43**, 1697 (1978).

Benzeneselenenyl bromide and chloride, 5, 518–522; **6**, 459–460; **7**, 286–287.

Addition to 1-alkenes.[1] The reaction of benzeneselenenyl bromide with 1-alkenes can be carried out under either kinetically or thermodynamically controlled conditions to give, after dehydrohalogenation, 2-phenylselenoalkenes (**2**) or 1-phenylselenoalkenes (**4**), respectively.[1]

$$
RCH{=}CH_2 + C_6H_5SeBr
$$

$$
\xrightarrow{\text{THF, } -78°} \underset{\mathbf{1}}{\overset{SeC_6H_5}{R\overset{|}{C}HCH_2Br}} \xrightarrow[\sim 90\%]{\text{(CH}_3)_3\text{COK, THF, } -78 \to 25°} \underset{\mathbf{2}}{\overset{SeC_6H_5}{R\overset{|}{C}{=}CH_2}}
$$

$$
\xrightarrow{\text{CH}_3\text{CN, } 25°} \underset{\mathbf{3}}{\underset{Br}{R\overset{|}{C}HCH_2SeC_6H_5}} \xrightarrow[\sim 90\%]{\text{(CH}_3)_3\text{COK, THF, } 25°} \underset{\mathbf{4}}{RCH{=}CHSeC_6H_5}
$$

The anti-Markownikoff adducts (**1**) on oxidation (H_2O_2, pyridine) form selenoxides that decompose to either vinyl or allyl bromides, depending on the structure of the R group:

$$
\underset{}{\overset{SeC_6H_5}{(CH_3)_3C\overset{|}{C}HCH_2Br}} \xrightarrow[51\%]{H_2O_2} \underset{H}{\overset{(CH_3)_3C}{}}C{=}C\underset{Br}{\overset{H}{}} + C_6H_5SeOH
$$

$$
\underset{}{\overset{SeC_6H_5}{(CH_3)_2CHCH_2\overset{|}{C}HCH_2Br}} \xrightarrow[67\%]{} \underset{(CH_3)_2CH}{\overset{H}{}}C{=}C\underset{H}{\overset{CH_2Br}{}}
$$

The selenoxides of the Markownikoff adducts (**3**) decompose to vinyl bromides (**4**); the yield is improved by addition of diisopropylamine.[2]

$$
\underset{\mathbf{3}}{\overset{Br}{R\overset{|}{C}HCH_2SeC_6H_5}} \xrightarrow[70-80\%]{\text{Oxid.}} \underset{\mathbf{4}}{\overset{Br}{R\overset{|}{C}{=}CH_2}}
$$

$$
\underset{}{65-75\%} \downarrow \begin{array}{l} \text{1) DMSO, AgPF}_6 \\ \text{2) N(C}_2\text{H}_5)_3 \end{array}
$$

$$
\underset{\mathbf{5}}{\overset{O}{R\overset{\|}{C}CH_2SeC_6H_5}}
$$

Oxidation of **3** by the method of Ganem and Boeckman (**5**, 264) leads to α-phenylseleno ketones (**5**).[3] A somewhat different route to these ketones has also been reported by Japanese chemists (equation I).[4] These ketones can be alkylated

(I) $CH_3(CH_2)_5CH{=}CH_2 \xrightarrow[95\%]{C_6H_5SeBr,\ C_2H_5OH}$

$$CH_3(CH_2)_5\underset{\underset{OC_2H_5}{|}}{C}HCH_2SeC_6H_5 \xrightarrow[85\%]{\substack{1)\ NaIO_4 \\ 2)\ \Delta}} CH_3(CH_2)_5\overset{\overset{O}{\|}}{C}CH_2SeC_6H_5$$

(II) $CH_3(CH_2)_5\overset{\overset{O}{\|}}{C}CH_2SeC_6H_5 \xrightarrow[71\%]{\substack{1)\ KOC(CH_3)_3 \\ 2)\ CH_3I}}$

$$CH_3(CH_2)_5\overset{\overset{O}{\|}}{C}{-}\underset{\underset{CH_3}{|}}{\overset{\overset{SeC_6H_5}{|}}{C}}H \xrightarrow[\substack{54\% \\ overall}]{\substack{1)\ NaIO_4 \\ 2)\ \Delta}} CH_3(CH_2)_5\overset{\overset{O}{\|}}{C}CH{=}CH_2$$

regioselectively as shown in equation (II) for a general synthesis of α,β-unsaturated ketones. The phenylseleno group can also be replaced by hydrogen by reaction with triethylamine and thiophenol.

Phenylselenolactonization. Two laboratories[5,6] have reported the conversion of γ,δ-unsaturated carboxylic acids into phenylselenolactones by reaction with this reagent under very mild conditions. One advantage is that the phenylselenenyl group can be used for further transformations.

Examples:

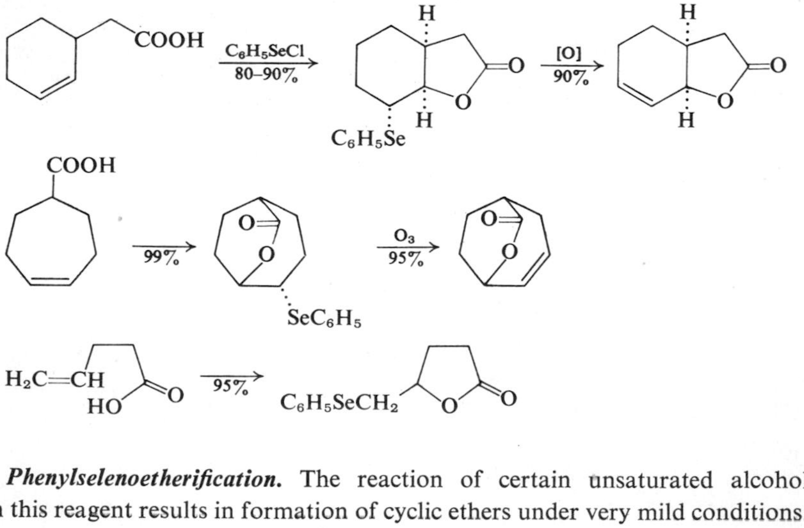

Phenylselenoetherification. The reaction of certain unsaturated alcohols with this reagent results in formation of cyclic ethers under very mild conditions.[7]

Examples:

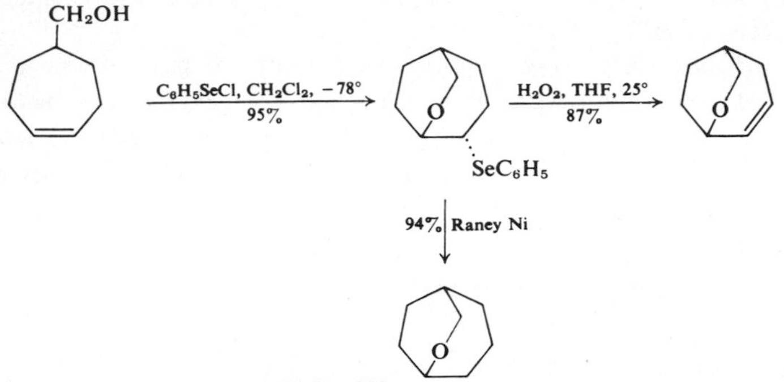

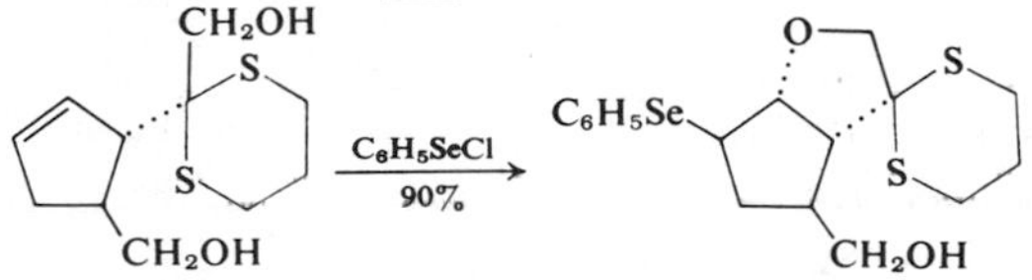

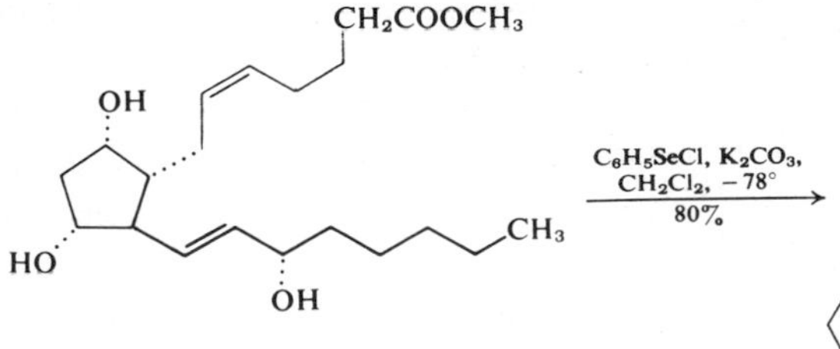

$C_6H_5SeCl, CH_2Cl_2, -78°$ / 95%

$H_2O_2, THF, 25°$ / 87%

94% Raney Ni

C_6H_5SeCl / 90%

$C_6H_5SeCl, K_2CO_3,$ $CH_2Cl_2, -78°$ / 80%

H_2O_2 / 95%

1

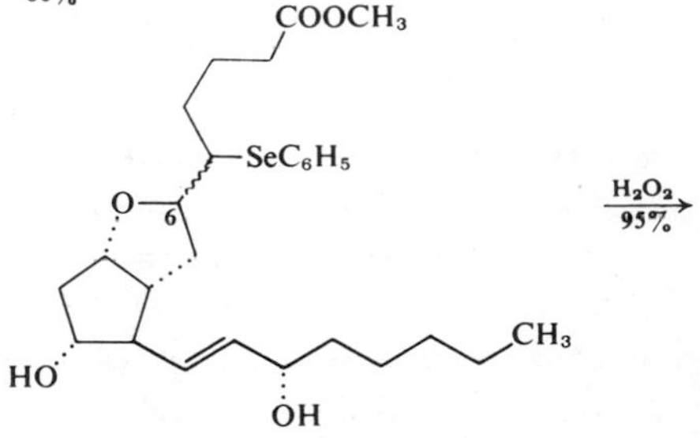

2 ($6\alpha/6\beta = 63:35$)

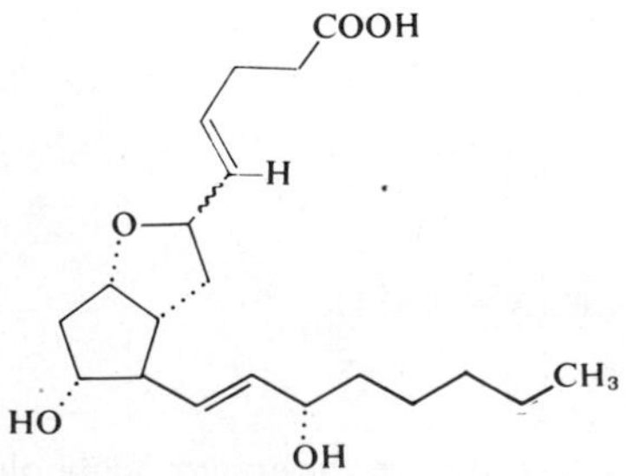

3 ($6\alpha/6\beta = 35:65$)

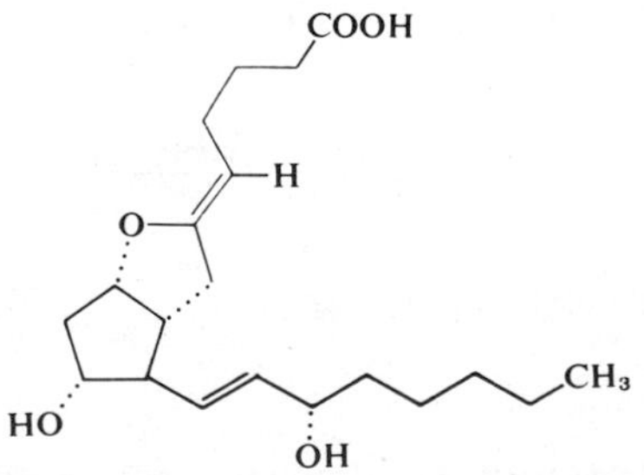

4

This reaction has been used in a synthesis of two isomers (3) of prostacyclin (PGI$_2$, 4), a major factor in blood platelet aggregation, from the prostaglandin PGF$_{2\alpha}$ methyl ester (1) by reaction with C$_6$H$_5$SeCl to give 2, as a mixture of diastereoisomers, in 75% yield.[8]

α-Phenylseleno carbonyl compounds. These compounds have generally been prepared by reaction of lithium enolates with benzeneselenenyl halides. They can also be prepared by reaction of trimethylsilyl enol ethers with benzeneselenenyl bromide (prepared *in situ* from bromine and diphenyl diselenide). Advantages of this method are regiospecificity and the mild conditions. Yields of products are generally in the range 75–90%.[9]

Review. Recent applications of organoselenium reagents have been reviewed in detail.[10,11]

[1] S. Raucher, *J. Org.*, **42**, 2950 (1977).
[2] *Idem, Tetrahedron Letters*, 3909 (1977).
[3] *Idem, ibid.*, 2261 (1978).
[4] T. Takahashi, H. Nagashima, and J. Tsuji, *ibid.*, 799 (1978).
[5] K. C. Nicolaou and Z. Lysenko, *Am. Soc.*, **99**, 3185 (1977); K. C. Nicolaou, Z. Lysenko, and S. Seitz, *Org. Syn.*, submitted (1978).
[6] D. L. J. Clive and G. Chittattu, *J.C.S. Chem. Comm.*, 484 (1977).
[7] K. C. Nicolaou and Z. Lysenko, *Tetrahedron Letters*, 1257 (1977).
[8] K. C. Nicolaou and W. E. Barnette, *J.C.S. Chem. Comm.*, 331 (1977); *Org. Syn.* submitted (1978).
[9] I. Ryu, S. Murai, I. Niwa, and N. Sonoda, *Synthesis*, 874 (1977).
[10] D. L. J. Clive, *Tetrahedron*, **34**, 1049 (1978).
[11] H. J. Reich, *Accts. Chem. Res.*, **12**, 22 (1979).

$$\overset{\text{O}}{\underset{\|}{}}$$

Benzeneseleninic acid, C$_6$H$_5$SeOH Mol. wt. 189.07, m.p. 122–124°. Diphenyl diselenide reacts with 3 moles of ozone to form benzeneseleninic anhydride, which reacts with water to form the seleninic acid.[1]

The acid can also be prepared by the reaction of diphenyl diselenide with H$_2$O$_2$.[2]

Acetoxyselenation. Benzeneseleninic acid in acetic acid adds to alkenes to form 2-acetoxyalkyl phenyl selenides in 80–90% yield. The actual reagent is probably C$_6$H$_5$SeOAc.[3]

Examples:

$$C_6H_5CH{=}CH_2 \xrightarrow[80\%]{} C_6H_5CH\overset{|}{\underset{\underset{\text{OAc}}{|}}{}}CH_2SeC_6H_5$$

Rearrangement of hydroxamic acids.[4] N-Arylbenzohydroxamic acids are rearranged by catalytic amounts of benzeneseleninic acid to p-hydroxybenzanilides

or N-benzoyl-*p*-quinonimines. A plausible mechanism is discussed in the paper.
Examples:

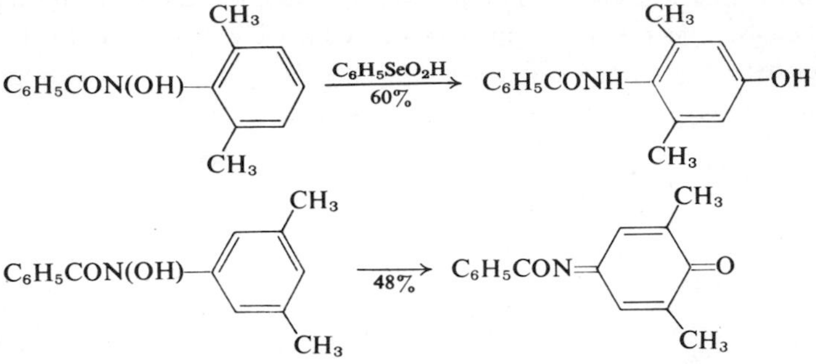

[1] G. Ayrey, D. Bernard, and D. T. Woodbridge, *J. Chem. Soc.*, 2089 (1962).
[2] J. D. McCullogh and E. S. Gould, *Am. Soc.*, **71**, 674 (1949).
[3] N. Miyoshi, Y. Takai, S. Murai, and N. Sonoda, *Bull. Chem. Soc. Japan*, **51**, 1265 (1978).
[4] T. Frejd and K. B. Sharpless, *Tetrahedron Letters*, 2239 (1978).

Benzeneseleninic anhydride (**1**), **6**, 240–241; **7**, 139.

Oxidation of phenols to hydroxydienones. A number of reagents effect this oxidation, but none favors formation of *o*-hydroxydienones over *p*-hydroxydienones. Barton et al.[1] have found that benzeneseleninic anhydride has some of the desired *ortho*-selectivity, particularly if the phenol is converted into the sodium phenolate prior to oxidation. Thus the sodium salt of 2,4-xylenol (**2**) under these conditions gives the *o*-hydroxydienone dimer(**3**) in 45% yield. The oxidation of **4** is of interest since it can be regarded as a model of ring A in tetracyclines.

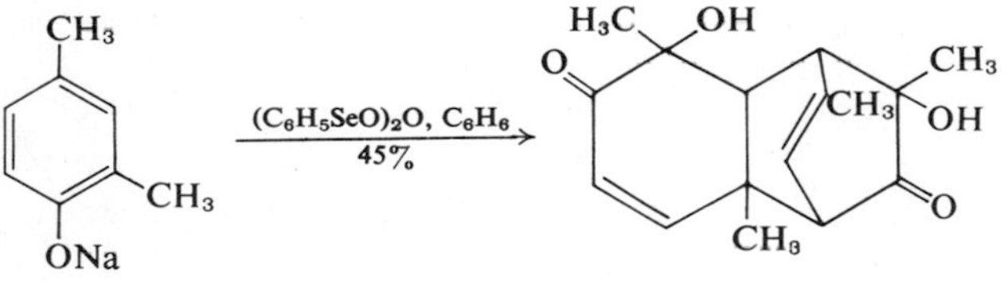

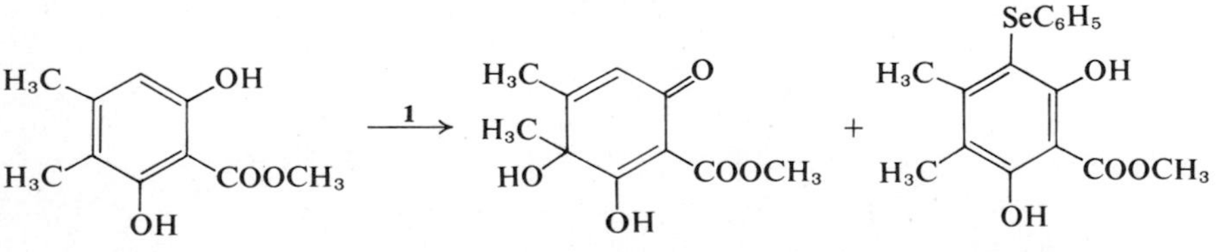

Phenylselenoimines; aminophenols.[2] In the presence of hexamethyldisilazane, this anhydride converts phenols into phenylselenoimines; as in the reaction without hexamethyldisilazane, there is a marked preference for *ortho*-substitution.

Benzene, CH_2Cl_2, or THF is used as the solvent. Since the products are re-duced by benzenethiol to aminophenols, the reaction can be used for amination of phenols. The products can also be reductively acetylated to aminophenol diacetates.

Oxidation of some nitrogen compounds.[3] The parent aldehydes are not regenerated on oxidation of the phenylhydrazones with this oxidant; instead acylazo compounds are formed (equation I). Aldehydes are regenerated, however, from tosylhydrazones or oximes with this reagent.

$$\text{(I)} \qquad RCH = NNHAr \xrightarrow[70-99\%]{(C_6H_5SeO)_2O} R\overset{\overset{\displaystyle O}{\|}}{C}\!\!-\!\!N = NAr$$

Miscellaneous dehydrogenations:

$$C_6H_5NHNHC_6H_5 \longrightarrow C_6H_5N = NC_6H_5 \ (99\%)$$

$$C_6H_5NHOH \longrightarrow C_6H_5N = O \ (89\%)$$

Oxidative cleavage of ethylene dithioketals. This reagent can be used for regeneration of ketones and aldehydes from 1,3-dithiolane derivatives at room temperature in THF or CH_2Cl_2 in yields of 60–90%. This method was successful in some cases where standard methods failed. An example is the conversion of **1** into the corresponding ketone, which could not be accomplished with various Hg(II) reagents or other reagents.[4]

1

Dehydrogenation of ketones.[5] This reagent is an attractive alternative, particularly to selenium dioxide, for dehydrogenation of ketosteroids. The by-product is diphenyl diselenide, which is easily removed and reoxidized to the anhydride. Moreover, the anhydride is fairly inert to double bonds.

Examples:

Thiones → ketones.[6] Xanthates, thioesters, thiocarbonates, thioamides, and thiones are converted into the parent carbonyl compounds when treated with this anhydride at 20° in THF. SeO_2 is not a satisfactory alternative for this reaction. Water and oxygen are not involved.

Examples:

[1] D. H. R. Barton, S. V. Ley, P. D. Magnus, and M. N. Rosenfeld, *J.C.S. Perkin I*, 567 (1977).

[2] D. H. R. Barton, A. G. Brewster, S. V. Ley, and M. N. Rosenfeld, *J.C.S. Chem. Comm.*, 147 (1977).

[3] D. H. R. Barton, D. J. Lester, and S. V. Ley, *ibid.*, 276 (1978); T. G. Back, *ibid.*, 278 (1978).

[4] D. H. R. Barton, N. J. Cussans, and S. V. Ley, *ibid.*, 751 (1977).

[5] D. H. R. Barton, D. J. Lester, and S. V. Ley, *ibid.*, 130 (1978).

[6] D. H. R. Barton, N. J. Cussans, and S. V. Ley, *ibid.*, 393 (1978).

Benzenesulfenyl chloride, 6, 30–32.

Chlorosulfenylation–dehydrochlorination.[1] This hygroscopic reagent is conveniently prepared and used in a methylene chloride solution by the reaction of NCS with thiophenol.[2] The coproduct succinimide does not interfere with the chlorosulfenylation of alkenes. The 1:1 adducts, β-chlorophenyl sulfides, are formed in high yield. The products are converted into allyl, vinyl, or dienyl sulfides on dehydrochlorination with DBU. The last example illustrates a route to α,β-unsaturated sulfones.

Examples:

$$CH_2=CHCH=CH_2 \xrightarrow[100\%]{} CH_2=CHCHCH_2SC_6H_5 \xrightarrow[81\%]{DBU} CH_2=CHCH=CHSC_6H_5$$
$$\underset{Cl}{|}$$

Lactonization. Unsaturated cycloalkanoic acids are converted by treatment with C_6H_5SCl into phenylsulfenyl lactones, which are useful intermediates to a number of products.[3]

Examples:

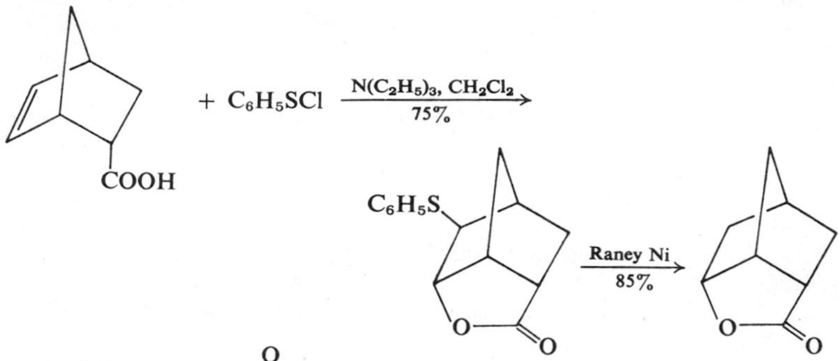

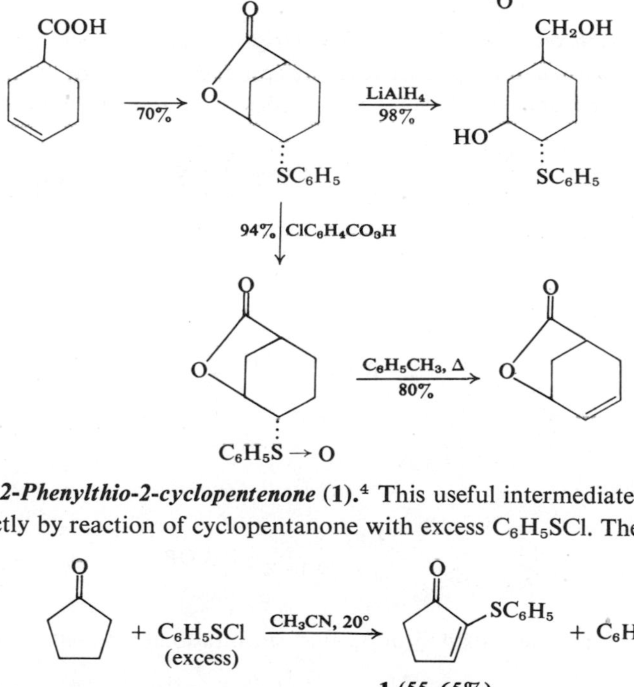

2-Phenylthio-2-cyclopentenone (1).[4] This useful intermediate can be obtained directly by reaction of cyclopentanone with excess C_6H_5SCl. The cyclopentenone

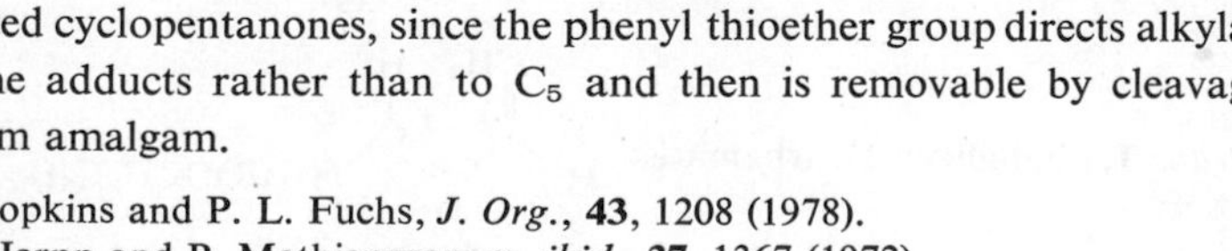

undergoes conjugate additions readily and hence is a useful precursor of 2,3-di-substituted cyclopentanones, since the phenyl thioether group directs alkylation to C_2 of the adducts rather than to C_5 and then is removable by cleavage with aluminum amalgam.

---

[1] P. B. Hopkins and P. L. Fuchs, *J. Org.*, **43**, 1208 (1978).
[2] D. N. Harpp and P. Mathiaparanam, *ibid.*, **37**, 1367 (1972).

[3] K. C. Nicolaou and Z. Lysenko, *J.C.S. Chem. Comm.*, 293 (1977).
[4] H. J. Monteiro, *J. Org.*, **42**, 2324 (1977).

1,3-Benzodithiolylium perchlorate (**1**). Mol. wt. 252.70, m.p. 182° (explosive), stable on storage.

Preparation[1,2]:

Aldehydes and ketones.[2] The salt reacts with Grignard reagents to form 2-substituted-1,3-benzodithioles (50–90% yield); the products are hydrolyzed by chloramine T and mercury(II) chloride to aldehydes. Ketones can be prepared

by introduction of another alkyl group; this substitution proceeds in only moderate yields.

[1] J. Nakayama, *Synthesis*, 38 (1975).
[2] I. Degani and R. Fochi, *J.C.S. Perkin I*, 1886 (1978).

Benzyl *trans*-**1,3-butadiene-1-carbamate**, (**1**).

Mol. wt. 203.24, m.p. 74–75°. This reagent is prepared in 49–56% overall yield by Curtius rearrangement of the acyl azide of *trans*-2,4-pentadienoic acid in the presence of benzyl alcohol.[1]

Pumiliotoxin C alkaloids. Alkaloids of this general type are *cis*-decahydroquinolines substituted at C_2 and C_5. Overman and Jessup[2] have devised a general synthesis for these alkaloids, shown in equation (I) for *dl*-pumiliotoxin C (**4**). The key step is a Diels-Alder reaction of **1**, as a synthetic equivalent of *trans*-1-amino-1,3-butadiene, with a β-alkyl-substituted *trans*-α,β-unsaturated aldehyde. Thus the dienecarbamate **1** reacts with *trans*-crotonaldehyde at 100° (2.5 hours) to form a single adduct (**2**) in 61% yield (10–15% of **1** is recovered). Wittig-

Horner reaction of **2** with the sodium salt of dimethyl 2-oxopentylphosphonate gives the enone **3** in 80% yield. Catalytic hydrogenation of **3** in acidic ethanol reduces both double bonds and unmasks the amine to give a $\Delta^{1,2}$-imine, which is reduced to **4**. The overall yield of **4** from **1** is > 50%.

[1] L. E. Overman, G. F. Taylor, and P. J. Jessup, *Tetrahedron Letters*, 3089 (1976); L. E. Overman, G. F. Taylor, C. B. Petty, and P. J. Jessup, *J. Org.*, **43**, 2164 (1978).
[2] L. E. Overman and P. J. Jessup, *Am. Soc.*, **100**, 5179 (1978).

Benzyl(chloro)bis(triphenylphosphine)palladium(II), $C_6H_5CH_2Pd[P(C_6H_5)_3]_2Cl$ (**1**). Mol. wt. 757.57, m.p. 140–144° dec., air stable. The yellow complex is prepared in 94% yield by reaction of benzyl chloride with $Pd[P(C_6H_5)_3]_4$ in benzene at 25° for 72 hours.[1,2]

Ketone synthesis.[3] Ketones can be prepared by reaction of acid chlorides and organotin compounds in HMPT catalyzed with **1** (equation I). The synthesis has

$$\text{(I)} \qquad RCOCl + R'_4Sn \xrightarrow[70-95\%]{1, \text{ HMPT}} RCOR' + R'_3SnCl$$

wide application. Most functional groups (even—CHO) can be tolerated; sterically hindered acid chlorides react normally; conjugate addition is not observed;

the reaction is truly catalytic. The actual catalyst is probably $Pd[P(C_6H_5)_3]_2$ formed by reduction of **1**.

[1] P. Fitton, J. E. McKeon, and B. C. Ream, *Chem. Commun.*, 370 (1969).
[2] K. S. Y. Lau, P. K. Wong, and J. K. Stille, *Am. Soc.*, **98**, 5832 (1976).
[3] D. Milstein and J. K. Stille, *ibid.*, **100**, 3636 (1978).

Benzylideneacetoneiron tricarbonyl, (**1**). Mol. wt. 302.06,

red crystals, m.p. 88–89°. The complex is prepared by the reaction under irradiation of benzylideneacetone with $Fe(CO)_5$ (60% yield).

Selective trapping of unstable dienes. This complex has the useful property of trapping the unstable diene tautomers of cyclooctatriene (**2**) and derivatives such as **5**. The kinetics of complex formation with unstable dienes has been studied and arguments have been advanced to account for the selectivity observed.[1]

[1] C. R. Graham, G. Scholes, and M. Brookhart, *Am. Soc.*, **99**, 1180 (1977).

Benzyltrimethylammonium hydroxide (Triton B), 1, 1252–1254; **5,** 29.

Cyclization. British chemists[1] have synthesized the novel fused β-lactam system of **4**, encountered in clavulanic acid (**6**),[2] by condensation of **1** and **2** to give the azetidinone **3**. Cyclization of **3** with Triton B results in a 1:1 mixture of **4** and **5**, epimeric at C_3. Actually **5** can be epimerized completely to the more stable **4** by treatment with DBN in $CHCl_3$.

This route has also been used for synthesis of a related compound containing three asymmetric centers (equation I).

[1] A. G. Brown, D. F. Corbett, and T. T. Howarth, *J.C.S. Chem. Comm.*, 359 (1977).
[2] T. T. Howarth, A. G. Brown, and T. J. King, *ibid.*, 266 (1976).

Birch reduction, 1, 54–56; **2,** 27–29; **3,** 19–20; **4,** 31–33; **5,** 30; **7,** 21.

   Reductive alkylation of α-tetralones. α-Tetralone (**1**) has been converted into **2** in 60% yield by Birch reduction with potassium and *t*-butyl alcohol in liquid ammonia at −78° followed by addition of lithium bromide and then methyl iodide.[1] Australian chemists[2] have found that this reductive alkylation is gener-

ally applicable to substituted α-tetralones. For successful results potassium and *t*-butyl alcohol are necessary for regiospecific protonation at C_6 and use of the lithium enolate is essential for regiospecific alkylation at C_9. By careful manipulation these workers were able to convert **3** into **4** in 93% yield.

   This reductive alkylation was used in a synthesis of **8**, which contains the 10-membered ring system of germacrane sesquiterpenes. Thus the tetralone **5** was converted by way of **6** into the tosylate **7**. On treatment with LDA in THF at 24°, **7** underwent a Grob-type fragmentation to give **8**.

[1] M. Narisada and F. Watanabe, *J. Org.*, **38**, 3887 (1973).
[2] J. M. Brown, T. M. Cresp, and L. N. Mander, *ibid.*, **42**, 3984 (1977).

Bis(acetonitrile)dichloropalladium(II), 7, 21–22.

Cyclization.[1] The last step in a total synthesis of ibogamine (**2**) involved cyclization of **1** with bis(acetonitrile)dichloropalladium(II) assisted by silver tetrafluoroborate and triethylamine. Work-up included reduction of an intermediate palladium species with $NaBH_4$. The alkaloid was obtained in 40–45% yield. This cyclization is related to the olefin arylation of Heck, catalyzed with "phenylpalladium acetate," $C_6H_5PdOCOCH_3$.[2]

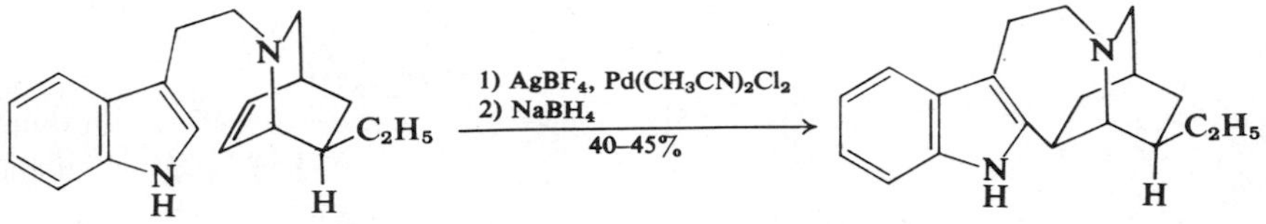

1) $AgBF_4$, $Pd(CH_3CN)_2Cl_2$
2) $NaBH_4$
40–45%

1   **2**

Alkylation of olefins. 1-Alkenes (and 1,2-disubstituted alkenes) can be alkylated primarily at the 2-position by stabilized carbanions in the presence of this Pd(II) complex and triethylamine. The reaction involves a palladium complex of the alkene followed by β-elimination of Pd(0) to give the unsaturated product, which can be hydrogenated if desired before work up.[3]

Examples:

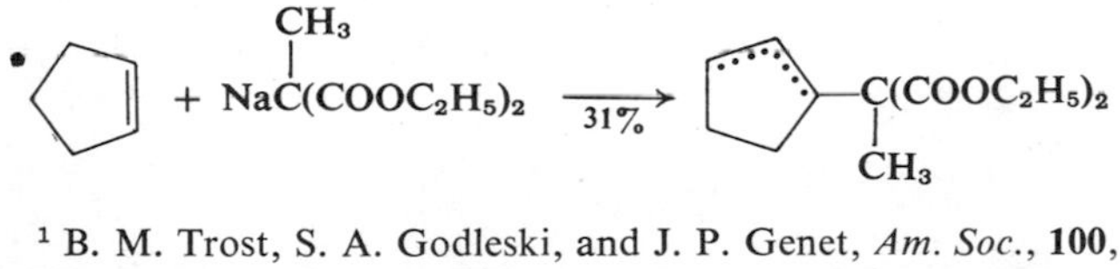

[1] B. M. Trost, S. A. Godleski, and J. P. Genet, *Am. Soc.*, **100**, 3930 (1978).
[2] R. F. Heck, *ibid.*, **90**, 5526 (1968); *Org. Syn.*, **51**, 17 (1971).
[3] T. Hayashi and L. S. Hegedus, *Am. Soc.*, **99**, 7093 (1977).

Bis[(−)-camphorquinone-α-dioximato]cobalt(II) hydrate (1). Mol. wt. 467.42, m.p. 240–241°. The chiral complex is prepared by the reaction of $CoCl_2 \cdot 6H_2O$ with the α-dioxime of (−)-camphorquinone.

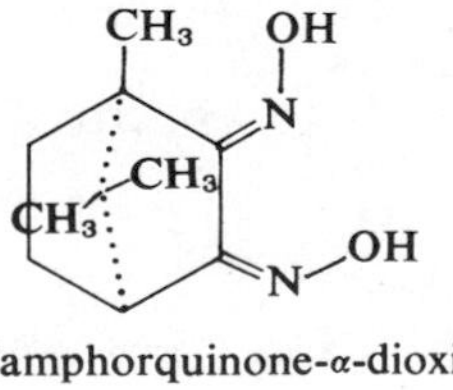

(−)-Camphorquinone-α-dioxime

Asymmetric cyclopropanation.[1] Optically active cyclopropanes can be prepared in optical yields as high as 88% by reaction of certain olefins with various diazo compounds in the presence of this catalyst. Chemical yields are usually 90–95%. However, the reaction is generally limited to terminal double bonds conjugated with an aryl or carbonyl group or with another double bond. The (1S)-enantiomer is always obtained in large excess.

Examples:

$$(C_6H_5)_2C{=}CH_2 + N_2CHCOOC_2H_5 \xrightarrow[95\%]{1}$$

(70% ee)

$$C_6H_5CH{=}CH_2 + N_2CHCOOCH_2C(CH_3)_3 \xrightarrow[87\%]{1}$$

(88% ee)

[1] A. Nakamura, A. Konishi, Y. Tatsuno, and S. Otsuka, *Am. Soc.*, **100**, 3443 (1978); A. Nakamura, A. Konishi, R. Tsujitani, and M. Kudo, *ibid.*, **100**, 3449 (1978).

Bis(cyclopentadienyl)methyltitanium, $(\eta\text{-}C_5H_5)_2TiCH_3$ (**1**). This complex is prepared *in situ* from Cp_2TiCl and 1 equiv. of CH_3Li in ether at $-78°$ (1 hour).

Cyclometalation of pyridines and quinolines.[1] The reagent (**1**) metalates 2-substituted pyridines, quinoline, and 8-methylquinoline to give purple crystalline compounds that can be isolated in about 25% yield. They probably have structures **2** and **3**. A typical reaction of these complexes with an electrophile is shown.

2 (R = CH$_3$, C$_6$H$_5$, CH=CH$_2$)

3 (R = H, CH$_3$)

$$\xrightarrow[\text{quant.}]{\substack{I_2\\-78 \to 20°}}$$

+ Cp$_2$TiI

4

[1] B. Klei and J. H. Teuben, *J.C.S. Chem. Comm.*, 659 (1978).

Bis(dicarbonylcyclopentadienyliron), $[C_5H_5Fe(CO)_2]_2$ (**1**). Mol. wt. 353.92, m.p. 194° dec. Supplier: Alfa. Preparation.[1]

Olefin complexes of dicarbonylcyclopentadienyliron.[2] The dimer **1** is converted into sodium dicarbonylcyclopentadienyl ferrate (**2**), (**5**, 610; **6**, 538–539) by reaction with 3% sodium amalgam in THF under N_2. This reagent reacts with methallyl chloride at 0° to form the complex **3**, which can be isolated and stored for prolonged periods at 20°. However, when **3** is heated (40°) with HBF_4 in CH_2Cl_2 it rearranges to dicarbonylcyclopentadienyl(isobutylene)iron tetrafluoro-

borate (**4**). This cationic olefin complex is stable at 0° under nitrogen indefinitely. When **4** is heated in the presence of an olefin (2–3 equiv.) ligand exchange leads to $C_5H_5Fe^+(CO)_2$(olefin) BF_4^- (**5**) if **5** is more stable thermodynamically than **4**.

[1] J. J. Eisch and R. B. King, *Organometal. Syn.*, **1**, 114 (1965).
[2] S.-B. Samuels, M. Rosenblum, and W. P. Giering, *Org. Syn.*, submitted (1977).

Bis-3-methyl-2-butylborane (Disiamylborane), Sia_2BH, **1**, 57–59; **3**, 22; **4**, 37; **5**, 39–41; **6**, 62.

1,3-Dienes. Negishi and Abramovitch[1] have found that reasonably hindered organoboranes are much more reactive to alkynyllithiums than an acetoxyl group is and have used this property for an efficient synthesis of (7E, 9Z)-dodecadien-1-yl acetate, a sex pheromone of the European grapevine moth (equation **I**).

[1] E. Negishi and A. Abramovitch, *Tetrahedron Letters*, 411 (1977).

Bis(2,4-pentanedionato)nickel [Nickel(II) acetylacetonate], 6, 417.

Ethers. This nickel derivative catalyzes the formation of ethers from an alcohol and an alkyl halide. The method fails in the case of phenols.[1]

$$R^1OH + R^2Cl(Br) \xrightarrow[65-90\%]{Ni(acac)_2} R^1OR^2 + HCl(Br)$$

Tetrasubstituted alkenes.[2] Methylmagnesium bromide in THF–benzene adds to 1-trimethylsilyl-1-octyne in the presence of a catalytic amount of this nickel compound and trimethylaluminum (1:1) to give the vinyl Grignard reagent **2**,

(I) $C_6H_{13}C\equiv CSi(CH_3)_3 + CH_3MgBr \xrightarrow{\substack{Ni(acac)_2, \\ Al(CH_3)_3}}$

which slowly isomerizes to **3** on standing. The rate of isomerization depends on the solvent. Other nickel species are ineffective as catalyst. As expected, **2** and **3** are highly reactive and react with aldehydes, ketones, carbon dioxide, and iodine.

[1] M. Yamashito and Y. Takegami, *Synthesis*, 803 (1977).
[2] B. B. Snider, M. Karras, and R. S. E. Conn, *Am. Soc.*, **100**, 4624 (1978).

Bis(2,4-pentanedionato)nickel–Diisobutylaluminum hydride.

Alkynylation of enones. The nickel catalyst obtained from Ni(acac)$_2$ and DIBAH catalyzes the conjugate addition of organoaluminum acetylides even to transoid enones. The acetylenic aluminum reagent is prepared according to Fried's procedure (**4**, 144) from a lithium acetylide and dimethylaluminum chloride. In all cases only 1,4-addition is observed.[1]

Examples:

[1] R. T. Hansen, D. B. Carr, and J. Schwartz, *Am. Soc.*, **100**, 2244 (1978).

Bis(N-*p*-toluenesulfonyl)sulfodiimide, TsN=S=NTs (1), **7**, 317. Mol. wt. 370.47, m.p. 48–50°, extremely sensitive to moisture.
Preparation[1,2]:

$$TsNH_2 + SOCl_2 \xrightarrow[69\%]{} TsN{=}S{=}O \xrightarrow[90\%]{Py} TsN{=}S{=}NTs + SO_2$$
$$1$$

Allylic amination (**7**, 317).[3] Allylic amination with this reagent was used in a synthesis of gabaculine (**2**), a metabolite of *S. toyocaenis* that inhibits an aminotransferase. The synthesis is outlined in equation (I). The overall yield of **2** from 3-cyclohexene-1-carboxylic acid was 23%.

[1] G. Kresze and W. Wucherpfenning, *Angew. Chem., Int. Ed.*, **6**, 149 (1967).
[2] T. Hori, S. P. Singer, and K. B. Sharpless, *J. Org.*, **43**, 1456 (1978).
[3] S. P. Singer and K. B. Sharpless, *ibid.*, **43**, 1448 (1978).

Bis(tri-*n*-butyltin) oxide, 6, 56–57; **7,** 26–27.

Oxidation of sulfides. This oxide in combination with bromine has been used for selective oxidation of a secondary alcohol in the presence of a primary alcohol (**7**, 26–27). This mild procedure is also applicable to oxidation of sulfides to sulfoxides without further oxidation to sulfones (equation I). The method can also be used to oxidize hydroxy sulfides to keto sulfoxides in one operation. The reagent also can be used for oxidation of sulfenamides to sulfinamides without formation of sulfonamides (equation II).[1]

(I) $RSR^1 + (Bu_3Sn)_2O + Br_2 \xrightarrow[70-90\%]{CH_2Cl_2,\ 20°} R\overset{\overset{\displaystyle O}{\|}}{S}R^1 + 2Bu_3SnBr$

(II) $C_6H_5SNR_2 \xrightarrow[80-92\%]{(Bu_3Sn)_2O,\ Br_2} C_6H_5\overset{\overset{\displaystyle O}{\|}}{S}NR_2$

Glycosyl esters.[2] β-D-Glycosyl esters can be prepared by tributylstannylation of the tetrabenzylglucopyranose (1) with bis(tri-*n*-butyltin) oxide followed by reaction with an acyl halide (equation I). This approach was used in a partial synthesis of the ester stevioside (1) from steviobioside (2). Stevioside is of interest because it is about 300 times as sweet as sugar.

(I)

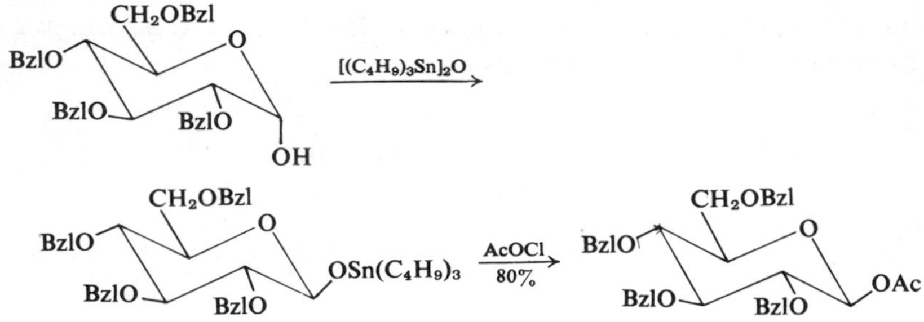

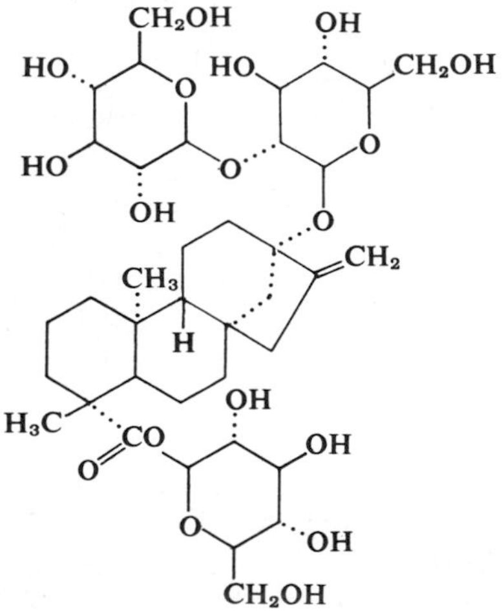

1

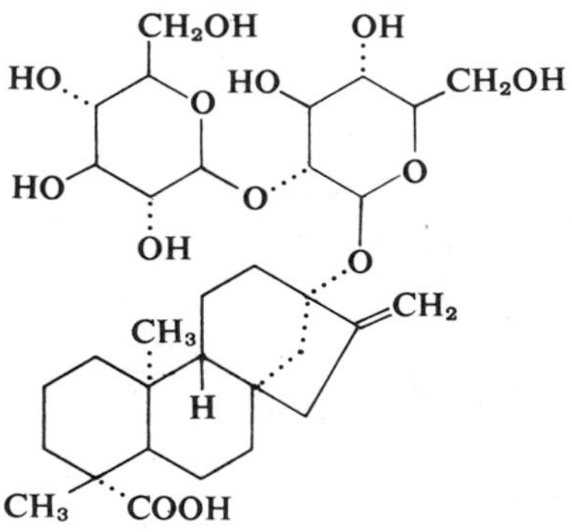

[1] Y. Ueno, T. Inoue, and M. Okawara, *Tetrahedron Letters*, 2413 (1977).
[2] T. Ogawa, M. Nozaki, and M. Matsui, *Carbohydrate Res.*, **60**, C7 (1978).

Bis(tri-*n*-butyltin) peroxide, $(n\text{-}C_4H_9)_3SnOOSn(C_4H_9\text{-}n)_3$ (1). Mol. wt. 612.05, somewhat unstable but can be stored in CH_2Cl_2 solution at $-20°$. The peroxide is prepared by reaction of tri(*n*-butyl)tin methoxide with 100% H_2O_2 at 0°.

Caution: All reactions with peroxides are potentially explosive.

Dialkyl peroxides. Triflates react with this peroxide at 20° to form dialkyl peroxides in surprisingly good yields. Ethers are often obtained as by-products.[1]
Examples:

$$(CH_3)_2CHOSO_2CF_3 + 1 \longrightarrow (CH_3)_2CHOOCH(CH_3)_2 + (CH_3)_2CHOCH(CH_3)_2$$
$$(79\%) \qquad\qquad (7\%)$$

The reaction has been modified to prepare 2,3-dioxabicyclo-[2.2.1]heptane (**3**),[2] the basic unit present in prostaglandin endoperoxides. Reaction of **2** with **1** under the conditions used above gave secondary products derived from **3**, which is unusually reactive. If the reaction is carried out *in vacuo* the desired peroxide **3**

can be isolated by use of a cold trap (−78°). When pure, **3** is stable to 70°. It rearranges to levulinaldehyde (80% yield) in the presence of Dabco (35°).

The same issue of the *Am. Soc.*[3] reports a different route to **3**, as shown in equation (I). After purification, **3** was obtained in crystalline form, m.p. 42–43.5°.

This synthesis is based on a synthesis of dioxetane developed by Kopecky *et al.*[4] (**6**, 287–288).

[1] M. F. Salomon and R. G. Salomon, *Am. Soc.*, **99**, 3500 (1977).
[2] R. G. Salomon and M. F. Salomon, *ibid.*, **99**, 3501 (1977); **101**, 4290 (1979).
[3] N. A. Porter and D. W. Gilmore, *ibid.*, **99**, 3503 (1977).
[4] K. R. Kopecky *et al.*, *Canad. J. Chem.*, **46**, 25 (1968); **53**, 1103 (1975).

Bis(trifluoroacetoxy)borane, $(CF_3OCO)_2BH$ (**1**). This reducing agent is formed as a complex with THF by addition of $BH_3 \cdot THF$ to TFA. The complex is not stable to prolonged storage, but is more stable if the THF is removed. The advantage of this new hydride is stability to acid.

Reduction of indoles to indolines.[1] This reduction requires an acidic medium and was chosen for a detailed study of this new reducing agent. Isolated **1** in TFA reduces **2** to **3** in 86% yield; but the same yield is attainable (in ~2 minutes) by the combination $BH_3 \cdot THF$ in TFA. Other somewhat less effective reagents for this reduction are $BH_3 \cdot$ pyridine in TFA, $BH_3 \cdot (CH_3)_2S \cdot THF$ in TFA, $NaBH_3CN$ in TFA, and catecholborane in TFA.

2   **3**

[1] B. E. Maryanoff and D. F. McComsey, *J. Org.*, **43**, 2733 (1978).

2,3-Bis(trimethylsilyloxy)-1,3-butadiene (1). Mol. wt. 230.45, b.p. 77–79°/10 mm.
  Preparation:

Diels-Alder diene. This diene undergoes ready cycloaddition reactions.[1]
  Examples:

[1] D. R. Anderson and T. H. Koch, *J. Org.*, **43**, 2726 (1978).

1,3-Bis(trimethylsilyloxy)cyclohexa-1,3-diene, (**1**).

Mol. wt. 256.48, b.p. 125°/12 mm. The diene is obtained in 77% yield by reaction of cyclohexane-1,3-dione with chlorotrimethylsilane, $ZnCl_2$, and $N(C_2H_5)_3$.

Diels-Alder reaction. This diene (also the 5-methyl and the 5,5-dimethyl derivatives) reacts with acrylonitrile to give, after hydrolysis, the *exo*-adduct predominantly (equation I).[1] This reaction is, therefore, another exception to the rule of *endo*-addition of Alder.[2] The *exo*-cycloaddition reaction of 1,3-bis(trimethylsiloxy)-5-methylcyclohexa-1,3-diene with acrylonitrile has been used for a

(I) $1 + \begin{matrix} CH_2 \\ \parallel \\ CHCN \end{matrix} \xrightarrow[]{170°} \xrightarrow[31\%]{HCl}$

2   3

stereoselective synthesis of pumiliotoxin C (**4**), a toxic alkaloid from certain frogs.[3]

4

[1] T. Ibuka, Y. Mori, T. Aoyama, and Y. Inubushi, *Chem. Pharm. Bull. Japan*, **26**, 457 (1978).

[2] K. Alder and G. Stein, *Angew. Chem.*, **50**, 510 (1937).

[3] T. Ibuka, Y. Mori, T. Aoyama, and Y. Inubushi, *Tetrahedron Letters*, 3169 (1976).

Bis(triphenylphosphine)copper tetrahydroborate, $[(C_6H_5)_3P]_2CuBH_4$ (**1**). Mol. wt. 602.96, stable white crystals, soluble in $HCCl_3$, C_6H_6, CH_3COCH_3, insoluble in ether, C_2H_5OH, and H_2O. The reagent is prepared by treatment of $CuSO_4·5H_2O$ or CuCl with excess triphenylphosphine; the product is then reduced with $NaBH_4$; yield 83%.[1]

Reduction of —COCl to —CHO.[2] Aliphatic and aromatic acid chlorides are reduced to aldehydes by 1 equiv. of the reagent (30 minutes, 20°). Yields are somewhat low unless triphenylphosphine (2 equiv.) is added. Acetone is the most suitable solvent. Yields of aldehydes are in the range 60–85%.

[1] J. M. Davidson, *Chem. Ind.*, 2021 (1964); S. J. Lippard and K. M. Melmed, *Inorg. Chem.*, **6**, 2223 (1967); S. J. Lippard and D. A. Ucko, *ibid.*, **7**, 1051 (1968).

[2] G. W. J. Fleet, C. J. Fuller, and P. J. C. Harding, *Tetrahedron Letters*, 1437 (1978); T. N. Sorrell and R. J. Spillane, *ibid.*, 2473 (1978).

9-Borabicyclo[3.3.1]nonane (9-BBN), 2, 31; **3,** 24–29; **4,** 41; **5,** 46–47; **6,** 62–64; **7,** 29–30.

Hydroboration. Hydroborations with this reagent have been reviewed[1]; Brown concludes that the regio- and stereoselectivity of 9-BBN surpass those of other hydroboration reagents. An example of the marked stereospecificity of hydroboration with 9-BBN is the reduction of 1-substituted cycloalkenes to *trans*-2-methylcycloalkanols (equation I). The same selectivity is shown in the hydroboration of 1,3-dimethylcycloalkenes (equation II). In fact this method is

(I)

(II)

excellent for preparation of cycloalkanols with two adjacent alkyl groups both *trans* to the hydroxyl group and for the corresponding cycloalkanones obtained on oxidation.[2]

Conjugation usually reduces the reactivity of dienes to hydroboration. Symmetrical acyclic 1,3-dienes can often be monohydroborated successfully with 9-BBN (equation III); unsymmetrical acyclic 1,3-dienes usually undergo monohydroboration without difficulty (equations IV and V).

(III) $(CH_3)_2C{=}CHCH{=}C(CH_3)_2$ $\xrightarrow[90\%]{\text{9-BBN, THF}}$ $(CH_3)_2C{=}CHCHCH(CH_3)_2$

(IV)

(V) $(CH_3)_3C{-}CH{=}CHCH{=}CH_2$ $\xrightarrow[99\%]{}$ $(CH_3)_3C{-}CH{=}CHCH_2CH_2{-}B$

Hydroboration of 1,3-cyclopentadiene with 9-BBN is too slow to be practicable, but monohydroboration of 1,3-cyclohexadiene is a useful reaction (equation VI). 9-BBN reacts with 1,3-cyclooctadiene to give predominantly dihydroboration products.[3]

(VI)

Reduction of enones. This borane is particularly useful for reduction of enones to the corresponding allylic alcohols. Complete details of use for this and other purposes are available.[4]

Examples:

$$C_6H_5CH{=}CHCHO \xrightarrow[99\%]{9\text{-BBN}} C_6H_5CH{=}CHCH_2OH$$

[1] H. C. Brown, R. Liotta, and L. Brener, *Am. Soc.*, **99**, 3427 (1977).
[2] L. Brener and H. C. Brown, *J. Org.*, **42**, 2702 (1977).
[3] H. C. Brown, R. Liotta, and G. W. Kramer, *ibid.*, **43**, 1058 (1978).
[4] S. Krishnamurthy and H. C. Brown, *ibid.*, **42**, 1197 (1977).

9-Borabicyclo[3.3.1]nonane–Pyridine,   Mol. wt. 201.12,

m.p. 76°, stable indefinitely under N_2. This reagent is prepared by reaction of 9-BBN with pyridine in pentane.[1]

Reduction of aldehydes.[2] This reagent reduces aldehyde groups with high selectivity in the presence of keto and many other groups (ester, lactone, nitrile, epoxide, halide, alkene, alkyne). The primary alcohol is isolated by addition of β-aminoethanol (equation I).

[1] H. C. Brown and S. U. Kulkarni, *Inorg. Chem.*, **16**, 3090 (1977).
[2] *Idem, J. Org.*, **42**, 4169 (1977).

Borane–Dimethyl sulfide (BMS), 4, 124, 191; **5**, 47.

This stable form of BH_3 is useful for the preparation of various borane reagents such as thexylborane, disiamylborane, catecholborane, and 9-BBN. The reagent is compatible with a wide variety of solvents.[1]

Reduction of enones. α,β-Unsaturated ketones are reduced by BMS in THF to allylic alcohols in good yields (65–95%). The reagent reduces anthraquinone to anthracene (70% yield).[2]

An improved route to 3-substituted 1,3-propanediols is formulated in equation (I) for conversion of pinacolone (1) into 4,4-dimethyl-1,3-pentanediol (3). The first step involves formylation of a methyl ketone by the method of Corey and Cane (4, 233). The product is then reduced in high yield by the nonbasic borane–dimethyl sulfide complex.[3]

(I) $(CH_3)_3CCOCH_3 + HCOOCH_3$ $\xrightarrow[74\%]{\text{NaH, ether}}$

　　　　1

$$(CH_3)_3C\overset{\overset{\displaystyle O}{\|}}{C}CH{=}CHONa \xrightarrow[76\%]{\substack{1)\ BH_3 \cdot (CH_3)_2S \\ 2)\ CH_3OH}} (CH_3)_3C\overset{\overset{\displaystyle OH}{|}}{C}HCH_2CH_2OH$$

　　　　　　　2　　　　　　　　　　　　　　　　　**3**

[1] H. C. Brown, A. K. Mandal, and S. U. Kulkarni, *J. Org.*, **42**, 1392 (1977).
[2] E. Mincione, *ibid.*, **43**, 1829 (1978).
[3] D. E. Pearson and J. D. Weaver, *Org. Prep. Proc. Int.*, **10**, 29 (1978).

Borane–Pyridine complex, 1, 963–964. Supplier: Aldrich.

Modification of tryptophan residues. Various indole derivatives are reduced to 2,3-dihydro compounds by borane–pyridine in ethanolic 20% hydrochloric acid.[1] This reduction fails completely with tryptophan itself and gives only low yields (15–55%) with N-acyltryptophans. However, use of CF_3COOH as solvent leads to the desired 2,3-dihydro compounds in yields of 85–95%. This is a useful method for selective modification of tryptophyl residues in peptides.[2] The imidazole group is partially decomposed under these conditions, but other groups encountered in peptides are stable.

Another useful method for modification of tryptophan groups is oxidation with DMSO and conc. HCl to form oxindole (2-hydroxytryptophan) groups. This reaction proceeds in >95% yield.[3]

Reduction of heterocycles. Pyridine–borane shows enhanced reducing properties in acetic acid, presumably owing to formation of pyridine–acetyl-borane, $Py\colon BH(OAc)_2$ or $Py\colon BH_2(OAc)$ (**1**). Thus **1** reduces quinoline as shown in equation (I). Some other reductions of heterocycles are: indole $\rightarrow$ indoline

(I)

　　　　　　　　　　　　　　　　(15%)　　　　　　(63%)

(86% yield); quinoxaline $\rightarrow$ 1,2,3,4-tetrahydroquinoxaline (95% yield).

Trimethylamine–borane in acetic acid is not effective for these reductions.[4]

[1] Y. Kikugawa, *J. Chem. Res. (S)*, 212 (1977).
[2] *Idem, ibid.*, 184 (1978).
[3] W. E. Savige and A. Fontana, *J.C.S. Chem. Comm.*, 599 (1976).
[4] Y. Kikugawa, K. Saito, and S. Yamada, *Synthesis*, 447 (1978).

Boron trifluoride etherate, 1, 70–72; **2**, 35–36; **3**, 33; **4**, 44–45; **5**, 54–55; **6**, 65–70; **7**, 31–32.

Aldol condensation of acetals; spiroannelation.[1] Japanese chemists have reported a new method of annelation that involves as the first step aldol condensation of acetals or ketals with the bissilylated succinoin **1**[2] mediated with BF_3 etherate or $TiCl_4$. The method is formulated in equation (I) for the reaction of the acetal of benzaldehyde. Owing to the ring strain of cyclobutanone, the aldol

adduct **2** rearranges (pinacolic) in TFA to 2-phenylcyclopentanedione (**3**).

Application to ketals provides an entry to 2,2-disubstituted cyclopentane-1,3-diones (equation II). Spirodiketones are also easily prepared in this way (equation III).

Catalysis of Diels-Alder reactions. Lewis acids are known to catalyze Diels-Alder reactions and also to influence the regioselectivity (**6**, 65–66). Trost *et al.*[3] have examined the effect of BF_3 etherate on the regioselectivity in the case of dienes containing both oxygen and sulfur substituents. For example, the thermal reaction of the diene **1** with methacrolein results in formation of **2** and **3** in a ratio of 13:1, with sulfur being the controlling element. In the catalyzed reaction the effect of sulfur is reinforced and the ratio of **2** and **3** is 50:1. The reaction of **1** with juglone (**4**) shows the opposite effect of acid catalysis. The uncatalyzed reaction results in formation of **5** and **6** in a ratio of 2:1, with sulfur again controlling the regioselectivity. However, in the catalyzed reaction only **6** is formed. In this case the oxygen function completely controls regiochemistry. Trost interprets these and similar results on the basis of two competing factors, odd electron density and polar factors. Sulfur should exert greater control than oxygen in the thermal reaction because it is less electronegative than oxygen. In the catalyzed reaction polar factors can become more important since the Lewis acid can coordinate with the dienophile and/or the diene and thus enhance or diminish electronic factors. In any case the observed effects can be used in a practical way.

[1] E. Nakamura and I. Kuwajima, *Am. Soc.*, **99**, 961 (1977).
[2] J. J. Bloomfield and J. M. Nelds, *Org. Syn.*, **57**, 1 (1977).
[3] B. M. Trost, J. Ippen, and W. C. Vladuchick, *Am. Soc.*, **99**, 8116 (1977).

Bromine, **3**, 34; **4**, 46–47; **5**, 55–57; **6**, 70–73; **7**, 33–35.

α-**Bromo ketones.** α-Bromo ketones are obtained on bromination of epoxides in CCl_4 under irradiation. Bromohydrins are formed when ether is the solvent.[1]

Examples:

$$n\text{-}C_4H_9\text{-epoxide} + Br_2 \xrightarrow[87\%]{h\nu,\ CCl_4} n\text{-}C_4H_9COCH_2Br$$

$$C_6H_5\text{-epoxide} \xrightarrow{85\%} C_6H_5COCH_2Br$$

$$\text{cyclohexene oxide} \xrightarrow{91\%} \text{2-bromocyclohexanone}$$

Cleavage of α-azido ketones. α-Azido ketosteroids undergo C—C bond cleavage on treatment with bromine in acetic acid at 20°. A typical reaction is formulated in equation (I).[2]

(I)

$$\xrightarrow[80\%]{Br_2,\ HOAc,\ 20°}$$

[1] V. Calò, L. Lopez, and D. S. Valentino, *Synthesis*, 139 (1978).
[2] T. T. Takahashi and J. Y. Satoh, *J.C.S. Chem. Comm.*, 409 (1978).

Bromine fluoride, BrF.

Vinyl fluorides. BrF, generated from NBA and HF in ether, adds to alkenes to form *vic*-bromofluoroalkanes in high yield. These products are dehydrobrominated by KOH either in DMSO (20–25°) or in triethylene glycol (150°).[1]

$$R^1\text{---}CH{=}C\begin{smallmatrix}R^2\\R^3\end{smallmatrix} \xrightarrow[60-95\%]{CH_3CONHBr,\ HF} R^1CH\text{---}\underset{F}{\underset{|}{C}}(\underset{R^3}{\overset{Br}{|}})R^2 \xrightarrow[60-85\%]{KOH} \underset{F}{\overset{R^1}{C}}{=}\underset{R^3}{\overset{R^2}{C}}$$

[1] L. Eckes and M. Hanack, *Synthesis*, 217 (1978).

5-Bromo-1,4-diphenyl-3-methylthio-*s*-triazolium bromide (1). Mol. wt. 427.17, m.p. 146–147° dec.

Preparation:

$$C_6H_5NHN{=}\underset{SCH_3}{\overset{|}{C}}NHC_6H_5 \xrightarrow[77\%]{HCOOH,\ KBr} \text{(triazolium)} \xrightarrow[2)\ N(C_2H_5)_3\ (76\%)]{1)\ Br_2\ (85\%)} \text{(1)}$$

1

Conversion of —CH₂NH₂ into —CHO.[1] This transformation can be carried out as formulated for conversion of 1-propylamine into propanal (equation I).

$$(I) \quad C_3H_7NH_2 + 1 \xrightarrow[91\%]{2N(C_2H_5)_3, \, -2HBr} C_3H_7N{=}\underset{\underset{C_6H_5}{|}}{\overset{\overset{C_6H_5}{|}}{N}}\overset{SCH_3}{\underset{N}{\diagdown}} \xrightarrow[89\%]{C_2H_5OOCN{=}NCOOC_2H_5}$$

$$CH_3CH_2CH{-}N{=}\underset{\underset{\underset{COOC_2H_5}{|}}{C_2H_5OOC}}{\overset{\overset{C_6H_5}{|}}{N}} \xrightarrow{HCl} CH_3CH_2CHO + H_2N{-}\overset{SCH_3}{\underset{C_6H_5}{N}}\, Cl^-$$

$$(48\%)$$

[1] G. Doleschall, *Tetrahedron Letters*, 2131 (1978).

N-Bromosuccinimide, 1, 78–80; **2**, 40–42; **3**, 34–36; **4**, 49–53; **5**, 65–66; **6**, 74–76; **7**, 37–40.

Anti-Markownikoff hydrobromination of alkenes. This reaction can be carried out by hydrosilylation of an alkene followed by conversion to an organopentafluorosilicate, in which the C—Si bonds are highly reactive to halogens. The method can also be used for preparation of alkenyl halides (last example).[1]

Examples:

$$n\text{-}C_6H_{13}CH{=}CH_2 + HSiCl_3 \xrightarrow{H_2PtCl_6} [RCH_2CH_2SiCl_3] \xrightarrow[88.5\%]{KF}$$

$$K_2[RCH_2CH_2SiF_5] \xrightarrow[77\%]{NBS} n\text{-}C_8H_{17}Br$$

$$C_6H_5C{\equiv}CH \xrightarrow[57\%]{} \underset{H}{\overset{C_6H_5}{\diagup}}C{=}C\underset{Br}{\overset{H}{\diagup}}$$

Allylic bromination. Ikota and Ganem[2] have reported a stereospecific synthesis of racemic *trans*-3,4-dihydroxy 3,4-dihydrobenzoic acid (**5**), a metabolite of shikimic acid. The starting material was 1,4-dihydrobenzoic acid, which was converted into the bromolactone **1** as formulated. This substance was resistant to most allylic oxidizing reagents, but reacted slowly with NBS to give the ene dibromide **2**. This product was converted by reaction with sodium acetate into a 1:1 mixture of the allylic acetates **3** and **4**. Hydrolysis of 3 led to the desired **5**. The conversions **1 → 2 → 3** and **4** occur on the β-side of the bicyclic ketone. This stereoselectivity was used in a conversion of the undesired acetate **4** into **3**. Acidic hydrolysis of **4** gave the corresponding allylic alcohol (**6**). This was converted into the mesylate (7), which underwent a clean S_N2' reaction with lithium acetate to give 3. Thanks to this recycling, the overall yield of 5 from the dihydrobenzoic acid was 17%.

An earlier synthesis of 5 reported by Chiasson and Berchtold[3] is shown in equation (I).

Oxidation of trimethylsilyl ethers.[4] Trimethylsilyl ethers[5] of primary alcohols are converted by NBS under irradiation of a sunlamp into esters; under these conditions and in the presence of pyridine the ethers of secondary alcohols are converted into ketones.

Examples:

$$CH_3(CH_2)_5OSi(CH_3)_3 \xrightarrow[80\%]{\substack{NBS, \\ h\nu, 0°}} CH_3(CH_2)_4COO(CH_2)_5CH_3$$

$$C_6H_5CH_2OSi(CH_3)_3 \xrightarrow[48\%]{} C_6H_5CHO$$

$$C_6H_5\underset{\substack{| \\ OSi(CH_3)_3}}{CH}CH_3 \xrightarrow[76\%]{\substack{NBS, Py, \\ h\nu, 20°}} C_6H_5\underset{\substack{\| \\ O}}{C}CH_3$$

$$CH_3(CH_2)_5\underset{\substack{| \\ OSi(CH_3)_3}}{CH}CH_3 \xrightarrow[55\%]{} CH_3(CH_2)_5\underset{\substack{\| \\ O}}{C}CH_3$$

One interesting application of this reaction is the preparation of mixed esters by oxidation of silyl ethers in the presence of aldehydes.

Examples:

$$CH_3CH_2OSi(CH_3)_3 + CH_3(CH_2)_8CHO \xrightarrow[83\%]{\substack{NBS, \\ h\nu, 0°}} CH_3(CH_2)_8COOCH_2CH_3$$

$$CH_3CH_2OSi(CH_3)_3 + C_6H_5CHO \xrightarrow[45\%]{} C_6H_5COOCH_2CH_3$$

$$(CH_3)_3COSi(CH_3)_3 + CH_3(CH_2)_4CHO \xrightarrow[42\%]{} CH_3(CH_2)_4COOC(CH_3)_3$$

This mixed ester synthesis fails with unsaturated substrates.

[1] K. Tamao, J. Yoshida, M. Takahashi, H. Yamamoto, T. Kakui, H. Matsumoto, A. Kurita, and M. Kumada, *Am. Soc.*, **100**, 290 (1978).

[2] N. Ikota and B. Ganem, *ibid.*, **100**, 351 (1978).

[3] B. A. Chiasson and G. A. Berchtold, *ibid.*, **96**, 2898 (1974).

[4] H. W. Pinnick and N. H. Lajis, *J. Org.*, **43**, 371 (1978).

[5] M. E. Jung, *ibid.*, **41**, 1479 (1976).

α-Bromovinyltrimethylsilane (1). Mol. wt. 179.14, b.p. 56.7°/67 mm., 26–27°/ 15 mm.

Preparation[1,2]:

$$Cl_3SiCH=CH_2 \xrightarrow[82\%]{Br_2,\ h\nu} \underset{Br}{\overset{Cl_3Si}{\diagdown}}C=CH_2 \xrightarrow[70\%]{CH_3MgI} \underset{Br}{\overset{(CH_3)_3Si}{\diagdown}}C=CH_2$$

1

Terminal allenes.[2] Aldehydes and ketones can be converted into terminal allenes by the sequence formulated in equation (I).

(I)   **1** $\xrightarrow[\substack{75-80\%}]{\substack{1)\ t\text{-BuLi,}\\ \text{ether, }-78°\\ 2)\ R^1COR^2}}$ $\underset{OH}{\overset{R^2\ Si(CH_3)_3}{R^1-C-C=CH_2}}$ $\xrightarrow{SOCl_2,\ CCl_4}$ $\left[\underset{Cl}{\overset{R^2\ Si(CH_3)_3}{R^1-C-C=CH_2}} \right.$

$+ \underset{R^2\ \ \ \ \ CH_2Cl}{\overset{R^1\ \ \ \ \ Si(CH_3)_3}{C=C}} + \underset{R^1\ \ \ \ \ CH_2Cl}{\overset{R^2\ \ \ \ \ Si(CH_3)_3}{C=C}} \left. \right]$ $\xrightarrow[40-50\%]{\substack{(C_2H_5)_4NF,\\ DMSO}}$ $\underset{R^2}{\overset{R^1}{\diagdown}}C=C=CH_2$

[1] A. Ottolenghi, M. Fridkin, and A. Zilkha, *Canad. J. Chem.*, **41**, 2977 (1963).
[2] T. H. Chan, W. Mychajlowskij, B. S. Ong, and D. N. Harpp, *J. Org.*, **43**, 1526 (1978).

1-*t*-Butoxycarbonyl-4-dimethylaminopyridinium tetrafluoroborate (1). Mol. wt. 310.1, m.p. 91°, soluble in water. This reagent is prepared[1] in high yield by the reaction of di-*t*-butyl dicarbonate[2] with 4-dimethylaminopyridinium tetrafluoroborate in ethyl acetate:

$$(CH_3)_3COC-O-C-OC(CH_3)_3 + (CH_3)_2N-\overset{+}{\underset{}{\bigcirc}}NHBF_4^- \xrightarrow{95\%}$$

$$(CH_3)_2N-\bigcirc\overset{+}{N}-\overset{O}{C}-OC(CH_3)_3 \quad BF_4^-$$

1

BOC-Amino acids and peptides.[1] The reagent reacts with sodium salts of amino acids (or peptides) in water to form the BOC derivatives in high yield.

[1] E. Giublé-Jampel and M. Wakselman, *Synthesis*, 772 (1977).
[2] B. M. Pope, Y. Yamamoto, and D. S. Tarbell, *Org. Syn.*, **57**, 45 (1977).

(2S,4S)-N-*t*-Butoxycarbonyl-4-diphenylphosphino-2-diphenylphosphinomethylpyrrolidine (BPPM) (1).

Preparation from L-hydroxyproline[1]:

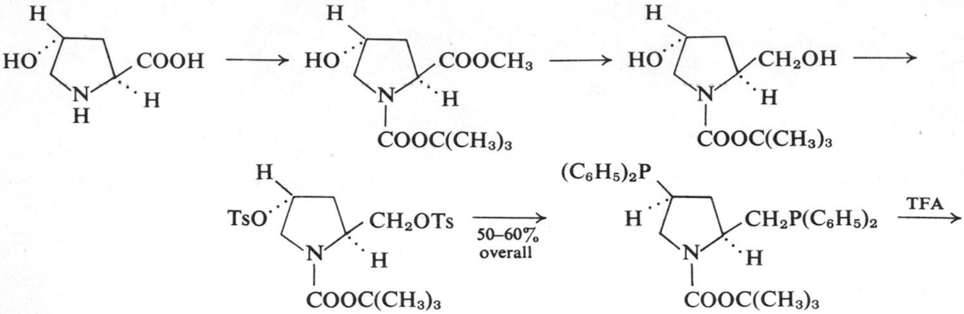

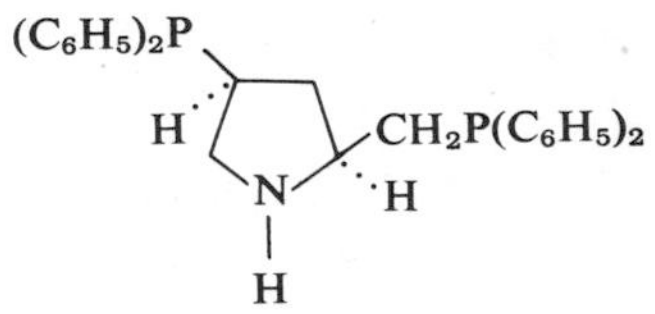

2 (PPM)

Asymmetric hydrogenation. BPPM is an effective ligand for asymmetric hydrogenation of unsaturated acids of type **3** with the rhodium catalyst [Rh(1,5-hexadiene)Cl]$_2$. Optical yields of 83–91% can be obtained when triethylamine is

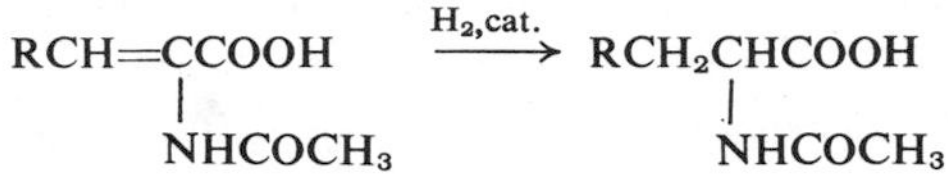

$$RCH{=}CCOOH \xrightarrow{H_2,cat.} RCH_2CHCOOH$$

$$\underset{NHCOCH_3}{|} \qquad\qquad \underset{NHCOCH_3}{|}$$

3   **4**

added. This effect of base is not observed with the usual bisphosphine ligands such as DIOP (**4**, 273; **5**, 360–361) or with the closely related **2**.[1]

This asymmetric induction was used to obtain D-(−)-pantoyl lactone (**7**), the microbiological precursor to D-(+)-pantothenic acid (**8**). The key step, reduction

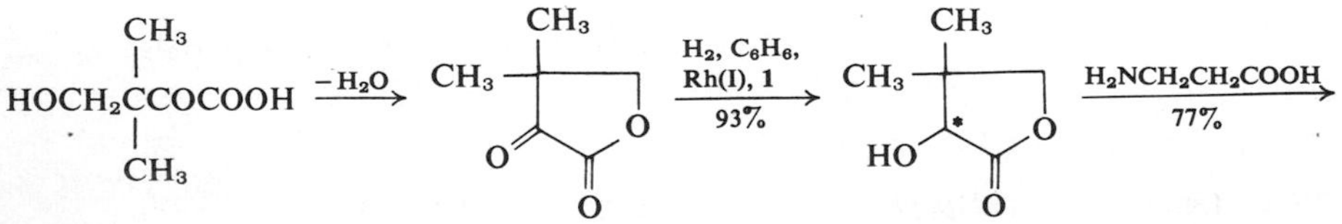

5   **6**   **7** (86.7% ee)

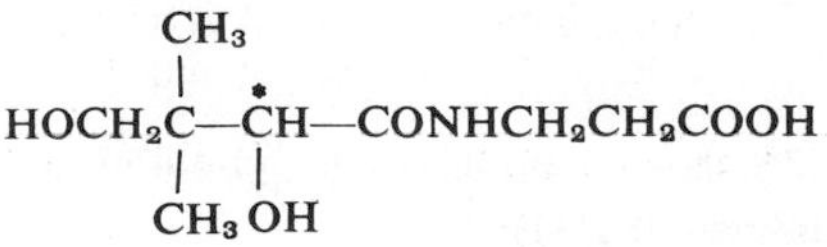

$$\underset{\underset{CH_3}{|}\ \ \underset{OH}{|}}{\overset{\overset{CH_3}{|}}{HOCH_2C-\overset{*}{C}H-CONHCH_2CH_2COOH}}$$

8

of **6** to **7**, was carried out using [Rh(1,5-cyclooctadiene)Cl]$_2$ with **1** as ligand in C$_6$H$_6$ at 30°.[2]

[1] K. Achiwa, *Am. Soc.*, **98**, 8265 (1976).
[2] I. Ojima, T. Kogure, T. Terasaki, and K. Achiwa, *J. Org.*, **43**, 3444 (1978).

t-**Butoxyfurane (1)**. Mol. wt. 140.18, b.p. 44°/16 mm.
    Preparation[1]:

γ-*Alkylidenebutenolides*.[2] A general route to this system from **1** is illustrated for the preparation of **3**. Previous routes to these compounds have been reviewed.[3]

[1] R. Sornay, J.-M. Meunier, and P. Fournari, *Bull. Soc.*, 990 (1971).
[2] G. A. Kraus and H. Sugimoto, *J.C.S. Chem. Comm.*, 30 (1978).
[3] Y. S. Rao, *Chem. Rev.*, **76**, 625 (1976).

t-**Butylchlorodimethylsilane, 4**, 57–58, 176–177; **5**, 74–75; **6**, 78–79.

   Oligoribonucleotides. Full details are available for the synthesis of oligoribonucleotides developed by the group of Ogilvie.[1] The success depends markedly on the availability of suitably protected monomers. This problem is not so important in the synthesis of oligodeoxynucleotides. In the case of adenosine and uridine nucleosides, the *t*-butyldimethylsilyl group is an excellent protecting group since it is compatible with the methoxytrityl group for protection of the 5′-hydroxyl group and with Letsinger's chlorophosphite procedure[2] for coupling of nucleosides to nucleotides. The silyl groups, regardless of their position, the methoxytrityl group, and the trichloroethyl phosphate protecting group are all removed within 30 minutes by tetra-*n*-butylammonium fluoride in THF at room temperature. Yields in the synthesis of dinucleotides are in the range 35–80%. Two trinucleotides were synthesized in 56 and 66% yields, respectively. Two tetranucleotides were prepared in yields of 57 and 40%.

[1] K. K. Ogilvie, S. L. Beaucage, A. L. Schifman, N. Y. Theriault, and K. L. Sadana, *Canad. J. Chem.*, **56**, 2768 (1978).

[2] The reagent is trichloroethylphosphorodichloridite, $CCl_3CH_2OP(O)Cl_2$; preparation: W. Gerrard, W. J. Green, and R. J. Phillips, *J. Chem. Soc.*, 1148 (1959).

t-**Butyl α-chloro-α-lithiotrimethylsilylacetate**, $\overset{\displaystyle Si(CH_3)_3}{\underset{\displaystyle Li}{Cl-C-COOC(CH_3)_3}}$ **(1)**.

Mol. wt. 228.72.

Preparation:

$$ClCH_2COOC(CH_3)_3 \xrightarrow[\text{2) ClSi(CH}_3)_3]{\text{1) LDA, THF}} \overset{\displaystyle Si(CH_3)_3}{ClCHCOOC(CH_3)_3} \xrightarrow[-78°]{\text{LDA, THF,}} \quad (1)$$

α-*Chloro-α,β-unsaturated esters*. The reagent reacts with aldehydes and ketones to form adducts, which on addition of thionyl chloride (**6**, 636–637) are converted to α-chloro-α,β-unsaturated esters in 20–55% yield. Both (Z)- and (E)-isomers are formed, with the former predominating in the eight examples reported.[1]

$$\overset{R^1}{\underset{R^2}{>}}C=O + 1 \xrightarrow[-78°]{\text{THF,}} \left[\overset{\displaystyle LiO \quad Cl}{\underset{\displaystyle R^2 \quad Si(CH_3)_3}{R^1-C-C-COOC(CH_3)_3}} \right] \xrightarrow[20-55\%]{SOCl_2} R^1R^2C=C\overset{\displaystyle Cl}{\underset{\displaystyle COOC(CH_3)_3}{}}$$

[1] T. H. Chan and M. Moreland, *Tetrahedron Letters*, 515 (1978).

t-**Butyl dilithioacetoacetate**, $LiCH_2COLiCHCOOC(CH_3)_3$ **(1)**. Mol. wt. 170.06. This dianion is generated preferably in two steps. *t*-Butyl acetoacetate is converted first into the monoanion with sodium hydride in THF and then into the dianion with *n*-butyllithium in hexane.[1]

Cleavage of α-hydroxy epoxides.[2] This reagent is similar to dilithioacetate (**7**, 113–114) in that it induces ring opening of *cis*-α-hydroxy epoxides (equation I). Of greater significance a protected form of the hydroxyl group can be used. Reaction of **1** with the α-methoxymethyloxy epoxide **2** is stereospecific and results

(I)

(II)

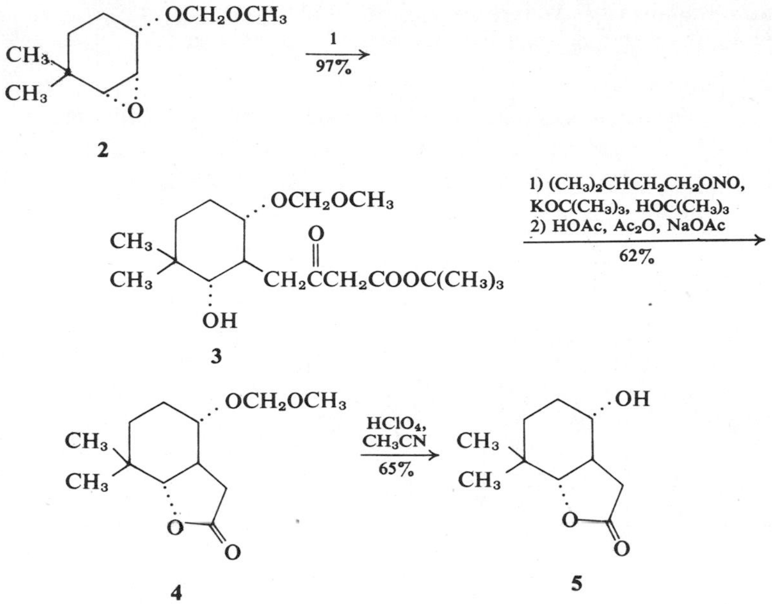

2

1) (CH₃)₂CHCH₂CH₂ONO, KOC(CH₃)₃, HOC(CH₃)₃
2) HOAc, Ac₂O, NaOAc
→
62%

3

HClO₄, CH₃CN
65%

4

5

(III)

65% overall

6

7

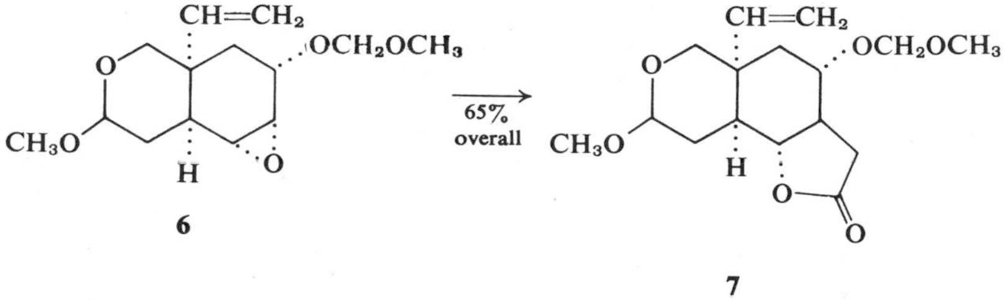

in **3** only. This is converted into the lactone alcohol **5** by a Beckmann degradation with loss of two carbon atoms (equation II). The same sequence is used to convert **6** into the lactone **7** in 65% overall yield (equation III).

In addition, *t*-butyl dilithioacetoacetate opens *trans*-α-hydroxy epoxides regioselectively (equation IV).

(IV)

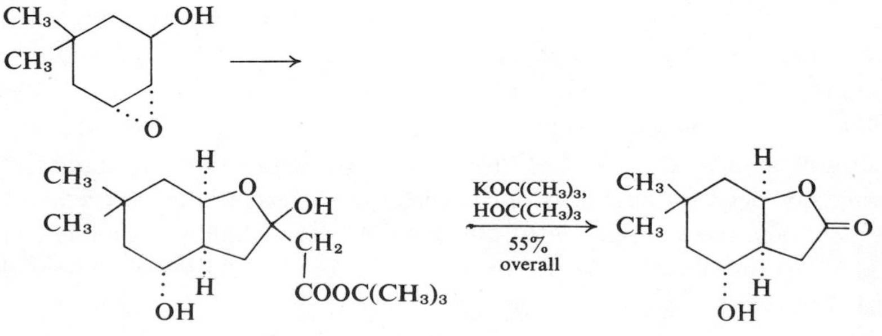

KOC(CH₃)₃, HOC(CH₃)₃
55% overall

[1] S. N. Huckin and L. Weiler, *Am. Soc.*, **96**, 1082 (1974).
[2] G. R. Kieczykowski, M. R. Roberts, and R. H. Schlessinger, *J. Org.*, **43**, 788 (1978).

t-**Butyl hydroperoxide, 1**, 88–89; **2**, 49–50; **3**, 77–78; **5**, 75–77; **6**, 81–82; **7**, 43–45.

Epoxidation of bishomoallylic alcohols. The Sharpless method of epoxidation (**5**, 75–76) of these alcohols is highly stereoselective. Thus reaction of **1** with *t*-butyl hydroperoxide catalyzed with $VO(acac)_2$ followed by treatment with acetic acid gives the tetrahydrofuranes **2** and **3** in the ratio 9:1.[1]

Remote epoxidation. Breslow and Maresca[2] have found that Sharpless epoxidation of olefins bearing hydroxyl groups (**5**, 76–77) is amenable to direction by a template (for an example of this method of control *see* **5**, 352–353). Thus the steroid ester **1** can be epoxidized in high yield by *t*-butyl hydroperoxide with catalysis by $Mo(CO)_6$. In contrast the ester **2** under the same conditions is recovered unchanged. Another striking example of the template effect is that only the 17,20-double bond of the ester **3** can be epoxidized by the Sharpless method. Ordinarily Δ^4- stenols are epoxidized readily. In these examples both the template and the steroid are rigid molecules.

A different picture emerges when this template effect is extended to flexible substrates such as all-*trans*-farnesol and all-*trans*-geranylgeraniol.[3] Thus the ester **4** under the conditions of Sharpless epoxidation is epoxidized to the 6,7-epoxide and the 10,11-epoxide in the ratio 40:60, with only negligible reaction at the allylic position. Similarly the ester **5** is converted into the 6,7-epoxide, the 10,11-epoxide, and the 14,15-epoxide in the ratio 11:36:53. In both cases there

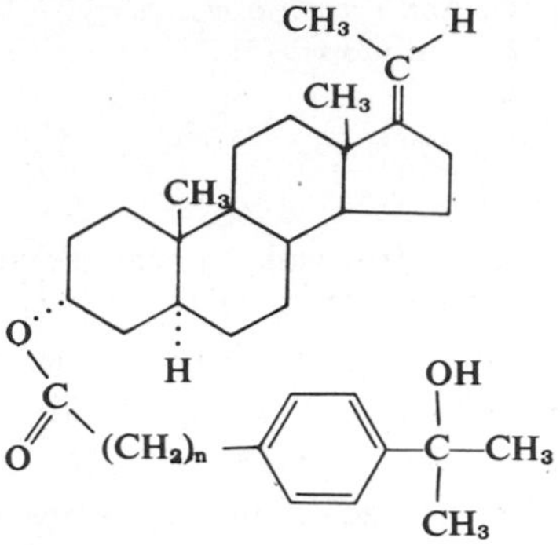

1 (n = 1)
2 (n = 2)

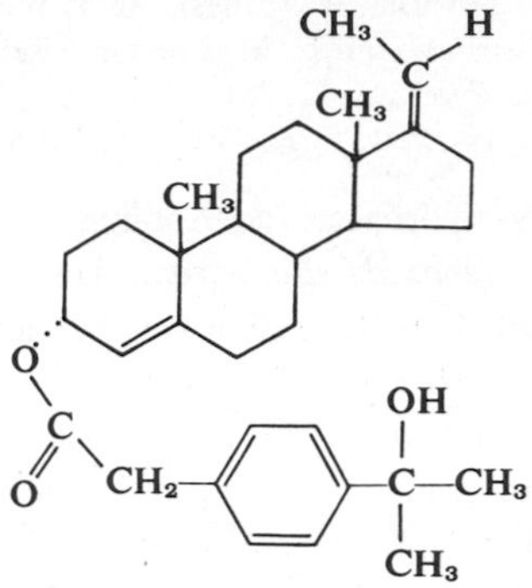

3

is a preference for attack at the end of the chain. The simplest explanation of these results is that the chain is extensively coiled and folded, even in benzene solution.

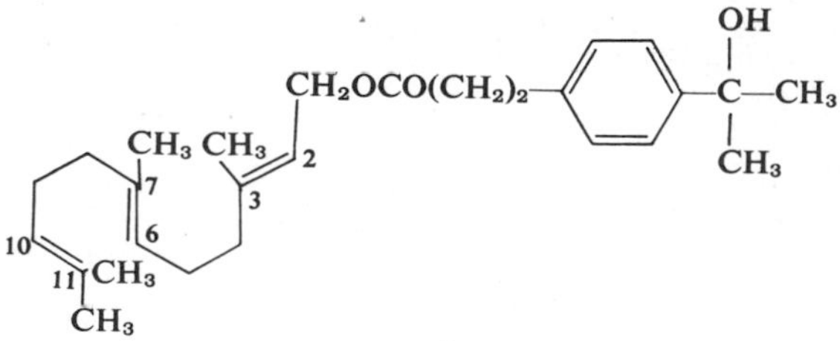

4

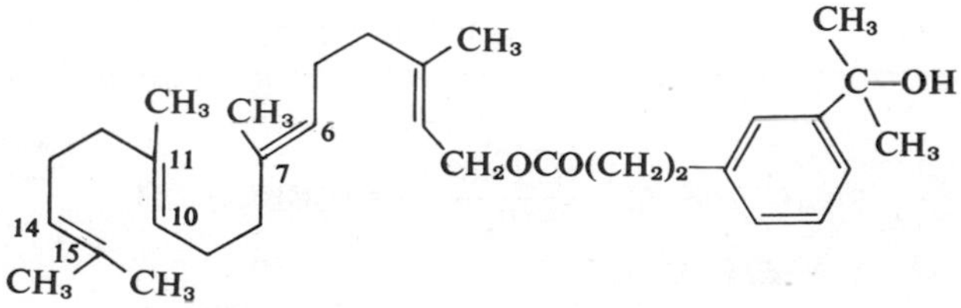

5

$\searrow$**CHNO$_2$** → $\searrow$**C=O.** This transformation can be carried out by conversion into the potassium nitronate followed by oxidation with *t*-butyl hydroperoxide catalyzed with VO(acac)$_2$ (**5**, 75–76). In some cases Mo(CO)$_6$ has proved to be a more effective catalyst.[4]

(45–85%)

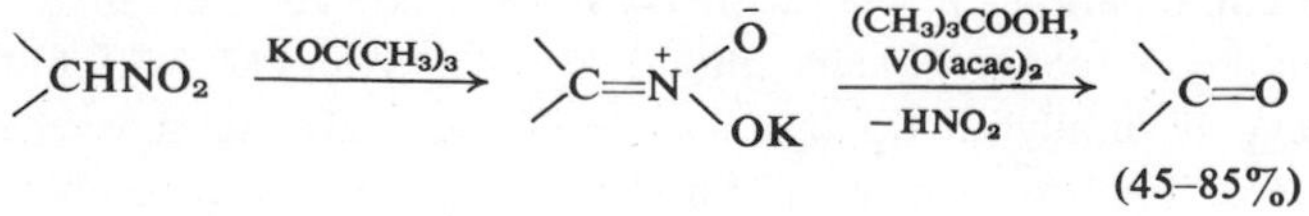

[1] T. Fukuyama, B. Vranesic, D. P. Negri, and Y. Kishi, *Tetrahedron Letters*, 2741 (1978).
[2] R. Breslow and L. M. Maresca, *Tetrahedron Letters*, 623 (1977).
[3] *Idem*, *ibid.*, 887 (1978).
[4] P. A. Bartlett, F. R. Green III, and T. R. Webb, *Tetrahedron Letters*, 331 (1977).

t-Butyl hydroperoxide–Alumina.

Selenoxide elimination. The yield of alkenes from alkyl phenyl selenoxides and alkyl methyl selenoxides under usual conditions (30% H_2O_2, O_3, $^1O_2$) tends to be rather low because of formation of the original selenide. Much higher yields are obtained if the selenides are oxidized with *t*-butyl hydroperoxide (4 equiv.) in the presence of basic alumina (8 equiv.) in THF at 55°. No epoxidation is observed under these conditions. A less satisfactory method is ozonization in CH_2Cl_2 in the presence of 1–3 equiv. of triethylamine.[1]

Examples:

$$C_8H_{17}CH_2CH_2SeC_6H_5 \xrightarrow[\substack{O_3, N(C_2H_5)_3: 75\% \\ t\text{-BuOOH,} \\ Al_2O_3: 86\%}]{O_3: 30\%} C_8H_{17}CH=CH_2$$

$$C_8H_{17}CH_2CH_2SeCH_3 \xrightarrow[\substack{O_3, N(C_2H_5)_3: 23\% \\ t\text{-BuOOH,} \\ Al_2O_3: 83\%}]{} C_8H_{17}CH=CH_2$$

$$\underset{\underset{SeCH_3}{|}}{C_{10}H_{21}CH_2CHCH_3} \xrightarrow[84\%]{\substack{t\text{-BuOOH,} \\ Al_2O_3}} C_{10}H_{21}CH_2CH=CH_2 + C_{10}H_{21}CH=CHCH_3$$
$$65:35$$

Oxidation-elimination of β-hydroxy selenides.[2] *t*-Butyl hydroperoxide combined with alumina is also more effective than H_2O_2 or H_2O_2–Al_2O_3 for selenoxide elimination of β-hydroxy selenides (equation I).

$$\text{(I)} \quad R'CH_2CH(SeR)_2 \xrightarrow[50-78\%]{\substack{1)\ n\text{-BuLi} \\ 2)\ HCHO}} \underset{\underset{SeR}{|}}{R'CH_2CHCH_2OH} \xrightarrow[50-95\%]{\substack{t\text{-BuOOH,} \\ Al_2O_3}} R'CH=CHCH_2OH$$

$$(R = CH_3, C_6H_5)$$

[1] D. Labar, L. Hevesi, W. Dumont, and A. Krief, *Tetrahedron Letters*, 1141 (1978).
[2] *Idem*, *ibid.*, 1145 (1978).

t-Butyl hydroperoxide–Selenium dioxide, $(CH_3)_3COOH$–SeO_2.

Allylic oxidation. Selenium dioxide has been widely used for allylic oxidation of olefins, but this reaction is often troublesome because of formation of selenium-containing by-products and colloidal selenium, which are not easily eliminated. The combination of SeO_2 with H_2O_2 to reoxidize selenium species to SeO_2 has been used to convert olefins into diols and epoxides (**2**, 362), but this method is not general for allylic oxidations. Umbreit and Sharpless[1] have effected allylic oxidations with an excess of *t*-butyl hydroperoxide (90%) in combination with stoichiometric or catalytic (1.5–2%) amounts of SeO_2. Sluggish oxidations can be catalyzed by a carboxylic acid such as salicylic acid. In general, yields of allylic alcohols are at least comparable to, and generally superior to, those obtained with SeO_2 alone.

Examples:

[1] M. A. Umbreit and K. B. Sharpless, *Am. Soc.*, **99**, 5526 (1977).

***n*-Butyllithium**, **1**, 95–96; **2**, 51–53; **4**, 60–63; **5**, 78; **6**, 85–91; **7**, 45–47.

Reaction with cyclic epoxides. The reaction of epoxides of medium-size rings with base (lithium di-*n*-alkylamides) has been extensively studied by Cope and by Crandall (**1**, 610–611; **2**, 247). Two products are usually obtained: an allylic alcohol formed by β-elimination and a bicyclic alcohol derived from a carbene intermediate formed by α-elimination. Boeckman[1] has now found that the carbene pathway is highly favored if *n*-butyllithium (3 equiv.) is used as base and if the reaction is conducted at low temperatures, −78 to 25° in the case of *cis*-cyclooctene oxide (**1**) and *cis*-cyclodecene oxide. The oxide of cyclododecene,

however, under these conditions is converted mainly into 2-cyclododecenol. The oxide of cyclohexene follows another course: the main products are cyclohexanone (56%) and the allylic alcohol 2-cyclohexenol (29%). The mechanism of ketone formation is not known.

Cyclization of o-bromophenylalkanoic acids. In connection with the problem of preparation of substituted anthraquinones required in synthesis of anthracyclines, Whitlock *et al.*[2] have developed an alternative to Friedel-Crafts cycloacylation that can accommodate various functional groups. An example is the preparation of 1-indanone (**2**) by treatment of 3-(*o*-bromophenyl) propionic acid (**1**) with *n*-butyllithium at − 80° for 5 minutes. The cyclization was shown to

be generally applicable (except for *o*-bromophenylacetic acid) and was used for the preparation from **3** of the more complicated system **4**, of interest as a possible intermediate for synthesis of anthracycline antibiotics.

Cleavage of 1,4-benzodioxane. Treatment of 1,4-benzodioxane (**1**) with *n*-butyllithium in ether at 0° results in formation of the monovinyl ether of catechol (**2**).[3]

[1] R. K. Boeckman, Jr., *Tetrahedron Letters*, 4281 (1977).
[2] R. J. Boatman, B. J. Whitlock, and H. W. Whitlock, Jr., *Am. Soc.*, **99**, 4822 (1977).
[3] A. C. Ranade and S. Jayalakshmi, *Chem. Ind.*, 234 (1978).

n-Butyllithium–Nickel(II) bromide.

Arylation and vinylation of lithium ester enolates. Addition of *n*-butyllithium to suspensions of nickel(II) bromide in THF produces a species that catalyzes substitution of lithium ester enolates (prepared with LDA in THF at − 78°) by aryl and vinyl halides with retention of configuration. Lithium enolates of α,β-unsaturated esters undergo substitution in this reaction at the γ-position (last example). The mechanism of this reaction is not certain.[1]

Examples:

$$LiCH_2COOC(CH_3)_3 + C_6H_5I \xrightarrow[73\%]{\substack{NiBr_2, \\ n\text{-}C_4H_9Li}} C_6H_5CH_2COOC(CH_3)_3$$

$$\underset{\underset{CH_3}{|}}{\overset{\overset{CH_3}{|}}{Li}CCOOC_2H_5} + \underset{H}{\overset{C_6H_5}{>}}C=C\underset{H}{\overset{Br}{<}} \xrightarrow{50\%} \underset{H}{\overset{C_6H_5}{>}}C=C\underset{\overset{|}{H}}{\overset{CCOOC_2H_5}{<}}$$

$$\underset{\underset{CH_3}{|}}{\overset{\overset{CH_3}{|}}{Li}CCOOC_2H_5} + \underset{H}{\overset{C_6H_5}{>}}C=C\underset{Br}{\overset{H}{<}} \xrightarrow{57\%} \underset{H}{\overset{C_6H_5}{>}}C=C\underset{\underset{CH_3}{\overset{|}{CCOOC_2H_5}}}{\overset{H}{<}}$$

$$LiCH_2CH=CHCOOC_2H_5 + C_6H_5I \xrightarrow{60\%} C_6H_5CH_2CH=CHCOOC_2H_5$$

[1] A. A. Millard and M. W. Rathke, *Am. Soc.*, **99**, 4833 (1977).

n-Butyllithium–Potassium *t*-butoxide, 5, 552.

α-*Methylene*-γ-*lactones*. A new route to these lactones involves conversion of methallyl alcohol to the dianion with the complex of potassium *t*-butoxide and *n*-butyllithium. The dianion reacts with aldehydes or ketones to give diols, which are oxidized by activated MnO_2 to α-methylene-γ-lactones (equation I). Unfortunately the yields in the first step are rather low.[1]

(I)

[1] R. M. Carlson, *Tetrahedron Letters*, 111 (1978).

n-Butyllithium–Tetramethylethylenediamine, 2, 403; 3, 284; 4, 485–489; 5, 678–680; 7, 47–48.

Lithium o-lithiobenzylate.[1] *n*-Butyllithium in combination with TMEDA converts benzyl alcohol into lithium *o*-lithiobenzylate (**1**) in about 70% yield. The new reagent (**1**) reacts readily with electrophiles as formulated.

*Metalation of N-phenyl-*m*-anisamide; anthrone synthesis.* A new anthrone synthesis involves metalation of **1** with 2 equiv. of *n*-BuLi and TMEDA in THF (−78 to −10°). The dianion **2** condenses with the aldehyde **3** to form the phthalide **4**, which is converted into the anthrone **7** by classical reactions. This route is valuable for synthesis of unsymmetrical anthraquinones.[2]

Ene-type reaction.[3] Treatment of the diene (**1**) with *n*-BuLi and TMEDA followed by quenching with water leads to 1,2-dimethyl-3-isopropenylcyclopentane (**2**). The reaction is a polar equivalent of the ene reaction.

Dehydrogenation (*cf.* **4**, 485). Gausing and Wilke[4] have reported two high-yield preparations of cyclooctatetraene. Reaction of 1,5-cyclooctadiene (**1**) and *n*-butyllithium·TMEDA in a ratio of 1:3 gives the crystalline compound **2** in practically quantitative yield. Cyclooctatetraene (**3**) is obtained from **2** by oxidation with $CdCl_2$. The intermediate **2** can also be obtained from 1,3,6-cyclooctatriene, but not from 1,3-cyclooctadiene or 1,3,5-cyclooctatriene.

For preparative purposes the German chemists prefer metalation of **1** with phenylsodium[5] (3 equiv.) in pentane followed by reaction with dry oxygen. The yield by this sequence is 78.5%.

1,5-Cyclooctadiene (**1**) has also been converted by potassium at 97° into a dianion, which on air oxidation is converted into the tetraene (**3**) (30% yield).[6]

[1] N. Meyer and D. Seebach, *Angew. Chem., Int. Ed.*, **17**, 521 (1978).
[2] J. E. Baldwin and K. W. Bair, *Tetrahedron Letters*, 2559 (1978).
[3] J. H. Edwards and F. J. McQuillin, *J.C.S. Chem. Comm.*, 838 (1977).
[4] W. Gausing and G. Wilke, *Angew. Chem., Int. Ed.*, **17**, 371 (1978).
[5] J. F. Nobis and L. F. Moormeier, *Ind. Eng. Chem.*, **46**, 539 (1954).
[6] W. J. Evans, A. L. Wayda, C.-W. Chang, and W. M. Cwirla, *Am. Soc.*, **100**, 333 (1978).

sec-Butyllithium–Tetramethylethylenediamine.

Metalation of a methyl ester. The methyl group of an aromatic methyl ester (**1**) can be metalated by treatment with *sec*-butyllithium and TMEDA at −75 to −98° for about 2 hours. The carbanion **a** is useful for a number of reactions with electrophiles as shown in the formulation.[1]

[1] P. Beak and B. G. McKinnie, *Am. Soc.*, **99**, 5213 (1977).

t-**Butyllithium, 1,** 96–97; **5,** 79–80.

Cyclic vinyl ether carbanions. Cyclic vinyl ethers can be metalated by *t*-butyl-lithium at − 78 to 5° if the concentration of THF is kept to a minimum (compare metalation of methyl vinyl ether, **6,** 372). The resulting carbanions react with alkyl iodides and carbonyl compounds to form adducts in about 50–75% yields. The products are useful for further transformations.[1]

Examples:

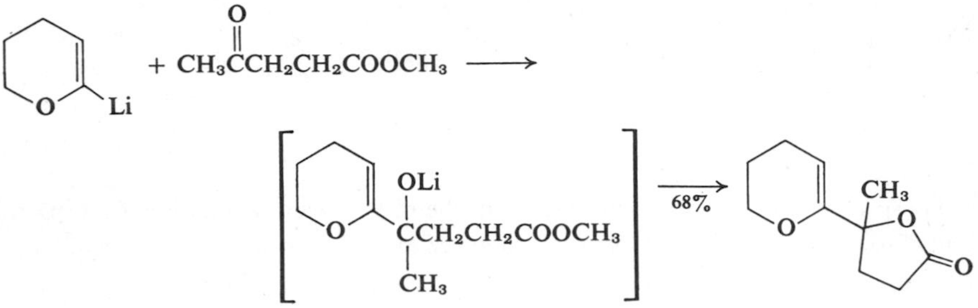

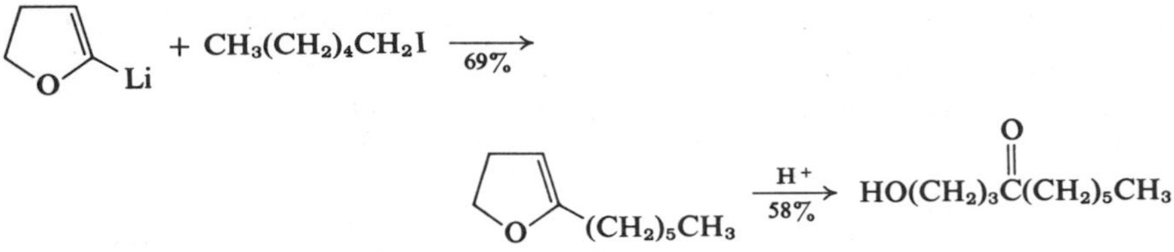

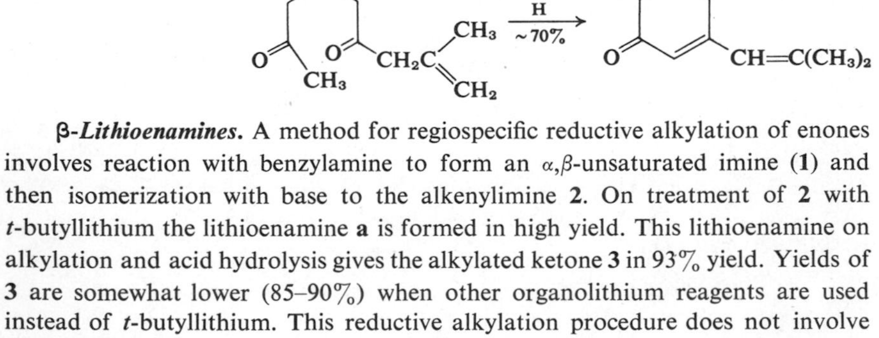

β-*Lithioenamines.* A method for regiospecific reductive alkylation of enones involves reaction with benzylamine to form an α,β-unsaturated imine (**1**) and then isomerization with base to the alkenylimine **2**. On treatment of **2** with *t*-butyllithium the lithioenamine **a** is formed in high yield. This lithioenamine on alkylation and acid hydrolysis gives the alkylated ketone **3** in 93% yield. Yields of **3** are somewhat lower (85–90%) when other organolithium reagents are used instead of *t*-butyllithium. This reductive alkylation procedure does not involve the more usual Birch reduction but rather isomerization (**1** → **2**) followed by trapping of (**a**).[2]

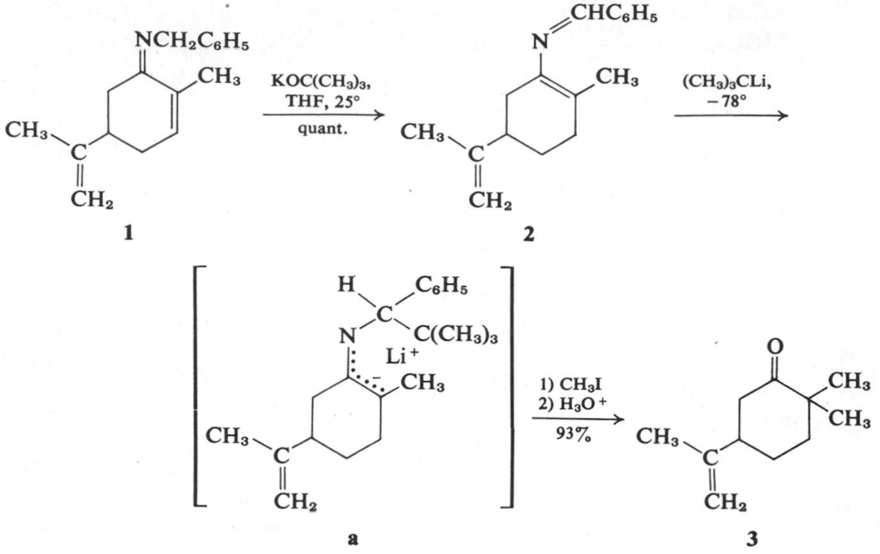

Other examples:

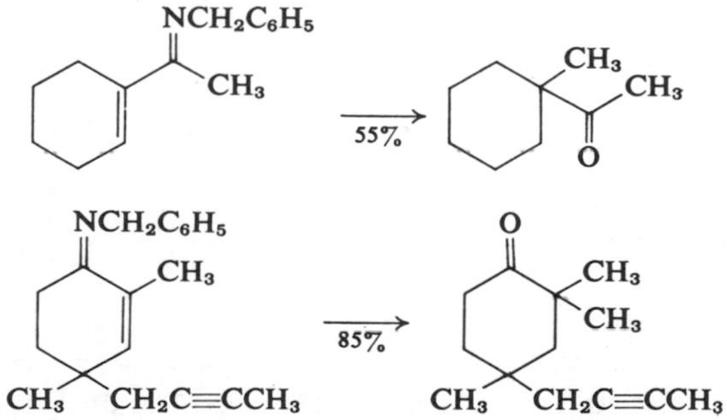

β-Lithioenamines can also be prepared from β-bromoenamines by exchange with *n*- or *t*-butyllithium at − 70°. They are also prepared in one pot from enamines by bromination in THF at − 60° followed by addition of the base. These substances react vigorously with various electrophiles at low temperatures without polysubstitution or substitution on nitrogen. The products are convertible into the corresponding carbonyl compounds.[3]

Examples:

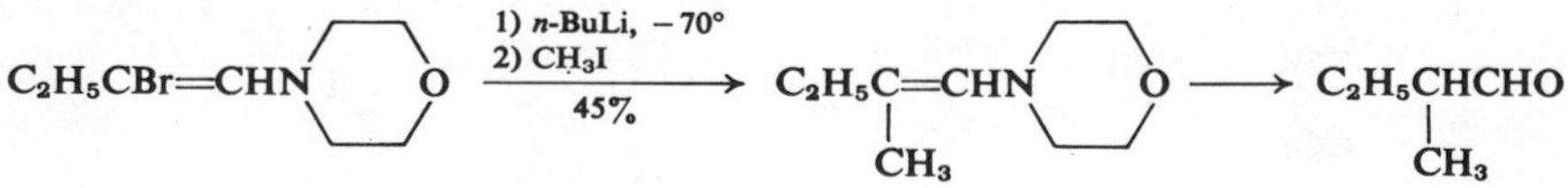

$$C_6H_5C{=}CHBr \xrightarrow[40\%]{\begin{array}{c}1)\ n\text{-BuLi}\\2)\ CH_3I\end{array}} C_6H_5C{=}CHCH_3 \longrightarrow C_6H_5CCH_2CH_3$$

(with $N(CH_3)_2$ on the first two structures and $\stackrel{O}{\|}$ on the last)

Lithioalkyl phenyl thioethers. Although thioanisole and dimethyl sulfide can be α-metalated by *n*-butyllithium and either Dabco or TMEDA (**2**, 324, 403), this metalation is not general for alkyl aryl sulfides. For this purpose *t*-butyllithium in combination with HMPT is generally useful.[4]

Examples:

$$(CH_3)_2CHCH_2SC_6H_5 \xrightarrow[\text{HMPT}]{t\text{-}C_4H_9Li,} \left[(CH_3)_2CHCHSC_6H_5 \right] \xrightarrow[86\%]{(CH_3)_3SiCl} (CH_3)_3SiCHSC_6H_5$$

(with $\overset{Li}{|}$ over the bracketed species and $\overset{CH(CH_3)_2}{|}$ over the product)

$$(CH_3)_2CHCH_2SC_6H_5 \xrightarrow[\text{HMPT}]{t\text{-}C_4H_9Li,} \cdots \xrightarrow{56\%}$$

$$CH_3OCH_2(CH_2)_2CHCH_2SC_6H_5 \xrightarrow[\text{HMPT}]{t\text{-}C_4H_9Li,} \xrightarrow[92\%]{(CH_3)_3SiCl} CH_3OCH_2(CH_2)_2CHCHSC_6H_5$$

(with $\overset{CH_3}{|}$ over the reactant carbon and $\overset{CH_3}{|}$ and $\underset{Si(CH_3)_3}{|}$ on the product)

[1] R. K. Boeckman, Jr., and K. J. Bruza, *Tetrahedron Letters*, 4187 (1977).
[2] P. A. Wender and M. A. Eissenstat, *Am. Soc.*, **100**, 292 (1978); P. A. Wender and J. M. Schaus, *J.Org.*, **43**, 782 (1978).
[3] L. Duhamel and J.-M. Poirier, *Am. Soc.*, **99**, 8356 (1977).
[4] T. M. Dolak and T. A. Bryson, *Tetrahedron Letters*, 1961 (1977).

t-**Butyl methyl iminodicarboxylate potassium salt,** $KN{\overset{\underset{\displaystyle +\ -}{}}{{<}{\begin{smallmatrix}COOC(CH_3)_3\\COOCH_3\end{smallmatrix}}}}$ (**1**). Mol. wt. 213.28, stable indefinitely at room temperature, sparingly soluble in DMF or DMSO.

Preparation:

$$\underset{CONH_2}{\overset{COOC(CH_3)_3}{|}} \xrightarrow[83\%]{\begin{array}{c}Pb(OAc)_4,\ THF,\\CH_3OH\end{array}} HN{<}{\begin{smallmatrix}COOC(CH_3)_3\\COOCH_3\end{smallmatrix}} \xrightarrow{KOH} \mathbf{1}$$

Gabriel synthesis. The main drawback to the classical Gabriel synthesis[1] is difficulty in cleavage of the phthaloyl protective group. The reagent (**1**) can serve as a substitute for potassium phthalimide. The methoxycarbonyl group is cleaved by mild alkali, and thus allows direct synthesis of *t*-butoxycarbonylamino acids (equation I).[2]

$$\text{(I)} \quad BrCH_2COOC_2H_5 + 1 \xrightarrow{\text{DMF, 20°}} C_2H_5OOCCH_2N \begin{array}{c} COOC(CH_3)_3 \\ \\ COOCH_3 \end{array} \xrightarrow[\text{71\% overall}]{\text{NaOH, 20°}}$$

$$(CH_3)_3CO\overset{\overset{\displaystyle O}{\|}}{C}-NHCH_2COOH$$

[1] S. Gabriel, *Ber.*, **20**, 2224 (1887).
[2] J. D. Elliott and J. H. Jones, *J.C.S. Chem. Comm.*, 758 (1977); C. T. Clarke, J. D. Elliott, and J. H. Jones, *J.C.S. Perkins I*, 1088 (1978).

C

Calcium amalgam, Ca/Hg. The amalgam is prepared by treating calcium shot with mercuric chloride in DMF and HMPT.

Alkene synthesis from vic-dinitro compounds. In the original synthesis (**4**, 460–461), sodium sulfide (or sodium thiophenoxide) was used as the reducing agent. This method, however, does not tolerate various functional groups: ester, keto, cyano, or ether. This defect can be remedied by use of calcium amalgam, which reduces nitro groups more readily than other common functional groups.[1]

Examples:

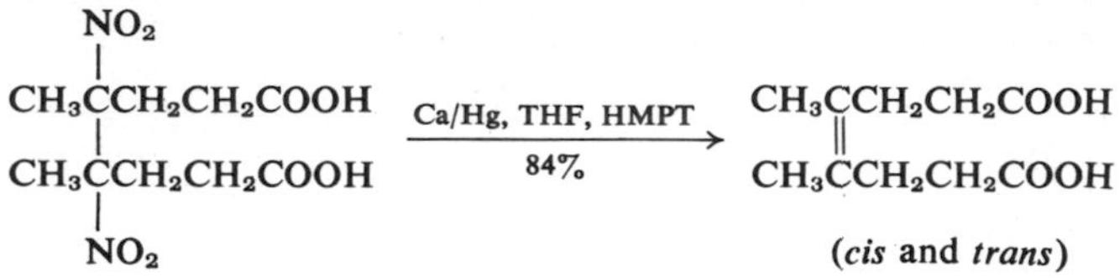

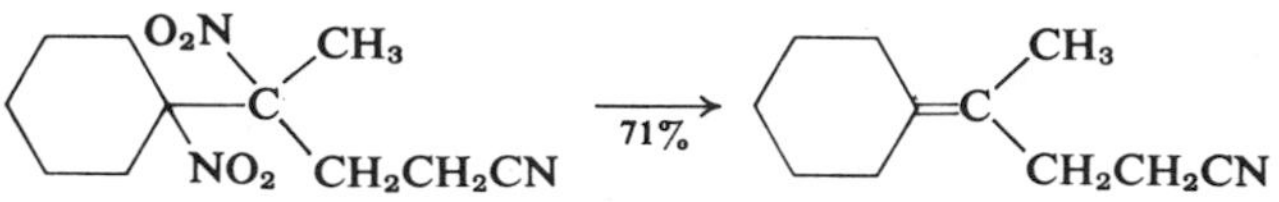

[1] N. Kornblum and L. Cheng, *J. Org.*, **42**, 2944 (1977).

(+)-Camphanic acid,

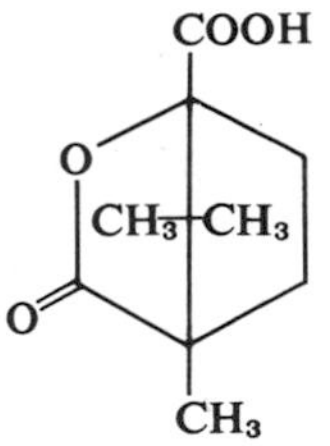

(R*COOH) (**1**). Mol. wt. 198.22, m.p. 201°.

Preparation.[1]

Optically active hydrobenzoins. Optically active hydrobenzoins **4a** and **4b** can be obtained by reaction of the racemic oxide of *trans*-stilbene (**2**) with (+)-camphanic acid to form two *threo*-esters, **3a** and **3b**. These are hydrolyzed to the two enantiomeric forms of hydrobenzoin, **4a** and **4b**.[2]

The method is general. A variation is epoxidation of a stilbene derivative with *m*-chloroperbenzoic acid in the presence of excess **1** to obtain the two monocamphanates.

74

$$C_6H_5\cdots\overset{\overset{\displaystyle R^*OCO}{|}}{\underset{\underset{\displaystyle H}{|}}{C}}-\overset{\overset{\displaystyle H}{|}}{\underset{\underset{\displaystyle OH}{|}}{C}}\cdots C_6H_5 + C_6H_5\cdots\overset{\overset{\displaystyle H}{|}}{\underset{\underset{\displaystyle R^*OCO}{|}}{C}}-\overset{\overset{\displaystyle OH}{|}}{\underset{\underset{\displaystyle H}{|}}{C}}\cdots C_6H_5$$

2   3a   3b

NaOH, C₂H₅OH   NaOH, C₂H₅OH

$$C_6H_5\cdots\overset{\overset{\displaystyle HO}{|}}{\underset{\underset{\displaystyle H}{|}}{C}}-\overset{\overset{\displaystyle H}{|}}{\underset{\underset{\displaystyle OH}{|}}{C}}\cdots C_6H_5 \qquad C_6H_5\cdots\overset{\overset{\displaystyle H}{|}}{\underset{\underset{\displaystyle HO}{|}}{C}}-\overset{\overset{\displaystyle OH}{|}}{\underset{\underset{\displaystyle H}{|}}{C}}\cdots C_6H_5$$

4a [SS(−)]   4b [RR(+)]

[1] O. Aschan, *Ber.*, **28B**, 922 (1895); J. Bredt, *Ann.*, **395**, 39 (1913).
[2] M.-J. Brienne and A. Collet, *J. Chem. Res. (M)*, 772 (1978).

o-Carbethoxybenzyl phenyl sulfoxide (1). Mol. wt. 288.36, oil.

Preparation:

1-Hydroxy-2,3-disubstituted naphthalenes.[1] The anion of 1 forms conjugate addition products (2) with α,β-unsaturated ketones and esters. These sulfoxides

are unstable and slowly decompose to naphthalene derivatives (3) on standing. This conversion is carried out more readily by heating the reaction mixture containing 2 to reflux. Yields of 3 are in the range 30–70%, being highest when R¹ is an alkyl group.

This naphthalene synthesis proceeds in somewhat higher yields when the phthalide sulfone **4** is used in place of **1**. In this reaction, 1,4-dihydroxynaphthalenes (**5**) are obtained. These products can be converted into naphthoquinones or into the more stable dimethyl ethers.

(II)

4

1) LDA, THF, −78°
2) R¹CH=CHCOR²
− C₆H₅SO₂H

5 (30–85%)

[1] F. R. Hauser and R. P. Rhee, *J. Org.*, **43**, 178 (1978).

Carboethoxycyclopropyltriphenylphosphonium tetrafluoroborate, **5**, 90–91; **6**, 93–94.

Spiroannelation. The complete paper on the reaction of the enolates of α-formylcycloalkanones with this reagent to form spiro[4.5]decanes (**6**, 93–94) has been published.[1] The formulas for β-vetivone (**6**) and α-vetispirene (**7**) in **6**, 94 should be corrected as shown.

6 (β-vetivone)   **7 (α-vetispirene)**

[1] W. G. Dauben and D. J. Hart, *Am. Soc.*, **99**, 7307 (1977).

Carbon monoxide, **2**, 60, 204; **3**, 41–43; **5**, 96; **7**, 53.

Brown's ketone synthesis (**2**, 148–149). Brown's method for conversion of dienes into cyclic ketones via carbonylation of trialkylboranes[1] has been used in a synthesis of (±)-3-methoxy-1,3,5(10)-estratriene (**4**).[2] The key step is hydroboration of the diene **1** with thexylborane to form **2**, which is immediately treated with carbon monoxide and then oxidized to give the ketone **3**. This is cyclized to the $\Delta^{9,11}$-estratetraene and then reduced to give **4**.

[1] H. C. Brown, *Organic Synthesis via Boranes*, Wiley, New York, 1975, pp. 127–130.
[2] T. A. Bryson and W. E. Pye, *J. Org.*, **42**, 3214 (1977).

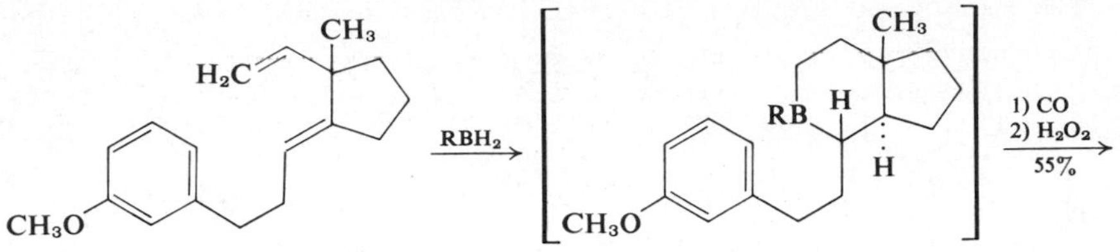

1

2 [R = CH(CH₃)CH(CH₃)₂]

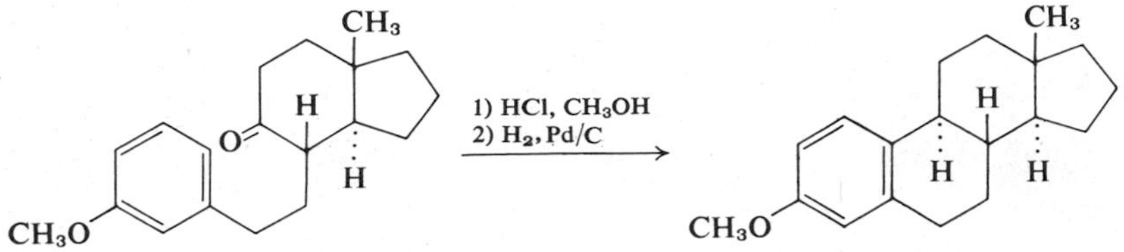

3   4

N,N′-Carbonyldiimidazole, 1, 114–116; **2,** 61; **5,** 97–98; **6,** 97.

Thiol and selenol esters. These useful intermediates in the synthesis of complex molecules can be prepared in good to high yields by activation of carboxylic acids with N,N′-carbonyldiimidazole or N,N′-carbonyldi-1,2,4-triazole (**1,** 116; **2,** 61). The products, without isolation, react with thiols or selenols to form the esters (equation I).[1]

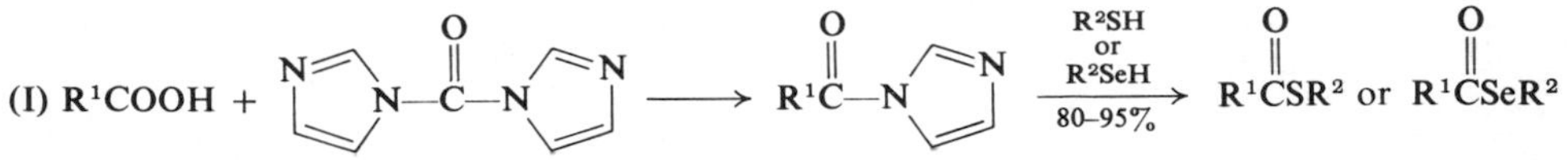

(I) R^1COOH +   →   → R^1CSR^2 or R^1CSeR^2

[1] H.-J. Gais, *Angew. Chem., Int. Ed.,* **16,** 244 (1977).

2-Carboxy-1-methoxycarbonylethyltriphenylphosphorane (1). Mol. wt. 392.38, m.p. 144°.

Preparation[1]:

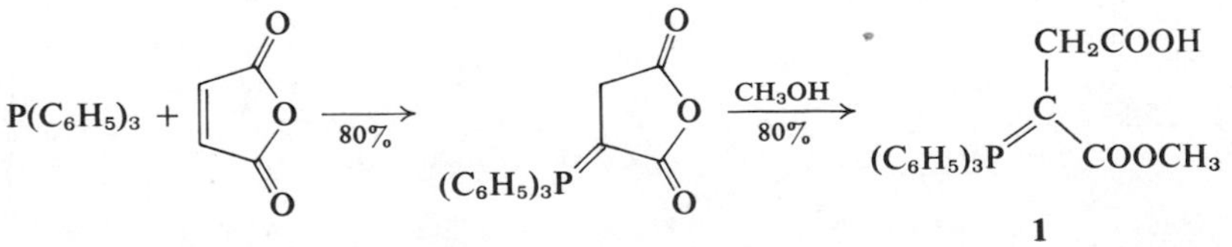

$P(C_6H_5)_3$ +

1

Wittig reactions. The *t*-butyl ester of **1** has been used for a Wittig reaction in a new synthesis of (±)-lysergic acid (**4**) from the aldehyde **2.**[2]

CHO

$+ (C_6H_5)_3P=C$ with CH$_2$COOC(CH$_3$)$_3$ and COOCH$_3$

$\dfrac{\text{1) } \Delta}{\text{2) TFA}}$ 70%

C_6H_5CON

2

COOCH$_3$

COOH

C_6H_5CON

3

Several steps →

COOH

N—CH$_3$

HN

4

McMurry and Donovan[3] have used **1** directly in a general synthesis of β-alkyl-idenebutyrolactones (**5**) from aldehydes (equation I). The reaction proceeds with both aromatic and aliphatic aldehydes to form the intermediate **a**, which is reduced and then treated with acid to give the lactone **5**. As expected, these

(I)   RCHO + **1** ⟶ $\left[\begin{array}{c} \text{—COOH} \\ \text{RCH} \diagdown \text{COOC}_2\text{H}_5 \end{array} \right]$ $\dfrac{\text{1) NaAl(C}_2\text{H}_5)_2\text{H}_2}{\text{2) TsOH, C}_6\text{H}_6}$ 65–78%

a

RCH= (lactone) $\dfrac{\text{Al}_2\text{O}_3}{70\text{–}80\%}$ RCH$_2$ (butenolide) $\dfrac{(i\text{-C}_4\text{H}_9)_2\text{AlH}}{85\text{–}95\%}$ RCH$_2$ (furane)

5   **6**   **7**

lactones are easily isomerized to butenolides (**6**). Transformation of **6** to furanes (**7**) is accomplished by reduction with diisobutylaluminum hydride (**1**, 262; **2**, 191; **6**, 200).

This sequence was used to synthesize the naturally occurring furane perillene (**8**) from the sensitive aldehyde $(CH_3)_2C=CHCH_2CHO$.

$(CH_3)_2C=CHCH_2CH_2$ (furane)

8

[1] R. F. Hudson and P. A. Chopard, *Helv.*, **46**, 2178 (1963); A. F. Cameron, F. D. Duncanson, A. A. Freer, V. W. Armstrong, and R. Ramage, *J.C.S. Perkin II*, 1030 (1975).
[2] V. W. Armstrong, S. Coulton, and R. Ramage, *Tetrahedron Letters*, 4311 (1976).
[3] J. E. McMurry and S. F. Donovan, *ibid.*, 2869 (1977).

Caro's acid (Peroxysulfuric acid), 1, 118–119; **3**, 43; **6**, 97–98.

Selective N-oxidations. A few years ago Kyriacou[1] reported that H_2SO_4–60% H_2O_2 is useful for oxidation of polyhalogenated diazines, which usually resist oxidation. Dow chemists[2] concluded that the actual reagent is Caro's acid, which is indeed the case; they used this reagent for oxidation of chloropyrazines and chloroquinoxalines. In general the oxidation proceeds only to the mono oxide, and the nitrogen nearer to the halogen substituent is the one more prone to attack.

Examples:

(55%)   (22%)

[1] D. Kyriacou, *J. Heterocyclic Chem.*, **8**, 697 (1971).
[2] C. E. Mixan and R. G. Pews, *J. Org.*, **42**, 1869 (1977).

Catecholborane (1,3,2-Benzodioxaborole), 4, 25, 69–70; **5**, 100–101; **6**, 33–34, 98; **7**, 54–55.

Selective reductions.[1] Catecholborane can be used for several selective reductions. Although it reduces both aldehydes and ketones, it can preferentially reduce the former group. It reduces alkenes only slowly, but alkynes relatively easily. It reduces nitriles, but very slowly, in contrast to BH_3.

Deoxygenation (**6**, 98; **7**, 54). Kabalka's method for deoxygenation of a carbonyl group has been used in a synthesis of pachydictyol-A (**3**), a diterpene hydroazulene alcohol from brown marine algae, from **1**, readily available from α-santonin.[2]

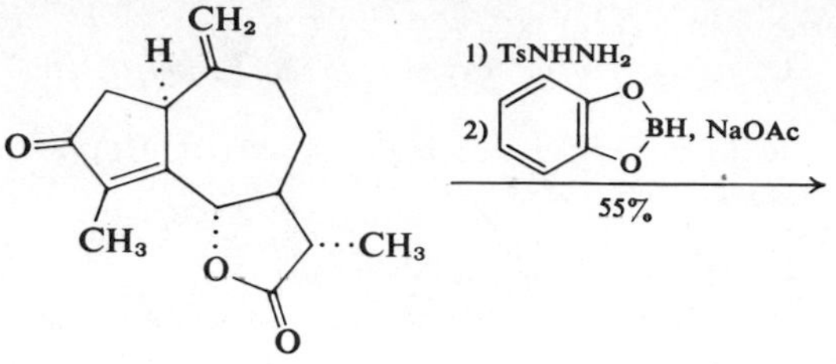
1) TsNHNH₂
2) BH, NaOAc
55%

1

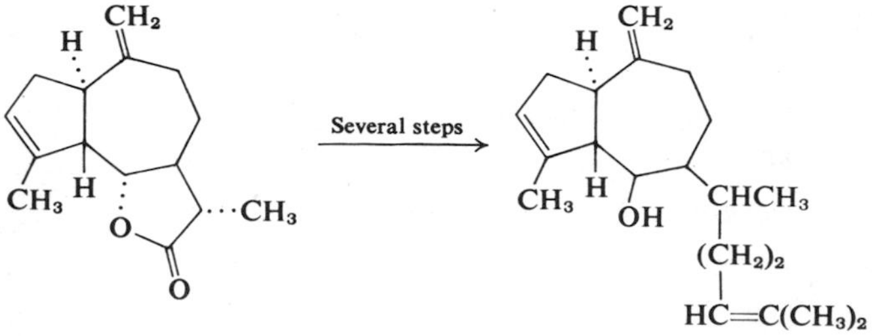
Several steps

2 **3**

[1] G. W. Kabalka, J. D. Baker, Jr., and G. W. Neal, *J. Org.*, **42**, 512 (1977).
[2] A. E. Greene, *Tetrahedron Letters*, 851 (1978).

Ceric ammonium sulfate [Ammonium cerium(IV) sulfate], 7, 56.

Oxidation of arenes. CAS in a mixture of dilute H_2SO_4 and acetonitrile oxidizes polycyclic arenes to quinones.[1]

Examples:

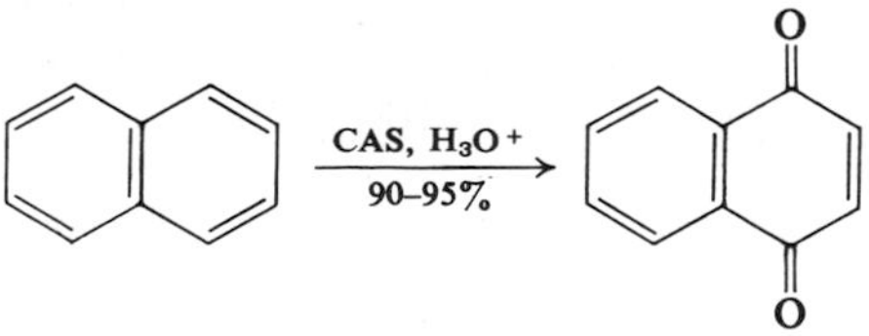
CAS, H₃O⁺
90–95%

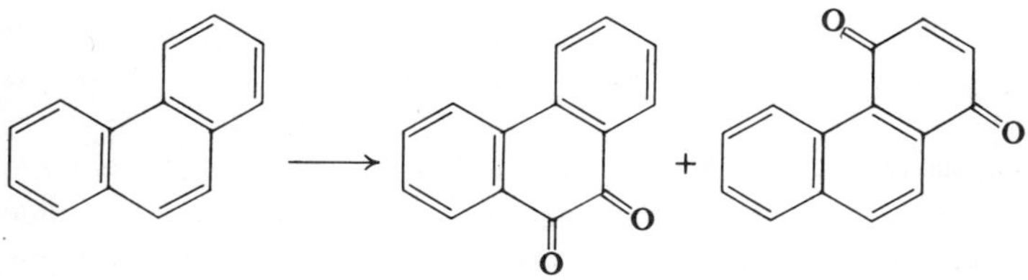

(60%) (15%)

(45%)   (26%)

In the case of naphthalene and substituted derivatives, a 1,2-shift has been observed.[2] Oxidation of the deuterium labeled naphthalene **1** gives three products, **2**–**4**. Thus deuterium migrates to some extent from the 1- to the 2-position.

1   2   3   4

Similar migration of bromine and phenyl substituents has also been observed in oxidation of substituted naphthalenes.

[1] M. Periasamy and M. V. Bhatt, *Synthesis*, 330 (1977).
[2] *Idem, Tetrahedron Letters*, 2357 (1977).

Cesium carbonate; cesium bicarbonate, Cs_2CO_3; $CsHCO_3$, **7,** 57.

Amino acid esters. Esters of protected amino acids and peptides can be prepared by conversion to the cesium salt by titration to neutrality with aqueous $CsHCO_3$ or Cs_2CO_3. After evaporation to dryness, the neutral salt is esterified by reaction with an alkyl halide in DMF. The reaction proceeds without observable racemization.[1]

$$RX + R'COOCs \xrightarrow[70-90\%]{DMF} R'COOR + CsX$$

[1] S. S. Wang, B. F. Gisin, D. P. Winter, R. Makofske, I. D. Kulesha, C. Tzougraki, and J. Meienhofer, *J. Org.*, **42**, 1286 (1977).

Cesium fluoride, 7, 57–58.

Transesterification. Mixed trialkyl phosphates can be prepared by a general transesterification reaction depicted in equation (I). The cesium fluoride can be replaced by tetra-*n*-butylammonium fluoride, but there is no reaction in the

$$(\mathrm{I}) \quad (Cl_3CCH_2O)_2\overset{\displaystyle O}{\overset{\|}{P}}OR \xrightarrow[85-100\%]{R^1OH,\ CsF,\ 20°} Cl_3CCH_2\overset{\displaystyle O}{\overset{\|}{P}}OR \xrightarrow[85-100\%]{R^2OH,\ CsF,\ 80°} R^2O\overset{\displaystyle O}{\overset{\|}{P}}OR$$

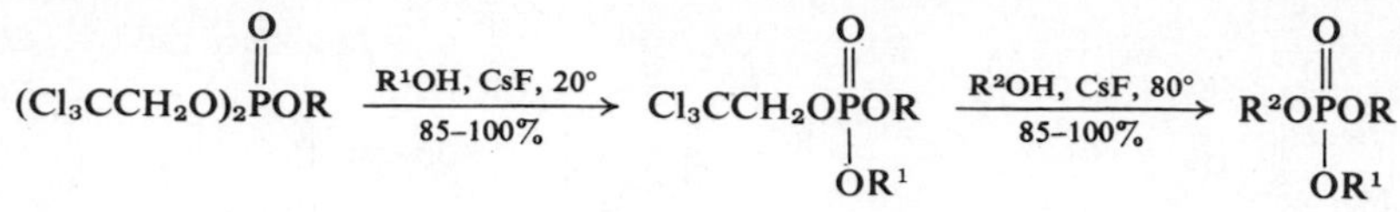

$$\underset{\textbf{1}}{} \qquad\qquad \underset{\textbf{2}}{\overset{\displaystyle OR^1}{}} \qquad\qquad \underset{\textbf{3}}{\overset{\displaystyle OR^1}{}}$$

absence of F^-. *t*-Butyl alcohol does not react with **1**, even at 80°; hence it can be used as solvent.

The method can be used for preparation of cyclic phosphates from diols and of nucleotide triesters.[1]

[1] K. K. Ogilvie, S. L. Beaucage, N. Theriault, and D. W. Entwistle, *Am. Soc.*, **99**, 1277 (1977).

Chloral, CCl_3CHO, **1**, 122; **3**, 45; **6**, 100–101.

Ene reactions.[1] Chloral has been found to undergo the ene reaction with mono-, 1,1-di-, and 1,2-disubstituted alkenes. The ene reaction in general requires temperatures of about 100° to proceed and is usually not subject to acid catalysis; however, in the chloral reaction, several Lewis acids have been found to be effective catalysts: $FeCl_3$, $AlCl_3$, $SnCl_4$.

Examples:

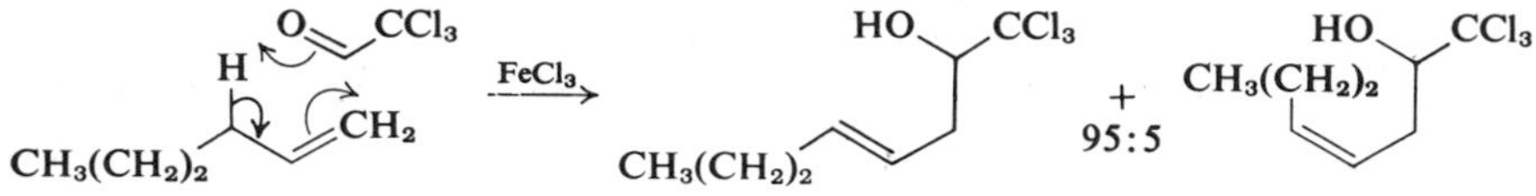

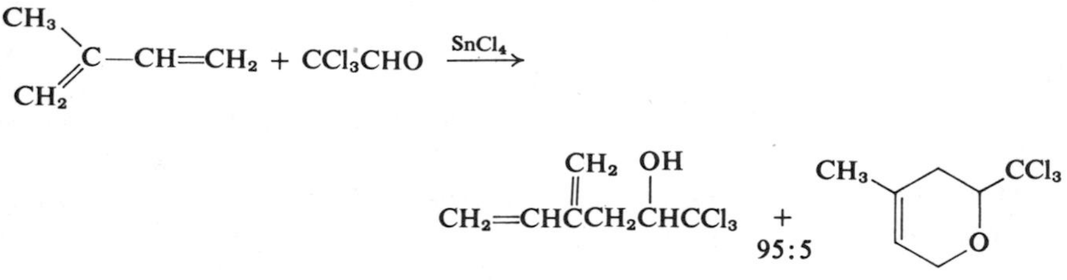

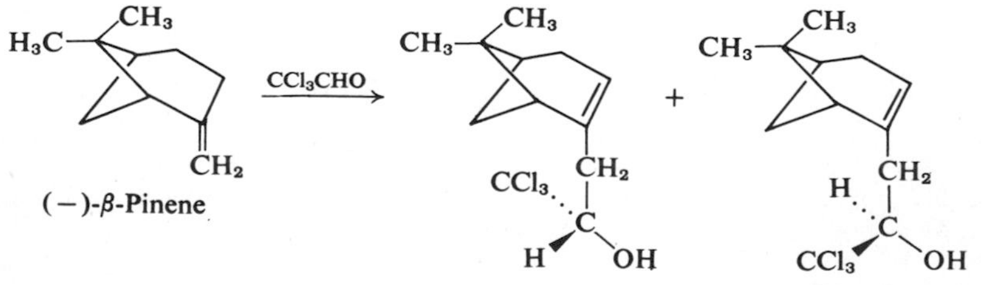

The catalyst also affects the stereoselectivity. In the thermal reaction of $(-)$-β-pinene at 95°, the ratio of the products **1** and **2** is 18:82; when the reaction is catalyzed by $SnCl_4$ the ratio is 92:8. Adduct **1** is the sole product in the reaction catalyzed by $TiCl_4$.

Since the 1-hydroxy-2,2,2-trichloroethyl unit of the ene products is convertible into —CH(OH)COOH, —CHOHCH₃, and other groups, this ene reaction could have synthetic utility.

[1] G. B. Gill and B. Wallace, *J.C.S. Chem. Comm.*, 380, 382 (1977).

Chloramine T, **4**, 75, 445–446; **5**, 104; **7**, 58.

Review.[1] The reaction of N-halogeno-N-metallo reagents with organic functional groups has been reviewed (191 references). The chemistry of N-chloro-N-sodiocarbamidates (ROCNClNa⁺)

$$RO\overset{O}{\overset{\|}{C}}N^-ClNa^+$$

is also covered.

[1] M. M. Campbell and G. Johnson, *Chem. Rev.*, **78**, 65 (1978).

Chlorine–Chlorosulfonic acid.

α-*Chlorination of carboxylic acids.* Carboxylic acids are chlorinated in the α-position by chlorine in the presence of chlorosulfonic acid as the catalyst for enolization. Yields are improved by added oxygen. No solvent is necessary for the lower acids; 1,2-dichloroethane is a suitable solvent for the higher acids. An example is the chlorination of ε-benzoylaminocaproic acid (**1**).[1]

$$C_6H_5CONH(CH_2)_4CH_2COOH \xrightarrow[79\%]{Cl_2,\ ClSO_3H,\ O_2,\ ClCH_2CH_2Cl} C_6H_5CONH(CH_2)_4\underset{Cl}{CHCOOH}$$

1

[1] Y. Ogata, T. Harada, K. Matsuyama, and T. Ikejira, *J. Org.*, **40**, 2960 (1975); Y. Ogata and T. Sugimoto, *Chem. Ind.*, 538 (1977); Y. Ogata, T. Sugimoto, and M. Inaishi, *Org. Syn.*, submitted (1977).

μ-Chlorobis(η^5-cyclopentadienyl)(dimethylaluminum)-μ-methylenetitanium (1).

Mol. wt. 284.62, reddish-orange crystals. This methylene-bridged compound is prepared by the reaction shown in equation (I).

$$(I) \quad Cp_2TiCl_2 + 2Al(CH_3)_3 \xrightarrow[-CH_4]{C_6H_5CH_3,\ 20°} Cp_2Ti\underset{Cl}{\overset{CH_2}{\diagup\diagdown}}Al(CH_3)_2 + Al(CH_3)_2Cl$$

1 (80–90%)

Olefin homologation.[1] The methyl groups of **1** exchange with some aluminum alkyls and halides, but the methylene group is unreactive. However, **1** reacts with ethylene (toluene, 20°, 18 hours) to form propylene in 32% yield. Reactions of this type are generally improved by addition of trimethylamine. This olefin homologation may involve the transient species **a** formed by CH_2—Al dissociation (equation II).

(II) $1 \rightleftharpoons$ $\left[\begin{array}{c} Cp_2Ti{=}CH_2 \\ | \\ ClAl(CH_3)_2 \end{array}\right]$ $\xrightarrow[32\%]{CH_2=CH_2}$ $CH_3CH{=}CH_2$

a

Another reaction of **1** of interest is that with certain carbonyl groups such as cyclohexanone (equation III).

(III)   ⬡=O + **1** $\xrightarrow[65\%]{C_6H_5CH_3, \ -15°}$ ⬡=CH_2

[1] F. N. Tebbe, G. W. Parshall, and G. S. Reddy, *Am. Soc.*, **100**, 3611 (1978).

Chlorobis(cyclopentadienyl)hydridozirconium(IV), **6**, 175–179; **7**, 101–102.

1,4-Enones. The vinylzirconium compounds obtained from acetylenes and this reagent (**6**, 177–178) do not react with α,β-unsaturated ketones. However, in the presence of nickel(II) acetylacetonate conjugate addition occurs to give, after hydrolysis, 1,4-enones in generally good yields, at least in the case of complexes derived from terminal acetylenes. Two typical examples are formulated in equation (I). A zirconium O-enolate is a probable intermediate.[1]

(I) $(CH_3)_3CC{\equiv}CH \xrightarrow{(C_5H_5)_2HZrCl}$

Vinyl iodides. One step in the preparation of a synthon for one fragment of the antibiotic erythronolide B involved conversion of **1** to **2**. This conversion was accomplished in high yield by hydrozirconation (C_6H_6, 43°) followed by iodination (CCl_4, 25°). Only one iodoalkene was formed.[2]

$$
\underset{1}{\underset{\substack{\text{CH}_3 \quad \text{CH}_2\text{CH}_3}}{\overset{\text{CH}_3}{\underset{\text{H}}{\text{C}}}\overset{}{\underset{}{\text{C}}}\equiv\text{C}-\text{CHOSi(CH}_3)_2\text{C(CH}_3)_3}}
\xrightarrow[84\%]{\substack{1)\ (\text{C}_5\text{H}_5)_2\text{HZrCl} \\ 2)\ \text{I}_2 \\ 3)\ \text{H}_2\text{O}}}
\underset{2}{\underset{\text{CH}_2\text{CH}_3}{\overset{\text{CH}_3 \quad \text{I}}{\text{C}=\text{C}}}\ \text{CHOSi(CH}_3)_2\text{C(CH}_3)_3}
$$

Cross-coupling of acetylenes with aryl halides; (E)-alkenylarenes. (E)-Alkenyl-zirconium compounds (**1**), obtained by hydrozirconation of terminal alkynes, react smoothly with aryl bromides and iodides in the presence of 10 mole % of $\text{Ni}[\text{P(C}_6\text{H}_5)_3]_4$ to form (E)-alkenylarenes (**2**) in yields generally of 70–95%.[3]

$$
\underset{1}{\underset{\substack{\text{H} \qquad \text{ClZrCp}_2}}{\overset{\substack{\text{R} \qquad \text{H}}}{\text{C}=\text{C}}}}
+ \text{ArBr(I)}
\xrightarrow[70\text{–}95\%]{\text{Ni}[\text{P(C}_6\text{H}_5)_3]_4,\ \text{THF},\ 20°}
\underset{2}{\underset{\substack{\text{H} \qquad \text{Ar}}}{\overset{\substack{\text{R} \qquad \text{H}}}{\text{C}=\text{C}}}}
$$

1,3-Dienes. The (E)-1-alkenylzirconium compounds, obtained by hydro-zirconation of 1-alkynes (**6**, 177–178), undergo cross coupling with alkenyl halides to form conjugated dienes (*cf.* the similar reaction of alkenylalanes with alkenyl halides, **7**, 95–96). In the present synthesis tetrakis(triphenylphosphine)-palladium (**6**, 571–573) can usually serve as catalyst. Nickel catalysts are less effective than Pd catalysts in this case.[4]

Examples:

$$
\underset{\substack{\text{Cl}}}{\underset{\substack{\text{H} \qquad \text{ZrCp}_2}}{\overset{\substack{n\text{-C}_5\text{H}_{11} \qquad \text{H}}}{\text{C}=\text{C}}}}
+
\underset{\substack{\text{H} \qquad \text{C}_4\text{H}_9\text{-}n}}{\overset{\substack{\text{I} \qquad \text{H}}}{\text{C}=\text{C}}}
\xrightarrow[91\%]{\text{Pd}(0)}
\underset{\substack{\text{H} \qquad \text{C}_4\text{H}_9\text{-}n}}{\overset{\substack{n\text{-C}_5\text{H}_{11} \qquad \text{H}}}{\text{C}=\text{C}\ \text{C}=\text{C}}}
$$

$$
\xrightarrow{77\%}
$$

$$
\xrightarrow{60\%}
$$

This cross coupling can also be used with alkenylzirconium compounds derived from internal alkynes if metal salts of Zn or Cd, such as $ZnCl_2$, are added in catalytic amounts. For example, the reaction of **1** with **2** under the above conditions fails, but in the presence of 0.2 equiv. of $ZnCl_2$ it results in formation of **3** in 72% yield (2 hours).[5]

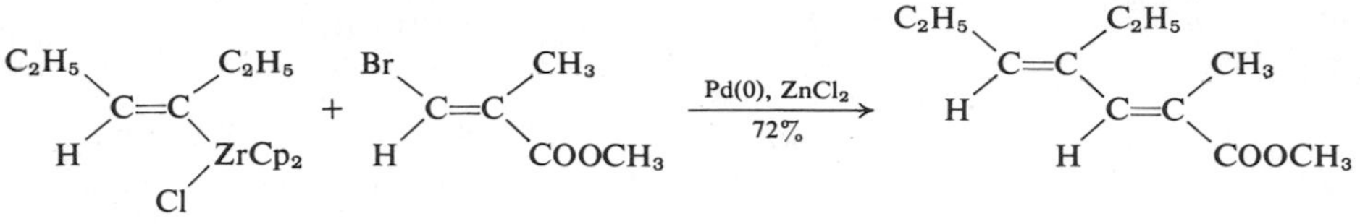

Transmetalation. Carr and Schwartz[6] have found that an organic group bound to zirconium can be transferred to aluminum. Thus reaction of **1**, obtained by hydrozirconation of alkenes or alkynes, with $AlCl_3$ results in a color change from bright yellow to orange as an organoaluminum dichloride is formed. On addition of an acid chloride, a ketone (**2**) is formed rapidly. This new ketone synthesis proceeds in high yield from alkyl- and alkenylaluminum dichlorides.

Examples:

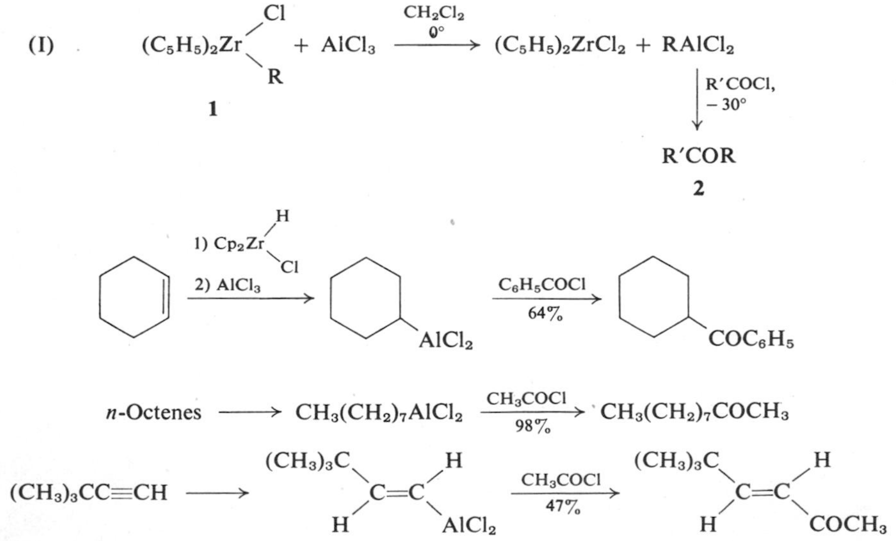

Alkenylzirconium complexes can also undergo transmetalation from Zr to Cu to form alkenylcopper(I) complexes, which are useful as synthetic intermediates. Since alkenylcopper(I) complexes are thermally unstable, dienes are obtained when the reaction is carried out at 0–20°, with formation of a copper mirror. The copper(I) complexes can also be converted into ate complexes and used for conjugate addition.[7]

Example:

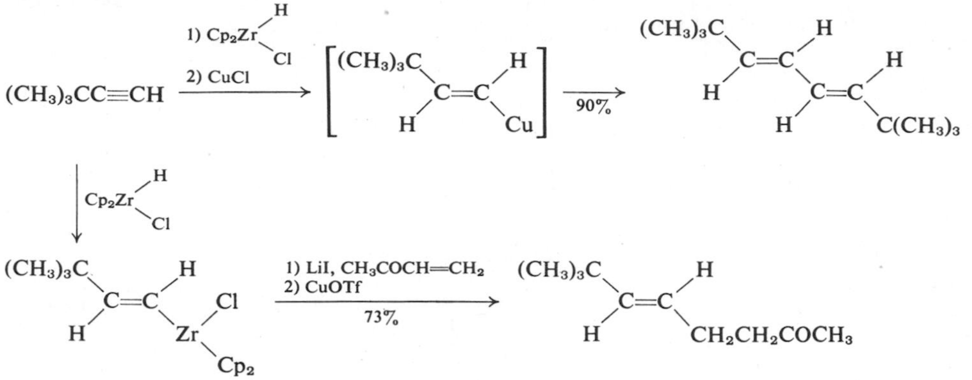

[1] M. J. Loots and J. Schwartz, *Am. Soc.*, **99**, 8045 (1977).
[2] E. J. Corey, E. J. Trybulski, L. S. Melvin, Jr., K. C. Nicolaou, J. A. Secrist, R. Lett, P. W. Sheldrake, J. R. Falck, D. J. Brunelle, M. F. Haslanger, S. Kim, and S. Yoo, *ibid.*, **100**, 4618 (1978).
[3] E. Negishi and D. E. Van Horn, *ibid.*, **99**, 3168 (1977).
[4] N. Okukado, D. E. Van Horn, W. L. Klima, and E. Negishi, *Tetrahedron Letters*, 1027 (1978).
[5] E. Negishi, N. Okukado, A. O. King, D. E. Van Horn, and B. I. Spiegel, *Am. Soc.*, **100**, 2254 (1978).
[6] D. B. Carr and J. Schwartz, *ibid.*, **99**, 638 (1977).
[7] M. Yoshifuji, M. J. Loots, and J. Schwartz, *Tetrahedron Letters*, 1303 (1977).

1-Chlorocarbonylbenzotriazole, 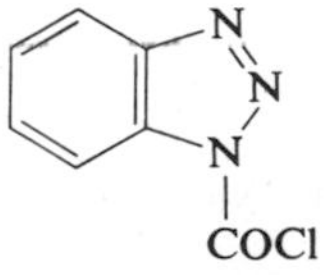 (**1**). Mol. wt. 181.58. The

reagent is prepared by the reaction of benzotriazole with phosgene.

Carbamates. The reagent reacts with an alcohol to form a stable 1-alkoxy-carbonyltriazole (**2**), which reacts with amines to form carbamates (**3**).[1]

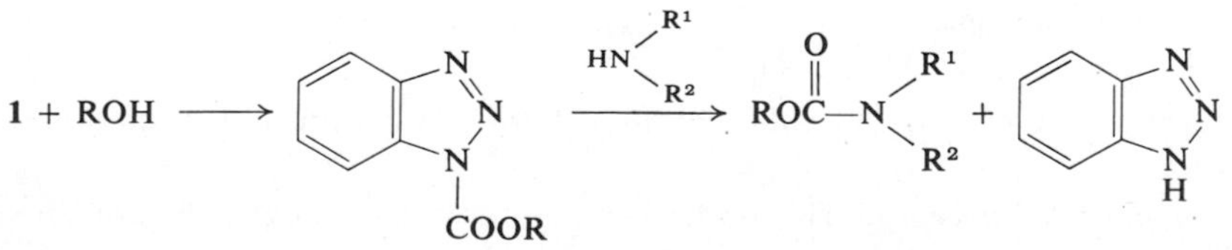

[1] I. Butula, L. Čurković, M. V. Proštenik, V. Vela, and F. Zorko, *Synthesis*, 704 (1977).

Chlorocyanoketene, $\underset{CN}{\overset{Cl}{\diagdown}}C{=}C{=}O$ **(1).** Mol. wt., 101.50.

Preparation *in situ* from mucochloric acid:

$$\xrightarrow{C_6H_6,\ \Delta}\ \mathbf{1} + N_2 + CH_3OCHO$$

The same method can be used to obtain bromo- and iodocyanoketene.[1]

β-*Lactams.*[2] These halocyanoketenes and also alkylcyanoketenes undergo [2 + 2] cycloaddition to formimidates to form β-lactams (2-azetidinones), all of which have the same stereochemistry as **2**.[3]

2

[1] H. W. Moore, L. Hernandez, and A. Sing, *Am. Soc.*, **98**, 3728 (1976).

[2] D. M. Kunert, R. Chambers, F. Mercer, L. Hernandez, Jr., and H. W. Moore. *Tetrahedron Letters*, 929 (1978).

[3] R. Chambers, D. Kunert, L. Hernandez, Jr., F. Mercer, and H. W. Moore, *ibid.*, 933 (1978).

2-Chloro-1,3-dithiane,   **(1).** Mol. wt. 154.69, m.p. 50° dec. The reagent can be prepared by chlorination of 1,3-dithiane with sulfuryl chloride or NCS.

Ketene dithioacetals.[1] This reagent (**1**) can be used for preparation of ketene dithioacetals by reaction with triphenylphosphine or triethyl phosphite to form **2** or **5**. The former reagent (**2**) can be used to form ketene dithioacetals (**4**) from aldehydes; the latter reagent (**5**) can be used with either aldehydes or ketones.

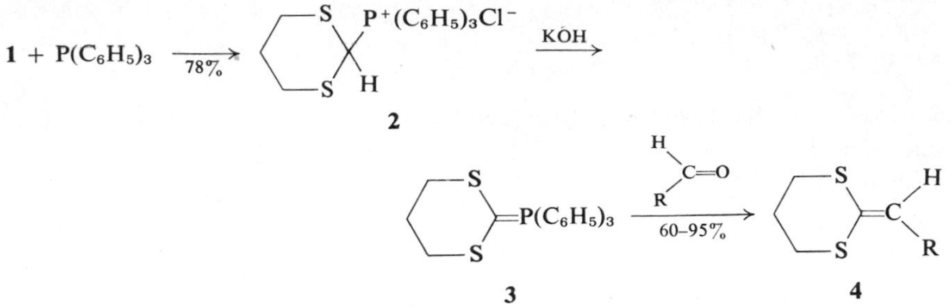
1 + P(C6H5)3
78%
P+(C6H5)3Cl−
KOH
2
=P(C6H5)3
60–95%
3
4

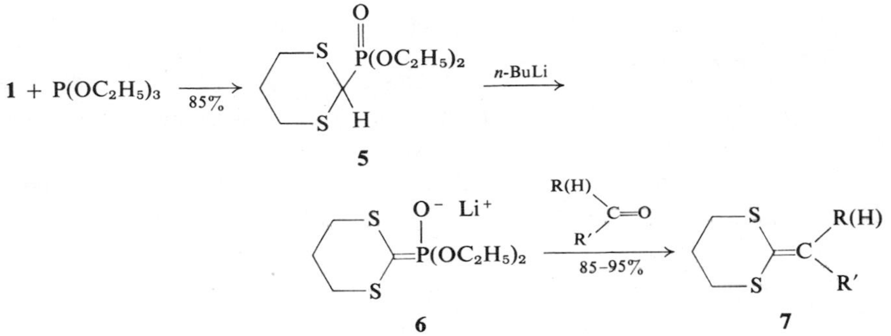
1 + P(OC2H5)3
85%
P(OC2H5)2
n-BuLi
5
O− Li+
=P(OC2H5)2
85–95%
6
7

Formylation.[2] Aliphatic aldehydes can be converted into protected malondialdehydes by alkylation of the morpholinoenamines with 2-chloro-1,3-dithiane (equation I). Cycloalkanones can be formylated in the same way (equation II); acidic substances can be alkylated directly (equation III).

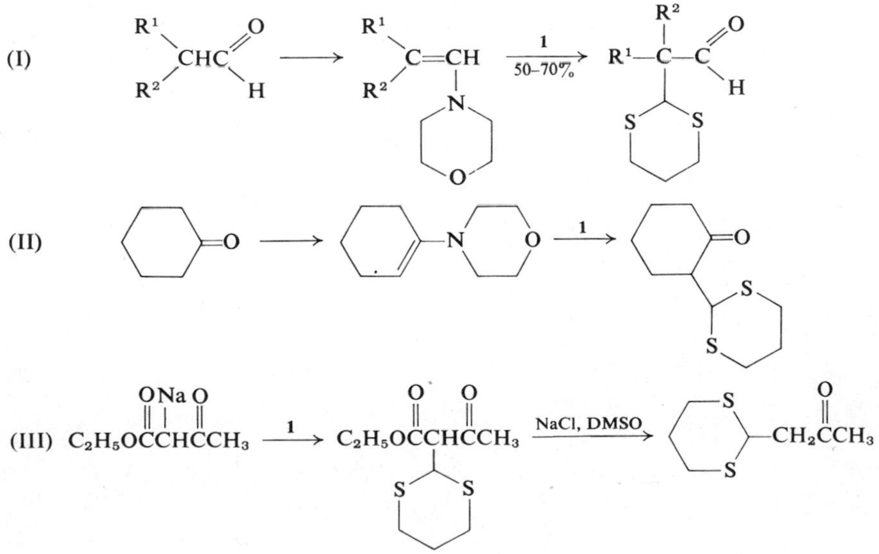
(I)
50–70%
(II)
(III) C2H5OCCHCCH3
C2H5OCCHCCH3
NaCl, DMSO
CH2CCH3

[1] C. G. Kruse, N. L. J. M. Broekhof, A. Wijsman, and A. van der Gen, *Tetrahedron Letters*, 885 (1977).
[2] E. C. Taylor and J. L. La Mattina, *ibid.*, 2077 (1977).

2-Chloro-3-ethylbenzoxazolium tetrafluoroborate (1). Mol. wt. 269.44, m.p. 131° dec. Supplier: Fluka.

Preparation:

Alkyl chlorides.[1] The reagent (1) reacts with alcohols in the presence of tetraethylammonium chloride and triethylamine to form alkyl chlorides in yields of 75–95%. The reaction proceeds with inversion ordinarily, but retention was observed in the reaction of two carbohydrates, perhaps because of an anomeric effect.

Examples:

Isocyanides.[2] The reagent (**1**) in combination with 2 equiv. of triethylamine converts formamides to isocyanides at room temperature in 70–85% yield (equation I).

(I) $R(Ar)NHCHO + 1 + 2N(C_2H_5)_3 \xrightarrow{CH_2Cl_2}$

$$RNC + \qquad\qquad =O + (C_2H_5)_3N \cdot HCl + (C_2H_5)_3N \cdot HBF_4$$

$$\text{(70–85\%)}$$

1,2-Dichloroalkanes.[3] The reagent in combination with tetraethylammonium chloride and triethylamine converts epoxides into 1,2-dichloro compounds at room temperature (12–48 hours).

Examples:

$$C_6H_5CH-CH_2 \xrightarrow{81\%} C_6H_5CHClCH_2Cl$$

Isocyanates.[4] Methyl thiocarbamates react with **1** and triethylamine at 20° to form isocyanates in 65–85% yield (equation I).

(I) $R(Ar)NH\overset{\text{S}}{\overset{\|}{C}}OCH_3 + 1 + N(C_2H_5)_3 \xrightarrow{CH_2Cl_2}$

$$R(Ar)N{=}C{=}O + \qquad\qquad =S + CH_3Cl + (C_2H_5)_3\overset{+}{N}HBF_4{}^-$$
$$\text{(65–85\%)}$$

Decarbonylation of α-hydroxy carboxylic acids.[5] These acids are converted into ketones on reaction with **1** and triethylamine (2 equiv.) at room temperature; yield 65–90%.

Example:

$$(CH_3)_2C\underset{COOH}{\overset{OH}{<}} + 1 + 2N(C_2H_5)_3 \xrightarrow[-CO, -H_2O]{CH_2Cl_2}$$

$$(CH_3)_2C{=}O + \underset{(71\%)}{} \quad \text{[benzoxazolone ring]} {=}O + (C_2H_5)_3\overset{+}{N}HBF_4{}^- + (C_2H_5)_3\overset{+}{N}HCl^-$$

[1] T. Mukaiyama, S. Shoda, and Y. Watanabe, *Chem. Letters*, 383 (1977).
[2] Y. Echigo, Y. Watanabe, and T. Mukaiyama, *ibid.*, 697 (1977).
[3] *Idem, ibid.*, 1013 (1977).
[4] *Idem, ibid.*, 1345 (1977).
[5] T. Mukaiyama and Y. Echigo, *ibid.*, 49 (1978).

Chloroform, $CHCl_3$. Mol. wt. 119.39, b.p. 61–62°.

α-*Methoxy carboxylic acids*. These acids can be prepared by reaction of aliphatic aldehydes with chloroform in the presence of NaH in THF as basic catalyst (exothermic reaction) for about 3 hours at 0° followed by addition of CH_3OH containing dissolved NaOH.[1]

$$RCHO + CHCl_3 \xrightarrow{NaH, THF}$$

$$\left[\underset{OH}{RCHCCl_3} \xrightarrow{NaOH} \underset{O}{RCH{-}C(Cl)_2} \xrightarrow{CH_3OH} \underset{OCH_3}{RCHCOCl}\right] \xrightarrow[50-60\%]{H_2O} \underset{OCH_3}{RCHCOOH}$$

[1] E. L. Compere, Jr. and A. Shockravi, *J. Org.*, **43**, 2702 (1978).

2-[(Chloroformyl)oxy]ethyltriphenylphosphonium chloride,

$$(C_6H_5)_3\overset{+}{P}CH_2CH_2O\overset{O}{\overset{\|}{C}}Cl, \quad \underset{Cl^-}{} \text{ Peoc-Cl (1). Mol. wt. 404.9, m.p. 136°.}$$

Preparation:

$$(C_6H_5)_3\overset{+}{P}CH_2CH_2OHCl^- \xrightarrow[95\%]{COCl_2} 1$$

Protection of amino acids. The reagent (1) reacts with amino acid esters in chloroform–methylene chloride at −15° in the presence of 1 equiv. of triethyl-amine to give the protected derivatives **2**. The blocking Peoc group is split by

$$\underset{Cl^-}{(C_6H_5)_3\overset{+}{P}CH_2CH_2O\overset{O}{\overset{\|}{C}}{-}NH\overset{R}{\overset{|}{C}}HCOOR'}$$

2

1 N NaOH in CH_3OH–water. On the other hand, the group is exceptionally stable to acid treatment.[1]

One interesting property of this protective group is that it increases the solubility of the amino acid in water. Thus it has been possible to synthesize a Peoc-tripeptide ester in reasonable yield in aqueous solution, using a water-soluble carbodiimide and 1-hydroxybenzotriazole as condensing agents.[2]

[1] H. Kunz, *Ber.*, **109**, 2670 (1976); *idem*, *Ann.*, 1674 (1976).
[2] *Idem*, *Angew. Chem., Int. Ed.*, **17**, 67 (1978).

3-Chloro-1-lithio-3-methylbut-1-yne, $Cl\!-\!\overset{\displaystyle CH_3}{\underset{\displaystyle CH_3}{C}}\!-\!C\!\equiv\!CLi$ (**1**). Mol. wt. 108.50.

The reagent is prepared by reaction of 3-chloro-3-methyl-1-butyne[1] with methyllithium in ether at $-75°$.[2]

3-Methylbuta-1,2-dienylidene. This carbene (**2**) is formed by decomposition of **1** at temperatures above $-75°$. It reacts as expected to cyclopropylate olefins (equation I).

(I) $1 \xrightarrow{\; >-75° \;} (CH_3)_2C\!=\!C\!=\!C\!: \xrightarrow[52\%]{\; C_6H_5CH=CH_2 \;} (CH_3)_2C\!=\!C\!=\!\!\!\triangleleft^{\,C_6H_5}$

         2

The five-carbon carbene has been used in a synthesis of the monoterpene artemisia ketone (**5**).[2] Reaction of the sulfide **3** with **2** gives a cumulenic ylide (**a**), which decomposes to a mixture of four isomeric products (54% yield). The major product (38% yield) is an allenic thioether (**4**) formed by a [2.3]sigmatropic shift. On hydrolysis **4** is converted into the terpene **5**.

The carbene undergoes insertion into S—S and S—Li bonds. Examples of these reactions are formulated in equations (II) and (III).[3]

(II) $2 \xrightarrow[81-84\%]{(RS)_2} (CH_3)_2C{=}C{=}C(SR)_2 \xrightarrow[93-95\%]{SiO_2}$ $H_2C \overset{CH_3}{\underset{}{\diagdown}} {=} \overset{SR}{\underset{SR}{\diagup}}$

(III) $2 \xrightarrow{C_6H_5SLi} (CH_3)_2C{=}C{=}C \overset{SC_6H_5}{\underset{Li}{\diagdown}} \xrightarrow[83\%]{C_4H_9I}$

$(CH_3)_2C{=}C{=}C \overset{SC_6H_5}{\underset{C_4H_9\text{-}n}{\diagdown}} \xrightarrow{HgCl_2,\ CH_3CN}$ $\overset{CH_3}{\underset{CH_3}{\diagdown}} C{=}CH \overset{C_4H_9\text{-}n}{\underset{O}{\diagup}}$

[1] G. H. Hennion and A. P. Boiselle, *J. Org.*, **26**, 725 (1961).
[2] D. Michelot, G. Linstrumelle, and S. Julia, *Syn. Comm.*, **7**, 95 (1977).
[3] J.-C. Clinet and S. Julia, *J. Chem. Res.* (*M*), 1714 (1978).

Chloromethyl methyl sulfide, 6, 109–110.

Protection of carboxylic acids.[1] Carboxylic acids can be converted into methylthiomethyl esters in high yield by reaction of the potassium salt with chloromethyl methyl sulfide in refluxing benzene containing catalytic amounts of sodium iodide and 18-crown-6. The ester group is stable to mild reducing agents and aqueous base. Deprotection is effected by reaction with $HgCl_2$ in refluxing acetonitrile–water.

Protection of phenols.[2] Aryl methylthiomethyl ethers can be obtained in 90–95% yield by reaction of sodium phenoxides with chloromethyl methyl sulfide in HMPT (20°, 16 hours). Deprotection is accomplished (90–95% yield) with $HgCl_2$ in refluxing CH_3CN–H_2O (10 hours).

[1] L. G. Wade, Jr., J. M. Gerdes, and R. P. Wirth, *Tetrahedron Letters*, 731 (1978).
[2] R. A. Holton and R. G. Davis, *ibid.*, 533 (1977).

$\overset{O}{\overset{\uparrow}{}}$

Chloromethyl phenyl sulfoxide, $C_6H_5SCH_2Cl$. Mol. wt. 174.66, m.p. 35–36°. The sulfoxide is obtained in > 70% yield by oxidation of chloromethyl phenyl sulfide with *m*-chloroperbenzoic acid.[1]

Chloromethyl ketones. The ketones can be prepared by reaction of an aldehyde with lithio chloromethylphenyl sulfoxide in THF at $-78°$ to form a β-hydroxy-α-chloro sulfoxide. Pyrolysis of the product in xylene (160°) leads to a chloromethyl ketone.[2]

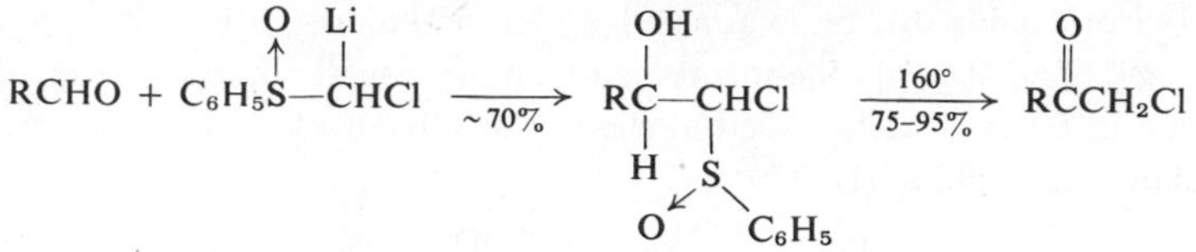

$$\underset{\text{O Li}}{\text{RCHO} + C_6H_5S-\overset{\uparrow}{C}HCl} \xrightarrow{\sim 70\%} \underset{\substack{H \quad S \\ O \quad C_6H_5}}{R\overset{OH}{\underset{|}{C}}-CHCl} \xrightarrow[75-95\%]{160°} \underset{O}{R\overset{\parallel}{C}CH_2Cl}$$

α,β-*Unsaturated aldehydes.* The β-hydroxy-α-chloro sulfoxides, formed as shown above from carbonyl compounds, can be converted into α-epoxy sulfoxides by treatment with base. When heated in toluene these products decompose to α,β-unsaturated aldehydes. An example is formulated in equation (I).[3]

(I)

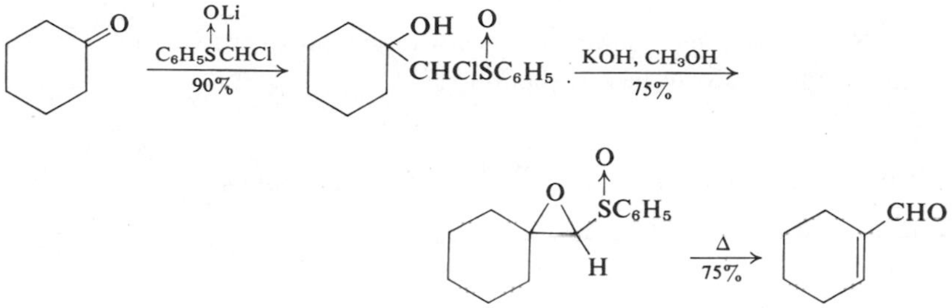

[1] T. Durst, *Am. Soc.*, **91**, 1034 (1969).
[2] V. Reutrakul and W. Kanghae, *Tetrahedron Letters*, 1225 (1977).
[3] *Idem, ibid.*, 1377 (1977).

2-Chloro-1-methylpyridinium iodide, I⁻ (1). Mol. wt. 255.49, m.p. 200°

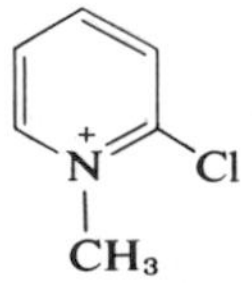

dec., hygroscopic. Suppliers: Aldrich, Fluka.

Reactions. Mukaiyama and collaborators[1] have reported a number of useful transformations mediated by **1** in the presence of base. Thus esters are formed in 80–95% yield from acids and alcohols in the presence of 1 equiv. of **1** and 2 equiv. of triethylamine (equation I).[1]

(I)

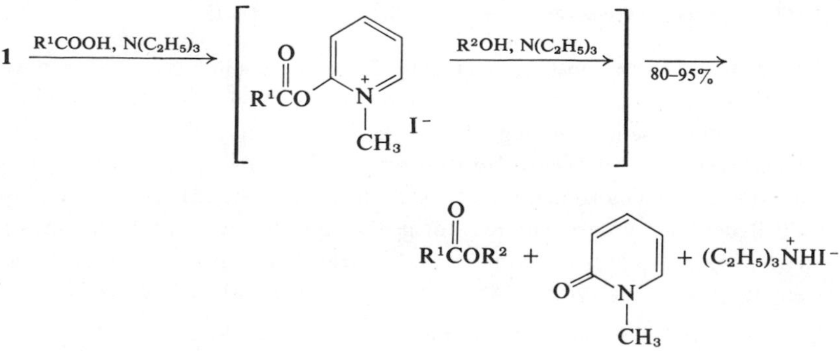

ω-Hydroxy acids can be lactonized under similar conditions. Yields are low in the case of eight- and nine-membered lactones; smaller and larger lactones are obtained in 60–90% yield. Acetonitrile is the solvent of choice. The reaction is formulated in equation (II).[2]

(II) $HO(CH_2)_nCOOH + 1 \xrightarrow{2N(C_2H_5)_3}$

$(n = 5, 10, 11, 14)$

(60–90%)

N,N'-Disubstituted thioureas are converted into carbodiimides by reaction with **1** and 2 equiv. of base (equation III).[3]

(III) $R^1NHCNHR^2 + 1 \xrightarrow{2N(C_2H_5)_3} R^1N{=}C{=}NR^2 +$

Pyridyl sulfides can be prepared by reaction of alcohols with 1-methyl-2-fluoropyridinium tosylate and base followed by reaction with pyrid-2-thione (equation IV).[4]

(IV) $+ ROH \xrightarrow{N(C_2H_5)_3}$ $\xrightarrow[65-90\%]{, N(C_2H_5)_3}$

[1] T. Mukaiyama, M. Usui, E. Shimada, and K. Saigo, *Chem. Letters*, 1045 (1975).
[2] *Idem, ibid.*, 49 (1976).
[3] T. Shibanuma, M. Shiono, and T. Mukaiyama, *ibid.*, 575 (1977).
[4] T. Mukaiyama, S. Ikeda, and S. Kobayashi, *ibid.*, 1159 (1975).

Chloromethyltriphenylphosphonium iodide (1). Mol. wt. 438.69. The salt is obtained as a white precipitate (60–70% yield) by reaction of triphenylphosphine and chloroiodomethane in refluxing THF.

Chloromethylenetriphenylphosphorane **(2)**.[1] Treatment of **1** with potassium *t*-butoxide forms chloromethylenetriphenylphosphorane **(2)**. Chloromethylation of aldehydes and ketones with reagent prepared in this way results in 1-chloro-1-alkenes [(E)- and (Z)-isomers] in 60–95% yield. If excess base is used, aromatic aldehydes give arylacetylenes, ArCHO $\longrightarrow$ ArC≡CH (40–90% yield).

[1] S. Miyano, Y. Izumi, and H. Hashimoto, *J.C.S. Chem. Comm.*, 446 (1978).

$$(C_6H_5)_3\overset{+}{P}CH_2Cl\bar{I} \xrightarrow{\text{KOC(CH}_3)_3, \text{ HOC(CH}_3)_3} (C_6H_5)_3P{=}CHCl$$

$$\mathbf{1} \qquad\qquad\qquad\qquad\qquad \mathbf{2}$$

m-**Chloroperbenzoic acid, 1,** 135–139; **2,** 68–69; **3,** 45–50; **6,** 110–114; **7,** 62–64.

Caution: An explosive decomposition of excess reagent has been reported during work-up of a reaction mixture.[1]

α-*Hydroxy ketones.* Oxidation of the enol methyl ether of cyclohexanone (**1**) with *m*-chloroperbenzoic acid in methanol results in formation of adipoin dimethyl acetal (**2**) in 81% yield. The epoxy ether **a** is undoubtedly the intermediate. The same oxidation of 2,3-dihydro-γ-pyrane (**3**) yields **4** and **5** in the ratio 9:1. This method therefore can be used for α-hydroxylation of ketones.[2]

Protection of γ-*lactones.*[3] γ-Lactones can be converted in about 90% yield into protected lactols by reduction with diisobutylaluminum hydride (**1,** 261; **2,** 140) followed by reaction with an alcohol in the presence of BF_3 etherate. Regeneration of the lactone can be effected in high yield in one step by reaction with 1.1 equiv. of *m*-chloroperbenzoic acid in CH_2Cl_2 containing BF_3 etherate. This regeneration step gives low yields of δ-lactones from δ-lactols.

Baeyer-Villiger oxidation. The Baeyer-Villiger oxidation of **1** to **2** was used in one step in a total synthesis of (±)-sarracenin (**3**), an iridoid monoterpene. The oxidation was carried out in CH_2Cl_2 in the presence of sodium bicarbonate, which approximately doubles the rate of reaction.[4] This method is credited to Dr. P. S. Stotter (private communication).

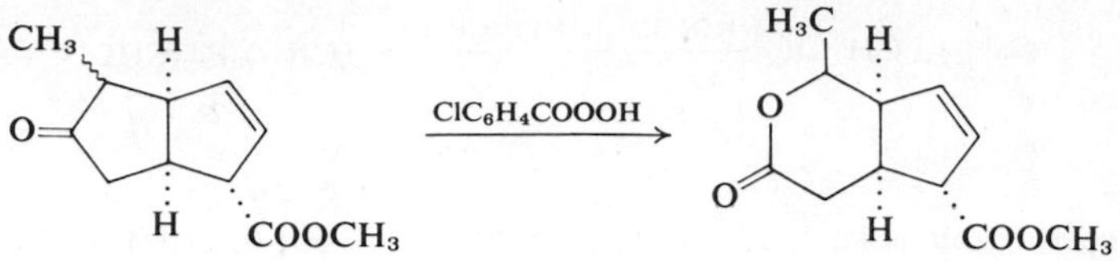

1

2

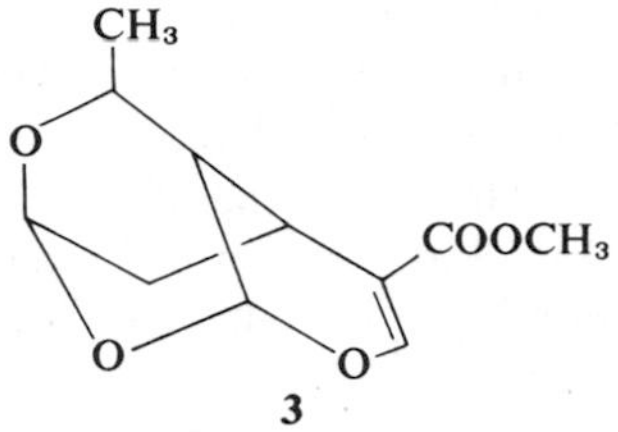

3

Primary alcohols. 1-Alkenes can be converted into primary alcohols by the sequence formulated in equation (I), which involves hydrosilylation, conversion to a pentafluorosilicate, and finally cleavage with *m*-chloroperbenzoic acid (*cf.*, N-Bromosuccinimide, this volume). The choice of solvent is important in this

(I) n-$C_6H_{13}CH{=}CH_2$+ $HSiCl_3$ $\xrightarrow{H_2PtCl_6}$ $C_6H_{13}CH_2CH_2SiCl_3$ $\xrightarrow{KF}$

$K_2[C_8H_{17}SiF_5]$ $\xrightarrow[82\%]{ClC_6H_5CO_3H}$ n-$C_8H_{17}OH$

reaction, yields being the highest in DMF. When benzene is the solvent the yield is only *ca.* 30%, but addition of 18-crown-6 improves the yield to 70%. Use of peracetic acid or *t*-butyl hydroperoxide results mainly in formation of octyltrifluorosilane.

The method gives low yields of secondary alcohols when applied to internal alkenes.[5]

Other examples:

$CH_3OOC(CH_2)_8CH{=}CH_2$ $\xrightarrow{77\%}$ $CH_3OOC(CH_2)_8CH_2CH_2OH$

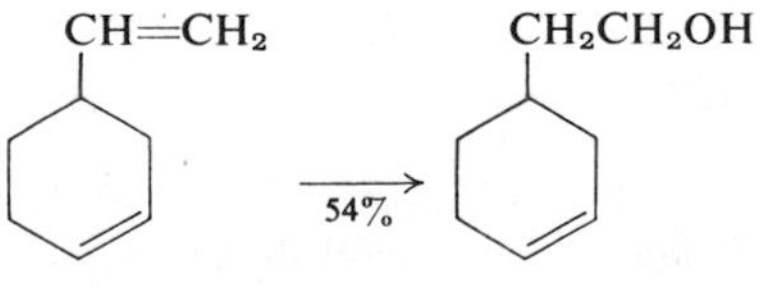

Oxidation of alkyl iodides. Alkyl iodides are inert to O_3, HIO_4, and H_2O_2, but are oxidized by *m*-chloroperbenzoic acid in CH_2Cl_2 or CCl_4. Primary alkyl iodides are converted into the corresponding alcohols in good yield; an example is shown in equation (I). On the other hand, alkyl iodides containing electron-attracting substituents at the α-position are converted into unsaturated products;

(I) $\quad\quad\quad$ $C_6H_5CH_2CH_2I \xrightarrow[73\%]{ClC_6H_4CO_3H} C_6H_5CH_2CH_2OH$

examples are shown in equations (II) and (III). The stereochemistry of this elimination was established in another system as *syn*, as in elimination of amine

(II) $\quad$ $\underset{\displaystyle \overset{\displaystyle\parallel}{O}}{C_2H_5OC}\underset{\displaystyle I}{CH}CH_2(CH_2)_6CH_3 \xrightarrow[81\%]{ClC_6H_4CO_3H} \underset{\displaystyle \overset{\displaystyle\parallel}{O}}{C_2H_5OC}CH{=}CH(CH_2)_6CH_3$

(III) $\quad$ $C_6H_5\overset{\displaystyle\overset{O}{\parallel}}{\underset{\displaystyle\underset{O}{\parallel}}{S}}{-}\overset{\displaystyle CH_3}{\underset{\displaystyle I}{C}}{-}CH_3 \xrightarrow[87\%]{} C_6H_5\overset{\displaystyle O}{\underset{\displaystyle\parallel}{S}}{-}C\overset{\displaystyle CH_2}{\underset{\displaystyle CH_3}{\diagup}}$

oxides, sulfoxides, and selenoxides. Reich and Peake[6] propose that iodoso compounds are the intermediates in this oxidation of alkyl iodides.

Oxidation of hydroxyl groups (6, 110–111). Cella[7] has extended this reaction of peracids catalyzed by 2,2,6,6-tetramethylpiperidine hydrochloride (TMP·HCl) to an efficient preparation of epoxy ketones. Thus unsaturated alcohols can be oxidized in one step to α,β-, β,γ-, and other epoxy ketones (yield 60–90%).

Oxidation of saturated secondary alcohols with about 4 equiv. of *m*-chloroperbenzoic acid catalyzed by TMP·HCl results in esters or lactones via a Baeyer-Villiger reaction.

Examples:

$$C_6H_5CH_2CHOHCH_3 \xrightarrow[90\%]{ClC_6H_4CO_3H,\ TMP\cdot HCl} C_6H_5CH_2O\overset{\displaystyle\overset{O}{\parallel}}{C}CH_3$$

$$(C_6H_5)_2CHOH \xrightarrow[53.5\%]{} C_6H_5O\overset{\displaystyle\overset{O}{\parallel}}{C}C_6H_5$$

Oxidation of an indole to a hydroxyindoline. Of a number of oxidizing reagents, only *m*-chloroperbenzoic acid was found to oxidize the indole **1** to the hydroxyindoline **2**. This oxidation is extremely stereospecific and occurs exclusively from the α-side of the indole, the β-side being blocked by the eight-membered ring. The hydroxyindoline rearranges under acidic conditions to the tetrahydro derivative (**3**) of austamide, a toxic metabolite of *Aspergillus ustus*.[8]

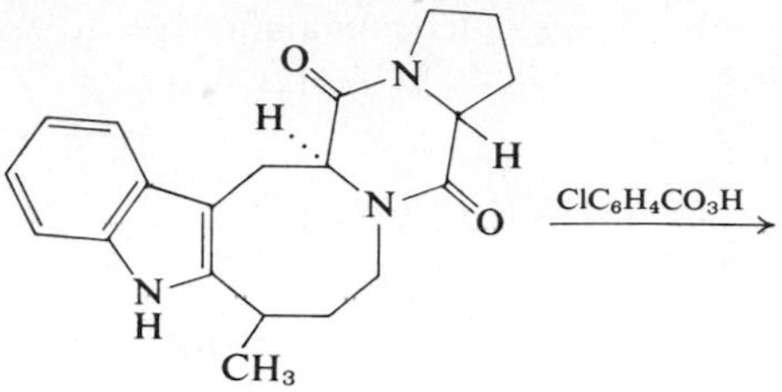

1

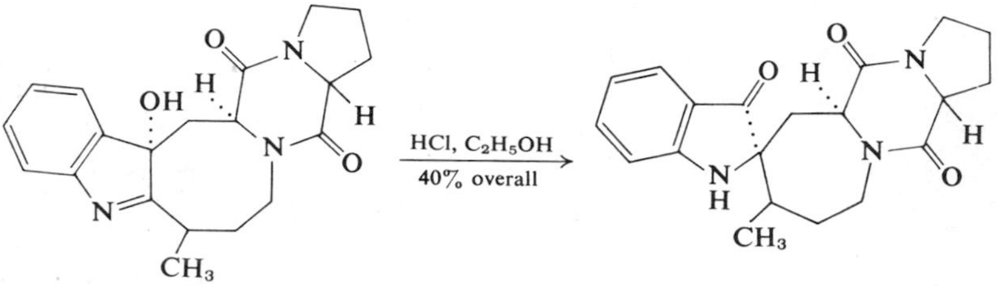

2

3

Reaction with a β,γ-unsaturated ketone. Attempted Baeyer-Villiger reaction of **1**, readily available from methyl ricinoleate, with *m*-chloroperbenzoic acid under usual conditions unexpectedly leads to the known furanoid fatty acid (**3**), although in only 25% yield. When the epoxidation is conducted under milder conditions, the expected epoxide (**2**) is obtained readily. When this epoxide is refluxed in chloroform containing trace amounts of hydrogen chloride, **3** is obtained in 80% yield. This reaction may be relevant to biosynthesis of natural furanoid fatty acids such as **3**.[9]

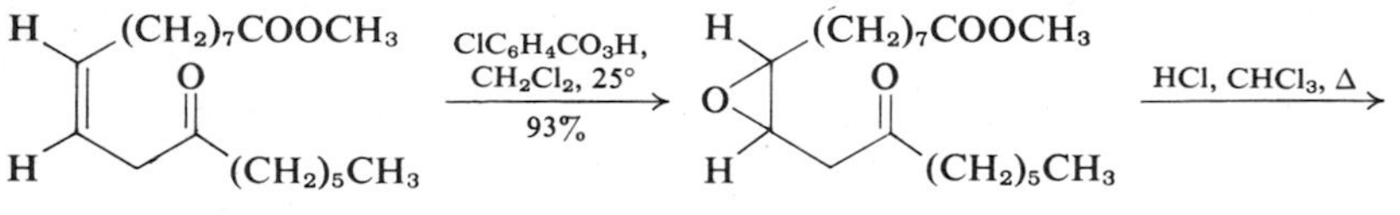

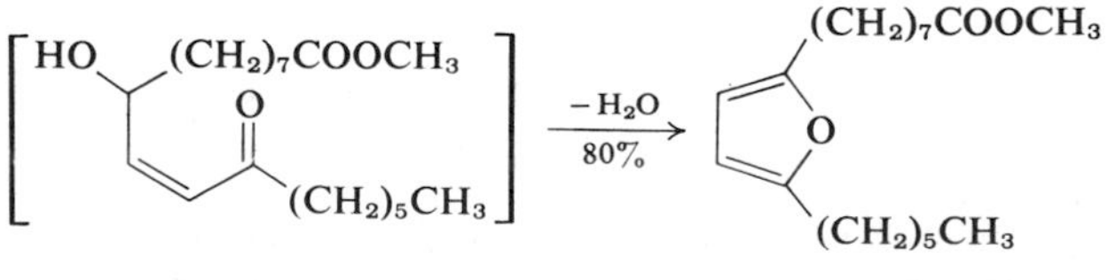

α-Hydroxy enones; α-acetoxy enones. These useful synthetic intermediates can be prepared by *m*-chloroperbenzoic acid oxidation of 2-trimethylsilyloxy-1,3-dienes followed by desilylation (equation I).[10]

(I)

$$1) ClC_6H_4COOOH$$
$$2) (C_2H_5)_4NF$$

$$1) ClC_6H_4COOOH$$
$$2) (C_2H_5)_4NF, Ac_2O$$

Arene oxides. The literature contains scattered references to formation of arene oxides by peracid oxidation of polycyclic arenes. One difficulty in attaining reasonable yields is that the acid by-product can induce rearrangement of the oxides. Griffin *et al.*[11] circumvent this difficulty by use of a two-phase system composed of aqueous sodium bicarbonate–methylene chloride. Under these conditions phenanthrene oxide was prepared in 59% yield using *m*-chloroperbenzoic acid. The oxidation becomes more complex with higher homologs of phenanthrene, but pyrene-4,5-oxide and chrysene-5,6-oxide were isolated in 14 and 9% yields, respectively. These oxides are deoxygenated to the parent arene by thiourea.

Peracid oxidation of naphthalene leads to *anti*-1,2;3,4-naphthalene dioxide (**2**, 15–20% yield).[12] A similar oxidation of 2-methylnaphthalene gives both *syn*- and *anti*-dioxides. Oxidation of anthracene results mainly in the quinone, but 9,10-diphenylanthracene can be oxidized to a dioxide of still uncertain structure.

$$\xrightarrow[15-20\%]{RCOOOH}$$

1   **2**

β-Lactams.[13] Peracid oxidation in the presence of pyridine of iminium salts (**2**), prepared as shown from azetidine-2-carboxylic acids (**1**), results in formation of β-lactams (**3**), possibly via peresters (**a**). The reaction is slow in the absence of a base.

[1] W. W. Brand, *Chem. Eng. News*, June 12, 88 (1978).
[2] A. A. Frimer, *Synthesis*, 578 (1977).
[3] P. A. Grieco, T. Oguri, and Y. Yokoyama, *Tetrahedron Letters*, 419 (1978).
[4] J. K. Whitesell, R. S. Matthews, and A. M. Helbling, *J. Org.*, **43**, 785 (1978).
[5] K. Tamao, T. Kakui, and M. Kumada, *Am. Soc.*, **100**, 2268 (1978).
[6] H. J. Reich and S. L. Peake, *ibid.*, **100**, 4888 (1978).

[7] J. A. Cella, J. P. McGrath, J. A. Kelley, O. ElSoukkary, and L. Hilpert, *J. Org.*, **42**, 2077 (1977).

[8] A. J. Hutchinson and Y. Kishi, *Tetrahedron Letters*, 539 (1978).

[9] S. Ranganathan, D. Ranganathan, and M. M. Mehrotra, *Synthesis*, 838 (1977).

[10] G. M. Rubottom and J. M. Gruber, *J. Org.*, **43**, 1599 (1978).

[11] K. Ishikawa, H. C. Charles, and G. W. Griffin, *Tetrahedron Letters*, 427 (1977).

[12] K. Ishikawa and G. W. Griffin, *Angew. Chem., Int. Ed.*, **16**, 171 (1977).

[13] H. H. Wasserman and A. W. Tremper, *Tetrahedron Letters*, 1449 (1977).

5-Chloro-1-phenyltetrazole (**1**), **2**, 319–320; **4**, 377; **7**, 64.

OH — H.[1] A recent synthesis of (−)-codeine and (−)-morphine required as one step the selective removal of the 2-hydroxyl group of the morphinan **2**. This was accomplished by etherification with the tetrazole compound to give **3** as the major product, easily obtained free from the isomeric ether by crystallization. Hydrogenolysis of **3** gave (−)-1-N-formylnordihydrothebainone (**4**), a known precursor to the morphine alkaloids.

[1] H. C. Beyerman, J. van Berkel, T. S. Lie, L. Maat, J. C. M. Wessels, H. H. Bosman, E. Buurman, E. J. M. Bijsterveld, and H. J. M. Sinnige, *Rec. trav.*, **97**, 127 (1978).

N-Chlorosuccinimide, 1, 139; **2**, 69–70; **5**, 127–129; **6**, 115–118; **7**, 65.

Oxidative decarboxylation of α-methylthio carboxylic acids (**6**, 396). Complete details for this reaction have been published.[1] The reaction affords a method for transformation of $\text{\raisebox{0pt}{>}}$CHCOOH into $\text{\raisebox{0pt}{>}}$C=O (equation I). The most difficult step usually is generation of the dianion of the acid. It is also possible to sulfenylate an ester, which is then hydrolyzed to the acid by KOH. Either NCS or

(I)

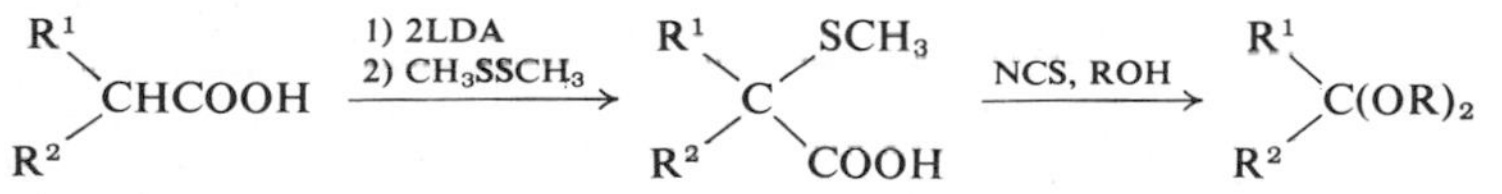

$$\underset{R^2}{\overset{R^1}{>}}\text{CHCOOH} \quad \xrightarrow[\text{2) CH}_3\text{SSCH}_3]{\text{1) 2LDA}} \quad \underset{R^2}{\overset{R^1}{>}}\underset{\text{COOH}}{\overset{\text{SCH}_3}{\text{C}}} \quad \xrightarrow{\text{NCS, ROH}} \quad \underset{R^2}{\overset{R^1}{>}}\text{C(OR)}_2$$

$NaIO_4$ in an absolute alcohol can be used as oxidant, since neither reagent oxidizes sulfides to sulfones.

Examples:

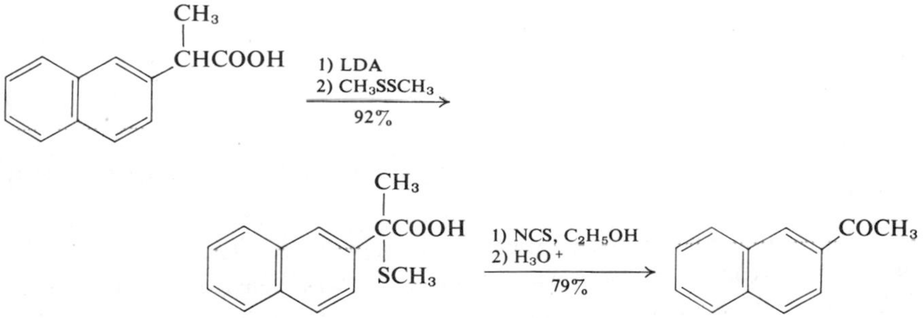

This reaction was used in one step in a synthesis of juvabione (**2**), a juvenile hormone mimic, from perillaldehyde (**1**).

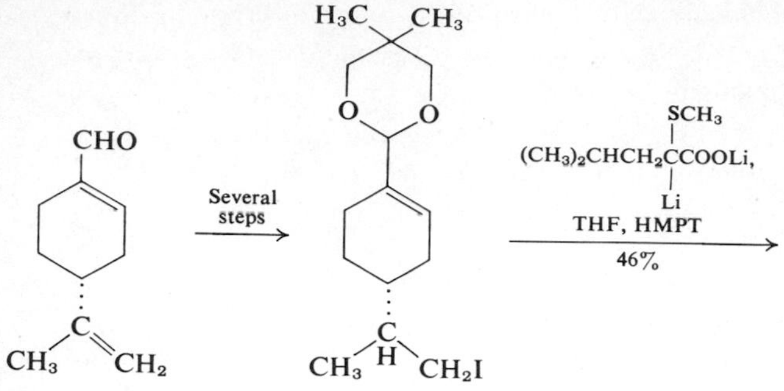

1

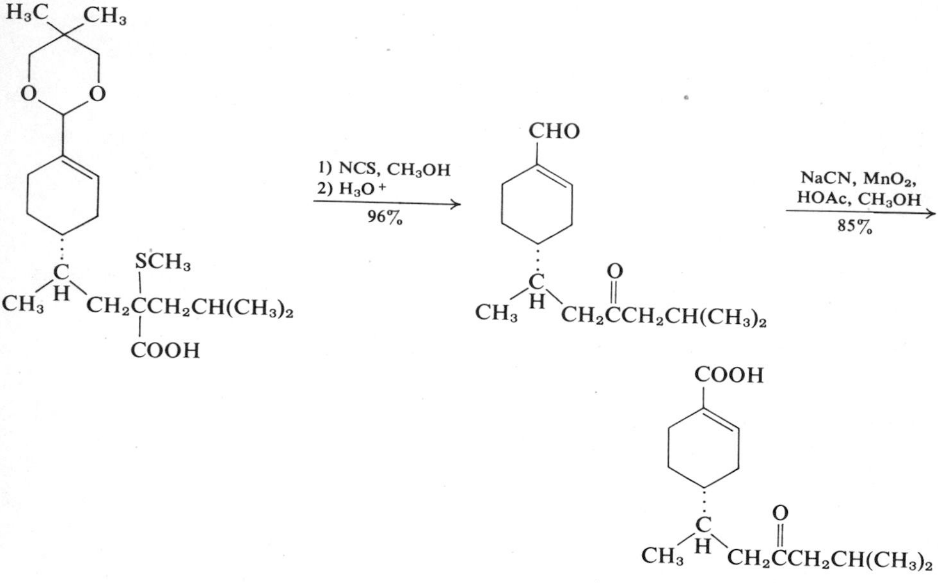

2

(1-Chloro-1-alkenyl)silanes. These useful synthetic intermediates can be prepared by reaction of NCS with (Z)-(1-alumino-1-alkenyl)silanes in ether at −20° (equation I). Use of chlorine in this reaction results in low yields, but iodine can be used to prepare (1-iodo-1-alkenyl)silanes. The corresponding bromo compounds are prepared by use of Br_2 and pyridine. In all three cases the (E)-isomer is obtained. The (Z)-isomers are obtained by isomerization of (E)-isomers with bromine in the presence of light. Both (E)- and (Z)-isomers undergo metal exchange with *n*-butyllithium in ether at −70°.[2]

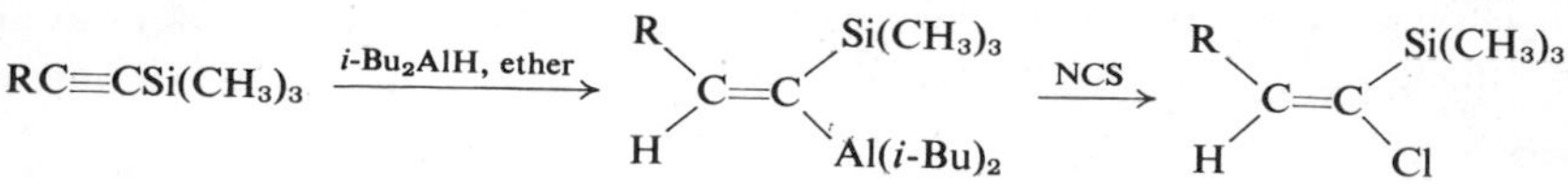

α-*Aryl-α′,α′-dichloro ketones.* A typical route to compounds of this type is formulated in equation (I). This method is not as successful for preparation of the corresponding dibromo ketones, since these compounds are readily converted to monobromo ketones on attempted purification by exchange with HBr present in the crude product.[3]

(I) $C_6H_5COCH_2CH_3 + C_6H_5NH_2$ $\xrightarrow[98\%]{TsOH, C_6H_6}$ $\underset{\overset{\|}{N}}{C_6H_5\overset{\displaystyle N^{C_6H_5}}{C}CH_2CH_3}$ $\xrightarrow[quant.]{2NCS, CCl_4, 20°}$

$\underset{\overset{\|}{N}}{C_6H_5\overset{\displaystyle N^{C_6H_5}}{C}C(Cl)_2CH_3}$ $\xrightarrow[81\% \text{ overall}]{H_3O^+}$ $C_6H_5\overset{O}{\overset{\|}{C}}C(Cl)_2CH_3$

Isatins. 3-Methylthiooxindoles (1) (5, 114–115) can be converted into isatins (3) by chlorination with NCS followed by hydrolysis of the relatively unstable chlorinated intermediates (2).[4]

1    $\xrightarrow{NCS}$ 2 $\xrightarrow[40-80\% \text{ overall}]{\substack{HgO, BF_3 \cdot (C_2H_5)_2O, \\ H_2O, THF}}$ 3

[1] B. M. Trost and Y. Tamaru, *Am. Soc.*, **99**, 3101 (1977).
[2] G. Zweifel and W. Lewis, *J. Org.*, **43**, 2739 (1978).
[3] N. De Kimpe, R. Verhé, L. De Buyck, and N. Schamp, *Syn. Comm.*, **8**, 75 (1978).
[4] P. G. Gassman, B. W. Cue, Jr., and T.-Y. Luh, *J. Org.*, **42**, 1344 (1977).

Chlorosulfonyl isocyanate (CSI), 1, 117–118; **2,** 70; **3,** 51–53; **4,** 90–94; **5,** 132–136; **6,** 122; **7,** 65–66.

Reaction with cyclic 1,3-dienes. The general reaction is illustrated for cyclohexadiene (1). The primary product, formed in ∼10 minutes, is the β-lactam (2) resulting from [1,2]cycloaddition. On standing it rearranges first to 4 and then to 5, the products of [1,4]cycloaddition through oxygen and through nitrogen, respectively.[1]

These reactions are useful as routes to aza- and oxabicyclic systems.

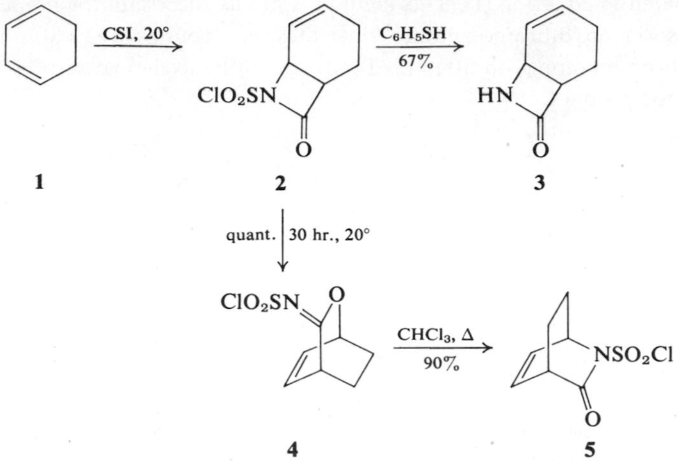

1   2   3

4   5

Reaction with indoles. The reagent reacts with indole or 2-substituted indoles (**1**) to form the derivatives **2**, which can be converted into indole-3-carboxamides (**3**) or into indole-3-carbonitriles (**4**). The compounds are versatile precursors to 3-substituted indoles.

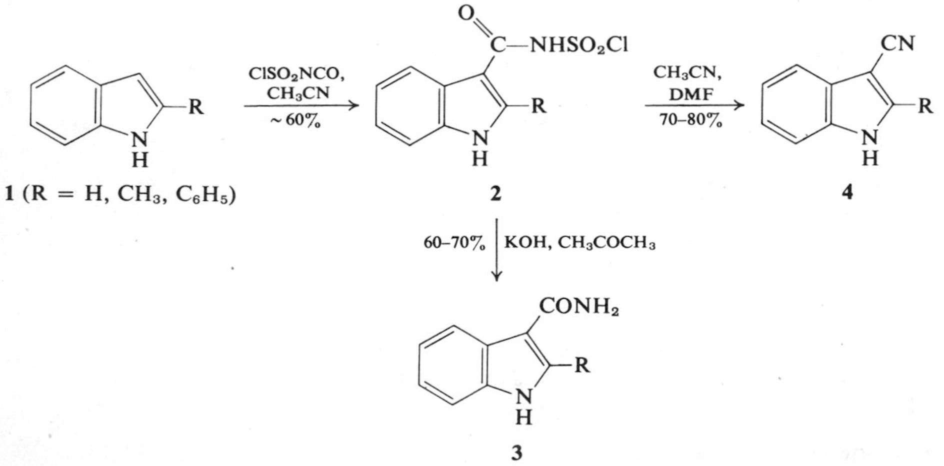

1 (R = H, CH₃, C₆H₅)   2   4

3

When the 3-position is substituted, the reagent displaces the H attached to nitrogen.[2]

[1] J. R. Malpass and N. J. Tweddle, *J.C.S. Perkin I*, 874 (1977).
[2] G. Mehta, D. J. Dhar, and S. C. Sur, *Synthesis*, 374 (1978).

Chlorotrimethylsilane, **1**, 1232; **2**, 435–438; **3**, 310–312; **4**, 537–539; **5**, 709–713; **6**, 626–628; **7**, 66–67.

Acyloin condensation (**2**, 435–436; **3**, 311–312; **4**, 537). An example of the preparation of cyclic acyloins with trapping of the intermediate enediol with $(CH_3)SiCl$ has been published in *Org. Syn.* (equation I).[1]

$$(I) \quad \begin{array}{c} CH_2COOC_2H_5 \\ | \\ CH_2COOC_2H_5 \end{array} \xrightarrow[78\%]{4Na,\ 4ClSi(CH_3)_3} \quad \begin{array}{c} OSi(CH_3)_3 \\ \\ OSi(CH_3)_3 \end{array} + 2C_2H_5OSi(CH_3)_3 + 4NaCl$$

$$71–86\% \Big| CH_3OH$$

$$\begin{array}{c} O \\ \\ OH \end{array} + 2CH_3OSi(CH_3)_3$$

The report contains a warning that explosions can occur unless the chlorotrimethylsilane is purified by distillation from calcium hydride under nitrogen. This procedure for preparation of succinoin is generally applicable to four to eight-membered acyloins and even to higher-membered rings if high dilution conditions are used.

β,γ-Unsaturated ketones.[2] Allylic alcohols can be converted into rearranged allylic ketones by a three-step process. The alcohol is first converted into an α-allyloxy ketone (**2**) by reaction with a diazoketone in the presence of BF_3 etherate or with 2-methoxyallyl bromide, $CH_2{=}C(OCH_3)CH_2Br$.[3] This product when treated with chlorotrimethylsilane and triethylamine in DMF undergoes Claisen rearrangement to an α-siloxy aldehyde, **a** → **3**. The final step is hydrolysis and oxidation to a ketone (**4**). Two examples are formulated in equations (I) and (II).

$$(I) \quad (CH_3)_2C{=}CHCH_2OH \xrightarrow[93\%]{\substack{C_6H_5COCHN_2, \\ BF_3\cdot(C_2H_5)_2O}} (CH_3)_2C{=}CHCH_2OCH_2\overset{O}{\overset{\|}{C}}C_6H_5 \xrightarrow[]{\substack{ClSi(CH_3)_3, \\ N(C_2H_5)_3}}$$

$$\mathbf{1} \qquad\qquad\qquad \mathbf{2}$$

$$\left[\begin{array}{c} OSi(CH_3)_3 \\ O \overset{}{\diagup} \diagdown C_6H_5 \\ \diagdown C(CH_3)_2 \end{array} \right] \xrightarrow[90\%]{} \begin{array}{c} H \\ O{=}\overset{}{\diagup} OSi(CH_3)_3 \\ H_2C \diagdown \overset{\diagup}{\diagdown} C_6H_5 \\ CH_3 \\ CH_3 \end{array} \xrightarrow[83\%]{\substack{CH_3OH,\ H_2O, \\ HIO_4}} CH_2{=}CH{-}\overset{CH_3}{\underset{CH_3}{\overset{|}{C}}}{-}COC_6H_5$$

$$\mathbf{a} \qquad\qquad \mathbf{3} \qquad\qquad\qquad \mathbf{4}$$

(II)

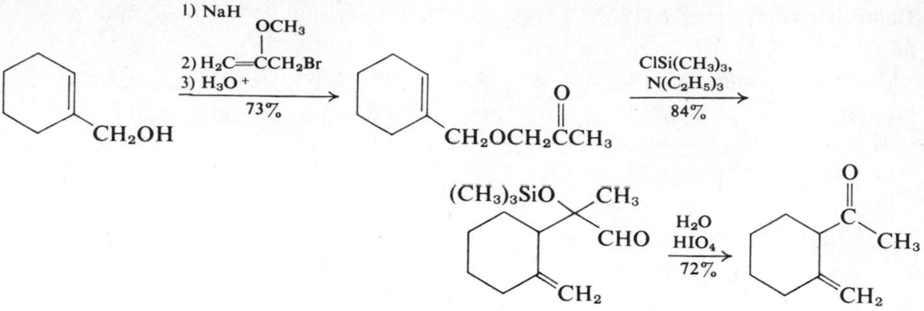

Addition to quinones.[4] In the presence of catalytic amounts of tetraethylammonium chloride or tetrafluoroborate, halotrimethylsilanes react with 1,4-benzoquinones (**1**) in CH_3CN to give, after hydrolysis, halohydroquinones (**2**) in yields of 75–95% (equation I). The intermediate adducts can be converted into disilyl ethers (**3**).

(I)

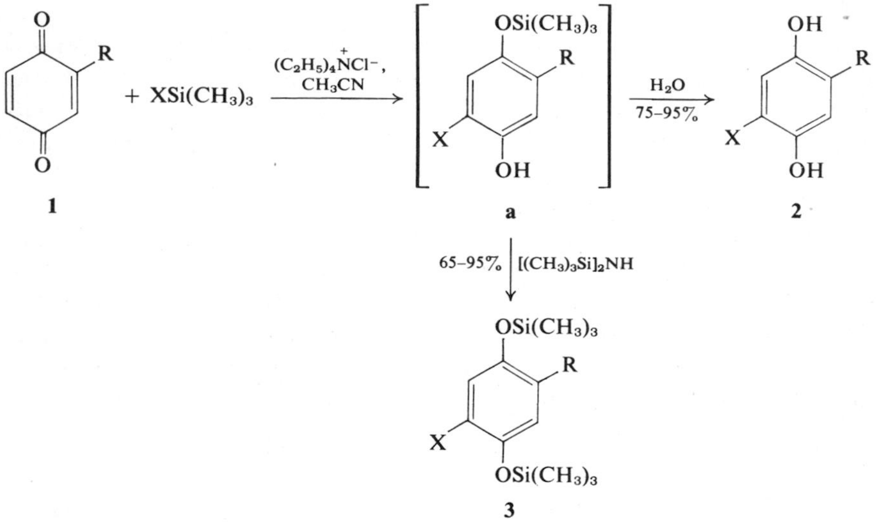

Dealkylation. Dialkyl phosphonates undergo dealkylation on treatment with chlorotrimethylsilane in the presence of sodium iodide in CH_3CN at 20°.[5]

(I) $$RP(O)(OR')_2 + 2(CH_3)_3SiCl + 2NaI \xrightarrow[\sim 100\%]{CH_3CN} RP(O)[OSi(CH_3)_3]_2 + 2R'I + 2NaCl$$

$$RP(O)[OSi(CH_3)_3]_2 \xrightarrow[80\text{–}85\%]{H_2O,\ 20^\circ} RP(O)(OH)_2$$

[1] J. J. Bloomfield and J. M. Nelke, *Org. Syn.*, **57**, 1 (1977).
[2] J. L. C. Kachinski and R. G. Salomon, *Tetrahedron Letters*, 3235 (1977).
[3] Prepared by reaction of 2-methoxypropene with NBS.
[4] L. L. Miller and R. F. Stewart, *J. Org.*, **43**, 3078 (1978).
[5] T. Morita, Y. Okamoto, and H. Sakurai, *Tetrahedron Letters*, 2523 (1978).

Chlorotris(triphenylphosphine)rhodium(I), **1**, 1252; **2**, 248–253; **3**, 325–329; **4**, 559–562; **5**, 736–740; **6**, 652–653; **7**, 68.

Aldehyde decarbonylation (**2**, 451; **3**, 327–329; **4**, 559). Suggs[1] has isolated the acylrhodium hydride **2** in 95% yield from the reaction of 8-quinolinecarboxy-aldehyde (**1**) with Wilkinson's catalyst in CH_2Cl_2. The hydride on refluxing in benzene is converted quantitatively into quinoline and $RhCl(CO)[P(C_6H_5)_3]_2$.

The tetrafluoroborate salt **3** can effect hydroacylation of 1-alkenes.[1]

Dimerization of α-hydroxy acetylenes. The dimerization of α-hydroxy acety-lenes with this catalyst to 1,4-disubstituted *trans*-vinylacetylenes was first descri-bed in 1968 (equation I).[2] The reaction has since been examined in detail.[3]

Triphenylphosphine is superior as the ligand to $(C_6H_5)_3N$, $(C_6H_5)_3As$, and $(C_6H_5)_3Sb$; substitution of halogen or methyl in the *ortho*- or *para*-positions of C_6H_5 is detrimental. Codimerizations were shown to be successful.

[1] J. W. Suggs, *Am. Soc.*, **100**, 640 (1978).
[2] H. Singer and G. Wilkinson, *J. Chem. Soc. A*, 849 (1968).
[3] H. J. Schmitt and H. Singer, *J. Organometal. Chem.*, **153**, 165 (1978).

Chromic acid–Silica gel, H_2CrO_4–SiO_2.

Oxidation. Chromic acid adsorbed on silica gel is useful for oxidation of primary and secondary alcohols to carbonyl compounds. Yields are generally 80–90%. In contrast, H_2CrO_4 adsorbed on alumina is ineffective.[1]

[1] E. Santaniello, F. Ponti, and A. Manzocchi, *Synthesis*, 534 (1978).

Chromic anhydride–3,5-Dimethylpyrazole complex, 5, 142.

Allylic oxidation. The oxidation of cholesteryl benzoate with this complex to the Δ^5-7-ketone is remarkably fast (complete in less than 30 minutes) and the yield of ketone is routinely 70–75%. The oxidation of Δ^6-cholestene-3β,5α-diol to the Δ^5-7-ketone is complete in less than 2 minutes and again clean. However, oxidation of Δ^6-cholestene-3β,5β-diol is much slower and leads to many by-products.[1] The paper suggests two possible mechanisms for these oxidations, both of which involve prior attack at the double bond.

McDonald and Suksamran[2] have reported use of the reagent (without details) for a benzylic oxidation.

[1] W. G. Salmond, M. A. Barta, and J. L. Havens, *J. Org.*, **43**, 2057 (1978).
[2] E. McDonald and A. Suksamran, *Tetrahedron Letters*, 4425 (1975).

Chromium carbonyl, 5, 142–143; **6,** 125–126; **7,** 71.

Ring expansion of arenes. French chemists have found this reaction is possible with the tricarbonylchromium complexes of arenes (**6,** 125). Thus treatment of **1** with the carbene C_6H_5CH: (**4,** 311) gives the tricarbonylchromium complex of

7-phenyl-1,3,5-cycloheptatriene (**2**) in 84% yield. When the arene is monosubstituted [C_2H_5, C_6H_5, OCH_3, $N(CH_3)_2$], 2- and 3-substituted derivatives of **2** are formed. Yields range from fair to good. Other halide carbene precursors can be used: $(C_6H_5)_2CHCl$ and $CH_2{=}CHCH_2Br$. Yields are rather low with the latter halide because of rearrangements encountered with vinylcarbenes. The *endo*-structure of the products has been confirmed by X-ray analysis.

[1] G. Simonneaux, G. Jaouen, R. Dabard, and M. Louër, *Nouv. J. Chem.*, **2**, 203 (1978).

Chromium(III) chloride–Lithium aluminum hydride. The reagent is prepared by reduction of 2 equiv. of anhydrous $CrCl_3$ with 1 equiv. of $LiAlH_4$ in THF at 0° and is presumably chromium(II) chloride. The advantage of this new chromium reagent is that it can be used in anhydrous solvents. The reagent is stable under an inert atmosphere in a refrigerator for more than a month.

Reduction of alkyl halides. The reduction of various halides with this new reagent has been studied. The allylic bromide **1** undergoes coupling mainly to the head-to-head dimer **2**.

$$(CH_3)_2C\!\!=\!\!CHCH_2Br \xrightarrow[70\%]{CrCl_2,\ THF} (CH_3)C\!\!=\!\!CHCH_2\text{-}CH_2CH\!\!=\!\!C(CH_3)_2$$

1   **2 (72%)**

$$+ (CH_3)_2C\!\!=\!\!CHCH_2\overset{\displaystyle CH_3}{\underset{\displaystyle CH_3}{C}}CH\!\!=\!\!CH_2 + CH_2\!\!=\!\!CH\overset{\displaystyle CH_3}{\underset{\displaystyle CH_3}{C}}\!\!-\!\!\overset{\displaystyle CH_3}{\underset{\displaystyle CH_3}{C}}CH\!\!=\!\!CH_2$$

3 (22%)   **4 (6%)**

Benzylic halides are also reduced to dimeric products: benzyl bromide → 1,2-diphenylethane (49% yield).[1]

The reduction of *gem*-dibromocyclopropanes is particularly interesting because it results in allenes as shown by the reduction of **5** and **7**.

5   **6**   **7**   **8**

Reduction of *dl*-1,2-dibromocyclododecane is nonstereospecific and gives both *cis*- and *trans*-cyclododecane (29 and 41%, respectively). α-Bromocyclododecanone is reduced to the ketone in 96% yield.[1]

Homoallylic alcohols. In the presence of this reagent allylic halides (or tosylates) add to aldehydes and ketones to form homoallylic alcohols. The more substituted γ-carbon of the allyl group becomes attached to the carbonyl carbon.[2]
Examples:

$$CH_3(CH_2)_5CHO + (CH_3)_2C\!\!=\!\!CHCH_2Br \xrightarrow[83\%]{} CH_3(CH_2)_5\overset{\displaystyle OH}{\underset{}{CH}}\!\!-\!\!\overset{\displaystyle CH_3}{\underset{\displaystyle CH_3}{C}}CH\!\!=\!\!CH_2$$

The reaction was used for a synthesis of yomogi alcohol (**1**).

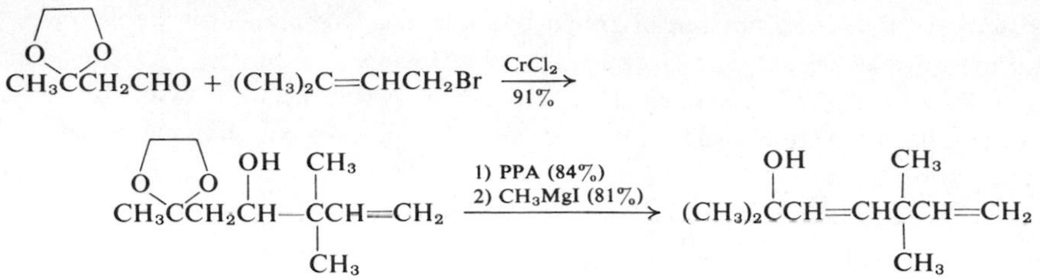

$$CH_3\overset{O\diagdown\diagup O}{C}CH_2CHO + (CH_3)_2C{=}CHCH_2Br \xrightarrow[91\%]{CrCl_2}$$

$$CH_3\overset{O\diagdown\diagup O}{C}CH_2\underset{}{CH}{-}\underset{CH_3}{\overset{CH_3}{C}}CH{=}CH_2 \xrightarrow[\text{2) } CH_3MgI\ (81\%)]{\text{1) PPA } (84\%)} (CH_3)_2\overset{OH}{C}CH{=}CH\underset{CH_3}{\overset{CH_3}{C}}CH{=}CH_2$$

1

The reaction is stereoselective; thus the reaction of benzaldehyde and crotyl bromide leads to the (SR, RS)-diastereomers (equation I).[3]

(I) 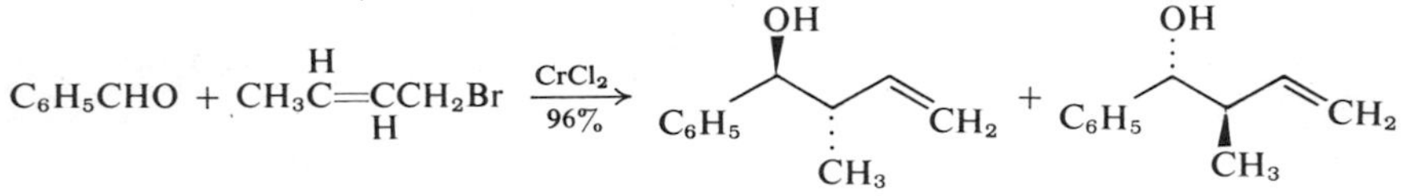 $C_6H_5CHO + CH_3\overset{H}{\underset{H}{C}}{=}CCH_2Br \xrightarrow[96\%]{CrCl_2}$

Another example:

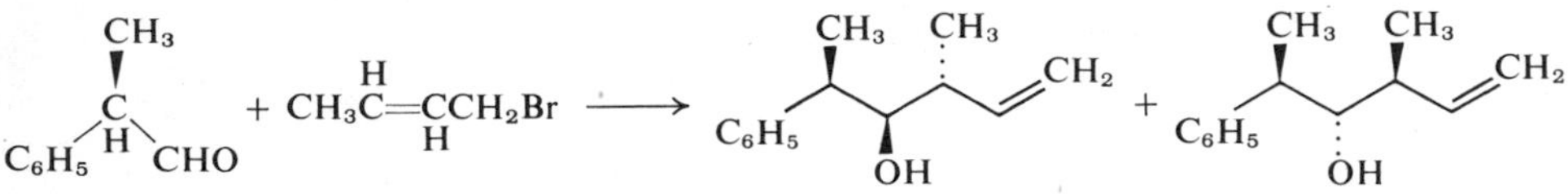

[1] Y. Okude, T. Hiyama, and H. Nozaki, *Tetrahedron Letters*, 3829 (1977).
[2] Y. Okude, S. Hirano, T. Hiyama, and H. Nozaki, *Am. Soc.*, **99**, 3179 (1977).
[3] C. T. Buse and C. H. Heathcock, *Tetrahedron Letters*, 1685 (1978).

Chromyl chloride, **1**, 151; **2**, 79; **3**, 62; **4**, 98–99; **5**, 144–145; **6**, 126–127.

Chemisorbed material. CrO_2Cl_2 absorbed on SiO_2–Al_2O_3 (Grace Davison No. 135) shows reduced activity and can be used for oxidation of alcohols to carbonyl compounds with yields comparable to those obtained with standard reagents but with the convenience shown by other chemisorbed reagents on an inert support. Esters, lactones, nitriles, and ethers are inert, but alkenes are oxidatively cleaved.[1]

Oxidation of alkenes. Oxidation of alkenes by CrO_2Cl_2 in CH_2Cl_2 at $-78°$ leads to epoxides and chlorohydrins as the primary products. The epoxides retain the geometry of the alkenes. The chlorohydrins are formed stereospecifically by *cis*-addition of OH and Cl to the double bond.[2] When the oxidation is carried out in the presence of acetyl chloride again at $-78°$ *vic*-chloro acetates are obtained in good yield. *cis*-Addition again predominates. In the case of unsymmetrical alkenes the chlorine is attached preferentially to the more substituted carbon atom.[3]

[1] J. San Filippo, Jr. and C. I. Chern, *J. Org.*, **42**, 2182 (1977).
[2] K. B. Sharpless, A. Y. Teranishi, and J. E. Bäckvall, *Am. Soc.*, **99**, 3120 (1977).
[3] J. E. Bäckvall, M. V. Young, and K. B. Sharpless, *Tetrahedron Letters*, 3523 (1977).

Cobalt(III) fluoride, CoF_3. Mol. wt. 115.93. Supplier: Alfa.

Oxidative cleavage of hydrazones and oximes. This reaction can be conducted with CoF_3 in $CHCl_3$ (reflux).[1]

(I)

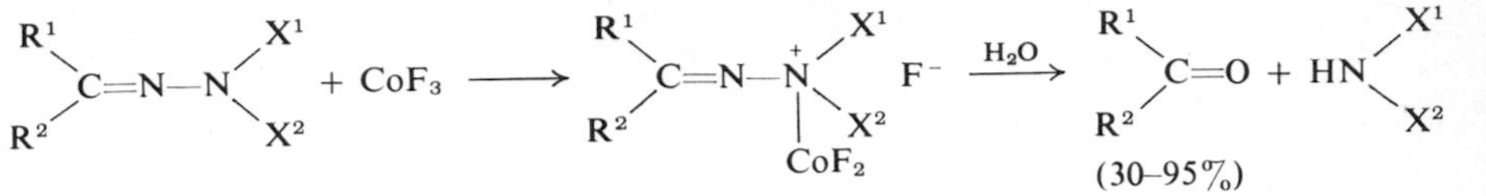

$$\underset{R^2}{\overset{R^1}{>}}C=N-N\underset{X^2}{\overset{X^1}{<}} + CoF_3 \longrightarrow \underset{R^2}{\overset{R^1}{>}}C=N-\overset{+}{N}\underset{\underset{CoF_2}{X^2}}{\overset{X^1}{<}} \quad F^- \xrightarrow{H_2O} \underset{R^2}{\overset{R^1}{>}}C=O + HN\underset{X^2}{\overset{X^1}{<}}$$

$$(30–95\%)$$

(II)

$$\underset{R^2}{\overset{R^1}{>}}C=N-OH \xrightarrow[2)\ H_2O]{1)\ CoF_3} \underset{R^2}{\overset{R^1}{>}}C=O$$

$$(25–65\%)$$

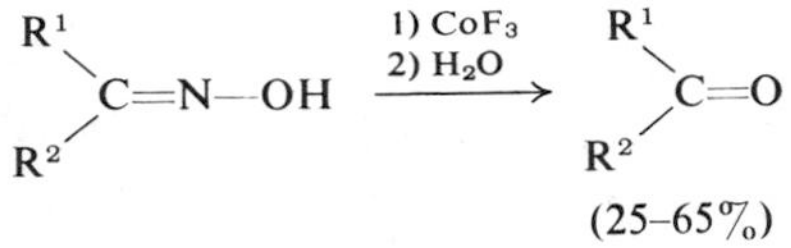

[1] G. A. Olah, J. Welch, and M. Henninger, *Synthesis*, 308 (1977).

Copper, 1, 157–158; **2**, 82–84; **3**, 63–65; **4**, 102–103; **5**, 146–148; **7**, 73–74.

Ullmann reaction. Brown and Robin[1] have found that two different aryl halides hindered by bulky substituents in the *ortho*-positions can be coupled by a modified Ullmann reaction. A large excess of copper powder (not activated) is necessary and the temperature should be lower than that which leads to self-coupling of the less reactive aryl halide. At optimal temperatures, usually 190–240°, the reaction can be completed in 20–30 minutes. In favorable cases yields of biphenyls as high as 75% can be obtained.

Example:

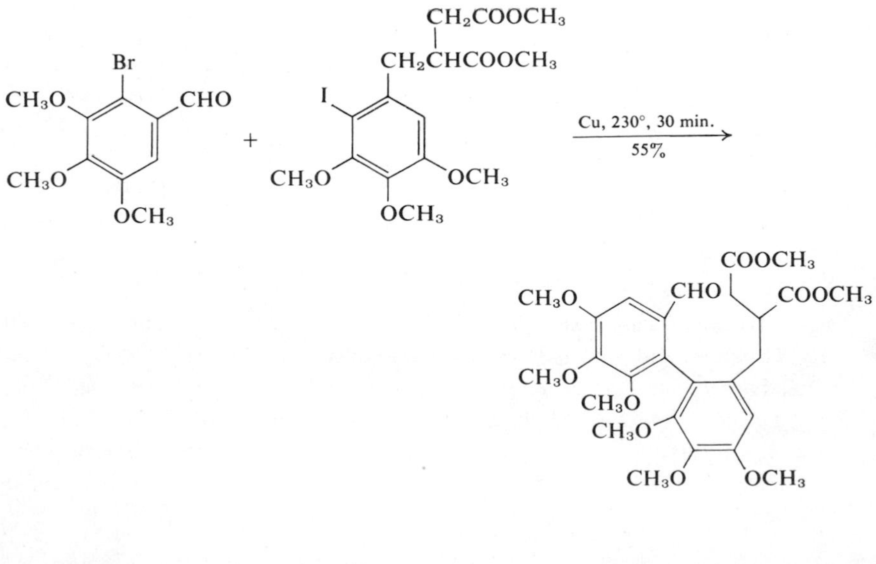

This version of the Ullmann reaction was used for synthesis of bisbenzocyclo-octadiene lignans related to steganacin. Thus picrostegane (**5**) was synthesized as shown in equation(I).

(I)

[1] E. Brown and J.-P. Robin, *Tetrahedron Letters*, 2015 (1977); *ibid.*, 3613 (1978).

Copper(II)–amine complexes. A typical complex is prepared from $Cu(NO_3)_2 \cdot 3H_2O$ (1 equiv.) and α-phenylethylamine (3 equiv.) in CH_3OH.

Oxidative phenolic coupling. Copper complexes have been used under aerobic conditions to oxidize phenols to quinones and to products of cleavage. Feringa and Wynberg[1] have now found that Cu(II)–amine complexes under anaerobic conditions can effect oxidative coupling of phenols. Some typical examples are formulated. The last example, coupling to dehydrogriseofulvin, is particularly interesting in that such a reaction may be involved in the biosynthesis of griseofulvins.

Examples:

[1] B. Feringa and H. Wynberg, *Tetrahedron Letters*, 4447 (1977).

Copper(II) acetate–Morpholine.

ortho-Hydroxylation of phenols. Some years ago Brockmann and Havinga[1] found that phenols undergo *ortho*-hydroxylation by oxygen in the presence of the copper(II) acetate complex with morpholine (soluble in ethanol). This reaction was used for preparation of *o*-quinones. For example, β-naphthol was converted under these conditions to 4-morpholino-1,2-naphthoquinone in 84% yield.

This reaction has been extended to the preparation of 5,6- and 7,8-dioxoisoquinolines[2] and is the key step in the synthesis[3] of the unusual ring system of the antibiotic mimosamycin (**6**), isolated from cultures of a strain of *S. lavendulae*.[4]

[1] W. Brockmann and E. Havinga, *Rec. trav.*, **74**, 937 (1955).
[2] Yu. S. Tsizin, *Khim. Geterotsikl. Soedin.* (*USSR*), 1253 (1974) [*C.A.*, **83**, 79198r (1975)].
[3] H. Fukumi, H. Kurihara, H. Hata, C. Tamura, H. Mishima, A. Kubo, and T. Arai, *Tetrahedron Letters*, 3825 (1977); H. Fukumi, H. Kurihara, and H. Mishima, *Chem. Pharm. Bull. Japan*, **26**, 2175 (1978).
[4] T. Arai, K. Gazawa, Y. Mikami, A. Kubo, and K. Takahashi, *J. Antibiotics* (*Tokyo*), **29**, 398 (1976).

Copper(I) bromide, 1, 165–166; **2**, 90–91; **3**, 67; **4**, 108; **5**, 163–164; **6**, 143–144; **7**, 79–80.

1,6-Addition of benzylmagnesium halides to a dienone.[1] *p*-Methoxybenzyl-magnesium bromide reacts with the dienone **1** to give the three possible products of addition. In the presence of CuBr only the 1,6-adduct **2** is formed. The product is cyclized by hot phosphoric acid to 5α- and 5β-2-oxopodocarpa-8,11,13-triene (**3**) in approximately equal amounts.

[1] B. R. Davis and S. J. Johnson, *J.C.S. Chem. Comm.*, **614** (1978).

Copper(I) bromide–Dimethyl sulfide, 6, 225.

Vinylcopper complexes. Use of this highly pure form of CuBr results in improved yields of vinylcopper complexes (**1**) from 1-alkynes (equation I). The reagents **1** were used in a new synthesis of functionalized trisubstituted alkenes (**2**) by conjugate addition to α,β-unsaturated carbonyl compounds (equation II). The probable stereochemistry of **2** is shown. The reactions are conducted in a one-flask procedure.[1]

(I) $R^1MgBr + CuBr \cdot S(CH_3)_2 \xrightarrow{\text{Ether, } -45°} R^1Cu[(CH_3)_2S]MgBr_2 \xrightarrow[\quad -25° \quad]{R^2C \equiv CH,}$

$(R^1 = C_2H_5, C_6H_{13}\text{-}n)$

$$\begin{array}{c} R^1 \\ \diagdown \\ C = CHCu[(CH_3)_2S] \cdot MgBr_2 \\ \diagup \\ R^2 \end{array}$$

1

(II) $1 + R^3CH = CHCR^4 \quad (C=O) \longrightarrow \xrightarrow[10-70\%]{H_2O} $

$$\begin{array}{c} R^1 \quad R^3 \quad O \\ \diagup\diagdown \quad \diagup \quad \| \\ R^2 \qquad\qquad\qquad R^4 \end{array}$$

2

Methylcopper (**4**, 334–335; **5**, 148; **6**, 377). Methylcopper itself is a very insoluble, polymeric solid and is less reactive than other alkylcopper compounds. However, Helquist and co-workers[2] have found that under conditions described above a fairly reactive complex (**1**) of methylcopper is formed that reacts, albeit slowly, with aliphatic 1-alkynes. These latter complexes such as **2**, react with

$$CH_3MgBr + (CH_3)_2S \cdot CuBr \xrightarrow[\quad -45° \quad]{\substack{\text{ether,} \\ (CH_3)_2S,}} CH_3Cu(CH_3)_2S \cdot MgBr_2$$

1

various electrophiles to give trisubstituted alkenes. The yields are improved by HMPT; 1,3-dienes are not formed if the temperature is low.

[1] A. Marfat, P. R. McGuirk, R. Kramer, and P. Helquist, *Am. Soc.*, **99**, 253 (1977).
[2] A. Marfat, P. R. McGuirk, and P. Helquist, *Tetrahedron Letters*, 1363 (1978).

1

$$n\text{-}C_8H_{13}C{\equiv}CH \downarrow$$

$$n\text{-}C_6H_{13}\overset{CH_3}{\underset{H}{\diagdown}}C{=}C\overset{O}{\underset{}{\diagdown}}CH_3 \xleftarrow[65\%]{CH_3COCl,\ HMPT} n\text{-}C_6H_{13}\overset{CH_3}{\underset{}{\diagup}}C{=}C\overset{CuS(CH_3)_2\cdot MgBr}{\underset{H}{\diagdown}} \xrightarrow[81\%]{BrCH_2CH=CH_2\ HMPT}$$

2

$$n\text{-}C_6H_{13}\overset{CH_3}{\underset{H}{\diagdown}}C{=}C{-}CH_2CH{=}CH_2$$

Copper(I) chloride, 1, 166–169; 2, 91–92; 3, 67–69; 4, 109–110; 5, 164–165; 6, 145–146; 7, 80–81.

Catalyst for oxygenation. The oxidative cleavage of *o*-phenylenediamine (**5**, 165) catalyzed by CuCl has been published.[1] Use of various oxidizing reagents in stoichiometric amounts (*e.g.* nickel peroxide, lead tetraacetate, and silver oxide) results in yields of less than 50%.

French chemists[2] have used the system CuCl–amine–O_2 for oxidation of alcohols to aldehydes or ketones. 1,10-Phenanthroline is more efficient than pyridine in this reaction.

Examples:

$$C_6H_5CH_2OH \xrightarrow[86\%]{O_2,\ CuCl,\ amine} C_6H_5CHO$$

$$C_6H_5CH{=}CHCH_2OH \xrightarrow[83\%]{} C_6H_5CH{=}CHCHO$$

$$C_6H_5CHOHCH_3 \xrightarrow[93\%]{} C_6H_5COCH_3$$

$$C_6H_5CH_2NH_2 \xrightarrow[90\%]{} C_6H_5C{\equiv}N$$

Decomposition of α-diazo ketones (**2**, 82–84; **3**, 63–65; **4**, 103; **5**, 41–42, 161, 164–165; **6**, 141). Copper-catalyzed decomposition of α-diazo ketones of type **1** results in formation of spiro[4.5]decanes (**2**) in high yield.[3]

1 (R = H, CH₃)   **a**   **2**

---

[1] J. Tsuji and H. Takayanagi, *Org. Syn.*, **57**, 33 (1977).
[2] C. Jallabert and H. Riviere, *Tetrahedron Letters*, 1215 (1977).
[3] C. Iwata, M. Yamada, Y. Shinoo, K. Kobayashi, and H. Okada, *J.C.S. Chem. Comm.*, 888 (1977).

Copper(I) chloride–Tetra-*n*-butylammonium chloride, $(C_4H_9)_4N^+CuCl_2^-$, soluble in acetone.

Rearrangement of a propargyl chloride. The rearrangement of the propargyl chloride **1** to the allene **2** catalyzed by CuCl has been conducted with this solubilized form.[1]

1 (α_D − 9.8°)   **2 (α_D − 94°)**

[1] O. I. Muscio, Jr., Y. M. Jun, and J. B. Philip, Jr., *Tetrahedron Letters*, 2379 (1978).

Copper(II) chloride, 1, 163; **2,** 84–85; **3,** 66; **4,** 105–107; **5,** 158–160; **6,** 139–141; **7,** 79.

Oxyselenation of alkenes. Use of 1 equiv. of cupric chloride or bromide as well as nickel(II) chloride or bromide allows the oxyselenation of alkenes by aryl or alkyl selenocyanates and an alcohol (or phenol). In the case of styrene, selenium becomes attached only to the terminal carbon atom.[1]

Examples:

$$n\text{-}C_4H_9CH{=}CH_2 + C_6H_5SeCN \xrightarrow[95\%]{CuCl_2,\ CH_3OH}$$

$$n\text{-}C_4H_9CH{-}CH_2 \qquad + \qquad n\text{-}C_4H_9CH{-}CH_2$$
$$\underset{CH_3O \quad SeC_6H_5}{|\qquad\quad|} \quad 82{:}18 \quad \underset{C_6H_5Se \quad OCH_3}{|\qquad\quad|}$$

[1] A. Toshimitsu, S. Uemura, and M. Okano, *J.C.S. Chem. Comm.*, 166 (1977).

Copper(II) chloride–Oxygen.

Chlorination of a flavanone. Naringenin (**1**) is selectively chlorinated at the 6- and 8-positions by cupric chloride (2 equiv.) in ethanol (or methanol) when

oxygen is bubbled through the solution. No reaction occurs in C_6H_6, ether, DMSO, or DMF.[1] Chlorination of flavanones has previously involved use of $CuCl_2$–LiCl in CH_3CN at $82°$[2] or of $ZnCl_2$–HCl.[3]

[1] Y. Takizawa and T. Mitsuhashi, *J. Heterocyclic Chem.*, **15**, 701 (1978).
[2] S. Uemura, H. Okazaki, A. Onoe, and M. Okano, *J.C.S. Perkin I*, 676 (1977).
[3] Y. Omote, Y. Takizawa, and N. Sugiyama, *J. Heterocyclic Chem.*, **8**, 319 (1971).

Copper hydride ate complexes. Semmelhack[1] has reviewed a number of these complexes, prepared usually from CuBr or CuI and various complex metal hydrides. They function as hydride transfer reagents, but show differences in behavior. One early complex of this type is hydrido(tri-*n*-butylphosphine)-copper(I)–HCu(PBu$_3$), prepared by Whitesides (**3**, 154). The structure or mode of reaction of the other related complexes is not clear.

The paper reports preparation of two new copper hydride complexes. The "lithium complex" (**1**) is prepared from CuBr and 2 equiv. of lithium trimethoxy-aluminum hydride (*cf.* **5**, 168); the "sodium complex" (**2**) is prepared from CuBr and 1 equiv. of sodium bis(2-methoxyethoxy)aluminum hydride. Both complexes are useful for selective reduction of the double bond of conjugated enones; **1** is more efficient for reduction of cyclohexenones and **2** is more efficient for reduction of acyclic enones. Aldehydes, ketones, and halides are also reduced; nitrile and ester units are inert. The effective stoichiometry of both reagents is consistent with the structures $LiCuH_2$ and $NaCuH_2$, but complex **1** is clearly different from a reagent assigned the structure $LiCuH_2$ by Ashby *et al.*[2]

[1] M. F. Semmelhack, R. D. Stauffer, and A. Yamashita, *J. Org.*, **42**, 3180 (1977).
[2] E. C. Ashby, T. F. Korenowski, and R. D. Schwartz, *J.C.S. Chem. Comm.*, 157 (1974).

Copper(I) iodide, 1, 169; **2**, 92; **3**, 69–71; **5**, 167–168; **6**, 147; **7**, 81–83.

Copper enolates of esters.[1] The lithium enolate of ethyl acetate (**1**, prepared with LDA in THF at $-78°$) reacts with propargyl bromide to give **2** as the only identified product (equation I). If the lithium enolate is first treated with copper(I)

$$\text{(I)} \quad HC\equiv CCH_2Br + CH_2\!\!=\!\!\overset{\overset{\displaystyle OLi}{|}}{C}OC_2H_5 \xrightarrow[12\%]{} HC\equiv CCH_2CH_2COOC_2H_5$$

$$\phantom{\text{(I)} \quad HC\equiv CCH_2Br + }\mathbf{1} \mathbf{2}$$

$$\text{(II)} \quad HC\equiv CCH_2Br + CH_2\!\!=\!\!\overset{\overset{\displaystyle OCu}{|}}{C}OC_2H_5 \xrightarrow[45\%]{} H_2C\!\!=\!\!C\!\!=\!\!CHCH_2COOC_2H_5$$

$$\phantom{\text{(II)} \quad HC\equiv CCH_2Br + }\mathbf{3} \mathbf{4}$$

iodide at $-78°$ for 30 minutes, a reactive species (**3**, "copper enolate") is formed that reacts with propargyl bromide to form the β-allenic ester **4**. The reaction can be used as a general route to 3,4-dienoic acids (equation III).

$$\text{(III)} \quad CH_3(CH_2)_4\overset{\overset{\displaystyle OSO_2CH_3}{|}}{C}HC\equiv CH + 3 \xrightarrow[76\%]{} CH_3(CH_2)_4CH\!\!=\!\!C\!\!=\!\!CHCH_2COOC_2H_5$$

$$\phantom{\text{(III)} \quad CH_3(CH_2)_4}\mathbf{5} \mathbf{6}$$

In the presence of sodium ethoxide, these dienoic acids can be isomerized in high yield to conjugated 2,4-dienoic acids. Surprisingly **6** is isomerized to a mixture of the (2E, 4Z)- and the (2E, 4E)-isomers in a 2:1 ratio. This mixture can be equilibrated mainly to the more stable (2E, 4E)-isomer (86%) by treatment with thiophenol in the presence of AIBN (**6**, 585–586). The ethyl (2E, 4Z)-decadienoate obtained by rearrangement of **6** is a component of the odoriferous principle of Bartlett pears.

Alkylation of allylic Grignard reagents. Alkylation of allylic Grignard reagents in the presence of CuI results in almost exclusive substitution at the α-position. The reaction is very slow in the absence of CuI.[2]

Examples:

$$CH_2\!\!=\!\!CHCH_2MgBr + CH_3(CH_2)_7I \xrightarrow[80\%]{\text{CuI,}\atop \text{ether, 0°}} CH_2\!\!=\!\!CH(CH_2)_8CH_3$$

$$(CH_3)_2C\!\!=\!\!CHCH_2MgBr + CH_3(CH_2)_6I \xrightarrow[80-90\%]{} (CH_3)_2C\!\!=\!\!CH(CH_2)_7CH_3$$

The last example is a new synthesis of geraniol. Note that alkyl tosylates react with allylic Grignard reagents, without CuI, to give mainly products of γ-substitution:

$$CH_3CH{=}CHCH_2MgBr \rightleftharpoons \underset{\underset{MgBr}{|}}{CH_3CHCH{=}CH_2} \xrightarrow[85\%]{n\text{-}C_7H_{15}OTs} \underset{n\text{-}C_7H_{15}}{\overset{CH_3}{\diagdown}}CHCH{=}CH_2$$

Alkylation of nitroarenes. Nitroarenes are alkylated by Grignard reagents with concomitant reduction of the nitro group in the presence of CuI in catalytic amounts.[3]

Examples:

[1] R. A. Amos and J. A. Katzenellenbogen, *J. Org.*, **43**, 555 (1978).
[2] F. Derguini-Boumechal, R. Lorne, and G. Linstrumelle, *Tetrahedron Letters*, 1181 (1977).
[3] G. Bartoli, R. Leardini, M. Lelli, and G. Rosini, *J.C.S. Perkin I*, 884 (1977); G. Bartoli, A. Medici, G. Rosini, and D. Tavernari, *Synthesis*, 436 (1978).

Copper(I) iodide–Triethyl phosphite complex, $CuI \cdot (C_2H_5O)_3P$. Mol. wt. 356.61, m.p. 110–111°. The complex is obtained in 88.6% yield by addition of CuI to a solution of triethyl phosphite in benzene. The mixture is stirred and heated until the CuI dissolves. The complex is obtained upon evaporation of the solvent under reduced pressure.[1]

Regiospecific Ullmann reaction.[2] Directed coupling of *o*-haloarylimines can be accomplished by conversion of one of the imines into a preformed copper reagent with this complex of CuI. An example is shown in equation (I).

This modified Ullmann reaction was used to synthesize the antileukemic agent ($\pm$)-steganacin (**1**), as outlined in equation (II).[3]

[1] Y. Nishizawa, *Bull. Chem. Soc. Japan*, **34**, 1170 (1961).
[2] F. E. Ziegler, K. W. Fowler, and P. Sanfer, *Am. Soc.*, **98**, 8282 (1976).
[3] F. E. Ziegler, K. W. Fowler, and N. D. Sinha, *Tetrahedron Letters*, 2767 (1978).

Copper(I) isopropenylacetylide, $CuC{\equiv}C{-}C\overset{\displaystyle CH_2}{\underset{\displaystyle CH_3}{\diagdown}}$ (**1**). Mol. wt. 128.63, m.p. 120–200° (slow decomp.). The reagent is prepared from isopropenylacetylene by treatment with copper sulfate and hydroxylamine hydrochloride in aqueous ethanolic ammonium hydroxide.

2-Isopropenylbenzofuranes.[1,2] Several natural benzofuranes substituted with an isopropenyl group have been synthesized by coupling of *o*-halophenols with this copper acetylide.

Examples:

Furocoumarins. The angular furocoumarin oroselone (**3**) has been obtained by reaction of the iodophenol **2** with **1** in pyridine. The linear isomer can be obtained by the same method.[3]

[1] F. G. Schreiber and R. Stevenson, *J.C.S. Perkin I*, 90 (1977).
[2] Idem, *Org. Prep. Proc. Int.*, **10**, 137 (1978).
[3] Idem, *J. Chem. Res.* (*M*), 1201 (1978).

Copper(I) oxide–Copper(II) nitrate, Cu_2O–$Cu(NO_3)_2$.

Phenols. Cohen and co-workers[1] have described complete details for conversion of diazonium salts to phenols by a redox procedure first published in 1967.[2] The classical method involves thermal decomposition in a strongly acidic medium and involves a cationic intermediate. In the newer method the diazonium salt is decomposed by copper(I) oxide in an aqueous solution containing excess copper(II) nitrate; a radical intermediate is involved. Yields by this method are at least equal to those obtained in the thermal reaction, and often higher.

[1] T. Cohen, A. G. Dietz, Jr., and J. R. Miser, *J. Org.*, **42**, 2053 (1977).
[2] A. H. Lewin and T. Cohen, *ibid.*, **32**, 3844 (1967).

Copper(II) sulfate, 1, 164–165; **2**, 89; **5**, 162–163; **6**, 141–142.

Acetonides (**2**, 89). Acetonides of *vic*-diols are generally prepared by condensation with acetone in the presence of anhydrous $CuSO_4$, which serves as a dehydrating reagent. Epoxides are also converted into acetonides under the same conditions. In this transformation the salt is also functioning as a Lewis acid for conversion of the epoxide into a *vic*-diol.

Copper(II) sulfate can also be used to convert carbonyl compounds into 1,3-dioxolane and 1,3-dioxane derivatives. Thus the reaction of benzaldehyde with ethylene glycol containing $CuSO_4$ gives the dioxolane derivative in 45% yield.[1]

Hydrolysis of α-dialkylamino nitriles. The hydrolysis of α-amino nitriles to carbonyl compounds has traditionally been conducted with either acids or strong bases, either of which can lead to undesired side reactions. Büchi *et al.*[2] find that cupric sulfate in aqueous CH_3OH greatly accelerates the hydrolysis and leads to higher yields. In the case of highly sensitive aldehydic products, K_2HPO_4 is added to control the pH to 5.0–5.5.

Examples:

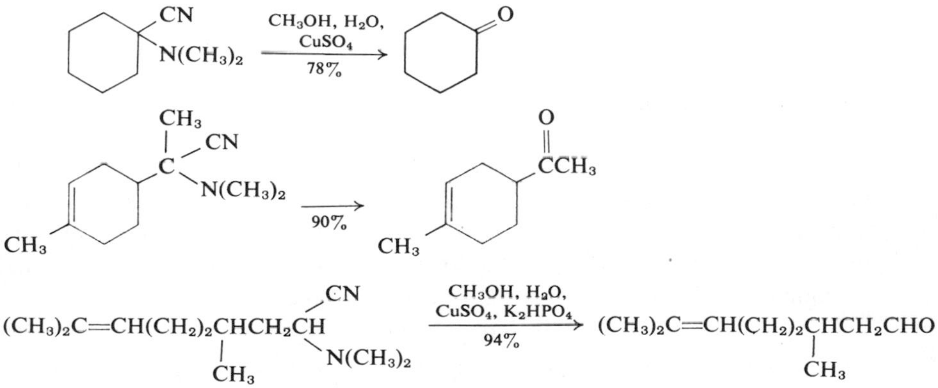

[1] R. P. Hanzlik and M. Leinwetter, *J. Org.*, **43**, 438 (1978).
[2] G. Büchi, P. H. Liang, and H. Wüest, *Tetrahedron Letters*, 2763 (1978).

Copper(I) trifluoromethanesulfonate [Copper(I) triflate], 5, 151–152; **6**, 130–133; **7**, 75–76.

Cohen *et al.*[1] have described a modified preparation that does not require special apparatus and hence is adaptable to preparation of various amounts of reagent. The yield for the preparation on a 45-g. scale is 60–75%. The report includes the preparation of (Z)-1-phenylthio-2-methoxy-1,3-butadiene (**2**) from 4,4-bis(phenylthio)-3-methoxy-1-butene (**1**) with elimination of thiophenol (**6**, 130–131). Diisopropylamine and a radical inhibitor are added to prevent polymerization of the diene **2**.

Fragmentation of five-and six-membered rings. Semmelhack and Tomesch[2] have reported a mild and efficient Grob-type fragmentation.[3] The reaction of 1,2-epoxycyclohexane (**1**) with the anion of bis(phenylthio)methane (**6**, 267) gives

the alcohol **2**. Treatment of **2** with $KOC(CH_3)_3$ or NaH results in recovery of **1** and products of decomposition. Addition of various metal salts does not improve matters, but use of *n*-butyllithium (to form the alkoxide) and this copper salt results in fragmentation to the aldehyde **3** in 92% yield. This product has both a free and a latent aldehyde group available for further elaboration. The fragmentation of the five-membered analog of **2** also proceeds satisfactorily under the same conditions (93% yield). For high yields 4 equiv. of the copper salt are necessary; indeed no fragmentation is observed when only 1 equiv. is used.

[1] T. Cohen, R. J. Ruffner, D. W. Shull, E. R. Fogel, and J. R. Falck, *Org. Syn.*, submitted (1977).
[2] M. F. Semmelhack and J. C. Tomesch, *J. Org.*, **42**, 2657 (1977).
[3] C. A. Grob and P. W. Schiess, *Angew. Chem., Int. Ed.*, **6**, 1 (1967).

Copper(II) trifluoromethanesulfonate, $Cu(OTf)_2$, **5**, 152; **7**, 76.

1,4-Diketones.[1] Lithium enolates of ketones are coupled to 1,4-diketones in the presence of copper(II) triflate (1 equiv.). Unlike the related oxidative coupling promoted by $CuCl_2$ (**6**, 139),[2] the present reaction is not limited to enolates of methyl ketones. Isobutyronitrile is used as cosolvent.

Example:

$$\underset{OLi}{\overset{|}{2C_6H_5COCH_2CH_3 \rightarrow 2C_6H_5C}=CHCH_3} \xrightarrow{2Cu(OTf)_2}$$

$$\underset{(80\%)}{C_6H_5\overset{O}{\overset{\|}{C}}-\overset{CH_3}{\overset{|}{CH}}-\overset{CH_3}{\overset{|}{CH}}-\overset{O}{\overset{\|}{C}}C_6H_5 + 2CuOTf + 2LiOTf}$$

Silyl enol ethers can also be coupled to 1,4-diketones by treatment with Cu-$(OTf)_2$ and Cu_2O in isobutyronitrile. Other copper salts are ineffective; the reaction proceeds in only low yields in solvents such as DMF or HMPT.
Example:

$$\underset{C_6H_5\overset{|}{C}=CH_2}{\overset{OSi(CH_3)_3}{\overset{|}{}}} \xrightarrow[55\%]{\overset{Cu(OTf)_2,}{Cu_2O}} C_6H_5\overset{O}{\overset{\|}{C}}CH_2CH_2\overset{O}{\overset{\|}{C}}C_6H_5$$

This oxidative dimerization of ketone enolates has been extended to an intramolecular oxidative coupling of dilithium enolates of 2,4-pentanediones to give 1,3-cyclopentanediones (equation I). Although yields are only moderate, the reaction is a useful route to compounds of this type and of spiro[4.n]-ring systems.[3]

(I)

$$\underset{\overset{|}{R^2}}{CH_3\overset{O}{\overset{\|}{C}}-\overset{R^1}{\overset{|}{C}}-\overset{O}{\overset{\|}{C}}CH_3} \xrightarrow[20-45\%]{\overset{1)\,2LDA,\,THF,\,-78°}{2)\,Cu(OTf)_2,\,i\text{-PrCN}}}$$

[1] Y. Kobayashi, T. Taguchi, and E. Tokuno, *Tetrahedron Letters*, 3741 (1977).
[2] Y. Ito, T. Konoike, T. Harada, and T. Saegusa, *Am. Soc.*, **99**, 1487 (1977).
[3] Y. Kobayashi, T. Taguchi, and T. Morikawa, *Tetrahedron Letters*, 3555 (1978).

Crotonic anhydride, $(CH_3CH=CHCO)_2O$. Mol. wt. 154.16, b.p. 246–248. This reagent is prepared by treatment of crotonic acid with DCC in THF.

Protection of hydroxyl groups as crotonate esters. Crotonate (and related) esters, prepared with the corresponding anhydrides, are deacylated by hydrazine hydrate in CH_3OH–Py at 20° in 30–45 minutes (equation I).[1] 4-Methoxycro-

(I)

tonyl esters ($CH_3OCH_2CH{=}CHCOR$) are cleaved particularly readily, but 4-methoxycrotonic acid is not readily accessible.

[1] R. Arentzen and C. B. Reese, *J.C.S. Chem. Comm.*, 270 (1977).

Crown ethers, 4, 142–145; **5,** 152–155; **6,** 133–137; **7,** 76–79.

Synthesis. A new general synthesis of crown ethers involves treatment of polyethylene glycols with sulfonyl chlorides in the presence of NaOH or KOH, which serves as the template. For example, 15-crown-5 was obtained in 50% isolated yield by this method (equation I). 18-Crown-6 is prepared in this way in 75% yield. Substituted crown ethers have also been prepared by this method.[1]

(I)   HO — O — O — O — O — OH $\xrightarrow{\text{TsCl, NaOH}}$ Dioxane

$$\left[HO\left(\diagup O \diagdown \right)_4 OSO_2C_6H_4CH_3 \right] \xrightarrow{50\%}$$

Decarboxylation. Addition of 18-crown-6 increases the rate of decarboxylation of alkali metal salts of carboxylic acids by 13- to 100-fold. Suitable solvents are THF, CH_3CN, Py, and DMSO. The rate is also correlated with the stability of the intermediate carbanion.[2]

Esters bearing an α-activating group undergo decarboalkoxylation in a one-pot reaction by treatment with KOH in benzene–ethanol and an equimolar amount of 18-crown-6. The reaction actually involves two steps (equation I). It can be carried out in ethanol–benzene, first at room temperature and then at reflux temperature. This method is useful in malonic ester syntheses. The crown ether can be recovered.[3]

$$(I)\quad R^1{-}\underset{R^2}{\overset{Y}{\underset{|}{\overset{|}{C}}}}COOC_2H_5 \xrightarrow[\text{crown ether}]{\text{KOH,}} R^1{-}\underset{R^2}{\overset{Y}{\underset{|}{\overset{|}{C}}}}COOK \xrightarrow[45{-}90\%]{80°} R^1{-}\underset{R^2}{\overset{Y}{\underset{|}{\overset{|}{C}}}}H$$

$(Y = COOH, COR^3, CN)$

Activation of potassium t-butoxide. 18-Crown-6 enhances the nucleophilicity and, to a lesser extent, the basicity of potassium *t*-butoxide in THF. For example in the presence of the crown ether, benzyl *t*-butyl ether can be prepared in high yield by the reaction shown in equation (I). This reaction is unsuccessful in DMSO, in which the basicity of potassium *t*-butoxide is enhanced.

(I) $KOC(CH_3)_3 + C_6H_5CH_2Cl \xrightarrow[\text{THF}]{\text{Crown ether,}} C_6H_5CH_2OC(CH_3)_3 + C_6H_5CH=CHC_6H_5$

(83%)   (<6%)

Two other useful reactions of potassium t-butoxide activated by 18-crown-6 are shown in equations (II) and (III).[4]

(II)

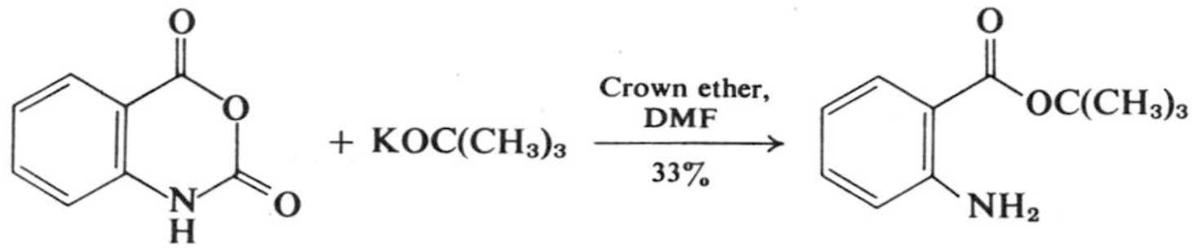

$+ KOC(CH_3)_3 \xrightarrow[\text{33%}]{\text{Crown ether,}}$

(III)

$\xrightarrow{\text{O}_2, KOC(CH_3)_3,\ \text{THF, crown ether}}$

$\xrightarrow{\text{100%}}$

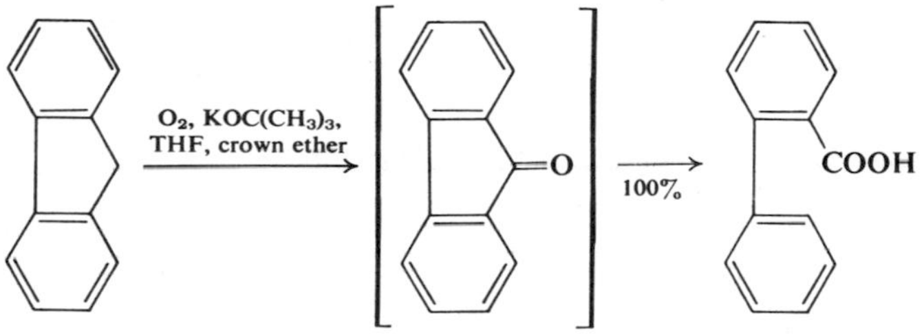

Metalation of arylmethanes.[5] Potassium hydride metalates compounds of medium acidity, but by itself does not metalate triphenylmethane. However, this and several other arylmethanes are metalated completely by KH in THF if 18-crown-6 is present. This was the first reported use of crown ethers to effect acid–base equilibria.

Bromochlorocarbene. Treatment of dibromochloromethane with aqueous NaOH and tetraalkylammonium salts results in formation of all three possible dihalocarbenes to about the same extent.[6] 18-Crown-6 shows somewhat more selectivity. Surprisingly, when dibenzo-18-crown-6 is used, only bromochlorocarbene is formed from dibromochloromethane. Thus *gem*-bromochlorocyclopropane derivatives can be prepared satisfactorily in 45–75% yield by the catalytic two-phase technique with this crown ether.[7]

KF-catalyzed Michael reactions. Fluoride ion solubilized by 18-crown-6 is an efficient catalyst for Michael reactions in aprotic solvents (C_6H_6, CH_2Cl_2, CH_3CN).[8]

Example:

$C_6H_5CH=CHCOC_6H_5 + CH_3NO_2 \xrightarrow[\text{94%}]{\begin{array}{c}\text{F}^-, 81°,\\ \text{CH}_3\text{CN}\end{array}} C_6H_5CH-CH_2COC_6H_5$

$\underset{\displaystyle CH_2NO_2}{|}$

Mixed biaryls.[9] The synthesis of biaryls by the Gomberg-Bachmann reaction[10] of aryldiazonium salts and arenes can be conducted to advantage under phase-transfer catalysis using potassium acetate to form a diazo anhydride. KOAc can be replaced by NaOH, KOH, or K_2CO_3, but with lower yields.

$$ArN_2{}^+BF_4{}^- \xrightarrow[-KBF_4]{\substack{KOAc, \\ crown\ ether}} [ArN{=}NOAc \rightarrow (ArN{=}N)_2O] \xrightarrow{-N_2}$$

$$ArN{=}N{-}O\cdot + Ar\cdot \xrightarrow{Ar'H} Ar{-}Ar'$$

Yields for the most part are in the range 60–80%, considerably higher than those reported earlier for this reaction (20–45%).

Reduction of nitroarenes. The reduction of nitroarenes to anilines by $Fe_3(CO)_{12}$ (**4**, 534) is catalyzed by this crown ether in the presence of KOH. The catalyzed reaction requires only half the amount of $Fe_3(CO)_{12}$. The yields are similar in both cases (60–85%).[11]

Review. Gokel and Durst[12] have reviewed the use of these macrocyclic polyethers as phase-transfer catalysts and as catalysts for ionic reactions (83 references).

[1] K. Ping-Lin, M. Miki, and M. Okahara, *J.C.S. Chem. Comm.*, 504 (1978).
[2] D. H. Hunter, M. Hamity, V. Patel, and R. A. Perry, *Canad. J. Chem.*, **56**, 104 (1978).
[3] D. H. Hunter and R. A. Perry, *Synthesis*, 37 (1977).
[4] S. A. DiBiase and G. W. Gokel, *J. Org.*, **43**, 447 (1978).
[5] E. Buncel and B. Menon, *Am. Soc.*, **99**, 4457 (1977).
[6] E. V. Dehmlow, M. Lissel, and J. Heider, *Tetrahedron*, **33**, 363 (1977).
[7] M. Fedoryński, *Synthesis*, 783 (1977).
[8] I. Belsky, *J.C.S. Chem. Comm.*, 237 (1977).
[9] S. H. Korzeniowski, L. Blum, and G. W. Gokel, *Tetrahedron Letters*, 1871 (1977).
[10] O. C. Dermer and M. T. Edmison, *Chem. Rev.*, **57**, 77 (1957).
[11] H. Alper, D. Des Roches, and H. des Abbayes, *Angew. Chem., Int. Ed.*, **16**, 41 (1977).
[12] G. W. Gokel and H. D. Durst, *Aldrichim. Acta*, **9**, 3 (1976).

Cryptates, **5**, 156; **6**, 137–138.

Review.[1] A review of cryptates (82 references) not only discusses known effects of inclusion complexes of this type but notes possible future uses. Thus a cryptate has been prepared that forms a complex with the highly toxic Cd^{2+} but not with Zn^{2+} and Ca^{2+}. Such cryptates should be useful in treatment of metal poisoning and in control of pollution. Some cryptates are very efficient phase-transfer catalysts (liquid to liquid). Isotope separation may be possible with cryptates.

[1] J.-M. Lehn, *Accts. Chem. Res.*, **11**, 49 (1978).

Cyanodiethylaluminum–Chlorotrimethylsilane. $(C_2H_5)_2AlCN$ is available from Alfa.

β-*Cyano silyl enol ethers.* These intermediates can be prepared from enones by conjugate addition of cyanodiethylaluminum followed by trapping of the aluminum enolate with chlorotrimethylsilane. The formulations indicate some useful transformations of these substances.[1]

Examples:

[1] M. Samson and M. Vandewalle, *Syn. Comm.*, **8**, 231 (1978).

Cyanogen bromide, 1, 174–176; **2**, 93; **4**, 110; **5**, 169–170; **6**, 148–149.

1,2-Iminocarbonates. The reaction of 1,2-glycols with *n*-butyllithium (2 equiv.) in THF and then with cyanogen bromide (1 equiv.) leads to 1,2-imino-carbonates in high yield. An example is the conversion of 1,2-O-isopropylidene-3-O-methyl-α-D-glucofuranose (**1**) into **2**. This compound has been converted to the 6-amino sugar **4** by ring scission with acetyl chloride to give **3** followed by cyclization (NaH), hydrolysis, and acetylation. The overall result is replacement of —OH by —NH$_2$.[1]

[1] D. H. R. Barton and W. Motherwell, *Nouv. J. Chim.*, **2**, 301 (1978).

Cyanogen chloride, ClCN. Mol. wt. 61.48, b.p. 13.8°. Preparation.[1]

Caution: Very toxic, similar to HCN.

Sulfonyl cyanides. The compounds can be prepared from sulfonyl chlorides by reduction to sodium sulfinates followed by reaction with cyanogen chloride.[2]

Example:

$$CH_3SO_2Cl \xrightarrow[\text{NaHCO}_3]{\text{Na}_2\text{SO}_3,} CH_3SO_2Na \xrightarrow[\substack{67-72\% \\ \text{overall}}]{\text{ClCN}} CH_3SO_2CN$$

[1] G. H. Coleman, R. W. Leeper, and C. C. Schulze, *Inorg. Syn.*, **2**, 90 (1946).
[2] M. S. A. Vrijland, *Org. Syn.*, **57**, 88 (1977).

Cyanogen iodide, ICN. Mol. wt. 152.92. The reagent is prepared *in situ* by reaction of iodine and mercury(II) cyanide.[1]

The iodide reacts with dithioacetals and -ketals as shown in equation (I). The products can be converted into α-ethylthiocarboxylic acids or nitriles.[1]

The reaction has recently been used to prepare 2-deoxyaldonitriles from aldose diethyl dithioacetals, as shown in equation (II) for the D-glucose derivative.[2]

[1] F. Pochat and E. Levas, *Tetrahedron Letters*, 1491 (1976).
[2] P. Herczegh, R. Bognar, and E. Timar, *Org. Prep. Proc. Int.*, **10**, 211 (1978).

$$
\begin{array}{ccc}
\text{CH(SC}_2\text{H}_5)_2 & \text{NCCH(SC}_2\text{H}_5) & \text{CH}_2\text{CN} \\
\mid & \mid & \mid \\
\text{—OAc} & \text{—OAc} & \text{—OAc}
\end{array}
$$

(II)   AcO— $\xrightarrow[96\%]{\text{ICN}}$ AcO— $\xrightarrow[34\%]{\text{Raney Ni,} \\ \text{CH}_3\text{COCH}_3,\ 70°}$ AcO—

AcO—   AcO—   AcO—

CH$_2$OAc   CH$_2$OAc   CH$_2$OAc

(two epimers)

Cyanomethyldiphenylphosphine oxide, $(C_6H_5)_2\overset{\text{O}}{\overset{\|}{P}}CH_2CN$ **(1)**. Mol. wt. 243.23. The reagent is prepared from $(C_6H_5)_2POC_2H_5$ and chloroacetonitrile.

α,β-*Unsaturated nitriles.* This phosphine oxide can be used as the Wittig-Horner reagent in a stereoselective synthesis of 2-alkenenitriles from aldehydes. The (E)-isomer **(2)** is formed with $>90\%$ selectivity from aromatic aldehydes, and with about 75% selectivity from aliphatic aldehydes.[1]

$$
\underset{R}{\overset{H}{>}}C=O + 1 \xrightarrow[80-95\%]{\substack{\text{KOC(CH}_3)_3, \\ \text{THF or DMF}}} \underset{R}{\overset{H}{>}}C=C\underset{H}{\overset{CN}{<}} + \underset{R}{\overset{H}{>}}C=C\underset{CN}{\overset{H}{<}} + (C_6H_5)_2\overset{\text{O}}{\overset{\|}{P}}OK
$$

(E)-2   (Z)-2

[1] A. Loupy, K. Sogadji, and J. Seyden-Penne, *Synthesis*, 126 (1977).

Cyanotrimethylsilane (Trimethylsilyl cyanide), 4, 542–543; **5,** 720–722; **6,** 632–633; **7,** 397–399.

Two other methods for the preparation have been published (equation I[1] and II[2]).

(I) $(CH_3)_3SiCl + AgCN \xrightarrow[74\%]{\Delta} (CH_3)_3SiCN$

(II) $(CH_3)_3SiCl + HCN \xrightarrow[70\%]{\substack{N(C_2H_5)_3, \\ \text{ether}}} (CH_3)_3SiCN$

[1] I. Ryu, S. Murai, T. Horiike, A. Shinonaga, and N. Sonoda, *Synthesis*, 154 (1978).
[2] B. Uznanski and W. J. Stec, *ibid.*, 154 (1978).

β-Cyclodextrin, 6, 151–152.

3-Allylation of 2-methylnaphthohydroquinone-1,4. Japanese chemists[1] have found that β-cyclodextrin can play a significant role in the allylation of 2-methyl-naphthohydroquinone-1,4. Thus condensation of allyl bromide with the hydro-quinone in dilute aqueous alkaline solution in the presence of β-cyclodextrin results in 2-methyl-3-allyl-1,4-naphthoquinone (equation I) in yields of 82–92%. The yields in the absence of β-cyclodextrin are in the range 13–19%. This synthesis is also applicable to the condensation with crotyl chloride. But yields are low when extended to a synthesis of vitamin K itself.

(I)   [structure: 2-methyl-1,4-dihydroxynaphthalene with OH, CH₃, OH]   + CH_2=$CHCH_2Br$ → β-Cyclodextrin, pH 9.0 (O₂)

[structure: quinone with CH₃ and CH₂CH=CH₂]   +   [structure: 2-methylnaphthoquinone with CH₃]

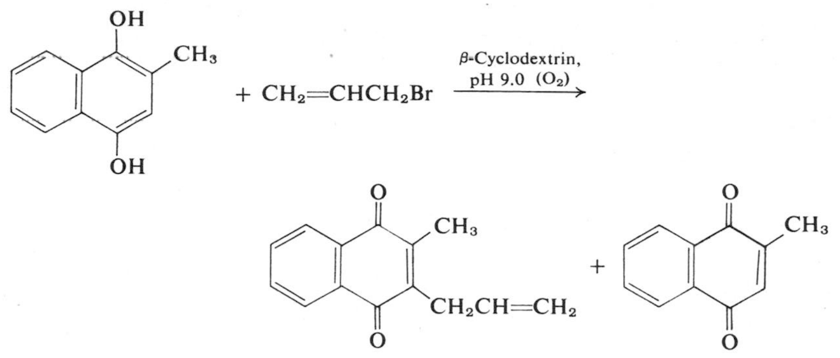

(82–92%)   (minor product)

This allylation-oxidation reaction has been extended to hydroquinone derivatives to give compounds related to α- and β-tocopherylquinones and to ubiquinone. Again β-cyclodextrin mimics allylquinone synthetase.[2]

Example:

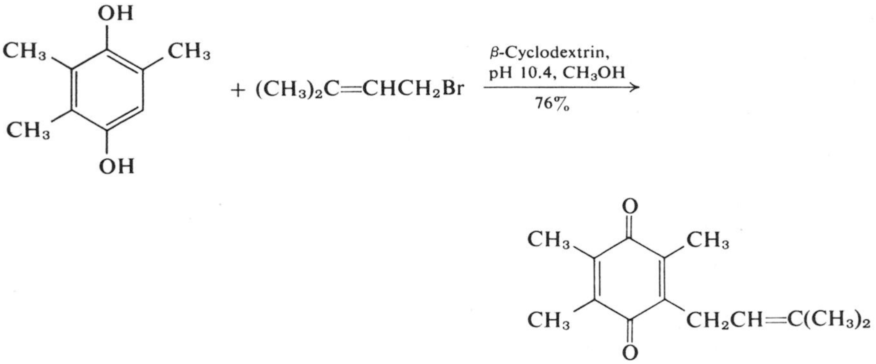

[structure: trimethylhydroquinone with OH, CH₃, CH₃, CH₃, OH]   + $(CH_3)_2C$=$CHCH_2Br$ → β-Cyclodextrin, pH 10.4, CH₃OH, 76%

[structure: quinone product with CH₃, CH₃, CH₃, CH₂CH=C(CH₃)₂]

Resolution of chiral sulfinyl compounds. Chiral sulfoxides can be partially resolved by inclusion in β-cyclodextrin, which is optically active. Better resolutions are obtained in the case of alkyl alkanesulfinates, RS(O)OR, when the sulfur atom is the only chiral center. The highest resolution was obtained with isopropyl methanesulfinate (68% optical purity). This method was used to obtain the first known optically active dialkylthiosulfinate, $(CH_3)_3CS(O)SC(CH_3)_3$, which is optically stable in refluxing benzene for 8 hours. It was obtained in 13.6% optical purity.[3]

Review.[4] Recent developments in applications of α- and β-cyclodextrin and of modified cyclodextrins have been reviewed (40 references).

[1] I. Tabushi, K. Fujita, and H. Kawakubo, *Am. Soc.*, **99**, 6456 (1977).

[2] I. Tabushi, Y. Kuroda, K. Fujita, and H. Kawakubo, *Tetrahedron Letters*, 2083 (1978).

[3] M. Mikołajczyk and J. Drabowicz, *Am. Soc.*, **100**, 2510 (1978).

[4] D. D. MacNicol, J. J. McKendrick, and D. R. Wilson, *Chem. Soc. Rev.*, **7**, 83 (1978).

N-Cyclohexylhydroxylamine, $\langle$ $\rangle$—NHOH. Mol. wt. 115.18, m.p. 102–103°. Supplier: Fluka.

α,β-*Unsaturated acids.* Gygax and Eschenmoser[1] have described several methods for conversion of *cis*-cyclohexene-4,5-dicarboxylic acid anhydride (**1**) into the methyl ester of 1,4-cyclohexadienecarboxylic acid (**5**). The first step in all cases involves conversion into the dioxoperhydro-1,2-oxazine **2**. Two methods are formulated for conversion of **2** into **5**. In fact, it is possible to effect the conversion in 60% yield without isolation of the intermediates. The transformation involves two fragmentations.

[1] P. Gygax and A. Eschenmoser, *Helv.*, **60**, 507 (1977).

(1,5-Cyclooctadiene)bis(methyldiphenylphosphine)iridium(I)   hexafluorophosphate, $(COD)Ir[PCH_3(C_6H_5)_2]_2{}^+ PF_6{}^-$ (**1**). Red solid, stable to air, soluble in CH_2Cl_2 at 0°.

Preparation[1]:

$$H_2IrCl_5 + COD \longrightarrow [IrHCl_2(COD)]_2 \xrightarrow[90-95\%\ overall]{NaOAc} [IrCl(COD)]_2 \xrightarrow{PCH_3(C_6H_5)_2}$$

$$IrCl(COD)[PCH_3(C_6H_5)_2] \xrightarrow{NH_4PF_6} \mathbf{1}$$

Isomerization of allylic alcohols. This cationic complex (**1**) is a very active catalyst for hydrogenation of alkenes, even of tri- and tetrasubstituted alkenes.[2] Since it can sometimes effect isomerization of the substrate as well, it was examined as a catalyst for isomerization of allylic alcohols to carbonyl compounds, a reaction that ordinarily requires fairly drastic conditions. After activation with hydrogen, **1** can effect this isomerization in THF at 20°. Primary and secondary allylic alcohols with monosubstituted double bonds are isomerized in quantitative yield at 20°. Alcohols with more substituted double bonds require more catalyst and/or higher temperatures (60°). Dismutation is not observed in this system.[3]

Examples:

$$CH_2{=}CHCH_2OH \xrightarrow[100\%]{\mathbf{1},\ 20°,\ \frac{1}{2}\ hr.} CH_3CH_2CHO$$

$$\xrightarrow{88\%} CH_3CH_2CH_2CHO$$

$$\xrightarrow[98\%]{18\ hr,\ 65°} CH_3CH_2CH_2C{=}O$$
$$\phantom{CH_3CH_2CH_2C{=}O}\ \ CH_3$$

Isomerization of allyl ethers.[4] This complex (**1**), after exposure to hydrogen, is an effective catalyst for isomerization of primary allyl ethers of types **2** and **4** to *trans*-propenyl ethers. Either THF or dioxane is a suitable solvent.

$$CH_2{=}CHCH_2OR \xrightarrow[95-98\%]{\mathbf{1},\ 20°}$$

[**2**, R = CH$_3$, C$_6$H$_5$,
C(CH$_3$)$_3$, C(CH$_3$)$_2$C$_6$H$_5$]

3

$$\xrightarrow{98\%}$$

4   **5**

Allyl ethers such as **6–8** are not isomerized.

6   **7**   **8**

[1] R. H. Crabtree and G. E. Morris, *J. Organometal. Chem.*, **135**, 395 (1977).
[2] R. H. Crabtree, H. Felkin, and G. E. Morris, *ibid.*, **141**, 205 (1977).
[3] D. Baudry, M. Ephritikhine, and H. Felkin, *Nouv. J. Chim.*, **2**, 355 (1978).
[4] Idem, *J.C.S. Chem. Comm.*, 694 (1978).

Cyclopentadiene, 1, 181–182.

Protection of $\diagdown C{=}C \diagdown$. A synthesis of the terpene alcohol (+)-ipsenol (**7**)

from itaconic anhydride (**1**) involves protection of the methylene group by a Diels-Alder reaction with cyclopentadiene. In the last step, the double bond is regenerated by a retro-Diels-Alder thermolysis at 450°.[1]

In an alternate synthesis of **7**[2] a similar double bond was protected as the Michael adduct with sodium thiophenoxide (**6**, 552).

[1] J. Haslouin and F. Rouessac, *Bull. soc.*, 1242 (1977).
[2] K. Mori, *Tetrahedron*, **32**, 1101 (1976).

Cyclopentadienylthallium(I), (**1**). Mol. wt. 269.47. Suppliers: Aldrich, Alfa.

Hexahydro-3,4,7-methenocyclopenta[a]pentalene. This hydrocarbon (**5**) was prepared originally (**5**, 230) in a multistep synthesis in an overall yield of about 7.5% from 9,10-dihydrofulvalene. It has now been prepared in one step by the reaction of 7-chloronorbornadiene (**2**) with TlCp (**1**) in dry diglyme; the yield of **5** is 8–12%.[1] Dihydro-*as*-indacenes are also formed from **3**. The reagent is a precursor to the anion of cyclopentadiene; in addition the metal ion may serve a catalytic role.

[1] M. A. Battiste and J. F. Timberlake, *J. Org.*, **42**, 176 (1977).

Cyclopropyl phenyl sulfide, (**1**). Mol. wt. 150.24, b.p. 62–63°/1.0 mm.

Preparation[1]:

$$BrCH_2CH_2CH_2Cl + C_6H_5SH \xrightarrow[95\%]{NaOH} C_6H_5SCH_2CH_2CH_2Cl \xrightarrow[75\%]{KNH_2} 1$$

Cyclobutanes, cyclobutenes, and cyclobutanones. This sulfide is metalated completely by *n*-butyllithium at 0°. The anion adds to aldehydes to give adducts in high yield. Yields of adducts are lower in the case of ketones because of enolization. Unlike diphenylsulfonium cyclopropylide (**4**, 211–214) the anion of **1** does not undergo conjugate addition to enones.[2]

The adducts can be rearranged to cyclobutenes on dehydration. For this purpose *p*-toluenesulfonic acid can be used, but the Burgess reagent (**4**, 227–228; **5**, 442–444) appears to be more useful. Ordinarily this reagent effects only dehydration; the presence of the sulfur substituent apparently promotes rearrangement.[3]
Examples:

$$1 + HCHO \xrightarrow[83\%]{n\text{-BuLi}} \quad \xrightarrow[47\%]{CH_3O_2CN^-SO_2N^+(C_2H_5)_3}$$

$$(CH_3)_2CH(CH_2)_3CHCH_2CH \xrightarrow{93\%} (CH_3)_2CH(CH_2)_3CHCH_2-$$

The adducts can also be rearranged to cyclobutanones; for this rearrangement three different conditions have been used, none of which is entirely general. These are outlined in the examples.
Examples:

$$\xrightarrow[84\%]{HBF_4,\ ether\ 20°}$$

$$\xrightarrow[90\%]{SnCl_4,\ CH_2Cl_2,\ 20°}$$

93:7

$$\xrightarrow[92\%]{p\text{-TsOH},\ C_6H_6,\ H_2O,\ \Delta}$$

[1] W. E. Truce, K. R. Hollister, L. B. Lindy, and J. E. Parr, *J. Org.*, **33**, 43 (1968).

[2] B. M. Trost, D. E. Keeley, H. C. Arndt, J. H. Rigby, and M. J. Bogdanowicz, *Am. Soc.*, **99**, 3080 (1977).

[3] B. M. Trost, D. E. Keeley, H. C. Arndt, and M. J. Bogdanowicz, *ibid.*, **99**, 3088 (1977).

1-Cyclopropyl-1-trimethylsilyloxyethylene   (**1**). Mol. wt.

156.30, b.p. 38–40°/12 mm. This reagent is prepared from cyclopropyl methyl ketone by reaction with LDA and $ClSi(CH_3)_3$ in 78% yield.

β-*Cyclopropylenones*. These substances can be prepared by Michael addition of **1** to enones followed by aldol condensation.[1]

Examples:

[1] P. A. Grieco and Y. Ohfune, *J. Org.*, **43**, 2720 (1978).

D

1,4-Diazabicyclo[2.2.2]octane–Bromine complex (Dabco·2Br$_2$). Mol. wt. 431.84, m.p. 155–160° dec., yellow, stable to light, air, water. This complex is prepared by admixture of the two compounds in CCl$_4$. It is very slightly soluble in a variety of solvents; however, in the presence of excess amine, solutions in CH$_3$CN or CH$_2$Cl$_2$ can be obtained.

Oxidation of alcohols.[1] The complex oxidizes benzyl alcohol to benzaldehyde in yields as high as 97%, but in general primary alcohols such as 1-butanol are oxidized only in low yields. However, secondary alcohols are oxidized to ketones in satisfactory yield (50–70%).

[1] L. K. Blair, J. Baldwin, and W. C. Smith, Jr., *J. Org.*, **42**, 1816 (1977).

1,5-Diazabicyclo[5.4.0]undecene-5 (DBU), **2**, 101; **4**, 16–18; **5**, 177–178; **6**, 158; **7**, 87–88.

Monoalkylation of tosylacetonitrile. Alkylation of TsCH$_2$CN ordinarily yields mainly dialkyl derivatives; selective monoalkylation of fairly stable carbanions is generally difficult. Japanese chemists[1] found that monoalkylation with primary alkyl halides can be effected in 90–95% yield if DBU is used as base with benzene as solvent. Considerable dialkylation is observed under phase-transfer conditions. Yields from secondary halides are somewhat lower.[1] The same conditions can be used for monoalkylation of methyl cyanoacetate and acetylacetone.[2]

[1] N. Ono, R. Tamura, R. Tanikaga, and A. Kaji, *Synthesis*, 690 (1977).
[2] N. Ono, T. Yoshimura, R. Tanikaga, and A. Kaji, *Chem. Letters*, 871 (1977).

Diborane, **1**, 199–207; **2**, 106–108; **3**, 76–77; **4**, 124–126; **5**, 184–186; **6**, 161–162; **7**, 89.

Hydroboration of an allylic alcohol. The reaction of the allylic alcohol **1**, prepared stereoselectively in five steps from (+)-*p*-menthene-1, with diborane followed by oxidation unexpectedly results in a mono alcohol (**2**) rather than a 1,2-glycol. Presumably the intermediate 1,2-borane borate undergoes elimination followed by rehydroboration. The product (**2**) was used in a synthesis of (−)-acorenone (**4**). (−)-Acorenone B (**7**) was also obtained from **1** as formulated. The conversions establish the absolute configuration of **4** and **7**.

Protection of t-*amines.* Recent examples of use of amine borane complexes include the intramolecular oxidative phenol coupling of **1** to **2**, a reaction that fails with the free amine.[2] Several spirodienone precursors to aporphine and

dibenzazonine alkaloids have been synthesized as borane complexes.[3] An amine complex has been used to markedly improve a Friedel-Crafts cyclization.[4] Thus **3** can be converted into **4** in 70% yield; the yield is much lower when the free amine is used. This reaction was the final step in the total synthesis of 14-hydroxy-morphinans such as **4**, the methyl ether of butorphanol.

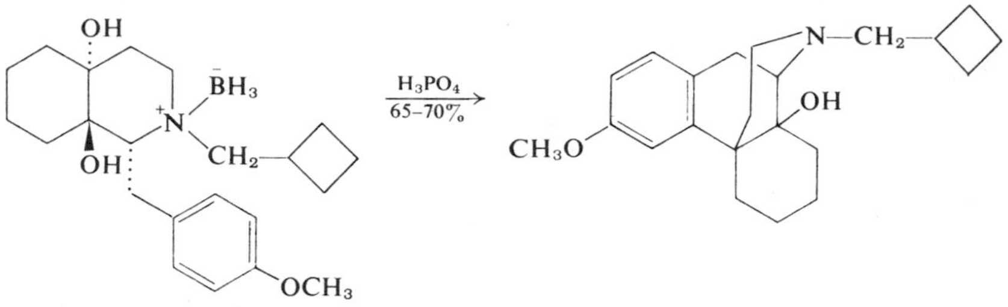

3   **4**

[1] M. Pesaro and J.-P. Bachmann, *J.C.S. Chem. Comm.*, 203 (1978).
[2] M. A. Schwartz, B. F. Rose, and B. Vishnuvajjala, *Am. Soc.*, **95**, 612 (1973).
[3] S. M. Kupchan and C.-K. Kim, *J. Org.*, **41**, 3210 (1976).
[4] I. Monković, C. Bachand, and H. Wong, *Am. Soc.*, **100**, 4609 (1978).

Dibromoborane–Dimethyl sulfide, $HBBr_2 \cdot S(CH_3)_2$. Mol. wt. 233.79, viscous liquid.
    Preparation:

$$BH_3 \cdot S(CH_3)_2 + 2S(CH_3)_2 + 2BBr_3 \longrightarrow 3BHBr_2 \cdot S(CH_3)_2$$

Hydroboration.[1] This reagent, in contrast to dichloroborane–dimethyl sulfide, hydroborates alkenes directly without necessity for an added Lewis acid such as BCl_3. Actually $HBBr_2 \cdot S(CH_3)_2$ would be expected to be less reactive than $HBCl_2 \cdot S(CH_3)_2$, since it should be more acidic. Irrespective of the theoretical interpretation of the unexpected activity, hydroboration with this reagent leads to synthesis of the previously unknown alkyldibromoboranes, obtained as the $S(CH_3)_2$ addition compounds. These complexes can be freed from dimethyl sulfide by distillation in the presence of BBr_3. The directive effects are similar to those observed with other borane reagents.

[1] H. C. Brown and N. Ravindran, *Am. Soc.*, **99**, 7097 (1977).

5,5-Dibromo-2,2-dimethyl-4,6-dioxo-1,3-dioxane (**1**). Mol. wt. 301.94, m.p. 75–76°.
    Preparation: by bromination of Meldrum's acid:

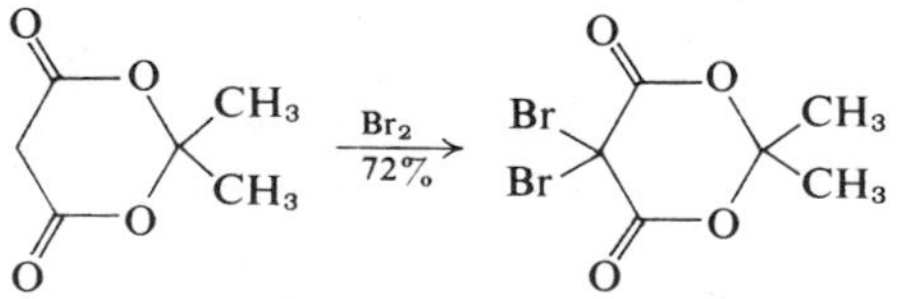

1

Bromination of carbonyl compounds.[1] Carbonyl compounds are brominated in the α-position by this reagent in a 2:1 molar ratio. In the case of saturated substrates, no catalyst is necessary. For enones, the rate can be enhanced by heat or addition of some HBr.

Examples:

$$2C_6H_5COCH_3 \xrightarrow{\text{1, Ether}} 2C_6H_5COCH_2Br + \quad (20\%)$$

$$(CH_3)_2CHCHO \xrightarrow[55\%]{} (CH_3)_2C(Br)CHO$$

$$C_6H_5CH{=}CHCOCH_3 \xrightarrow[75\%]{H^+} C_6H_5CH{=}CHCOCH_2Br$$

$$(CH_3)_2C{=}CHCOCH_3 \xrightarrow[45\%]{CCl_4, H^+, \Delta} (CH_3)_2C{=}CHCOCH_2Br$$

[1] R. Bloch, *Synthesis*, 140 (1978).

Di-*t*-butoxyacetylene, $(CH_3)_3CO{-}C{\equiv}C{-}OC(CH_3)_3$ (**1**). Mol. wt. 170.24, m.p. 8.5°, stable at 20° for some days.

Preparation[1]:

Reactions.[2] The triple bond of **1** is neither hydrogenated nor hydrated under usual conditions. It does react with dichlorocarbene, however; the adduct was used in a synthesis of deltic acid (**2**). The reagent was also used for a synthesis of squaric acid (**5**). When heated in benzene, **1** is converted into *t*-butoxyketene (**a**), which reacts with **1** to form **3**. Oxidation followed by hydrolysis gives **5**.

[1] M. A. Pericás and F. Serratosa, *Tetrahedron Letters*, 4433 (1977).
[2] *Idem, ibid.*, 4437 (1977).

$$1 \xrightarrow[\text{C}_6\text{H}_5\text{CH}_2\text{N}^+(\text{C}_2\text{H}_5)_3\text{Cl}^-]{\text{HCCl}_3, \text{NaOH}, \text{H}_2\text{O},} \left[\underset{(\text{CH}_3)_3\text{CO} \qquad \text{OC(CH}_3)_3}{\overset{\text{Cl} \quad \text{Cl}}{\triangle}} \right] \xrightarrow{13\text{–}35\%}$$

$$\underset{(\text{CH}_3)_3\text{CO} \qquad \text{OC(CH}_3)_3}{\overset{\text{O}}{\triangle}} \xrightarrow[\text{quant.}]{\text{CF}_3\text{COOH}} \underset{\text{HO} \qquad \text{OH}}{\overset{\text{O}}{\triangle}}$$

2

$$1 \xrightarrow[-\text{C}_4\text{H}_8]{\text{C}_6\text{H}_6, 80°} [\text{O}{=}\text{C}{=}\text{CHOC(CH}_3)_3] \xrightarrow[\text{quant.}]{1}$$

a

$$\underset{(\text{CH}_3)_3\text{CO} \qquad \text{OC(CH}_3)_3}{\overset{\text{O} \qquad \text{OC(CH}_3)_3}{\square}} \xrightarrow[83\%]{\text{NBS, CCl}_4}$$

3

$$\underset{(\text{CH}_3)_3\text{CO} \qquad \text{OC(CH}_3)_3}{\overset{\text{O} \qquad \text{O}}{\square}} \xrightarrow[\text{quant.}]{\text{CF}_3\text{COOH}} \underset{\text{HO} \qquad \text{OH}}{\overset{\text{O} \qquad \text{O}}{\square}}$$

4   **5**

Di-*t*-butyl dicarbonate, (BOC)$_2$O, **4**, 128; **7**, 91. The preparation has been published.[1]

Caution: Do not heat above 80°. Suppliers: Aldrich, Fluka.

[1] B. M. Pope, Y. Yamamoto, and D. S. Tarbell, *Org. Syn.*, **57**, 45 (1977).

2,6-Di-*t*-butyl-4-methylpyridine (1). Mol. wt. 205.34, m.p. 31–32°, b.p. 148–153°/95 mm.

Preparation[1]:

$$(\text{CH}_3)_3\text{CCOCl} \xrightarrow{\text{CF}_3\text{SO}_3\text{H}} [(\text{CH}_3)_3\text{CC}{\equiv}\text{O}^+\text{CF}_3\text{SO}_3^-] \xrightarrow[54\%]{(\text{CH}_3)_2\text{C}{=}\text{CH}_2}$$

$+ \text{H}_2\text{O}$

$$\xrightarrow[95\%]{\overset{\text{NH}_4\text{OH}, -40°}{\text{C}_2\text{H}_5\text{OH}}}$$

1

This hindered, nonnucleophilic base has been used in the alkylation of arenes with vinyl triflates. This reaction proceeds via vinyl cations.[2]

Example:

$$C_6H_6 + (C_6H_5)_2C{=}C\underset{C_6H_5}{\overset{OSO_2CF_3}{<}} \xrightarrow[54\%]{1, 150°} (C_6H_5)_2C{=}C(C_6H_5)_2$$

[1] A. G. Anderson and P. J. Stang, *J. Org.*, **41**, 3034 (1976).
[2] P. J. Stang and A. G. Anderson, *Am. Soc.*, **100**, 1520 (1978).

Di-*t*-butyl phosphorobromidate, $(C_4H_9O)_2\overset{\overset{\displaystyle O}{\|}}{P}Br.$ Mol. wt. 273.11, n_D 1.4490. Preparation.[1]

Selective phosphorylation.[2] One advantage of this reagent is that the *t*-butyl groups are removable by acid treatment. It reacts readily only with primary alcohols, but does react with alkoxides of both primary and secondary alcohols (equation I). However, it reacts significantly faster with alkoxides of primary

(I) $RONa + 1 \longrightarrow (t\text{-}C_4H_9O)_2\overset{\overset{\displaystyle O}{\|}}{P}OR \xrightarrow[50-80\%]{CF_3COOH, 20°} RO\overset{\overset{\displaystyle O}{\|}}{P}(OH)_2$

alcohols than with those of secondary alcohols so that selective phosphorylation of the former alcohols is possible.

[1] T. Gajda and A. Zwierzak, *Synthesis*, 243 (1976).
[2] *Idem, ibid.*, 623 (1977).

Dicarbonylcyclopentadienylcobalt, 5, 172–173; **6,** 153–154; **7,** 94–95.

Benzocyclobutenes; indans; tetralins.[1] These compounds can be obtained by cooligomerization of α,ω-diynes with substituted acetylenes; $CpCo(CO)_2$ is the most satisfactory catalyst. Yields of benzocycloalkenes obtained in this way are low unless bis(trimethylsilyl)acetylene is used (equation I) as the acetylene component.

(I)

Substitution in the diyne lowers the yield of the benzocyclobutene. Use of 1,6-heptadiyne or of 1,7-octadiyne in the reaction results in indans or tetralins. An example of extension of this synthesis to an anthraquinone is formulated in equation (II).

Steroid synthesis.[2] The cobalt-catalyzed cooligomerization of a 1,5-hexadiyne (**1**) with bis(trimethylsilyl)acetylene (**2**) has culminated in a synthesis of the estra-trienone **4**. The cyclization results in formation of the desired *trans, anti, trans* arrangement of the B, C, and D rings, probably for steric reasons. The reaction

represents the shortest known steroid synthesis. The precursor **1** can be prepared in 28% yield from 1,5-hexadiyne.

Review. Vollhardt[3] has reviewed the cooligomerization of acetylene catalyzed by transition metals for synthesis of complex molecules. The review includes some as yet unpublished syntheses of exotic compounds that would be difficult to prepare by classical reactions.

[1] R. L. Hillard III and K. P. C. Vollhardt, *Am. Soc.*, **99**, 4058 (1977).
[2] R. L. Funk and K. P. C. Vollhardt, *ibid.*, **99**, 5483 (1977).
[3] K. P. C. Vollhardt, *Accts. Chem. Res.*, **10**, 1 (1977); *Strem Chemiker*, **6**, 1 (1978).

Dicarbonylbis(triphenylphosphine)nickel, $[(C_6H_5)_3P]_2Ni(CO)_2$ (**1**), **1**, 61. Mol. wt. 639.29, m.p. 210–215° dec., light-yellow, air-stable solid. Additional supplier: Orgmet. Preparation.[1]

Oxidative bisdecarboxylation. This reaction using the organometallic reagent rather than lead tetraacetate was introduced by Trost and Chen.[2] These chemists found that in general thioanhydrides (prepared by treatment of anhydrides with Na_2S) are better substrates; yields of olefins tend to be low when either substrate contains β-hydrogens.

The reaction has been used to prepare norbornenes substituted at a bridgehead (equation I).[3]

(I)

[1] J. Chatt and F. A. Hart, *J. Chem. Soc.*, 1378 (1960).
[2] B. M. Trost and F. Chen, *Tetrahedron Letters*, 2603 (1971).
[3] G.L. Grunewald and D. P. Davis, *J. Org.*, **43**, 3074 (1978).

Di-μ-carbonylhexacarbonyldicobalt, $Co_2(CO)_8$, **1**, 224–225; **3**, 89; **4**, 139; **5**, 204–205; **6**, 172; **7**, 99–100.

Propargylation.[1] Propargylic alcohols can be converted into stabilized cations by reaction with $Co_2(CO)_8$ and then with tetrafluoroboric acid. The resulting cation propargylates activated aromatic compounds. The —$Co_2(CO)_6$ unit is removed by treatment with ferric nitrate.[2] An example is shown in equation (I). The steps can be conducted in sequence without purification of intermediates. The complex of propargyl alcohol itself reacts at both the *ortho-* and *para*-positions of anisole.

The same synthetic scheme has been used to alkylate β-dicarbonyl compounds such as benzoylacetone (equation II).[3]

(II)   HC≡CCH$_2$OH + CH$_3$COCH$_2$COC$_6$H$_5$ $\xrightarrow[\substack{-78 \to 0° \\ 90\%}]{\text{HBF}_4, \text{CH}_2\text{Cl}_2,}$
(CO)$_3$Co—Co(CO)$_3$

CH$_3$COCHCOC$_6$H$_5$ $\xrightarrow[\substack{\sim 95\%}]{\substack{\text{Fe(NO}_3)_3, \\ \text{C}_2\text{H}_5\text{OH}}}$ CH$_3$COCHCOC$_6$H$_5$
| |
CH$_2$C≡CH   CH$_2$C≡CH
|
Co$_2$(CO)$_6$

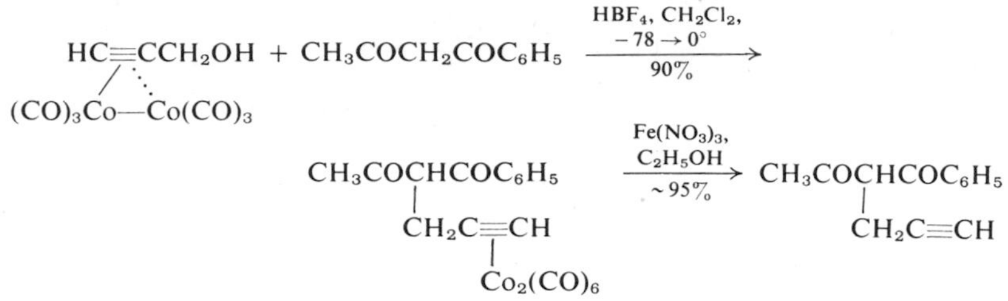

This reaction has been used to propargylate 2-acetylcyclohexanone (equation III).

(III)

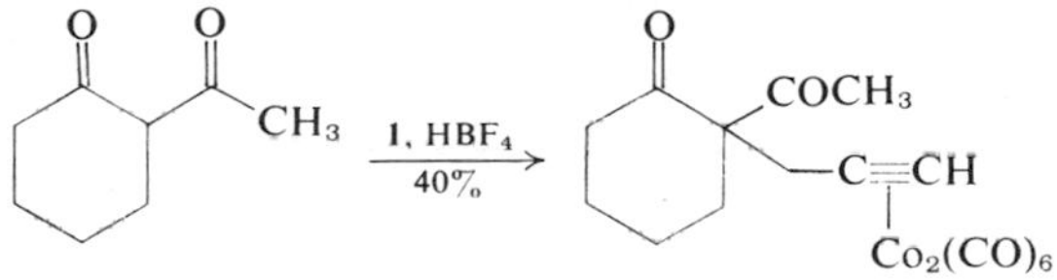

$\xrightarrow[40\%]{1, \text{HBF}_4}$

Conjugated enynes.[4] This system can be obtained by acid cleavage of an α'-acetylenic α-cyclopropyl alcohol, but this reaction is not stereoselective. However, (E)-enynes are obtained selectively when the triple bond is first complexed with Co$_2$(CO)$_8$.
Example:

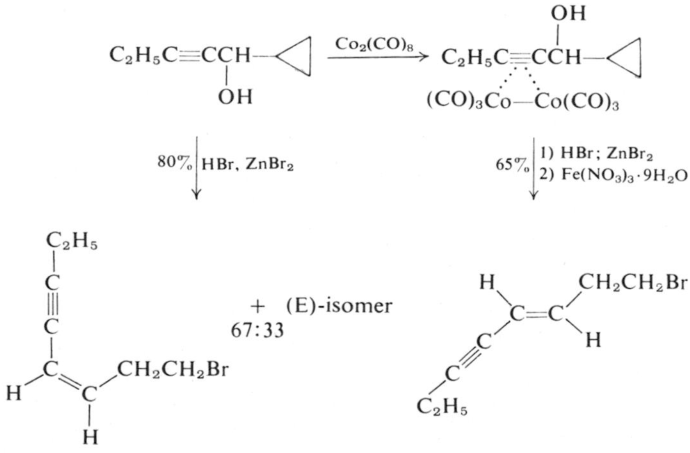

Δ²-Butenolides. Under phase-transfer conditions (cetyltrimethylammonium bromide), Co$_2$(CO)$_8$ catalyzes the reaction of carbon monoxide and methyl iodide with alkynes to form Δ^2-butenolides (equation I). When a 1,3-diene replaces the alkyne in this reaction, acylation occurs (equation II).[5]

(I) $C_6H_5C{\equiv}CH + CH_3I + CO$ $\xrightarrow{\begin{array}{c}Co_2(CO)_8,\ cat.,\\ NaOH,\ H_2O,\ C_6H_6\end{array}}$

(II) $\xrightarrow{54\%}$

Dehalogenation of α-bromo ketones. Treatment of α-halo ketones with 1 equiv. of $Co_2(CO)_8$ under phase-transfer conditions yields the dehalogenated ketone in high yield and traces of the 1,4-diketone. Yields of the methyl ketone are somewhat lower when the metal carbonyl is used in catalytic amounts.[6]

$RCCH_2Br + Co_2(CO)_8$ $\xrightarrow{\begin{array}{c}C_6H_5CH_2N(C_2H_5)_3{}^+Cl^-,\\ NaOH,\ H_2O,\ C_6H_6\end{array}}$ $RCCH_3 + RCCH_2CH_2CR$

(~95%)   (minor)

[1] R. F. Lockwood and K. M. Nicholas, *Tetrahedron Letters*, 4163 (1977).
[2] K. M. Nicholas and R. Pettit, *ibid.*, 3475 (1971).
[3] H. D. Hodes and K. M. Nicholas, *ibid.*, 4349 (1978).
[4] C. Descoins and D. Samain, *ibid.*, 745 (1976); C. Descoins, D. Samain, B. Lalanne-Casson, and M. Gallois, *Bull. soc.*, 941 (1977).
[5] H. Alper, J. K. Currie, and H. des Abbayes, *J.C.S. Chem. Comm.*, 311 (1978).
[6] H. Alper, K. D. Logbo, and H. des Abbayes, *Tetrahedron Letters*, 2861 (1977).

gem-Dichloroallyllithium (1), 5, 188–189.

Reaction with carbonyl compounds. A detailed study of the reaction of this reagent with various substrates has been published.[1] In most of the examples, the regioselectivity is strongly dependent on electronic factors rather than on steric factors. Several possible explanations for this unexpected regioselectivity are suggested.

The paper also reports some reactions of **1** with substrates other than aldehydes and ketones.

$CH_2{\cdots}\overset{-}{CH}{\cdots}CCl_2Li^+ + CO_2$ $\xrightarrow{91\%}$ $CH_2{=}CHCCl_2COOH$

$\xrightarrow[\text{low yield}]{1}$ $CH_2{=}CHCCl_2CH_2CH_2OH$

$HCOOCH_3$ $\xrightarrow{1 \atop 63\%}$ $CH_2{=}CHCCl_2CHO$

$$H_2O \xrightarrow[87\%]{1} HCCl_2CH=CH_2$$

$$CH_3I \xrightarrow[98\%]{1} CH_3CCl_2CH=CH_2$$

Cyclopentenones. Japanese chemists[2] have reported a new, general synthesis of cyclopentenones based on the reaction of this reagent with ketones as shown in equation (1) for the reaction of cyclohexanone with **1**, which leads to a dichloro homoallylic alcohol (**2**). On treatment with acid, **2** is converted into **a** and then

(1)

2

a   **b**   **3**

into the chloropentadienyl cation **b**, which undergoes ring closure followed by hydrolysis to give **3**. Several acids are effective for the conversion of **2** into **3**, but trifluoroacetic acid appears to be the most satisfactory.

The present procedure is applicable to both cyclic and acyclic ketones, but is particularly useful for cyclopentenone annelation.

[1] D. Seyferth, G. J. Murphy, and B. Mauzé, *Am. Soc.*, **99**, 5317 (1977).
[2] T. Hiyama, M. Shinoda, and H. Nozaki, *Tetrahedron Letters*, 771 (1978).

Dichlorobis(triphenylphosphine)palladium(II), **6**, 60–61; **7**, 95–96.

2-Alkynyl ketones.[1] 1-Alkynes react with some acyl chlorides in the presence of CuI and this palladium complex and triethylamine to form 2-alkynyl ketones. The reaction fails with acyl chlorides that react easily with the amine. Use of dimethylcarbamoyl chloride results in 2-alkynamides.

Examples:

$$C_6H_5C\equiv CH + C_6H_5COCl \xrightarrow[96\%]{[P(C_6H_5)_3]_2PdCl_2, \ CuI, \ N(C_2H_5)_3} C_6H_5C\equiv CCC_6H_5$$

$$n\text{-}C_4H_9C\equiv CH + (CH_3)_2CHCOCl \xrightarrow[61\%]{} n\text{-}C_4H_9C\equiv CCCH(CH_3)_2$$

$$C_6H_5C\equiv CH + (CH_3)_2NCOCl \xrightarrow[92\%]{} C_6H_5C\equiv CCN(CH_3)_2$$

[1] Y. Tohda, K. Sonogashira, and N. Hagihara, *Synthesis*, 777 (1977).

1,3-Dichloro-2-butene, 1, 214–215; **2**, 111–112.

Annelation. McKenzie[1] has reported a modification of the annelation procedure of Caine and Tuller (**2**, 111–112) that permits this dihalobutene to be used as a 3-pentenone equivalent. The acetylenic intermediate **2** is obtained as in the Caine method, but this is alkylated before hydration of the triple bond and cyclization to the final product (**3**).

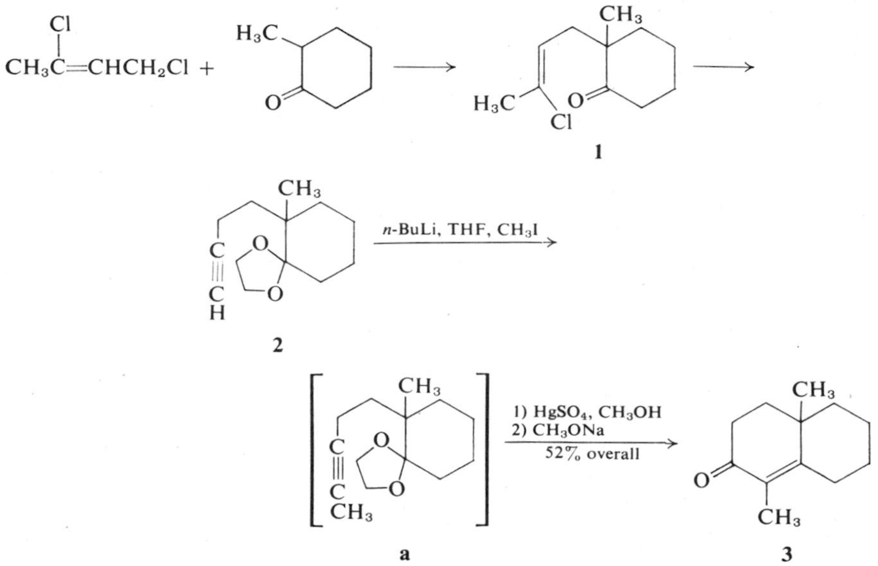

[1] T. C. McKenzie, *Synthesis*, 608 (1976).

Di-μ-chlorodichlorobis(pentamethylcyclopentadienyl)rhodium,
$\{Rh[\eta^5\text{-}C_5(CH_3)_5]_2Cl_2\}_2$ (**1**), red, crystalline solid. Preparation[1]:

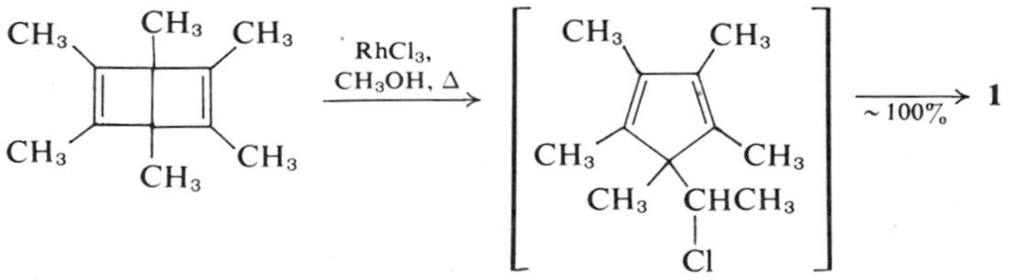

The complex also has been prepared from pentamethylcyclopentadiene, but this starting material is less accessible than hexamethyl-Dewar benzene. The complex $[Rh(C_5H_5)Cl_2]_n$ is known, but this is amorphous and insoluble in most organic solvents; moreover, the Rh—C_5H_5 bond is easily cleaved. The iridium analog of **1** is also known.

Hydrogenation catalyst.[2] The complex 1 is a very active catalyst for hydrogenation of olefins, particularly in a polar solvent such as 2-propanol and in the presence of a basic cocatalyst (usually triethylamine, but Na_2CO_3 is effective). It is about twice as active toward cyclohexene as Wilkinson's catalyst; the iridium dichloride complex corresponding to 1 is even more active than 1. Of greater significance, 1 is an effective homogeneous catalyst for hydrogenation of arenes to cyclohexanes. Intermediate cyclohexenes and cyclohexadienes have not been detected. In the 1-catalyzed hydrogenation of xylenes, *cis*-dimethylcyclohexanes are favored over the *trans*-isomers. A review has been published.[3]

Related catalyst. A related homogeneous rhodium catalyst, $RhHCl[\eta^6-C_6(CH_3)_6][P(C_6H_5)_3]$, has recently been reported to catalyze hydrogenation of benzene to cyclohexane, also without intermediate formation of cyclohexadienes or cyclohexene. However, this catalyst functions in the absence of an added base. It also has high catalytic activity for hydrogen transfer from secondary alcohols to alkenes and alkadienes and is recovered unchanged after the transfer.[4]

[1] J. W. Kang and P. M. Maitlis, *Am. Soc.*, **90**, 3259 (1968).
[2] M. J. H. Russell, C. White, and P. M. Maitlis, *J.C.S. Chem. Comm.*, 427 (1977).
[3] P. M. Maitlis, *Accts. Chem. Res.*, **11**, 301 (1978).
[4] M. A. Bennett, T.-N. Huang, A. K. Smith, and T. W. Turney, *J.C.S. Chem. Comm.*, 582 (1978).

2,3-Dichloro-5,6-dicyano-1,4-benzoquinone (DDQ), 1, 215–219; 2, 112–117; 3, 83–86; 4, 130–134; 5, 193–194; 6, 168–170; 7, 96–97.

Polycyclic dihydroarenes.[1] A novel route to conjugated dihydroarenes involves regiospecific hydrogenation of polycyclic arenes in the terminal ring over a 10% Pt/C catalyst to a tetrahydroarene followed by dehydrogenation with 1 equiv. of DDQ in refluxing benzene. The yields are surprisingly high.

Examples:

Dihydroarenes are important as precursors to diol epoxides, which are the probable biologically active forms of carcinogenic hydrocarbons.[2]

Dehydrogenation of 1,2-dibenzyloxytetrahydroarenes. In a recent synthesis of diol epoxides of carcinogenic arenes, Fu and Harvey[3] found that 1,2-dibenzyloxytetrahydroarenes are dehydrogenated in good yield by DDQ in refluxing dioxane. An example is formulated in equation (**I**). This method is superior to bromination-dehydrobromination; the NBS route to **2** from **1** failed completely.

(I)

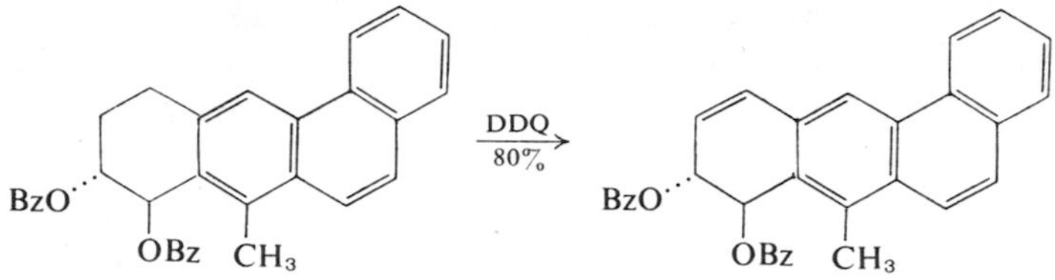

Oxidation of 1-alkylazulenes. 1-Alkylazulenes are oxidized by DDQ in aqueous acetone or dioxane to 1-acylazulenes. An alkyl group at C_2 is not oxidized under these conditions.[4]

Examples:

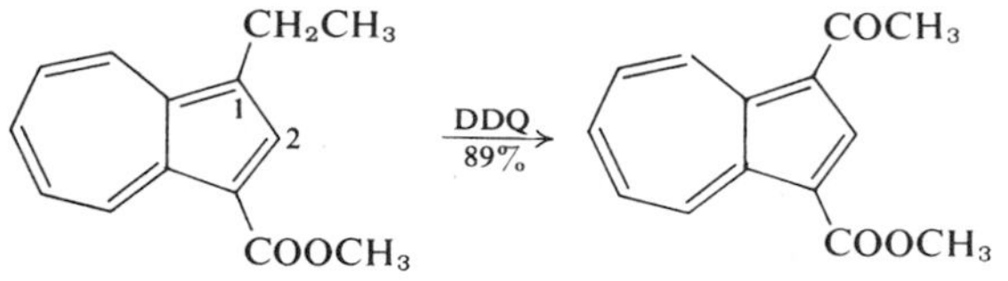

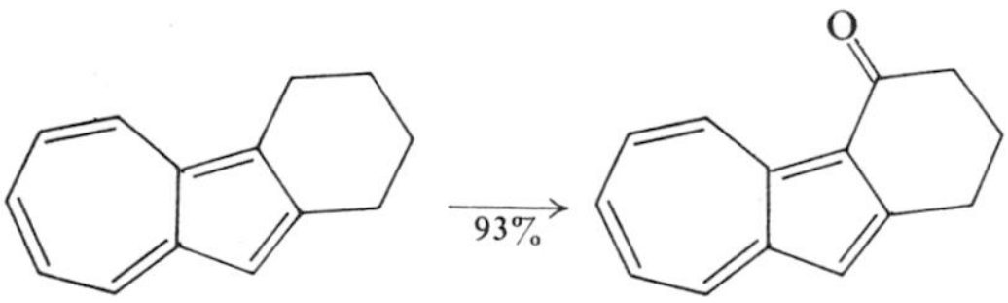

Kekulene.[5] The hydrocarbon **3** has a very complicated systematic name and since it can be regarded as a "super benzene," the name kekulene was advanced in 1965 even before the synthesis in 1978. The final steps involved photocyclo-dehydrogenation of **1**, which contains two *cis*-stilbene units, and dehydrogenation of **2** to **3** with DDQ. Because of the low solubility of **2**, 1,2,4-trichlorobenzene was used as solvent in a reaction conducted at 100° for 3 days. Kekulene is greenish-yellow and melts above 620°; the proton resonance spectrum argues against annulenoid aromaticity and favors a regular benzenoid system.

1    $\xrightarrow[70\%]{h\nu}$

2    $\xrightarrow[80\%]{DDQ}$ 3

Dehydrogenation of flavanones. Flavanones are oxidized to flavones in quantitative yield by DDQ. The reaction is slow in refluxing benzene (*ca.* 60 hours), but is complete within 5–12 hours when conducted in dioxane at reflux.[6]

[1] P. P. Fu, H. M. Lee, and R. G. Harvey, *Tetrahedron Letters*, 551 (1978).
[2] P. P. Fu and R. G. Harvey, *ibid.*, 415 (1977).
[3] *Idem, ibid.*, 2059 (1977).
[4] T. Ameniya, M. Yasunami, and K. Takase, *Chem. Letters*, 587 (1977).
[5] F. Diederick and H. A. Staab, *Angew. Chem., Int. Ed.*, **17**, 372 (1978).

[6] S. Matsuura, M. Iinuma, K. Ishikawa, and K. Kagei, *Chem. Pharm. Bull. Japan*, **26**, 305 (1978).

Dichloroketene, 1, 221–222; **2**, 118; **3**, 87–88; **4**, 134–135.

1,2-Cycloaddition. An improved method for *in situ* generation of dichloro-ketene involves treatment of trichloroacetyl chloride with zinc activated with $CuSO_4$ and with $POCl_3$. The role of $POCl_3$ seems to be that of complexing the $ZnCl_2$ formed in the reaction. Under these conditions adducts were obtained even from tri- and tetrasubstituted olefins.[1]

Note that adducts had been obtained previously from trisubstituted alkenes and dichloroketene generated with activated zinc.[2]

[1] L. R. Krepski and A. Hassner, *J. Org.*, **43**, 2879 (1978).
[2] P. W. Jeffs and J. Molina, *J.C.S. Chem. Comm.*, 3 (1973).

Dichloromethylenedimethylammonium chloride (Phosgene immonium chloride), 4, 135–138; **5**, 195–198; **6**, 170.

"Phosgeneiminium salts" (*PI salts*). Viehe[1] has reviewed heterocyclizations with these salts, which formally are related to phosgene and phosgeneimines (55 references, including unpublished results).

Reaction with carbodiimides. The adducts (**2**) from reaction of this phosgene-like salt with carbodiimides or N,N'-disubstituted ureas are useful for synthesis of several heterocycles (scheme I).[2]

Reaction with acyl chlorides.[3] A typical reaction is formulated in equation (I).

(I)

$$R—N{=}C{=}N—R$$
or
$$RNHCONHR$$

$$\xrightarrow[\sim 95\%]{\overset{Cl^-}{(CH_3)_2\overset{+}{N}{=}C(Cl)_2,\ 1}}$$

$$RN{=}\underset{Cl}{\overset{R}{\underset{|}{\overset{|}{C}}}}—\underset{Cl}{\overset{|}{\underset{|}{N}}}—C{=}\overset{+}{N}(CH_3)_2 Cl^-$$

2

30% | $C_6H_5NHNH_2$

70–80% | $HN{=}C\overset{C_6H_5}{\underset{NH_2}{}}$

70% | $H_2N\quad NH_2$ (benzene ring)

3   **4**   **5**

Scheme (I)

Reaction with nitriles.[4] A typical reaction is formulated in equation (I).

(I) $R'CH_2C{\equiv}N + 2(1)$ $\xrightarrow[\sim 25\%]{HCl,\ CHCl_3,\ \Delta}$

$$(CH_3)_2\overset{+}{N}\underset{Cl^-}{}{=}\underset{Cl}{\overset{R'}{\overset{|}{C}}}—\underset{Cl}{\overset{|}{C}}{=}\underset{Cl}{\overset{|}{C}}—C{=}N{-}C{-}N(CH_3)_2$$ $\xrightarrow[35-90\%]{\Delta}$ $(CH_3)_2N{-}$ (pyrimidine ring with Cl, Cl, R')

2   **3**

[1] H. G. Viehe, *Chem. Ind.*, 386 (1977).
[2] A. Elgavi and H. G. Viehe, *Angew. Chem., Int. Ed.*, **16**, 181 (1977).
[3] H. G. Viehe, B. Le Clef, and A. Elgavi, *ibid.*, **16**, 182 (1977).
[4] B. Stelander and H. G. Viehe, *ibid.*, **16**, 189 (1977).

Dichloromethylenetris(dimethylamino)phosphorane, $[(CH_3)_2N]_3P{=}CCl_2$ (**1**). Mol. wt. 246.11. This reagent is prepared *in situ* from hexamethylphosphorous triamide and bromotrichloromethane.

1,1-Dichloroalkenes. The reagent **1** reacts with aldehydes in CH_2Cl_2 at -20 to $-10°$ to form 1,1-dichloroolefins in 85–94% yield. Ketones react slowly if at all with this reagent. The reagent is a useful alternative to the more expensive combination of triphenylphosphine and CCl_4 (**4**, 551).[1]

$$RCHO + [(CH_3)_2N]_3P=CCl_2 \longrightarrow RCH=CCl_2 + [(CH_3)_2N]_3P=O$$
$$1$$

This method was used in a synthesis of 25-hydroxycholesterol (2) from stigmasterol.[2]

[1] W. G. Salmond, *Tetrahedron Letters*, 1239 (1977).
[2] W. G. Salmond, M. C. Sobala, and K. D. Maisto, *ibid.*, 1237 (1977).

2,3-Dichloro-1-propene, $H_2C=C(Cl)CH_2Cl$, 2, 120.

γ-Butyrolactones. Marshall and Flynn[1] have shown that the 2-chloro-2-propenyl group can serve as an acetic acid synthon. An example is the synthesis of the γ-butyrolactone formulated in equation (I), in which the chloropropenyl group is converted into an acyl chloride unit by ozonolysis. A more elaborate example of use of this synthon is formulated in equation (II).

(II)

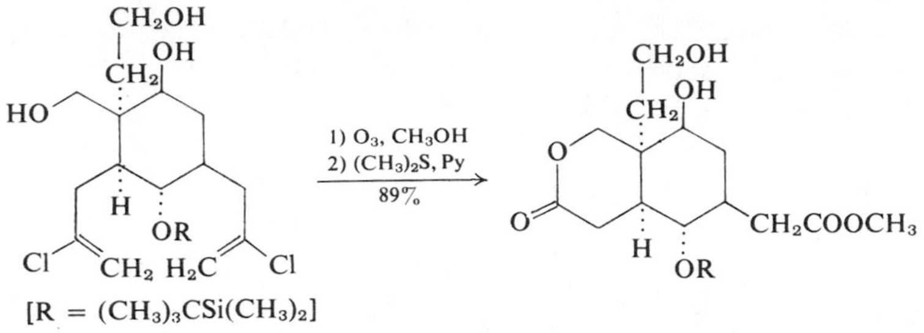

$$[R = (CH_3)_3CSi(CH_3)_2]$$

[1] J. A. Marshall and G. A. Flynn, *Syn. Comm.*, **7**, 417 (1977).

2,5-Dichlorothiophenium bismethoxycarbonylmethylide (1). Mol. wt. 283.13, stable crystalline solid.

Preparation[1]:

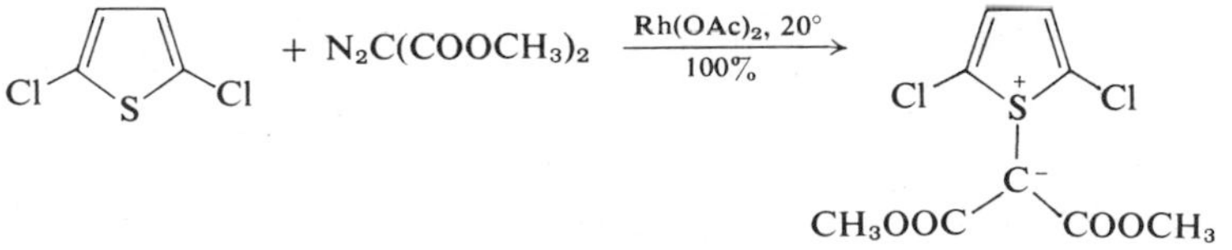

Bismethoxycarbonylcarbene. When **1** is warmed in 2,5-dichlorothiophene it rearranges in low yield to dimethyl 2,5-dichlorothiophene-3-malonate.[2] However, when **1** is heated with an alkene in the presence of rhodium(II) acetate or Cu-(acac)$_2$, bismethoxycarbonylcyclopropanes are obtained in 60–85% yield. Evidently **1** rearranges to bismethoxycarbonylcarbene or a derived metal carbenoid. This new reagent offers some advantages over diazomalonic esters: ease of handling, shorter reaction periods, and usually higher yields of products.[3]

Example:

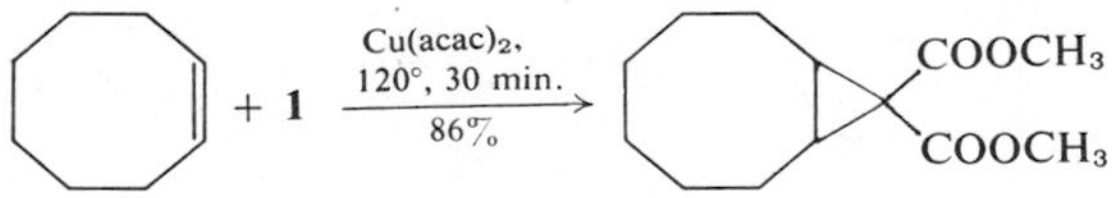

[1] R. J. Gillespie, J. Murray-Rust, P. Murray-Rust, and A. E. A. Porter, *J.C.S. Chem. Comm.*, 83 (1978).
[2] R. J. Gillespie, A. E. A. Porter, and W. E. Willmott, *ibid.*, 85 (1978).
[3] J. Cuffe, R. J. Gillespie, and A. E. A. Porter, *ibid.*, 641 (1978).

Dichlorotris(triphenylphosphine)ruthenium(II), 4, 564; **5,** 740–741; **6,** 654–655; **7,** 99.

Selective hydrogenations.[1] This is a satisfactory catalyst for selective hydrogenation of 1,5,9-cyclododecatriene, 1,5-cyclooctadiene, and several 1,6-dienes to monoenes.

Hydrogenation of aldehydes.[2] This complex is an effective catalyst for hydrogenation of aliphatic aldehydes to alcohols. Aryl aldehydes are reduced to benzyl alcohols. Keto and nitro groups are not reduced under the same conditions.

Claisen rearrangement of diallyl ethers. The Claisen rearrangement of allyl vinyl ethers[3] to give γ,δ-unsaturated aldehydes and ketones has been a valuable synthetic reaction (**3**, 300). Reuter and Salomon[4] have extended this rearrangement to unsymmetrical diallyl ethers, which are readily prepared by O-alkylation of allyl alcohols with allyl chlorides. These ethers have been found to rearrange to γ,δ-unsaturated aldehydes or ketones in the presence of $[(C_6H_5)_3P]_3RuCl_2$. The first step undoubtedly is isomerization of one allyl group to a vinyl group, a known reaction (**5**, 736), to give an allyl vinyl ether, which then undergoes Claisen rearrangement. The present work shows that this isomerization is sensitive to the number of substituents on the double bond; monosubstituted allyl ethers rearrange to vinyl ethers much more readily than di- or trisubstituted allyl ethers. Thus single products of rearrangement can often be obtained on Claisen rearrangement.

Examples:

γ-Butyrolactones. In the presence of this catalyst, trichloro- and dichloroacetic acid react with 1-alkenes to form γ-butyrolactones (equation I). The reaction is conducted in refluxing toluene (5 hours).[5]

[1] J. Tsuji and H. Suzuki, *Chem. Letters*, 1083 (1977).
[2] *Idem, ibid.*, 1085 (1977).
[3] D. J. Faulkner, *Synthesis*, 175 (1971).

[4] J. M. Reuter and R. G. Salomon, *J. Org.*, **42**, 3360 (1977).
[5] H. Matsumoto, T. Nakano, K. Ohkawa, and Y. Nagai, *Chem. Letters*, 363 (1978).

(I) 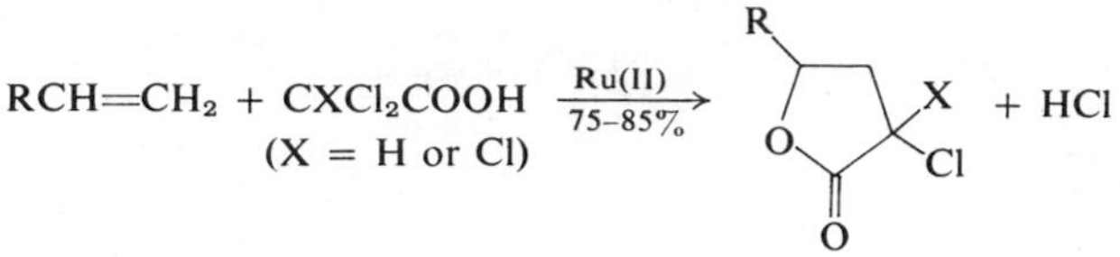 $RCH{=}CH_2 + CXCl_2COOH$ $\xrightarrow[75-85\%]{Ru(II)}$ $+ HCl$
           $(X = H \text{ or } Cl)$

N,N-Dichlorourethane (DCU), $C_2H_5OCONCl_2$ **(1)**. Mol. wt. 157.99, b.p. 74–76°/ 15 mm. The reagent is prepared by chlorination of urethane (80% yield).[1]

Selective chlorination.[2] DCU when activated by light converts cyclohexane to chlorocyclohexane in high yield. The reagent also converts adamantane into 1-chloroadamantane. A more useful reaction is chlorination of steroids at C_9 to give, after dehydrochlorination with $AgClO_4$, 9(11)-dehydro derivatives (equation I).

(I)

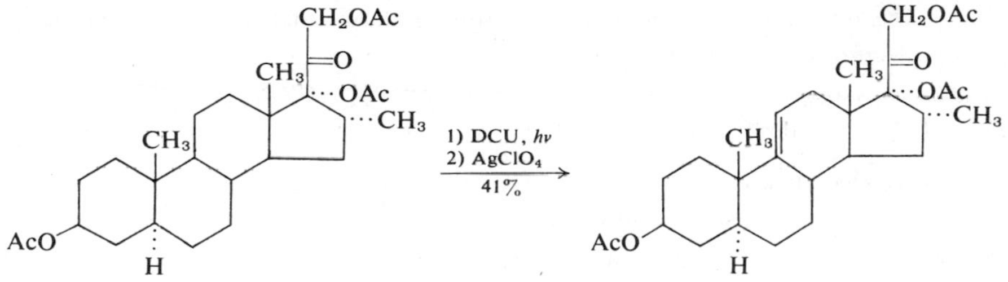

[1] T. A. Foglia and D. Swern, *J. Org.*, **31**, 3625 (1966); **33**, 766 (1968).
[2] Y. Mazur and Z. Cohen, *Angew. Chem., Int. Ed.*, **17**, 281 (1978).

Dicyanodimethylsilane, $(CH_3)_2Si(CN)_2$ **(1)**. Mol. wt. 110.20, m.p. 80–83°, b.p. 165–170°. Supplier: Petrarch Systems. Preparation.[1]

Reaction with β-diketones.[2] The reagent converts β-diketones into 5-cyano-2, 6-dioxa-1-silacyclohex-3-enes, from which the parent β-diketones can be recovered by treatment with methanol or AgF in THF (equation I).

(I)

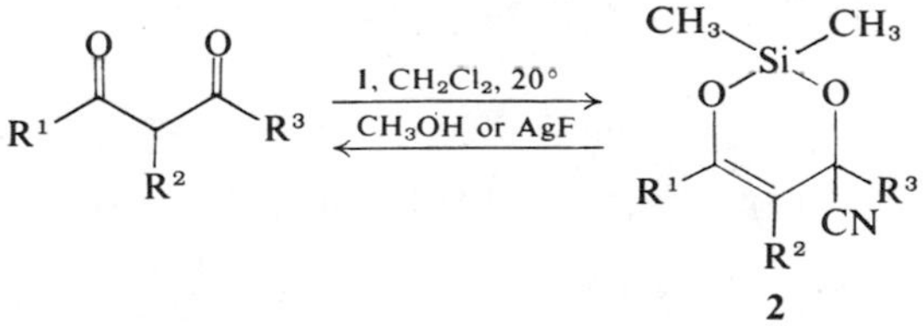

Examples:

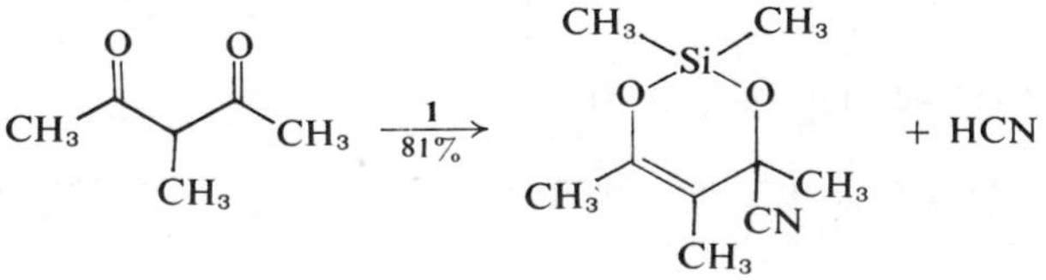

The reaction with juglone (last example) shows that **1** can also be used for protection of β-hydroxy ketones.

[1] J. J. McBride, Jr., and H. C. Beachall, *Am. Soc.*, **74**, 5247 (1952).
[2] I. Ryu, S. Murai, T. Horiike, A. Shinonaga, and N. Sonoda, *J. Org.*, **43**, 780 (1978).

Dicyclohexylborane, 3, 90–91; **4**, 141.

Alkanoic acids. Monohydroboration of 1-alkynyl(trimethyl)silanes (**2**) with dicyclohexylborane leads to 1-boryl-1-silylalkenes (**3**) almost exclusively. The usual alkaline peroxide oxidation of **3** gives alkanoic acids (**4**). Related reactions can be used to prepare α,β- and β,γ-unsaturated acids.[1]

Examples:

[1] G. Zweifel and S. J. Backlund, *Am. Soc.*, **99**, 3184 (1977).

Dicyclohexylcarbodiimide (DCC), 1, 231–236; **2**, 126; **3**, 91; **4**, 141; **5**, 206–207; **6**, 174; **7**, 100–101.

Ureas. In the presence of a tertiary amine (2 equiv.), a primary amine (2 equiv.) reacts with DCC (1 equiv.) and CO_2 (Dry Ice) at $-75°$ to form ureas (equation I).[1]

(I) $$2RNH_2 + C_6H_{11}N{=}C{=}NC_6H_{11} + CO_2 \xrightarrow[50-95\%]{N(C_2H_5)_3} RNH\overset{\overset{\displaystyle O}{\|}}{C}NHR$$

[1] H. Ogura, K. Takeda, R. Tokue, and T. Kobayashi, *Synthesis*, 394 (1978).

Dicyclohexylcarbodiimide–4-Dimethylaminopyridine, (DMAP, **5**, 260).

Esterification.[1] DCC has seen little use for esterification because of variable yields, which are satisfactory only in the case of phenols and thiophenols. Pyridine exerts a favorable effect; the more effective acylation catalyst DMAP is more useful. With this catalyst, esters and thioesters can be obtained generally in yields of 60–95%. Even the hindered 2,4,6-trimethylbenzoic acid can be converted in this way into the methyl ester in 74% yield. However, the method is subject to steric effects; for example, adamantanecarboxylic acid does not react with *t*-butyl alcohol under these conditions.

[1] B. Neises and W. Steglich, *Angew. Chem., Int. Ed.*, **17**, 522 (1978).

2,5-Diethoxy-1,4-benzoquinone (1). Mol. wt. 196.20, m.p. 183°.

The quinone is prepared by the reaction of benzoquinone and ethanol in the presence of zinc chloride (reflux).[1]

2,5-Dialkyl-1,4-benzoquinones (**3**). Moore *et al.*[2] have found that organolithium reagents undergo 1,2-addition to the carbonyl groups of **1**. The enol ether groups of the products are hydrolyzed to **3** by treatment with conc. H_2SO_4 in glyme. Two different lithium reagents can be used to give unsymmetrical disubstituted quinones. 2,5-Dialkyl-3,6-dichloro-1,4-benzoquinones can be

1   2   3

obtained in the same way from 2,5-dichloro-3,6-dimethoxy-1,4-benzoquinone.[3] The chloro groups of the products can be displaced by hydroxyl groups under hydrolytic conditions.

[1] E. Knoevenagel and C. Bückel, *Ber.*, **34**, 3993 (1901).
[2] H. W. Moore, Y. L. Sing, and R. S. Sidhu, *J. Org.*, **42**, 3320 (1977).
[3] K. Wallenfels and W. Draber, *Ber.*, **90**, 2819 (1957).

Diethylaluminum chloride, 4, 144–146.

Aldol synthesis.[1] A new synthesis of β-hydroxy ketones and esters involves regiospecific conversion of an α-bromo ketone or ester into an aluminum enolate by a coupled reaction with diethylaluminum chloride and zinc activated with copper(I) bromide in THF at $-20°$. This enolate adds to carbonyl compounds to give, after work-up, β-hydroxy ketones (equation I).

Examples:

$$BrCH_2COOC_2H_5 + C_6H_5CHO \xrightarrow[94\%]{} C_6H_5CH(OH)CH_2COOC_2H_5$$

An intramolecular version of this reaction has been used to prepare β-hydroxy macrolides (equation II).

[1] K. Maruoka, S. Hashimoto, Y. Kitagawa, H. Yamamoto, and H. Nozaki, *Am. Soc.*, **99**, 7705 (1977).

Diethylaluminum 2,2,6,6-tetramethylpiperidide (DATMP), 6, 181–183.

Review.[1] The superiority of this reagent for cleavage of epoxides is attributable to the fact that bonds from aluminum to electronegative atoms such as oxygen are unusually strong. DATMP is particularly reactive to trisubstituted epoxides, with which it reacts selectively to give only a secondary allylic alcohol. The required proton is supplied by an alkyl group attached to the ring. The reagent $(i\text{-}C_3H_7)_2N\text{---}Al(C_2H_5)_2$ is less selective than DATMP.

Oxetanes can also be cleaved by organoaluminum compounds, in particular, by diethylaluminum methylanilide, $\overset{CH_3}{\underset{C_6H_5}{\diagdown N}}\text{---}Al(C_2H_5)_2$ (1).[2] A higher tem-

perature and a longer reaction time are required, but still only a homoallylic alcohol is formed.

Examples:

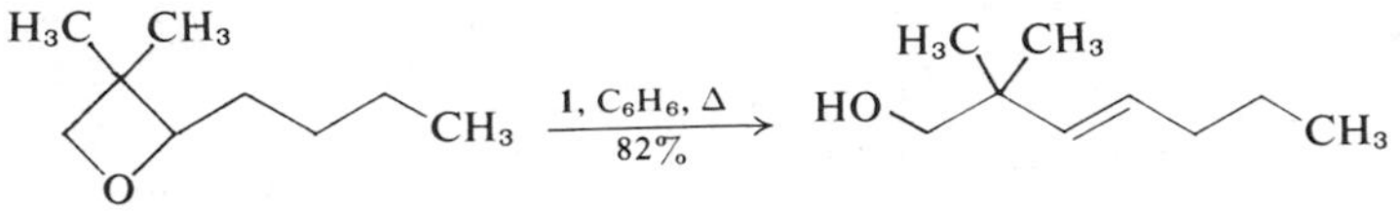

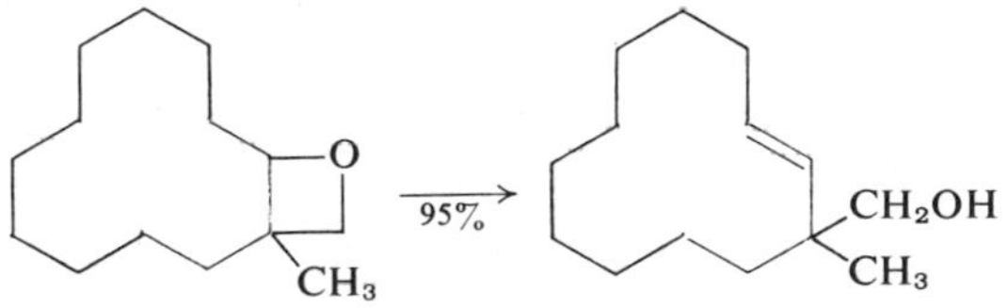

[1] H. Yamamoto and H. Nozaki, *Angew. Chem., Int. Ed.*, **17**, 169 (1978).
[2] Y. Kitagawa, A. Itoh, S. Hashimoto, H. Yamamoto, and H. Nozaki, *Am. Soc.*, **99**, 3864 (1977).

1-(Diethylamino)propyne, 2, 133–134; 3, 98; 5, 217–219; 7, 107–108.

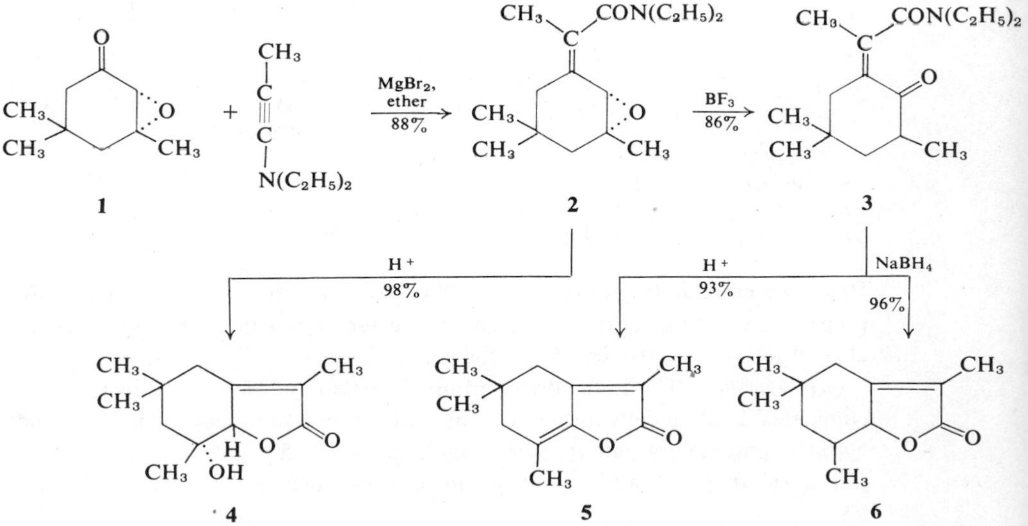

Reaction with an α-epoxy ketone.[1] The reagent reacts with the epoxy ketone **1** to give the amide **2** in high yield. The product can be converted into 3-methyl-2(5*H*)-furanones, **4–6**.

The reaction has been modified to provide a synthesis of methyl (R,S)-lichensterinic acid (**9**) from methyl α-ketopalmitate (**7**).[2]

Nitriles. The reagent reacts with benzaldoximes in acetonitrile (reflux) to form benzonitriles.[3]

[1] S. I. Pennanen, *Tetrahedron Letters*, 2631 (1977).
[2] *Idem, Heterocycles*, **9**, 1047 (1978).
[3] C. Bernhart and C.-G. Wermuth, *Synthesis*, 338 (1977).

Diethylaminosulfur trifluoride, 6, 183–184. Detailed directions are available for preparation of this fluorination reagent and for its reaction with *p*-nitrobenzyl alcohol to form *p*-nitrobenzyl fluoride (67% yield).[1]

Dehydration. The 11β-hydroxy group of 9α-fluorocorticosteroids is not easily eliminated by usual dehydrating reagents but reaction of 9α-fluorocorticosteroids with this potent fluorinating reagent results in Δ^{11}-9α-fluorosteroids (about 55% yield). The dihydroxyacetone side chain remains intact.[2]

OH → F. 7-Dehydro-25-hydroxycholesteryl acetate has been converted into the 25-fluoro derivative by reaction with diethylaminosulfur trifluoride (76% yield); the product was converted into 25-fluorovitamin D_3 by irradiation.[3]

The reagent is useful for preparation of 6-fluoro-6-deoxyhexoses as shown for the synthesis of 6-fluoro-6-deoxy-D-glucopyranose (equation 1). O-Acetyl groups are adequate for protection with this reagent.[4]

(I)

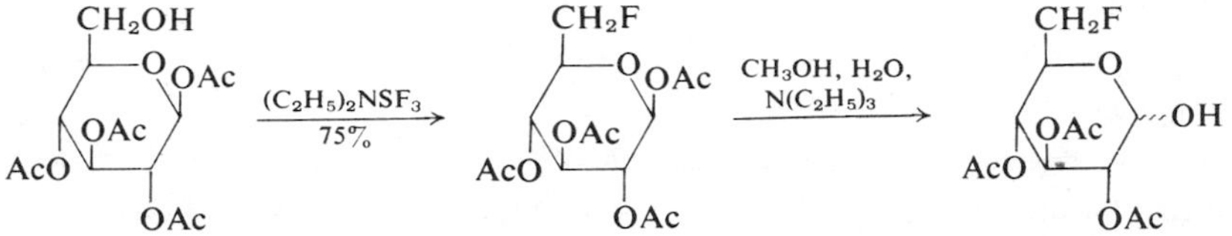

The reagent is also useful for conversion of carbonyl oxygen of suitably protected carbohydrates into a *gem*-difluoride. Yields range from 25 to 45%.[5]

[1] W. J. Middleton and E. M. Bingham, *Org. Syn.*, **57**, 50 (1977); *idem, ibid.*, **57**, 72 (1977).
[2] M. J. Green, H.-J. Shue, M. Tanabe, D. M. Yasuda, A. T. McPhail, and K. D. Onan, *J.C.S. Chem. Comm.*, 611 (1977).
[3] S. S. Yang, C. P. Dorn, and H. Jones, *Tetrahedron Letters*, 2315 (1977).
[4] M. Sharma and W. Korytnyk, *ibid.*, 573 (1977).
[5] R. A. Sharma, I. Kavai, Y. L. Fu, and M. Bobek, *ibid.*, 3433 (1977).

Diethyl *t*-butoxy(cyano)methylphosphonate (1). Mol. wt. 249.25, b.p. 116–118°/0.5 mm.

Preparation:

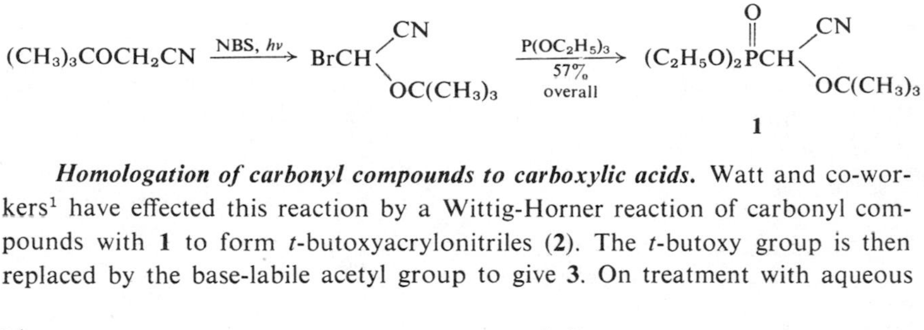

1

Homologation of carbonyl compounds to carboxylic acids. Watt and co-workers[1] have effected this reaction by a Wittig-Horner reaction of carbonyl compounds with **1** to form *t*-butoxyacrylonitriles (**2**). The *t*-butoxy group is then replaced by the base-labile acetyl group to give **3**. On treatment with aqueous

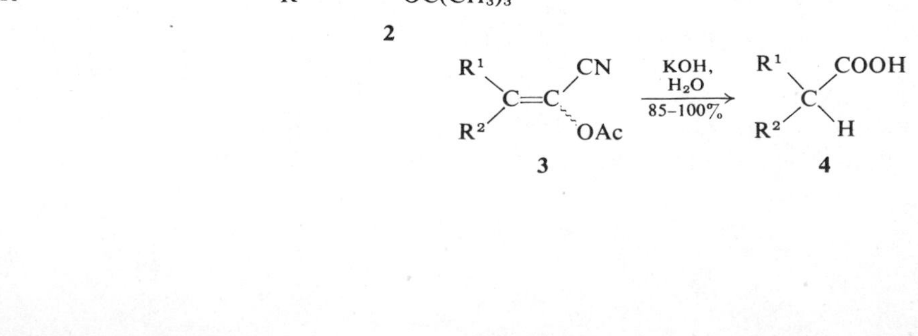

KOH in methanol the homologated acid **4** is obtained. Solvolysis with an alkoxide affords an ester; solvolysis with an amine affords an amide.[2]

[1] S. E. Dinizo, R. W. Freerksen, W. E. Pabst, and D. S. Watt, *J. Org.*, **41**, 2846 (1976).
[2] *Idem, Am. Soc.*, **99**, 182 (1977).

Diethyl carboxychloromethylphosphonate,

$$(C_2H_5O)_2\overset{\displaystyle O}{\underset{}{\overset{\|}{P}}}CH\overset{Cl}{\underset{COOH}{<}}$$

(1). Mol. wt.

230.58, oil. The reagent is prepared by carboxylation of the anion of diethyl chloromethylphosphonate (87–91% yield).

α-Chloro-α,β-unsaturated carboxylic acids. The dianion of **1**, prepared with *n*-butyllithium (THF, −65°), reacts with aldehydes and ketones to give, after hydrolysis, α-chloro-α,β-unsaturated acids (equation I).[1]

$$(I) \quad 1 \xrightarrow[\text{2) R}^1\text{COR}^2]{\text{1) 2}n\text{-BuLi}} \overset{R^1}{\underset{R^2}{>}}C=C\overset{Cl}{\underset{COOH}{<}} + (C_2H_5O)_2\overset{O}{\overset{\|}{P}}OH$$

$$(65\text{–}90\%)$$

[1] P. Savignac, M. Snoussi, and P. Coutrot, *Syn. Comm.*, **8**, 19 (1978).

Diethyl carboxymethylphosphonate (1). Mol. wt. 196.14, oil.

Preparation:

$$(C_2H_5O)_2\overset{O}{\overset{\|}{P}}H \xrightarrow[92\%]{\substack{\text{1) NaOC}_2\text{H}_5,\ \text{C}_2\text{H}_5\text{OH} \\ \text{2) CH}_3\text{I}}} (C_2H_5O)_2\overset{O}{\overset{\|}{P}}CH_3 \xrightarrow[75\text{–}85\%]{\substack{\text{1) }n\text{-BuLi, THF} \\ \text{2) CO}_2 \\ \text{3) H}_3\text{O}^+}} (C_2H_5O)_2\overset{O}{\overset{\|}{P}}CH_2COOH$$
$$\mathbf{1}$$

α,β-Unsaturated acids.[1] This reagent is more convenient than dibenzyl carboxymethylphosphonate[2] for synthesis of α,β-unsaturated acids from aldehydes or ketones.

Examples:

$$1 \xrightarrow[\text{THF, }-65°]{\text{LDA,}} (C_2H_5O)_2\overset{\overset{\displaystyle O}{\|}}{P}CHCOOLi \xrightarrow[96\%]{C_6H_5CHO} \overset{C_6H_5}{\underset{H}{>}}C=C\overset{H}{\underset{COOH}{<}}$$
$$\mathbf{2}$$

$$2 + \overset{CH_3}{\underset{C_2H_5}{>}}C=O \xrightarrow{83\%} \overset{CH_3}{\underset{C_2H_5}{>}}C=CHCOOH$$

$$(E/Z = 80:20)$$

[1] P. Coutrot, M. Snoussi, and P. Savignac, *Synthesis*, 133 (1978).
[2] G. A. Koppel and M. D. Kinnick, *Tetrahedron Letters*, 711 (1974).

Diethyl dibromomalonate, $Br_2C(COOC_2H_5)_2$. Mol. wt. 317.98, b.p. 108–110°/ 4 mm. This reagent is prepared by bromination of diethyl malonate by the method of Palmer and McWherter[1] (95% yield).

Bromination of a polyunsaturated carboxylic acid. Bromination of an acid of this type by the method of Rathke and Lindert (**4**, 306) proved unsatisfactory owing to low yields, but was accomplished by conversion of **1** into the α-anion with lithium N-isopropylcyclohexylamide, bromination with diethyl dibromo-malonate at $-78°$, and removal of the ester group. The pure acid **3** is fairly stable.[2]

$$CH_3(CH_2)_4{-}\overset{H}{(C}{=}\overset{H}{CCH_2)_3}{-}(CH_2)_4COOC(CH_3)_3 \xrightarrow[2)\ Br_2C(COOC_2H_5)_2]{1)\ LiICA}$$

$$\textbf{1}\ [RCH_2COOC(CH_3)_3]$$

$$RCHBrCOOC(CH_3)_3 \xrightarrow[27\%\ \text{overall}]{p\text{-TsOH}} RCHBrCOOH$$

$$\qquad\qquad\qquad \textbf{2} \qquad\qquad\qquad\qquad \textbf{3}$$

[1] C. S. Palmer and P. W. McWherter, *Org. Syn., Coll. Vol.*, **1**, 245 (1941).
[2] L. van der Wolf and H. J. J. Pabon, *Rec. trav.*, **96**, 72 (1977).

Diethylformamide diethyl acetal, $(C_2H_5)_2N{-}CH(OC_2H_5)_2$. This reagent is prepared in the same way as dimethylformamide diethyl acetal (**1**, 282); yield 60%.

Tertiary allenic amides. Diethylformamide acetals react with propargyl alcohols in refluxing xylene or dichlorobenzene (with removal of the alcohol) to form allenic amides in variable yields.[1]

Examples:

[1] K. A. Parker and J. J. Petraitis, *Tetrahedron Letters*, 4561 (1977).

O,O-Diethyl hydrogenphosphoroselenoate, $(C_2H_5O)_2\overset{O}{\overset{\|}{P}}{-}SeH$ (**1**). Mol. wt. 217.07; susceptible to air oxidation; store as sodium salt.
Preparation:

$$(C_2H_5O)_2PONa + Se \xrightarrow{C_2H_5OH} (C_2H_5O)_2\overset{O}{\overset{\|}{P}}SeNa \xrightarrow[90{-}97\%]{H^+} \textbf{1}$$

Deoxygenation of sulfoxides. The reagent reduces unhindered sulfoxides to sulfides in a yield of 75–95%.[1]

$$2(1) + R^1\overset{\overset{\displaystyle O}{\|}}{S}R^2 \xrightarrow{\text{CH}_2\text{Cl}_2} [(C_2H_5O)_2\overset{\overset{\displaystyle O}{\|}}{P}Se]_2 + R^1SR^2 + H_2O$$

[1] D. L. J. Clive, W. A. Kiel, S. M. Menchen, and C. K. Wong, *J.C.S. Chem. Comm.*, 657 (1977).

Diethyl ketomalonate, $O{=}C(COOC_2H_5)_2$ **(1)**. Mol. wt. 174.15, b.p. 208–210°. Supplier: Aldrich.

β,γ-*Unsaturated* δ-*lactones.* A new route to these lactones **(3)** involves a Diels-Alder reaction of this reagent with 1,3-dienes to give, after hydrolysis, the *gem*-dicarboxylic acids **(2)**. These can be converted into the lactones **(3)** by oxidative decarboxylation with lead tetraacetate, but yields are only about 20%. The Curtius degradation shown proved to be more efficient. The reagent **(1)** thus serves as an equivalent of carbon dioxide in a Diels-Alder reaction.[1]

[1] R. Bonjouklian and R. A. Ruden, *J. Org.*, **42**, 4095 (1977).

Diethyl phosphorobromidate, $(C_2H_5O)_2\overset{\overset{\displaystyle O}{\|}}{P}{\diagdown}_{Br}$ **(1)**. Mol. wt. 217.0. The reagent is prepared in high yield (92%) by bromination of triethyl phosphite with bromine at −30 to −25°; the ethyl bromide is removed *in vacuo*.

Amides; peptides. The reagent in combination with pyridine or 2,6-lutidine effects synthesis of amides and peptides in good yields. Racemization is slight under controlled conditions.[1]

$$1 + R^1COOH \xrightarrow[-15°]{\text{CH}_2\text{Cl}_2,\ \text{Py,}} R^1\overset{\overset{\displaystyle O}{\|}}{C}-O-\overset{\overset{\displaystyle O}{\|}}{P}(OC_2H_5)_2 \xrightarrow[85-95\%]{\substack{R^2NH_2, \\ -15 \to 20°}} R^1\overset{\overset{\displaystyle O}{\|}}{C}NHR^2$$

[1] A. Górecka, M. Leplawy, J. Zabrocki, and A. Zwierzak, *Synthesis*, 474 (1978).

Diethyl phosphorochloridate, 1, 248; **3**, 98; **5**, 217; **6**, 192.

Disubstituted acetylenes. β-Keto sulfones can be converted into acetylenes by conversion to enol phosphates followed by reduction with sodium amalgam.[1]
Examples:

[1] B. Lythgoe and I. Waterhouse, *Tetrahedron Letters*, 2625 (1978).

Diethyl trimethylsilyloxycarbonylmethylphosphonate (1). Mol. wt. 268.32, b.p. 93°/0.0005 mm.
Preparation:

$$BrCH_2COOSi(CH_3)_3 + P(OC_2H_5)_3 \xrightarrow[92\%]{95°} (C_2H_5O)_2\overset{\overset{O}{\|}}{P}CH_2COOSi(CH_3)_3$$
$$\mathbf{1}$$

α,β-*Unsaturated acids.*[1] These compounds can be prepared by a Wittig-Horner reaction of the anion of **1** with aldehydes or ketones (60–85% yield). This method was used to prepare Queen substance (last example).
Examples:

[1] L. Lombardo and R. J. K. Taylor, *Synthesis*, 131 (1978).

Diethylzinc–Methylene iodide, 1, 253; **2,** 134; **4,** 153.

Cyclopropanation of enol silyl ethers. Cyclopropanation of these substrates with the Simmons-Smith reagent has been reported by several laboratories (**4,** 588–589). Cyclopropanation can also be effected with diethylzinc–methylene iodide in ether or in *n*-pentane under controlled conditions (70–80% yields). This reagent can also be used to convert cyclic enol silyl ethers (**1**) to spiro ethers (**4**), as shown in equation (I). This reaction involves **2** and **3** as intermediates, but can be conducted in one pot.[1]

Chloroiodomethane can be used in place of CH_2I_2.[2]

[1] I. Ryu, S. Murai, and N. Sonoda, *Tetrahedron Letters*, 4611 (1977).
[2] S. Miyano, Y. Izumi, H. Fujui, and H. Hashimoto, *Synthesis*, 700 (1977).

Diimide, 1, 257–258; **2,** 139; **3,** 99–101; **4,** 154–155; **5,** 220; **6,** 795.

Generation.[1] As generated from hydrazine by the usual oxidizing reagents, diimide apparently consists of about equal amounts of the *cis*- and *trans*-forms. Since only the *cis*-isomer functions as a hydrogenating reagent, 2 moles of hydrazine are required for reduction of one double bond. Japanese chemists have now found that selenium can function as the oxidizing agent in the generation of diimide from hydrazine and that if the oxidation is conducted under N_2, only 1.3 equiv. of hydrazine is necessary for quantitative reduction of styrene to ethylbenzene. The implication is that only *cis*-diimide is generated. Actually if the reaction is conducted under oxygen, selenium can function catalytically.

$$NH_2NH_2 + 2Se \longrightarrow [NH{=}NH] + H_2Se_2 \xrightarrow{\frac{1}{2}O_2} 2Se + H_2O$$

Selective reduction of unsaturated endoperoxides. 2,3-Dioxabicyclo[2.2.1]-heptane (**3**) serves as a model of the prostaglandin endoperoxide (PGG₂), involved in the biosynthesis of prostaglandins and also of thromboxanes and PGI₂ (prostacyclin). One successful synthesis of **3** involved sensitized photooxygenation of cyclopentadiene to give the unstable heat-sensitive endoperoxide **2**, which was selectively reduced to **3** by diimide generated from potassium azodicarboxy-late with limited acetic acid; the reduction was carried out in methylene chloride at $-78°$. The overall yield was about 30%.[2]

1   2   3

This sequence has been used for preparation of other more elaborate endo-peroxides from dienes.[3]

[1] K. Kondo, S. Murai, and N. Sonoda, *Tetrahedron Letters*, 3727 (1977).
[2] W. Adam and H. J. Eggelte, *J. Org.*, **42**, 3987 (1977).
[3] W. Adam and I. Erden, *Angew Chem., Int. Ed.*, **17**, 210, 211 (1978); W. Adam, A. J. Bloodworth, H. J. Eggelte, and M. E. Loveitt, *ibid.*, **17**, 209 (1978).

Diisobutylaluminum hydride (DIBAL), 1, 260–262; **2**, 140–142; **3**, 101–102; **4**, 158–161; **5**, 224–225; **6**, 198–201; **7**, 111–113.

Reduction of allylic esters. Kreft[1] has reported that reduction of hindered allylic esters with lithium aluminum hydride can give anomalous products (equation I). However, reduction of the ester **1** with DIBAL proceeded normally to give the pure alcohol **2** in 81% yield.

(I)

1   2

HAl(*i*-Bu)₂
81%

+   +

3   4
49:29:31

Stereoselective synthesis of a homoallylic alcohol. The first steps in a simple stereoselective synthesis of the macrolide (R)-recifeiolide (**5**) involved conversion of 1-nonen-8-yne (**1**) into the lithium vinylalanate (**2**); this was then treated with (R)-methyloxirane to give the homoallylic alcohol (**6**, 199). This alcohol was converted by standard methods into the hydroxy carboxylic acid (**4**), which was then lactonized to **5** by the Corey method (**5**, 285–286; **6**, 246–247).[2]

$$CH_2\!\!=\!\!CH(CH_2)_5C\!\!\equiv\!\!CH \xrightarrow[\text{2) } n\text{-BuLi}]{\text{1) } HAl(i\text{-Bu})_2}$$

1

$$\left[\; CH_2\!\!=\!\!CH(CH_2)_5 \quad \substack{H \\ C=C} \quad Li^+ \\ Al^-(i\text{-Bu})_2 \\ Bu \;\right] \xrightarrow{55\%}$$

2

3

$$\xrightarrow{\text{Several steps}}$$

4

$$\xrightarrow[\text{2) AgBF}_4]{\text{1) } \big(\text{pyridyl-S}\big)_2,\ P(C_6H_5)_3}$$

5

[1] A. Kreft, *Tetrahedron Letters*, 1959 (1977).
[2] K. Utimoto, K. Uchida, M. Yamaya, and H. Nozaki, *ibid.*, 3641 (1977).

(−)-**Diisopinocampheylborane**, **1**, 262–263; **4**, 161–162; **6**, 202.

Asymmetric reduction of ketones. Highly pure reagent can be prepared by hydroboration of (+)-α-pinene either with borane–THF or with borane–dimethyl sulfide. The latter method is somewhat more convenient since borane–dimethyl sulfide is available commercially (Aldrich) and since reductions with reagent prepared in this way are more rapid than reductions with reagent prepared with BH_3–THF. Reagent prepared by either method reduces ketones of the type R^*COCH_3 to $R^*CHOHCH_3$ with optical yields of 9–37%.[1]

[1] H. C. Brown and A. K. Mandal, *J. Org.*, **42**, 2996 (1977).

N,N-Diisopropylformamide, $[(CH_3)_2CH]_2NCHO$. Mol. wt. 129.20, b.p. 83–86/11 mm. Supplier: Fluka.

α-Hydroxy amides.[1] This formamide is converted into the lithio derivative, diisopropylcarbamoyllithium (**1**), in nearly quantitative yield by *t*-butyllithium in diethyl ether–THF–pentane (4:4:1) (Trapp's mixture) at −95°. The carbamoyl-lithium reacts with aldehydes and ketones to form α-hydroxy amides (2). A reaction with one ester resulted in formation of an α-keto amide.

$$LiCON[CH(CH_3)_2]_2 + R^1COR^2 \xrightarrow[60-85\%]{} \begin{array}{c} R^1 \\ | \\ C-CON[CH(CH_3)_2]_2 \\ | \\ R^2 \quad OH \end{array}$$

1   **2**

[1] A. S. Fletcher, K. Smith, and K. Swaminathan, *J.C.S. Perkin I*, 1881 (1978).

2,6-Diisopropylphenoxymagnesium hydride,

$$\begin{array}{c} CH(CH_3)_2 \\ \hline \\ -OMgH \\ \hline \\ CH(CH_3)_2 \end{array}$$

. The hydride

exists as the dimer in THF. It is prepared by the reaction of magnesium hydride with bis(2,6-diisopropylphenoxy)magnesium in THF.

Reduction.[1] This hydride reduces alkyl iodides to the alkane in quantitative yield at 25° (24 hours); alkyl bromides, alkyl chlorides, and aryl iodides are relatively inert under these conditions. It reduces enones predominantly by 1,2-addition. It reduces 4-*t*-butylcyclohexanone quantitatively; the ratio of *cis*- to *trans*-alcohol is 83:17.[2] When magnesium hydride is used for this reduction more of the *trans*-alcohol is obtained, the ratio being 24:76.

[1] E. C. Ashby, J. J. Lin, and A. B. Goel, *J. Org.*, **43**, 1557 (1978).
[2] *Idem, ibid.*, **43**, 1560 (1978).

Dilithioacetate, $LiCH_2COOLi$ (1), **7**, 113–114.

Reactions with α-oxygenated epoxides (**7**, 113–114). The reaction of **1** with the α-oxygenated epoxides **2** and **5** followed by lactonization of the hydroxy acids thus formed results in the lactones **3** and **4**. In each case the epoxide is opened with inversion, but the two reactions differ in the site of attack, which seems to be controlled by the nature of the α-oxygen. When the α-oxygen function is a hydroxyl group, the nearest bond of the epoxide is opened (with inversion); when it is a trimethylsilyl ether, the remote epoxide bond is severed. These results do not agree with those obtained with diethyl(alkoxyethynyl)alanes (**7**, 104), where the stereochemical relation of the oxygen atom to the epoxide is the decisive factor.[1]

[1] S. Danishefsky, M.-Y. Tsai, and T. Kitakana, *J. Org.*, **42**, 394 (1977).

Dilithium tetrachlorocuprate, 4, 163–164; **5**, 226; **7**, 114.

Coupling of Grignard reagents with allylic acetates (**5**, 226). This reaction has been extended to a 1-acetoxy-2,4-diene for a synthesis of the sex pheromone of the codling moth (**3**).[1]

[1] D. Samain, C. Descoins, and A. Commerçon, *Synthesis*, 388 (1978).

Dilithium tetrachloropalladate, Li_2PdCl_4, 5, 413; **7**, 114–115.

Carbopalladation. Allylic amines and sulfides form palladium complexes in high yield with dilithium tetrachloropalladate (THF, 20°, 6–8 hours). These complexes react with certain carbanions at the β-position, a site that is not susceptible to nucleophilic displacement in allylic amines and sulfides. The complexes need not be isolated; these reactions are conducted in THF at room temperature by reaction of the allylic amine or sulfide with the carbanion in the presence of 1 equiv. of Li_2PdCl_4 (equations I and II). The products are the palladium complexes **1** and **3**. These are reduced by sodium borohydride or sodium cyanoborohydride

(II)

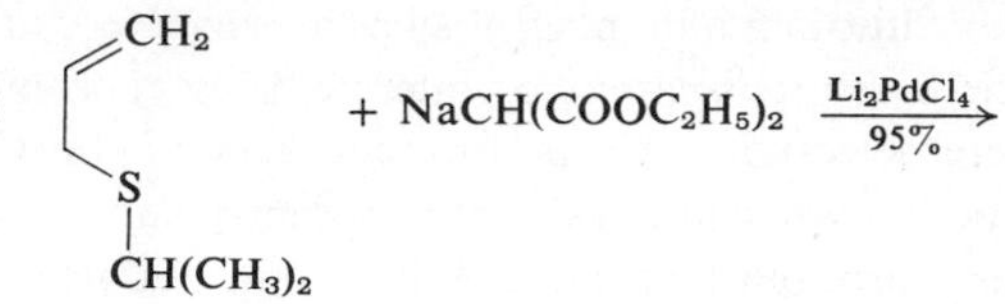

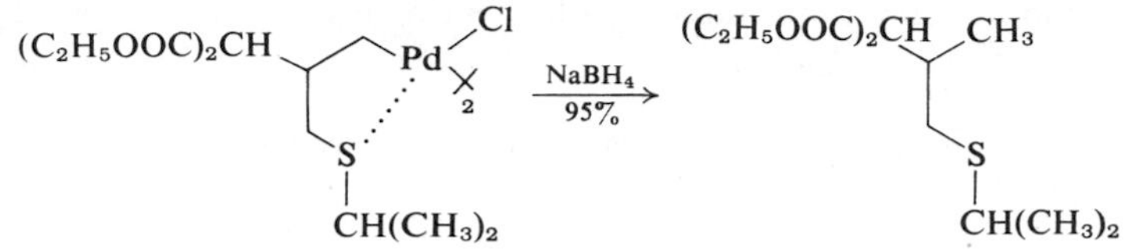

$$+ \text{NaCH(COOC}_2\text{H}_5)_2 \xrightarrow[95\%]{\text{Li}_2\text{PdCl}_4}$$

$$\xrightarrow[95\%]{\text{NaBH}_4}$$

3   4

to the γ-amino or γ-alkylthio esters **2** and **4**. Reduction can be carried out directly without isolation of the complexes.

Displacement of palladium in the complex **1** has been observed on addition to methyl vinyl ketone, which results in formation of the vinyl ketone **5** (equation III).[1]

(III) 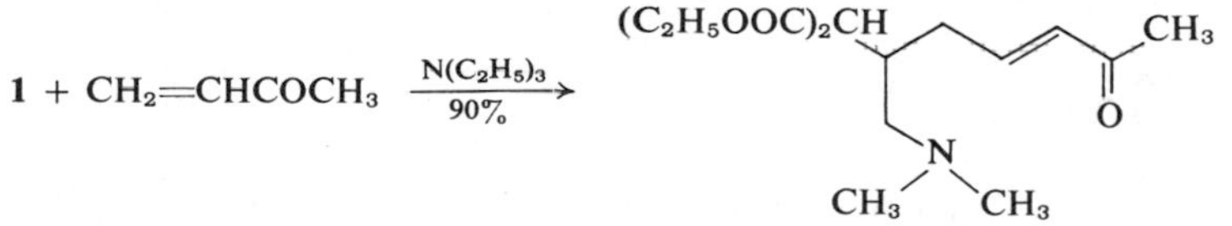 $1 + \text{CH}_2\text{=CHCOCH}_3 \xrightarrow[90\%]{\text{N(C}_2\text{H}_5)_3}$

5

Carbopalladation and alkoxypalladation are the key steps in a new synthesis of the Corey lactone diol (**10**) in about 35% overall yield from cyclopentadiene. The

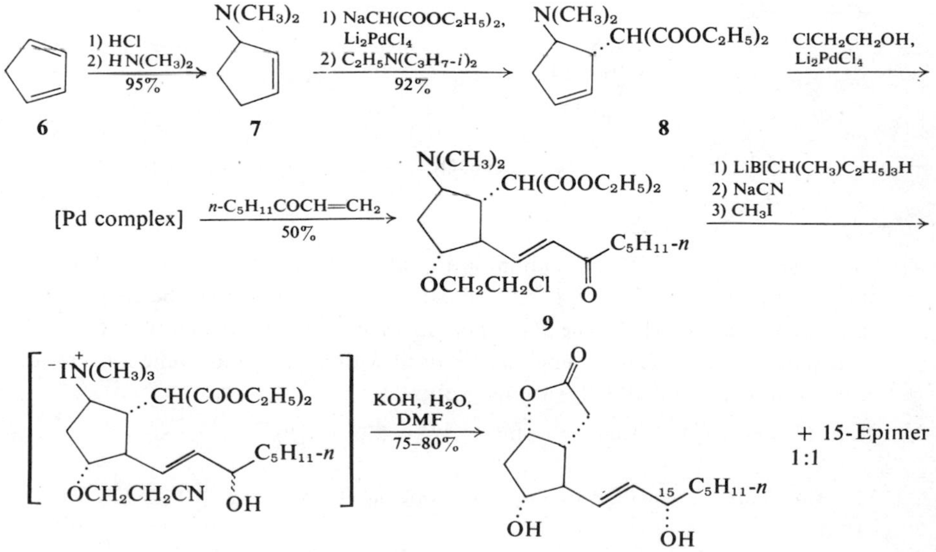

a   Scheme (I)   10

steps are outlined in scheme (1).[2] Although a number of steps are involved, most can be carried out in sequence without isolation of intermediates. The most striking feature is the high stereoselectivity, except for reduction of the C_{15}-ketone, a drawback in any of the known syntheses of **10**. Evidently both carbopalladation and alkoxypalladation proceed by *trans*-addition to the double bond and ketovinylation proceeds with retention of configuration of the group originally attached to palladium.

[1] R. A. Holton and R. A. Kjonaas, *Am. Soc.*, **99**, 4177 (1977).
[2] R. A. Holton, *ibid.*, **99**, 8083 (1977).

5,5-Dimethoxy-1,2,3,4-tetrachlorocyclopentadiene, 1, 270.

Cyclopentanones. Jung and Hudspeth[1] have devised a new cyclopentanone synthesis by three-carbon annelation of an alkene via a Diels-Alder reaction. They chose the dimethoxytetrachlorocyclopentadiene **1** as the diene component because it is remarkably reactive. The general method is outlined for methyl acrylate (**2**, equation I). The chlorine atoms of the adduct **3** are replaced by

hydrogen on reduction with sodium in liquid ammonia–ethanol to give **4**. Oxidation of the double bond of **4** is the most difficult step but can be effected with permanganate at pH 7. The final step involves bisdecarboxylation effected by warming in water. The method can be used with acyclic and cyclic alkenes and for alkenes substituted with various groups.

[1] M. E. Jung and J. P. Hudspeth, *Am. Soc.*, **99**, 5508 (1977).

1,1-Dimethoxy-3-trimethylsilyloxy-1,3-butadiene (1). Mol. wt. 202.32, b.p. 84–87°/10 mm.

Preparation:

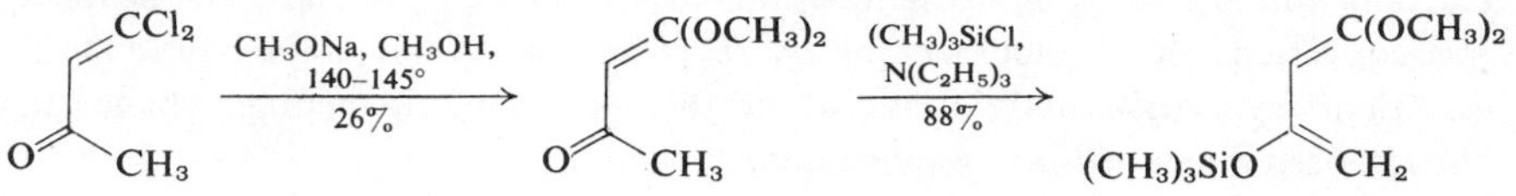

1

Diels-Alder reactions. The diene reacts with the juglone derivative **2** to give, after pyrolysis and desilylation, the anthraquinone **3**. Other regiospecific cyclo-additions are reported.[1]

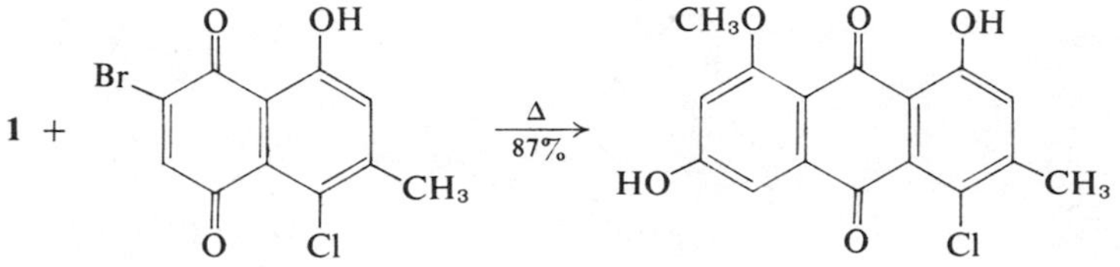

2   **3**

Actually **1** is a particularly reactive Diels-Alder diene; it is more reactive than 1-methoxy-3-trimethylsilyloxy-1,3-butadiene. In this cycloaddition reaction it functions as the equivalent of $^+COCH_2COCH_2{}^-$.[2]

Examples:

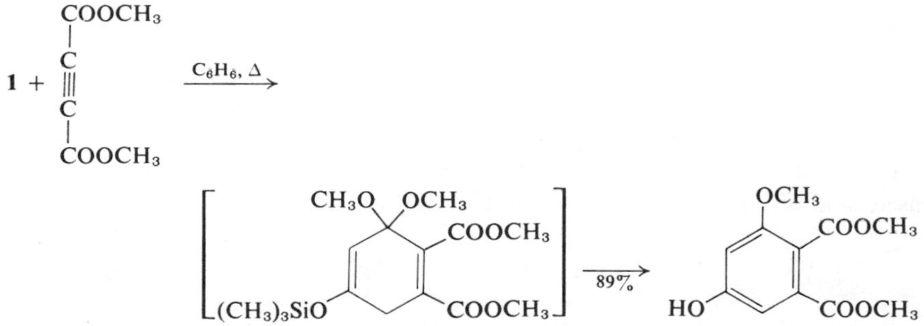

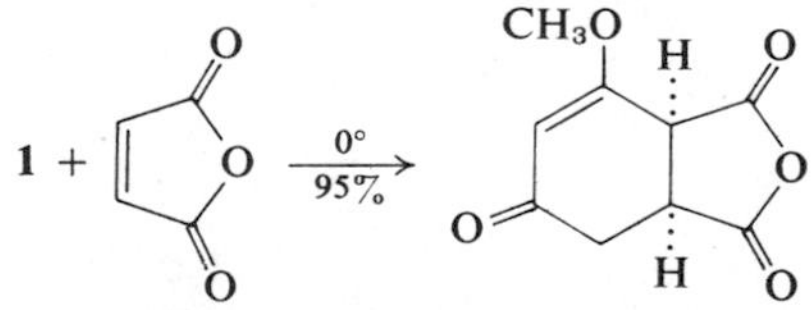

[1] J. Banville and P. Brassard, *J.C.S. Perkin I*, 1852 (1976).
[2] S. Danishefsky, R. K. Singh, and R. B. Gammill, *J. Org.*, **43**, 379 (1978).

N,N-Dimethylacetamide dialkyl acetals, 1, 271–272; **4,** 166–167; **5,** 226–227.
    Reaction with allylic alcohols (**1,** 271–272; **4,** 167, **5,** 226–227). A key step in a recent, stereocontrolled synthesis of natural thromboxane B₂ (**4**) from D-glucose

involved Claisen rearrangement of the allylic alcohol **1** to the dimethylamide **2** by reaction with N,N-dimethylacetamide dimethyl acetal at $25 \rightarrow 160°$. The product was converted into the iodolactone by reaction with iodine; this was reduced quantitatively by tri-*n*-butyltin hydride to the corresponding lactone, which had already been converted into **4** by standard reactions.[1]

Hernandez[2] has also reported a synthesis of **3** from **1** that differs from the Corey synthesis in that the orthoester Claisen rearrangement (**3**, 300–302; **6**, 607–608) is used to introduce the acetic acid side chain (equation I).

[1] E. J. Corey, M. Shibasaki, and J. Knolle, *Tetrahedron Letters*, 1625 (1977).
[2] O. Hernandez, *ibid.* 219 (1978).

3,3-Dimethylallylnickel bromide (1), 2, 291.

Prenylation. The next to last step in a total synthesis of the antibiotic diter-pene dictyolene (**2b**) involved introduction of the prenyl group. This was accom-plished by reaction of **1a** with 3,3-dimethylallylnickel bromide. Coupling of the mesylate **1b** with "(CH₃)₂C=CHCH₂Cu" resulted only in reduction of the mesylate group.[1]

1a (R = Ac, Y = I)   **2a** (R = Ac)
1b (R = Ac, Y = OMs)   **2b** (R = H)

[1] J. A. Marshall and P. G. M. Wuts, *Am. Soc.*, **100**, 1627 (1978).

(α,α-Dimethylallyl)trimethylsilane, CH_2=$CHCSi(CH_3)_3$ (**1**). Mol. wt. 142.31.

Preparation:

Prenylation.[1] The reagent (**1**) reacts with acetals in the presence of TiCl₄ or BF₃ etherate to give homoallylic ethers in high yield (equation I). The reaction of **1** with aldehydes and ketones is much slower and the expected homoallylic alcohols are usually minor products (if formed at all). The usual products are tetrahydrofuranes (equation II).

[1] A. Hosomi and H. Sakurai, *Tetrahedron Letters*, 2589 (1978).

(I) $1 + RCH(OR^1)_2$ $\xrightarrow[85-95\%]{\text{TiCl}_4, \text{CH}_2\text{Cl}_2}$ $(CH_3)_2C$=$CHCH_2CHR$
                                                                                         $|$
                                                                                       OR^1

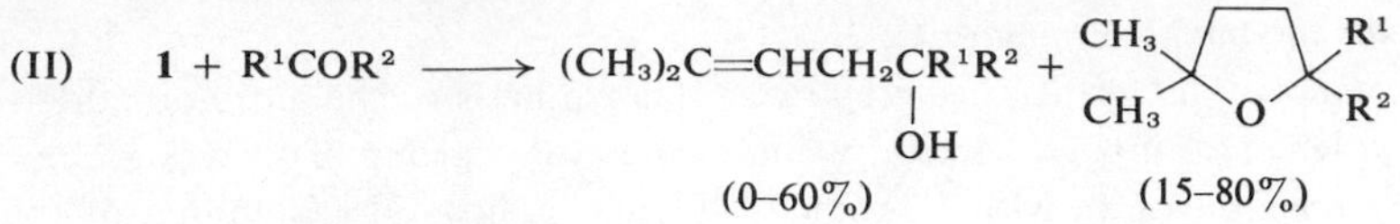

(II) $1 + R^1COR^2 \longrightarrow (CH_3)_2C{=}CHCH_2CR^1R^2 +$

$\underset{\text{OH}}{\phantom{(CH_3)_2C{=}CHCH_2CR^1R^2}}$

(0–60%)   (15–80%)

Dimethylaluminum amides, $(CH_3)_2AlNR^1R^2$. The reagents are prepared by addition of trimethylaluminum in hexane (Alfa) to a solution of ammonia or an amine in CH_2Cl_2 (N_2).

Amides. The reagents react with esters and lactones at 25–41° to give amides in yields of 75–100%.[1]

Examples:

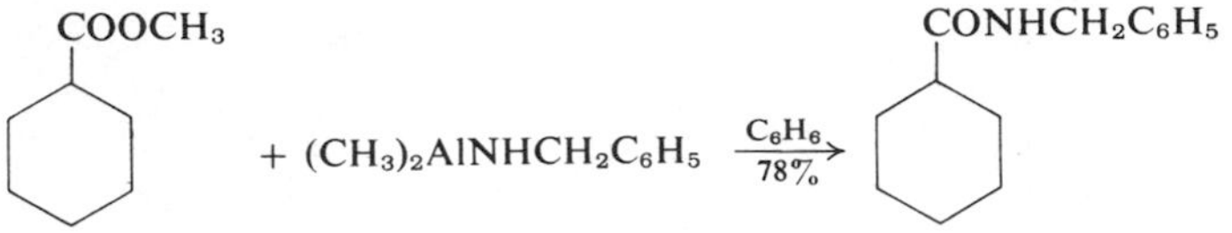

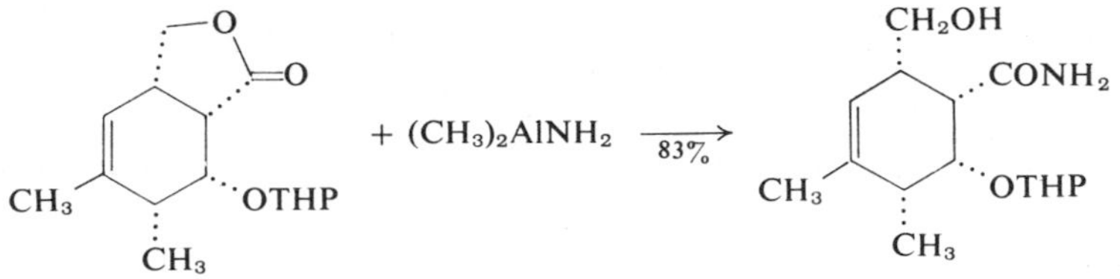

[1] A. Basha, M. F. Lipton, and S. M. Weinreb, *Tetrahedron Letters*, 4171 (1977); *idem*, *Org. Syn.* submitted (1978).

Dimethylaluminum methyl selenolate, $(CH_3)_2AlSeCH_3$ (**1**). Mol. wt. 151.04. The yellow reagent is prepared in toluene (reflux 2 hours) from trimethylaluminum and powdered selenium.

Methylselenol esters.[1] Methyl and ethyl carboxylic esters are converted into methylselenol esters by this reagent (0°–20°).

Examples:

$$CH_3(CH_2)_5COOCH_3 \xrightarrow[95\%]{1} CH_3(CH_2)_5COSeCH_3$$

2

$$C_6H_5COOCH_3 \xrightarrow[99\%]{1} C_6H_5COSeCH_3$$

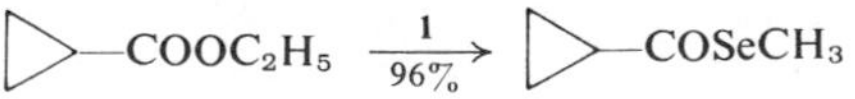

Selenol esters are more reactive than thiol esters and are useful as acyl-transfer reagents as shown by the conversion of **2** into an acid, an ester, and an amide.

$$2 \xrightarrow[\text{CH}_3\text{CN}]{\text{HgCl}_2,\ \text{CaCl}_2,} \begin{array}{l} \xrightarrow{\text{H}_2\text{O}} \text{CH}_3(\text{CH}_2)_5\text{COOH}\ (97\%) \\ \xrightarrow{\text{CH}_3\text{OH}} \text{CH}_3(\text{CH}_2)_5\text{COOCH}_3\ (88\%) \\ \xrightarrow{\text{H}_2\text{NC}_6\text{H}_{11}} \text{CH}_3(\text{CH}_2)_5\text{CONHC}_6\text{H}_{11}\ (88\%) \end{array}$$

Other reactions:

$$\xrightarrow[78\%]{1} \text{HO(CH}_2)_4\text{COSeCH}_3$$

$$\xrightarrow{1}$$

[1] A. P. Kozikowski and A. Ames, *J. Org.*, **43**, 2735 (1978).

4-Dimethylamino-3-butyne-2-one (1). Mol. wt. 111.14, oil, stable for prolonged periods only at $-50°$.

Preparation[1]:

$$(\text{CH}_3)_2\text{NCH}=\text{CHCOCH}_3 \xrightarrow[75\%]{\begin{array}{l}\text{1) Br}_2 \\ \text{2) KOC(CH}_3)_3\end{array}} (\text{CH}_3)_2\text{NC}\equiv\text{CCOCH}_3$$
$$\textbf{1}$$

Thiol(selenol) esters.[2] Activated enol esters (**2**) of carboxylic acids can be prepared as shown in equation (I). They can be converted into thiol and selenol esters (**3**).

(I) $\text{RCOOH} + \textbf{1} \longrightarrow$

$$\xrightarrow{90\text{–}95\%}$$

$$\textbf{a}$$

$$\xrightarrow[\substack{(\text{M = Li, Na, K};\\ \text{X = S, Se})}]{\text{M(XR}')}$$

$$(70\text{–}90\%)$$
$$\textbf{3}$$

$$\textbf{2}$$

[1] H.-J. Gais, K. Hafner, and M. Neuenschwander, *Helv.*, **52**, 2641 (1969).
[2] H.-J. Gais and T. Lied, *Angew. Chem., Int. Ed.*, **17**, 267 (1978).

(2-Dimethylaminoethyl)triphenylphosphonium   bromide, $(C_6H_5)_3P^+CH_2CH_2N$-$(CH_3)_2Br^-$ **(1)**. Mol. wt. 414.33, m.p. 196–199°. This Wittig reagent is prepared by treating a mixture of triphenylphosphine and 2-dimethylaminoethanol with 48% HBr or gaseous HBr (2 equiv.) to form the hydrobromides. These products are heated without solvent at 150–160°. When the melt is treated with aqueous K_2CO_3, **1** is obtained in 75–80% yield (equation I). The method is applicable to synthesis of related reagents.

(I) $(C_6H_5)_3P + HOCH_2CH_2N(CH_3)_2 \xrightarrow{\text{HBr}}$

$(C_6H_5)_3\overset{+}{P}HBr^- + HOCH_2CH_2\overset{+}{N}H(CH_3)_2Br^-$

$$75\text{–}80\% \left| \begin{array}{l} \text{1) 150–160°} \\ \text{2) } K_2CO_3,\ H_2O \end{array} \right. \downarrow$$

1

Aminoalkylation of carbonyl compounds.[1] The ylide **2** formed on treatment of **1** with *n*-butyllithium in THF, reacts with ketones to form aminoethylidene compounds (equation II).

(II)

2

In the reaction of **2** with aldehydes, both *cis*- and *trans*-products can be formed. However, in the case of the reaction with *p*-chlorobenzaldehyde, conditions favorable to formation of each isomer were found (equation III).

(III)

[1] A. Marxer and T. Leutert, *Helv.*, **61**, 1708 (1978).

(+)-(2S,3R)-4-Dimethylamino-3-methyl-1,2-diphenyl-2-butanol (Darvon alcohol), 5, 231.

Asymmetric reduction of ketones. Mosher observed in 1973 that ketones of the type ArCOR can be reduced by this alcohol (**1**) complexed with lithium aluminum hydride with high stereoselectivity (about 70% enantiomeric excess of one enantiomer in the case of $C_6H_5COCH_3$). He noted that the stereoselectivity decreased as the size of the R group increased and that stereoselectivity was slight with purely aliphatic ketones, RCOR'.[1]

Brinkmeyer and Kapoor[2] have now found that acetylenic ketones, RC≡CCOR', are reduced by $LiAlH_4$ and **1** at $-78°$ with the highest enantiomeric selectivity observed to date. Thus $CH_3C≡CCOCH_2CH(CH_3)_2$ is reduced to the corresponding (R)-alcohol with an enantiomeric excess of 82%. Similar asymmetric reductions were observed with seven other ketones of this type. Propargylic ketones are readily available by reaction of lithium acetylides with aldehydes followed by Jones oxidation of the propargylic alcohols.

Johnson et al.[3] have used this asymmetric reduction to effect an asymmetric total synthesis of 11α-hydroxyprogesterone (**6**) based on Johnson's method of polyene cyclization (**5**, 696). The key asymmetric step involved reduction of the ketone **2** to the alcohol **3** with the complex of **1** with lithium aluminum hydride. The yield of **3** was 93% with an enantiomeric ratio of (R) and (S) of 92:8. The

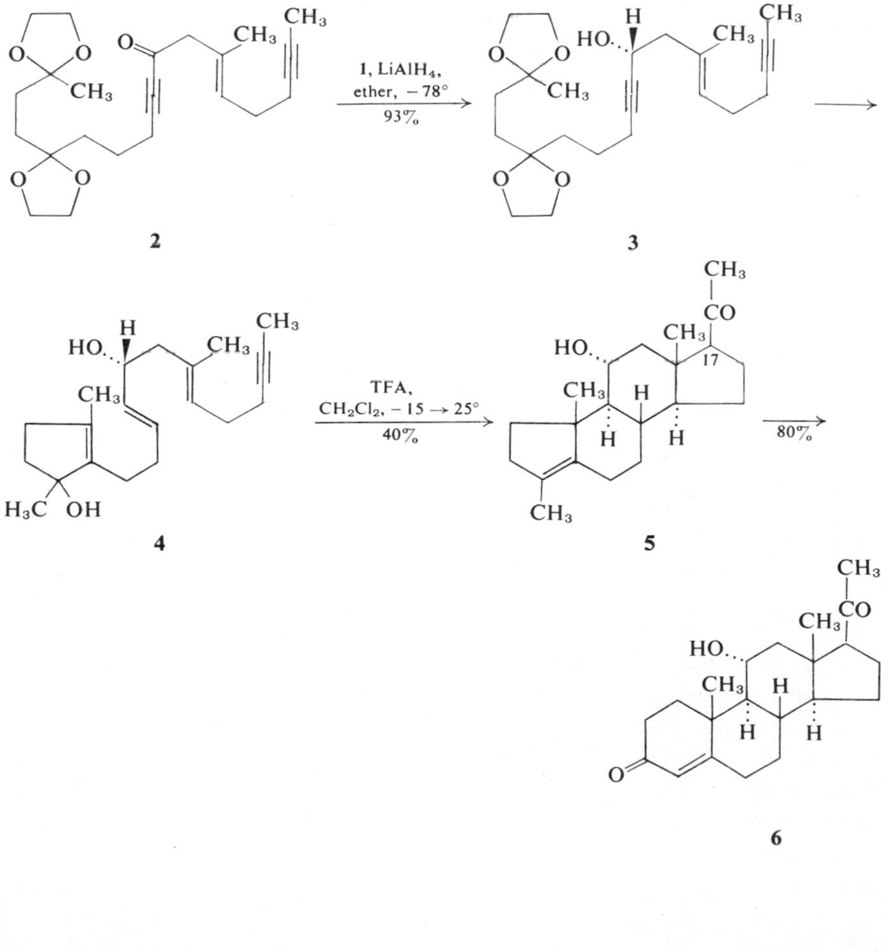

2   **3**

4   **5**   **6**

alcohol group of **3** is the precursor of the 11-hydroxy group of 11α-hydroxypro-gesterone. Remaining steps in the synthesis of **6** involved conversion of **3** into **4** as reported earlier for a synthesis of racemic **4**.[4] The optically active **4** was cyclized with trifluoroacetic acid as the catalyst to give **5**, which was converted by known methods into 11α-hydroxyprogesterone, $\alpha_D + 147°$. The ratio of **6** to enantio-**6**, was 92:8, the same as for **3** and enantio-**3**. Therefore the conversion of **3** into **6** proceeded with no perceptible racemization.

[1] J. D. Morrison and H. S. Mosher, *Asymmetric Organic Reactions*, American Chemical Society, 1976, pp. 204–205.
[2] R. S. Brinkmeyer and V. M. Kapoor, *Am. Soc.*, **99**, 8339 (1977).
[3] W. S. Johnson, R. S. Brinkmeyer, V. M. Kapoor, and T. M. Yarnell, *ibid.*, **99**, 8341 (1977).
[4] W. S. Johnson, S. Escher, and B. W. Metcalf, *ibid.*, **98**, 1039 (1976).

Dimethylchloromethylideneammonium chloride (Dimethylchloroforminium chloride), 1, 286–289; **3**, 116; **4**, 186; **5**, 251; **6**, 220.

Esterification. Stadler[1] recommends preparation of this reagent (**1**) in acetonitrile from DMF and oxalyl chloride because the by-products are gaseous CO and CO_2 (equation I). Addition of an acid to this suspension of **1** gives an

$$(I) \quad (CH_3)_2NCHO + ClOC{-}COCl \xrightarrow[-CO,\ -CO_2]{CH_3CN,\ 20°} (CH_3)_2\overset{+}{N}{=}C\diagdown\diagup\overset{Cl}{H}\ Cl^-$$

1

$$-HCl \Big| RCOOH$$

$$DMF + \begin{array}{c} \\ \overset{+}{N} \\ H\ Cl^- \end{array} + RCOOR' \xleftarrow[70-90\%]{R'OH,\ py.} \left[(CH_3)_2\overset{+}{N}{=}C\diagdown\diagup\overset{\overset{O}{\|}OCR}{H}\ Cl^- \right]$$

$$\qquad\qquad\qquad\qquad\qquad \textbf{2} \qquad\qquad\qquad\qquad\qquad \textbf{a}$$

activated form of the acid (**a**), which reacts rapidly with an alcohol or a phenol in the presence of pyridine to form an ester (**2**), usually in high yield. This method is suitable for esterification of amino acids N-protected with CBZ, CF_3CO, and toluenesulfonyl groups (55–90% yields).

[1] P. A. Stadler, *Helv.*, **61**, 1675 (1978).

Dimethylchloronium chloropentafluoroantimonate, $(CH_3)_2Cl^+SbF_5Cl^-$; **5**, 231–232. The salt is prepared from SbF_5 and an excess of CH_3Cl in SO_2.

Methylation of nitrogen.[1] N-Methylmethyleneaziridine (**1**) is not methylated by methyl fluorosulfonate in $CHCl_3$ at $-60°$ but rather is converted into a water-soluble polymer. Methylation is achieved by the dimethylchloronium salt in SO_2 at $-78°$. The product (**2**) shows remarkable thermal stability.

1   **2**

The aziridine (**1**) is also converted into an ion by $FSO_3H–SbF_5$ in SO_2.

[1] E. Jongejan, H. Steinberg, and Th. J. de Boer, *Rec. trav.*, **97**, 145 (1978).

Dimethyl diazomalonate, $N_2C(COOCH_3)_2$. Mol. wt. 158.11, b.p. 60–61°/2 mm. This reagent can be prepared in 50% yield by the reaction of dimethyl malonate with tosyl azide and diethylamine.[1]

   Reaction with enol ethers. Wenkert and co-workers[2] have examined the copper-catalyzed decomposition of this α-diazo compound in the presence of an enol ether of an aldehyde (**1**) and a ketone (**5**). In the first case, the expected cyclopropane ester (**2**) was obtained. This was reduced by lithium aluminum hydride to the diol, which cyclized to the hemiacetal (**3**) on exposure to acid. Collins oxidation of **3** gave the spiro-β-methylene-γ-lactone **4**.

1   **2**   **3**   **4**

5   **6**   **7**   **8**   **9**

Reaction of the diazo compound with **5** resulted in **6**. Lithium aluminum hydride reduction of the sodium enolate resulted in the acid-sensitive alcohol **7**. Treatment of **7** with acid led to the ketal **8** and then to the furane **9**. Reduction of the sodium enolate of **10** led to the α-diol **11**, which had been converted previously into the α-methylene-γ-lactone **12**.

[1] H. Lindemann, A. Wolter, and R. Groger, *Ber.*, **73**, 702 (1930); W. Ando, Y. Yagihara, S. Tozune, I. Imai, J. Suzuki, T. Toyama, S. Nakaido, and T. Migita, *J. Org.*, **37**, 1721 (1972).

[2] E. Wenkert, M. E. Alonso, B. L. Buckwalter, and K. J. Chou, *Am. Soc.*, **99**, 4778 (1977).

Dimethyl diazomethylphosphonate, 3, 113–114.

Diarylalkynes.[1] The anion (**1**) of this diazo compound, prepared with *n*-butyllithium (or potassium *t*-butoxide) in THF at low temperatures, converts diaryl ketones into diarylalkynes in high yield (equation I). Heteroaromatic ketones behave in the same way, but only very electrophilic aryl aldehydes

(I) $\underset{\mathbf{1}}{N_2\overset{\overset{Li\ \ \ O}{\diagdown\ \ \ \parallel}}{C}P(OCH_3)_2} + C_6H_5COC_6H_5 \xrightarrow[94\%]{} C_6H_5C{\equiv}CC_6H_5$

react. Dicarbonyl compounds give only low yields of alkynones owing to subsequent transformations (equation II). It is apparently impossible to prepare dialkylacetylenes in this way.

(II) $\mathbf{1} + C_6H_5COCOC_6H_5 \xrightarrow[25\%]{} C_6H_5C{\equiv}CCOC_6H_5$

Diazomethyltrimethylsilane, $(CH_3)_3SiCHN_2$, can be used in this synthesis with comparable results. A possible mechanism for this reaction involving loss of nitrogen with Wolff rearrangement and then loss of lithium trimethylsilanolate from the initial adduct is shown in equation (III).

[1] E. W. Colvin and B. J. Hamill, *J.C.S. Perkin I*, 869 (1977).

(III) $(C_6H_5)_2C{=}O + (CH_3)_3SiCN_2$ $\xrightarrow{}$ (with Li on the N)

$$\left[(C_6H_5)_2\overset{OLi}{\underset{N_2{}^+}{C{-}C}}{\overset{Si(CH_3)_3}{}} \right] \xrightarrow{-N_2} \left[\overset{LiO}{\underset{C_6H_5}{}}C{=}C\overset{Si(CH_3)_3}{\underset{C_6H_5}{}} \right] \xrightarrow{-(CH_3)_3SiOLi}$$

$$C_6H_5C{\equiv}CC_6H_5$$

Dimethyl disulfide, 5, 246–247; **6**, 217–218.

Reductive sulfenylation. Gassman et al.[1] have found that Stork's method for reductive alkylation of α,β-unsaturated ketones (**1**, 601–602) is applicable to reductive sulfenylation of cyclic α,β-unsaturated ketones and has the advantage of regiospecificity in the formation of α-methylthio ketones. Furthermore, the reaction proceeds with some preference for axial attack.

Examples:

(cis/trans = 55:45)

(cis/trans = 83:17)

Sulfenylation of amides.[2] The LDA and THF system appears to be the most useful one for α-monosulfenylation of N,N-disubstituted amides. Use of sodium amide in liquid ammonia results in polysulfenylation.

[1] P. G. Gassman, D. P. Gilbert, and S. M. Cole, *J. Org.*, **42**, 3233 (1977).
[2] P. G. Gassman and R. J. Balchunis, *ibid.*, **42**, 3236 (1977).

N,N-Dimethylformamide, 1, 278–281; **2**, 153–154; **3**, 115; **4**, 184; **5**, 247–249; **7**, 124.

Conjugated allenic aldehydes or ketones. Conjugated allenic aldehydes can be prepared in satisfactory yield by reaction of allenyllithium compounds with DMF (1.5 equiv.) in THF ($-70°$, 2 hours) (equation I). Conjugated allenic ketones are obtained when $RCON(CH_3)_2$ is used in the reaction (equation II).

(I)

$$\underset{CH_3}{\overset{CH_3}{>}}C=C=C\underset{Li}{\overset{H(C_4H_9\text{-}n)}{<}} + HCON(CH_3)_2 \xrightarrow[80-85\%]{} \underset{CH_3}{\overset{CH_3}{>}}C=C=C\underset{CHO}{\overset{H(C_4H_9\text{-}n)}{<}}$$

(II)

$$\underset{CH_3}{\overset{CH_3}{>}}C=C=C\underset{Li}{\overset{H}{<}} + (CH_3)_2CHCH_2CON(CH_3)_2 \xrightarrow{71\%}$$

$$\underset{CH_3}{\overset{CH_3}{>}}C=C=C\underset{\overset{|}{\underset{O}{CCH_2CH(CH_3)_2}}}{\overset{H}{<}}$$

Ipsenol (**3**), the pheromone of bark beetles, has been synthesized by conjugate addition of lithium divinyl cuprate to the allenic ketone (**1**).[1]

$$(CH_3)_2CHCH_2\overset{\overset{O}{\|}}{C}-CH=C=CH_2 + (CH_2=CH)_2CuLi \xrightarrow[95\%]{Ether, -60°}$$
$$\quad\quad\quad\quad\quad \mathbf{1}$$

$$(CH_3)_2CHCH_2\overset{\overset{O}{\|}}{C}CH_2\overset{\overset{CH=CH_2}{|}}{C}=CH_2 \xrightarrow[95\%]{\substack{NaBH_4, \\ CH_3OH-H_2O}} (CH_3)_2CHCH_2\overset{\overset{OH}{|}}{C}HCH_2\overset{\overset{CH=CH_2}{|}}{C}=CH_2$$
$$\quad\quad\quad\quad \mathbf{2} \quad\quad\quad\quad\quad\quad\quad\quad\quad\quad\quad \mathbf{3}$$

Isoxazoles. Isoxazoles can be prepared by reaction of the dianions of *syn*-oximes with amides. The synthesis of 3,4-tetramethyleneisoxazole (**2**) has been

formulated.[2] This synthesis is an extension of one by Hauser *et al.*[3] The method is suitable for various modifications.

[1] J. C. Clinet and G. Linstrumelle, *Nouv. J. Chim.*, **1**, 373 (1977).
[2] G. N. Barber and R. A. Olofson, *J. Org.*, **43**, 3015 (1978).
[3] C. F. Beam, M. C. D. Dyer, R. A. Schwarz, and C. R. Hauser, *ibid.*, **35**, 1806 (1970).

N,N-Dimethylformamide dimethyl acetal, **1**, 281–282; **2**, 154; **3**, 115–116; **4**, 184–185; **6**, 221–222.

α-*Alkylation of amino acids.*[1] One method to accomplish this reaction is conversion of the amino acid **1** into the amidino ester **2** by reaction with dimethylformamide dimethyl acetal. This derivative can be deprotonated at the α-position. The anion is readily alkylated (**3**), but reacts rather sluggishly with aldehydes or ketones. Alkylated amino acids (**4**) are obtained by hydrolysis with conc. HCl.

Rearrangement of allylic alcohols to β,γ-*unsaturated amides*[2] (**1**, 271–272; **4**, 167; **5**, 253). This reaction, which involves a [2,3]sigmatropic rearrangement, occurs with complete transfer of chirality. Thus the reaction of the (R, Z)-allylic alcohol **1** with N,N-dimethylformamide dimethyl acetal gives as the only product the optically active (E)-β,γ-unsaturated amide **2**, shown to have the (R)-configuration. However, the (S, E)-isomer of **1** also rearranges mainly to **2**, with only a trace of the (S, Z)-isomer. These results suggest that both rearrangements proceed through a five-membered cyclic transition state such as **a** in the case of **1** and by a doubly suprafacial [2,3] sigmatropic process to give **2**.

Similar results were observed with other allylic alcohols. This rearrangement was used to obtain the optically active C_{14}-carbon unit **3**, which is an important intermediate in vitamin E synthesis, originally prepared from natural phytol.

3 ($\alpha_D + 9.4$)

The same laboratory has also shown that the Claisen rearrangement also proceeds with almost complete chiral transmission.[3]

Alkenes from **vic,cis-diols.** This conversion can be effected in a two-step, one-pot process. The diol is converted into the 1-dimethylamino(methylene) acetal by reaction with DMF dimethyl acetal. The product is converted into a trimethylalkylammonium salt, which is converted into an alkene when heated.[4] Examples:

The second example illustrates the conversion of diethyl D-tartrate into diethyl fumarate.

[1] J. J. Fitt and H. W. Gschwend, *J. Org.*, **42**, 2639 (1977).
[2] K.-K. Chan and G. Saucy, *ibid.*, **42**, 3828 (1977).
[3] K.-K. Chan, N. Cohen, J. P. DeNoble, A. C. Specian, Jr., and G. Saucy, *ibid.*, **41**, 3497 (1976).
[4] S. Hanessian, A. Bargiotti, and M. LaRue, *Tetrahedron Letters*, 737 (1978).

N,N-Dimethylhydrazine, 1, 289–290; **2**, 154–155; **3**, 117; **5**, 254; **6**, 223; **7**, 126–130.

Monoalkylation of α,β-*unsaturated ketones* (*see also* **4**, 300). Alkylation of enamines of enones usually gives low yields of C-alkylated products because of competing N-alkylation. An expedient is to convert the N,N-dimethylhydrazone of the enone into the metalloenamine by a strong base (sodium hydride or LDA usually). This derivative undergoes monoalkylation in reasonable yields. The final step is regeneration of the enone system.[1]

Example:

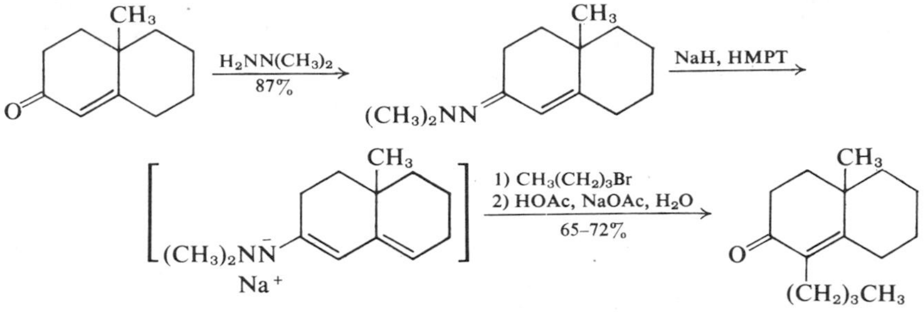

α,β-*Unsaturated aldehydes*. The reaction of α-bromo aldehydes containing a β-hydrogen with dimethylhydrazine in the presence of K_2CO_3 gives α,β-unsaturated hydrazones, convertible by acid hydrolysis into the corresponding aldehydes.[2]

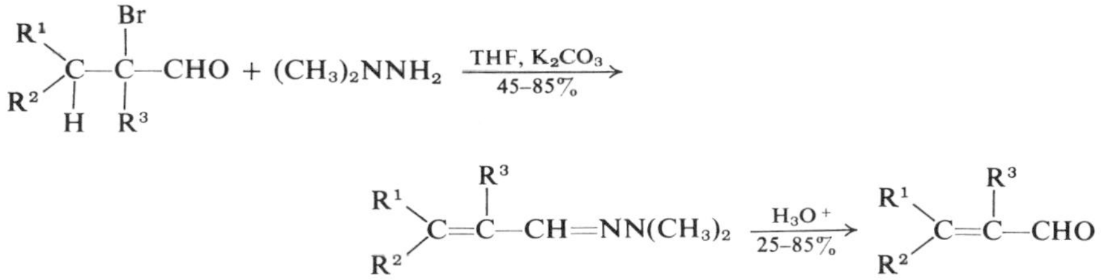

[1] G. Stork and J. Benaim, *Org. Syn.*, **57**, 69 (1977).
[2] L. Duhamel and J.-Y. Valnot, *Compt. rend.* (*C*), **286**, 47 (1978).

Dimethyl malonate, $CH_2(COOCH_3)_2$. Mol. wt. 132.12, b.p. 180–181°.

α-*Methylene*-γ-*lactones*.[1] A new route to these substances involves the reaction of dimethyl sodiomalonate with an epoxide to form *trans*-α-alkoxy-carbonyl-γ-lactones, which are then converted by the method of Miller and Behare (**5**, 314)[2] into α-methylene-γ-lactones. An example is formulated in equation (I).

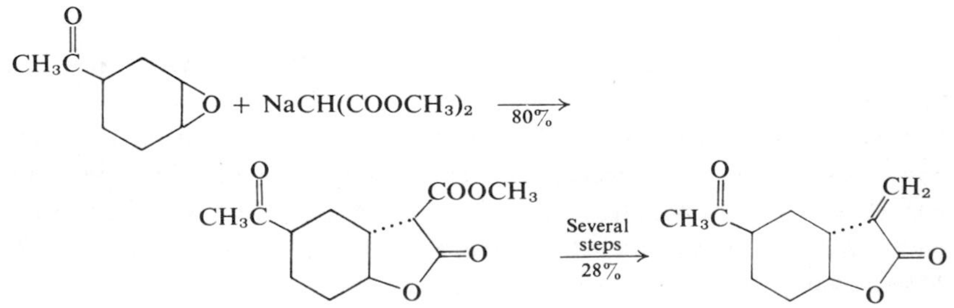

[1] H. Marschall, F. Vogel, and P. Weyerstahl, *Ann.*, 1557 (1977).
[2] R. B. Miller and E. S. Behare, *Am. Soc.*, **96**, 8102 (1974).

Dimethyl(methylene)ammonium salts, 3, 114–115; **4,** 186–187; **7,** 130–132.

1-Alkenes. Reaction of a Grignard reagent or an alkyllithium in ether (refluxing) with dimethyl(methylene)ammonium iodide results in formation of an alkyldimethylamine, which can be isolated as the methyl iodide salt. The amine is converted into a 1-alkene by oxidation to the oxide followed by pyrolysis at 160°.[1]
Examples:

$$CH_3(CH_2)_7MgBr + CH_2{=}N^+(CH_3)_2I^- \xrightarrow[91\%]{}$$

$$CH_3(CH_2)_7CH_2N(CH_3)_2 \xrightarrow[87\%]{H_2O_2,\ 160°} CH_3(CH_2)_6CH{=}CH_2$$

$$C_6H_5CH_2MgBr \xrightarrow[85\%]{} C_6H_5CH_2CH_2N(CH_3)_2 \xrightarrow[83\%]{} C_6H_5CH{=}CH_2$$

Mannich reaction. Detailed directions are available for the preparation of the trifluoroacetate salt (**1**) by the reaction of bis(dimethylamino)methane with trifluoroacetic acid. The reagent (**1**) was used to effect the regioselective Mannich reaction of 3-methyl-2-butanone (**2**) to give the less substituted possible product **3**.[2]

$$(CH_3)_2CHCOCH_3 + (CH_3)_2\overset{+}{N}{=}CH_2CF_3CO_2^- \xrightarrow[49-58\%]{-10 \to 145°}$$

  2   **1**

$$(CH_3)_2CHCOCH_2CH_2N(CH_3)_2$$

  3

[1] J. L. Roberts, P. S. Borromeo, and C. D. Poulter, *Tetrahedron Letters*, 1299 (1977).
[2] M. Gaudry, Y. Jasor, T. Bui Khac, and A. Marquet, *Org. Syn.*, submitted (1978).

Dimethyloxosulfonium methylide, 1, 315–318; **2,** 171–73; **3,** 125–127; **4,** 197–199; **5,** 254–257; **7,** 133.

Tropolones. A new tropolone synthesis involves in the first step reaction of dimethyloxosulfonium methylide with the quinone monoketal **1** to form the cyclopropyl ketone **2**. Reaction of **2** with isopropylmagnesium bromide followed by dehydration and deketalization gives **3**. On treatment with base, the enolate undergoes ring opening to form, after silylation, the dihydrotropolone derivative **4**. This product is converted into β-dolabrin (**5**) by oxidation followed by demethylation.

A similar approach resulted in the synthesis of **8**, an intermediate in two previous syntheses of colchicine (**9**). For this purpose the cyclopropyl ketone **2** was allowed to react with the Grignard reagent **6** to afford **7**. When **7** was treated with trifluoroacetic acid the dihydrotropolone **8** was formed in high yield by way of two isolable intermediates.[1]

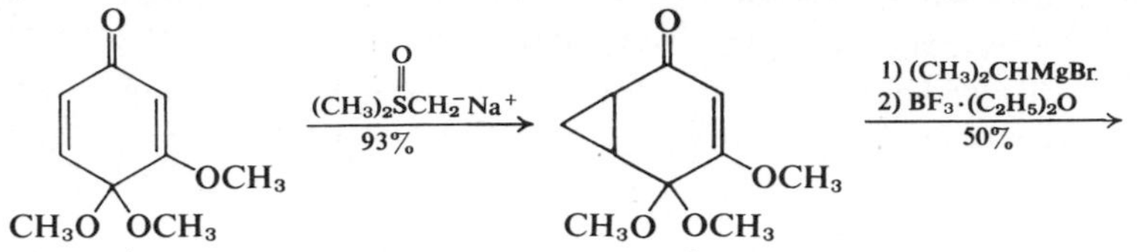

The reaction scheme showing compounds 1 through 9.

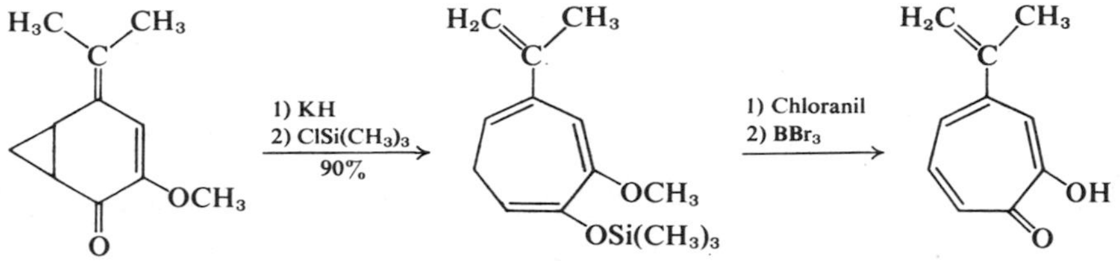

Compound 1 → 2:
$(CH_3)_2\overset{O}{S}CH_2^- Na^+$, 93%

Compound 2 →:
1) $(CH_3)_2CHMgBr$
2) $BF_3 \cdot (C_2H_5)_2O$, 50%

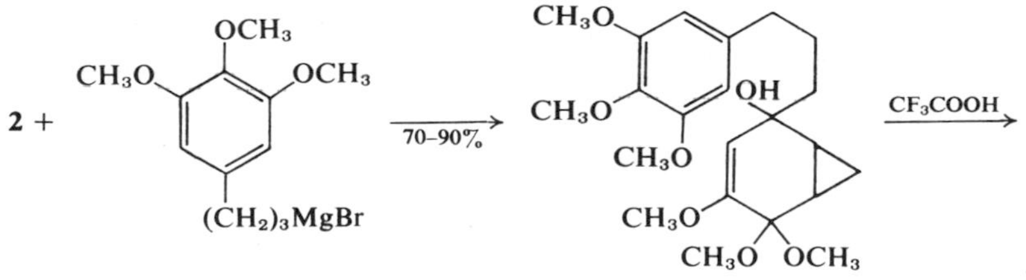

Compound 3 → 4:
1) KH
2) $ClSi(CH_3)_3$, 90%

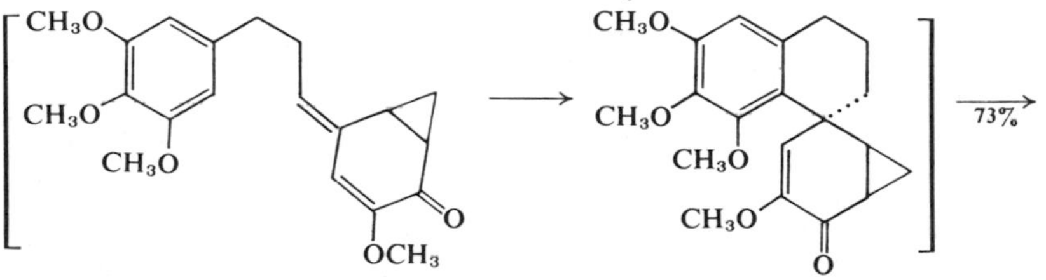

Compound 4 → 5:
1) Chloranil
2) BBr_3

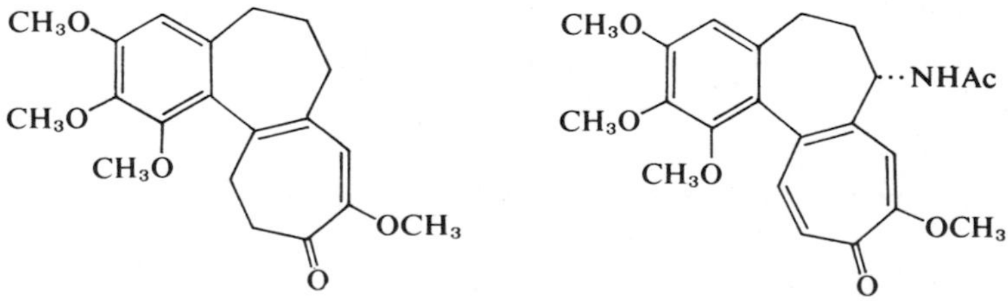

2 + 6 → 7:
70–90%

7 →:
CF_3COOH

a → b → 8, 9:
73%

Cycloalkenylcyclopropanes. Substances of this type have been prepared by reaction of β-chlorocycloalkenones with dimethyloxosulfonium methylide to form an intermediate that reacts with an electrophilic olefin to give a cyclopropane derivative. Yields range from 20 to 80%; the ratio of *cis*- to *trans*-isomers is about 25:75.[2]

Example:

(*cis*- and *trans*-isomers, 35%)

[1] D. A. Evans, D. J. Hart, and P. M. Koelsch, *Am. Soc.*, **100**, 4593 (1978).
[2] J.-C. Chalchat, R. Garry, A. Michet, and R. Vessière, *Compt. rend.* (*C*), **286**, 329 (1978).

$$CH_3$$
Dimethylphenylsilyllithium, C_6H_5SiLi (**1**), 7, 133.
$$CH_3$$

Protection of enones. This silyllithium reagent (2 equiv.) in the presence of CuI (1 equiv.) adds to the β-position of α,β-unsaturated ketones at −23° in 65–99% yield.[1] The original α,β-unsaturated keto group can be regenerated by bromination followed by desilylation with base (70–80% yield).[2] In the case of cyclic ketones, CuBr$_2$ regenerates enones directly. With open-chain ketones, β-bromo ketones are obtained, but treatment with NaHCO$_3$ regenerates the enone.

Examples:

Still[3] has used trimethylsilyllithium for this conjugate addition; but this reagent is less reactive.

[1] D. J. Ager and I. Fleming, *J.C.S. Chem. Comm.*, 177 (1978).
[2] I. Fleming and J. Goldhill, *ibid.*, 176 (1978).
[3] W. C. Still, *J. Org.*, **41**, 3063 (1976).

(+)-**trans-2,5-Dimethylpyrrolidine,**   (1). Mol. wt. 99.17,

α_D + 10.6°. The reagent is prepared by reduction of the N-aminopyrrolidine[1] followed by resolution with (−)-mandelic acid.[2]

Asymmetric alkylation of cyclohexanone. Yamada *et al.*[3] have reported asymmetric induction in the alkylation of enamines derived from proline esters, but typical optical yields are somewhat low (10–30%). Optical yields of 80–93% have now been observed in alkylations of the enamine of cyclohexanone derived from **1**.[2] The higher yields obtained from enamines derived from **1** are attributed to the C_2-axis of symmetry in **1**.

[1] P. B. Dervan and T. Uyehara, *Am. Soc.*, **98**, 2003 (1976).
[2] J. K. Whitesell and S. W. Felman, *J. Org.*, **42**, 1663 (1977).
[3] M. Kitomoto, K. Hiroi, S. Terashima, and S. Yamada, *Chem. Pharm. Bull. Japan*, **22**, 459 (1974).

Dimethyl selenoxide, CH_3SeCH_3. Mol. wt. 125.03, m.p. 94°. The substance is prepared by oxidation of $(CH_3)_2SeBr_2$ with Ag_2O in CH_3OH.[1]

Oxidations.[2] Dimethyl selenoxide is more effective than dimethyl sulfoxide (**4**, 194; **6**, 227) for oxidation of trivalent phosphorus and thiocarbonyl compounds. It oxidizes acyclic P(III) compounds to oxides (with nearly complete inversion of configuration); cyclic P(III) compounds are oxidized with retention of configuration. The same stereochemical result obtains in oxidations with DMSO.

Thio- and selenophosphoryl compounds are oxidized readily to phosphine oxides in yields of over 90%.

Thiocarbonyl groups are oxidized to carbonyl groups at room temperature. Simple thioureas give ureas in 80–85% yield. This reaction is useful for oxidation of thiouracils to uracils in 80–90% yield. This oxidation permits selective modification of the 4-thiouracil groups in transfer RNA.

[1] R. Paetzold, V. Linder, G. Boechmann, and P. Reich, *Z. Anorg. Allg. Chem.*, **352**, 295 (1967).
[2] M. Mikołajczyk and J. Suczak, *J. Org.*, **43**, 2132 (1978).

Dimethylsulfonium methylide, 1, 314–315; **2**, 169–171; **3**, 124–125; **4**, 196–197.

Oxiranes (**1**, 315–316). The original method of Corey has been modified to provide a one-pot process.[1] Dimethylsulfonium methylide is prepared *in situ* by reaction of dimethyl sulfide with dimethyl sulfate in CH_3CN followed by addition of base (sodium methoxide). Addition of the carbonyl compound leads to the oxirane and dimethyl sulfide, which can be recycled (equation I). Yields are generally higher than those obtained by the original method.

$$(CH_3)_2S + (CH_3)_2SO_4 \longrightarrow [(CH_3)_3S^+ \cdot CH_3SO_4{}^-] \xrightarrow[\substack{2)\ R_1R_2C=O}]{1)\ CH_3ONa} \begin{array}{c} R^1 \\ R^2 \end{array}\!\!\triangle\!\!{}^O + (CH_3)_2S$$

$$(60\text{–}90\%)$$

Addition to 9,10-anthraquinone. The reagent **1**, prepared in DMSO from trimethylsulfonium iodide and NaH, reacts with anthraquinone (**2**) (1 equiv.) to form the monoepoxide **3** in 95% yield (crude). The reaction with 2 equiv. of **1** gives the diepoxide **4** in 94% yield (crude). The crude **4** travels as a single spot in TLC; it is reduced by L-Selectride (**4**, 312–313; **6**, 348) to **5**.[2]

$$\begin{array}{ccc} \textbf{2} & \textbf{1} & \textbf{3} \end{array}$$

$$\begin{array}{cc} \textbf{4} & \textbf{5} \end{array}$$

[1] T. Kutsuma, I. Nagayama, T. Okazaki, T. Sakamoto, and S. Akaboshi, *Heterocycles*, **8**, 397 (1977).
[2] T. J. McCarthy, W. F. Connor, and S. M. Rosenfeld, *Syn. Comm.*, **8**, 379 (1978).

Dimethyl sulfoxide, 1, 296–310; **2**, 157–158; **3**, 119–123; **4**, 192–194; **5**, 263–266; **6**, 225–229; **7**, 133–135.

Aldehyde synthesis. α-Bromo selenides, prepared in quantitative yield from vinyl selenides,[1] are converted into aldehydes in about 85% yield when treated with anhydrous DMSO for a brief time. If the reaction is allowed to stand α-seleno aldehydes are obtained by reaction of the aldehyde with the benzeneselenenyl bromide.[2]

$$
\begin{array}{c}
R^1 \\
\diagdown \\
\diagup \quad C=C \diagdown \\
R^2 \qquad\qquad SeC_6H_5
\end{array}
\xrightarrow{HBr}
R^2-\overset{\overset{R^1}{|}}{\underset{\underset{H}{|}}{C}}-\overset{\overset{H}{|}}{\underset{\underset{SeC_6H_5}{|}}{C}}-Br
\xrightarrow{DMSO}
$$

$$
\left[R^2-\overset{\overset{R^1}{|}}{\underset{\underset{H}{|}}{C}}-\overset{\overset{H}{|}}{\underset{\underset{SeC_6H_5}{|}}{C}}-O-\overset{+}{S}(CH_3)_2 \quad Br^- \right]
\xrightarrow{10\ min.}
R^2-\overset{\overset{R^1}{|}}{\underset{\underset{H}{|}}{C}}-CH{=}O + C_6H_5SeBr + (CH_3)_2S
$$

Pfitzner-Moffatt oxidation (**1**, 304–307; **2**, 162; **3**, 121). The oxidation of the alcohol **1** under the conditions of Moffatt, using DMSO, DCC, TFA, and pyridine, was accompanied by dehydration to the conjugated diene. Substitution of TFA and pyridine by 85% phosphoric acid gave **2** in 80% yield. This product was converted by conventional methods into *trans*-damascone (**5**).[3]

¹ W. Dumont, M. Sevrin, and A. Krief, *Angew. Chem., Int. Ed.*, **16**, 541 (1977).
² *Idem, Tetrahedron Letters*, 183 (1978).
³ H.-J. Liu, H.-K. Hung, G. L. Mhehe, and M. L. D. Weinberg, *Canad. J. Chem.*, **56**, 1368 (1978).

Dimethyl sulfoxide–Iodine, 5, 347; **6**, 295.

Diaryl α-diketones. Certain activated methylene compounds are oxidized by DMSO in the presence of catalytic amounts of iodine and sulfuric acid. The method is suitable for preparation of diaryl di- and triketones.[1]

Examples:

$$C_6H_5COCH_2C_6H_5 \xrightarrow{\text{DMSO, I}_2,\text{ H}^+} C_6H_5COCOC_6H_5 + \underset{\underset{C_6H_5}{|}}{C_6H_5COCHC_6H_5}$$

$$\text{(40\%)} \qquad \text{(1\%)}$$

(with SCH₃ group: $C_6H_5COCH(SCH_3)C_6H_5$)

$$C_6H_5COCH_2COC_6H_5 \xrightarrow[89\%]{\text{[O], H}_2\text{O}} \underset{\underset{OH}{|}}{\overset{\overset{OH}{|}}{C_6H_5COCCOC_6H_5}}$$

[1] N. Furukawa, T. Akasaka, T. Aida, and S. Oae, *Perkin I*, 372 (1977).

Dimethyl sulfoxide–Oxalyl chloride, DMSO–(COCl)₂.

Oxidations of alcohols. Omura and Swern[1] have reported results of an extensive study of oxidation of alcohols of various types by DMSO in combination with an "activator." Of the previously known methods based on DMSO or DMS, the DMSO–pyridine·sulfur trioxide reagent of Parikh and Doering (**2**, 394) and the DMS–NCS reagent of Corey and Kim (**4**, 87–90) are clearly the best, particularly for oxidation of less hindered alcohols. (A hindered alcohol such as isoborneol is oxidized in high yield by a variety of reagents.) Albright (**5**, 266) had briefly mentioned use of DMSO and several acid halides and anhydrides. Omura and Swern studied such activators in more detail and found that thionyl chloride (SOCl₂) and sulfuryl chloride (SO₂Cl₂) are both fairly efficient; but oxalyl chloride, (COCl)₂, is a more satisfactory adjunct to DMSO. Alcohols in general, primary, secondary, allylic, benzylic, bicyclic, and hindered, are oxidized to carbonyl compounds in yields usually in the range 80–100%. Oxalyl chloride is preferred to DMSO–TFAA (**7**, 126)[2] because it is less costly and less toxic and because it is superior for oxidation of primary alcohols.

[1] K. Omura and D. Swern, *Tetrahedron*, **34**, 1651 (1978); A. J. Mancuso, S.-L. Huang, and D. Swern, *J. Org.*, **43**, 2480 (1978).
[2] S. L. Huang, K. Omura, and D. Swern, *Synthesis*, 297 (1978).

N,N-Dimethylthiocarbamoyl chloride, 2, 173–174; **3**, 127–128; **4**, 202.

α,β-Unsaturated aldehydes by bishomologation of carbonyl compounds. This reaction can be carried out by conversion of a carbonyl compound into an allylic thionocarbamate, which undergoes [3.3] sigmatropic rearrangement to an allylic thiocarbamate. This product can be sulfenated at the α-position; the final step involves hydrolysis to the α,β-unsaturated aldehyde. The reaction is illustrated for conversion of hexanal into (E)-2-octenal. In other cases (E)- and (Z)- mixtures have been obtained.

$$CH_3(CH_2)_4CHO + CH_2{=}CHMgBr \longrightarrow$$

The same sequence can be carried out by addition of an alkyllithium to an α,β-unsaturated aldehyde to form an allylic alkoxide, which can then be converted to an α,β-unsaturated ketone in the same way. The process is illustrated for conversion of acrolein to (E)-2-heptenal.[1]

$$CH_2{=}CHCHO + n\text{-}C_4H_9Li \longrightarrow$$

[1] T. Nakai, T. Mimura, and A. Ari-izumi, *Tetrahedron Letters*, 2425 (1977).

N,N-Dimethyl-α-trimethylsilylacetamide, $(CH_3)_3SiCH_2CON(CH_3)_2$. Mol. wt. 159.31, b.p. 68–72°/3 mm. The reagent is prepared from N,N-dimethylacetamide by deprotonation with LDA (THF, $-78°$) followed by reaction with chlorotrimethylsilane and triethylamine (-78 to 20°); yield 78%.[1]

p-*Quinone methide ketals.* p-Quinone methides are not readily available; the most direct route, Wittig olefination of p-quinones, is not generally satisfactory. On the other hand, p-quinone ketals are available by several methods, such as the reaction of p-alkoxy phenols with thallium(III) nitrate (7, 363). Evans and co-workers[2] have converted p-quinone ketals into p-quinone methide ketals by reaction with the anion (1) of N,N-dimethyl-α-trimethylsilylacetamide[3] in THF at -70 to 20°. The products (3) are stable for several months in a refrigerator

Scheme (I)

(argon), but decompose on attempted chromatography. Three useful reactions of these new substances are shown in scheme (I).

The reaction of **2** with **1** served as a model for synthesis of the *Amaryllidaceae* alkaloid cherylline (**11**).[2] The key step was the reaction of **2** with the α-silyl acetamide **7** (equation I).

A shorter route to **11** involves reaction of the phosphorane **12** with **2**, as shown in equation (II). The phosphorane route to quinone methide ketals is widely applicable.

[1] P. F. Hudrlik, D. Peterson, and D. Chow, *Syn. Comm.*, **5**, 359 (1975).
[2] D. J. Hart, P. A. Cain, and D. A. Evans, *Am. Soc.*, **100**, 1548 (1978).
[3] See also R. P. Woodbury and M. W. Rathke, *J. Org.*, **43**, 1947 (1978).

(II)

12

13

3,5-Dinitrobenzoyl *t*-butyl nitroxyl, $3,5-(NO_2)_2C_6H_3CON[C(CH_3)_3]O$. Mol. wt. 282.24, m.p. 100° dec. This reagent is prepared by oxidation of the corresponding hydroxamic acid, $3,5-(NO_2)_2C_6H_3CON[C(CH_3)_3]OH$, with alkaline $K_3Fe(CN)_6$.

Oxidation. This nitroxyl (**1**) is similar to Fremy's salt [**1**, 940; $(KSO_3)_2NO\cdot$] in that it oxidizes phenols and hydroquinones to quinones, usually in satisfactory yields. Secondary alcohols are oxidized to ketones in high yield; but the aldehydes formed from primary alcohols are subject to further oxidation.[1]

Examples:

$$C_6H_5CHOHCHOHC_6H_5 \xrightarrow[89\%]{1} C_6H_5COCOC_6H_5$$

$$C_6H_5CH_2OH \xrightarrow[65\%]{1} C_6H_5CHO$$

$$C_6H_5CH=CHCH_2OH \xrightarrow[83\%]{1} C_6H_5CH=CHCHO$$

$$C_6H_5CHOHCH_3 \xrightarrow[90\%]{1} C_6H_5COCH_3$$

[1] S. A. Hussain, T. C. Jenkins, and M. J. Perkins, *Tetrahedron Letters*, 3199 (1977).

Dinitrogen tetroxide, 1, 324–329; **2**, 175–176; **3**, 130; **4**, 202–203.

Thionitrites. Aryl and alkyl thionitrites, RSNO, can be prepared by reaction of thiols with N_2O_4 in CCl_4 or ether at about $-10°$ (quant. yields). Thionitrites decompose at about 0°. The reaction with thiols, sulfinic acids, and amines is formulated.[1]

[1] S. Oae, D. Fukushima, and Y. H. Kim, *J.C.S. Chem. Comm.*, 407 (1977).

$$R^1SH$$

$$\downarrow N_2O_4, -10°$$

$$R^1SSO_2R^3 \xleftarrow{R^3SO_2H} [R^1SNO] \xrightarrow{R^2SH} R^1SSR^2$$

$$\downarrow HN{<}{\overset{R^4}{\underset{R^5}{}}}$$

$${\overset{R^4}{\underset{R^5}{}}}{>}N{-}N{=}O + R^1SSR^1$$

Dinitrogen tetroxide–Iodine, N_2O_4–I_2.

Reaction with 1,5-, 1,6-, and 1,7-dienes.[1] The reaction of N_2O_4–I_2 with these dienes results in formation of the monoadducts **2** as the major products. The

$$CH_2{=}CH(CH_2)_nCH{=}CH_2 \xrightarrow[80-95\%]{\substack{N_2O_4, I_2, \\ ether, 0°}} O_2NCH_2{-}\overset{I}{C}H(CH_2)_nCH{=}CH_2$$

$$\textbf{1} (n = 2, 3, 4) \qquad\qquad\qquad\qquad \textbf{2}$$

products are useful for synthesis of ω-nitroalkenes (**3**), ω-unsaturated aldehydes (**4**), and ω-unsaturated nitrile oxides (**5**).

$$CH_2{=}CHCH_2\overset{\overset{CH_3}{|}}{\underset{\underset{CH_3}{|}}{C}}CH_2CH_2CHO$$

4

$$\uparrow 77\% \;\; CH_3ONa, CH_3OH; H_3O^+$$

$$O_2NCH_2{-}\overset{I}{C}HCH_2\overset{\overset{CH_3}{|}}{\underset{\underset{CH_3}{|}}{C}}CH_2CH{=}CH_2 \xrightarrow[72\%]{\substack{NaBH_4, \\ CH_3OH}} CH_2{=}CHCH_2\overset{\overset{CH_3}{|}}{\underset{\underset{CH_3}{|}}{C}}CH_2CH_2CH_2NO_2$$

2a $C_6H_5NCO, N(C_2H_5)_3$ **3**

$$\left[CH_2{=}CHCH_2\overset{\overset{CH_3}{|}}{\underset{\underset{CH_3}{|}}{C}}CH_2CH_2C{\equiv}N \to O \right] \xrightarrow{91\%} \text{(structure 6)}$$

5   **6**

[1] V. Jäger and H. J. Günther, *Angew. Chem., Int. Ed.*, **16**, 246 (1977).

Dioxobis(*t*-butylimido)osmium(VIII);   Oxotris(*t*-butylimido)osmium(VIII).   *Cf.* Trioxo(*t*-butylimido)osmium (VIII), **6**, 641–642.

These new osmium complexes have been prepared as shown in equations (I) and (II).

(I) $OsO_4 + 2(C_6H_5)_3P{=}NC(CH_3)_3 \xrightarrow[63\%]{CH_2Cl_2}$

1 (yellow, m.p. 119–120°)

(Ia)

$+ (C_6H_5)_3P{=}NC(CH_3)_3 \xrightarrow[82\%]{CH_2Cl_2} \mathbf{1}$

(II) $OsO_4 + 3(n{\text-}C_4H_9)_3P{=}NC(CH_3)_3 \xrightarrow[45\%]{}$

2 (orange-red, m.p. 114–116°)

Both **1** and **2** react with monosubstituted and *trans*-disubstituted alkenes to form complexes that give *vic*-diamines on reductive cleavage (equation III). The reaction involves stereospecific *cis*-addition of the reagent.[1]

(III)

[1] A. O. Chong, K. Oshima, and K. B. Sharpless, *Am. Soc.*, **99**, 3420 (1977).

Dioxygentetrakis(triphenylarsine)rhodium(I) hexafluorophosphate or perchlorate, $RhO_2[As(C_6H_5)_3]_4{}^+A^-(A^- = PF_6{}^-$ or $ClO_4{}^-)$. This rhodium dioxygen complex is prepared by reaction of chloro(4,5-cyclooctadiene)rhodium(I) dimer (Strem) with triphenylarsine in the presence of oxygen (air) followed by addition of NH_4PF_6 or $LiClO_4$.[1]

RCH=CH₂ → RCOCH₃. Methyl ketones are obtained in about 85% yield by reaction of terminal alkenes with this complex under heterogeneous conditions (no solvent). The reaction is selective; for example, only 1-octene-7-one is obtained from 1,7-octadiene. The oxygen of the product is derived exclusively from **1**.[2]

[1] L. M. Haines, *Inorg. Chem.*, **10**, 1685 (1971).
[2] F. Igersheim and H. Mimoun, *J.C.S. Chem. Comm.*, 559 (1978).

Diperoxo-oxohexamethylphosphoramidomolybdenum(VI), 4, 203–204, **5,** 269–270, **7,** 136.

ArBr → ArOH. This conversion can be carried out by reaction of arylmagnesium bromides with $MoO_5 \cdot Py \cdot HMPT$ in THF at -78 to $10°$; yield 67–89% (four examples).[1]

Oxidation of dihydropyrane. Oxidation of dihydropyrane (1) with $MoO_5 \cdot$ HMPT in CH_2Cl_2 yields the cleavage product 2 in high yield, but reaction in methanol gives *cis*- and *trans*-2-methoxy-3-hydroxytetrahydropyrane, (3 and 4). These two products evidently are formed from the expected oxide, which is probably also the precursor of 2.[2]

Enolate hydroxylation (5, 269–270). Complete details for hydroxylation of enolates of ketones, esters, and lactones with $MoO_5 \cdot Py \cdot HMPT$ have been published. The peroxide is light sensitive, but can be stored for months in a refrigerator when shielded from light. It does not seem to be shock sensitive. A major advantage of this method of hydroxylation is the formation of the acyloin corresponding to the kinetic enolate. Thus 1 is converted only into 2 and 3 in the ratio 7:1.

Hydroxylation of methyl ketones in this way is usually not satisfactory because of the competing aldol condensation. Direct oxygenation is superior to this method in the case of enolates of branched esters (**6**, 427–428).[3]

[1] N. J. Lewis, S. Y. Gabhe, and M. R. DeLaMater, *J. Org.*, **42**, 1479 (1977).
[2] A. A. Frimer, *J.C.S. Chem. Comm.*, 205 (1977).
[3] E. Vedejs, D. A. Engler, and J. E. E. Telschow, *J. Org.*, **43**, 188 (1978).

Diphenylcarbodiimide, $C_6H_5N{=}C{=}NC_6H_5$. Mol. wt. 194.23, b.p. 110–112°, 0.2 mm., stable for several weeks at 0°.

Preparation[1]:

$$2C_6H_5NCO \xrightarrow[82-93\%]{} C_6H_5N{=}C{=}NC_6H_5 + CO_2$$

Cleavage of amides. The 7-amide group of cephalosporins as well as the 6-amide group of penicillins can be cleaved without concurrent cleavage of the β-lactone ring or other reactions by a two-step process. The substrate **1** is first treated with oxalyl chloride in benzene at 20° to form an oxamic acid (**2**), which is then treated with the carbodiimide in CH_2Cl_2 at 25° to give the amine **3**. Dicyclohexylcarbodiimide is not useful for this reaction because it gives in addition to **3** by-products containing dicyclohexyl groups.[2]

[1] T. W. Campbell and J. J. Monagle, *Org. Syn.*, *Coll. Vol.*, **5**, 501 (1973).
[2] M. Shiozaki, N. Ishida, K. Iino, and T. Hiraoka, *Tetrahedron Letters*, 4059 (1977).

Diphenyldi(1,1,1,3,3,3-hexafluoro-2-phenyl-2-propoxy)sulfurane, 4, 205–207; **5,** 270–272.

Epoxides. (**5**, 270–272). Eschenmoser and Eugster[1] have prepared the chiral diol **1** (5R, 6R-5, 6-dihydro-β, β-carotene-5, 6-diol) as bright violet leaflets. Because of the severe steric hindrance **1** has the solubility properties of a hydrocarbon. On treatment of **1** with the sulfurane **3**, the chiral epoxide **2** is obtained.

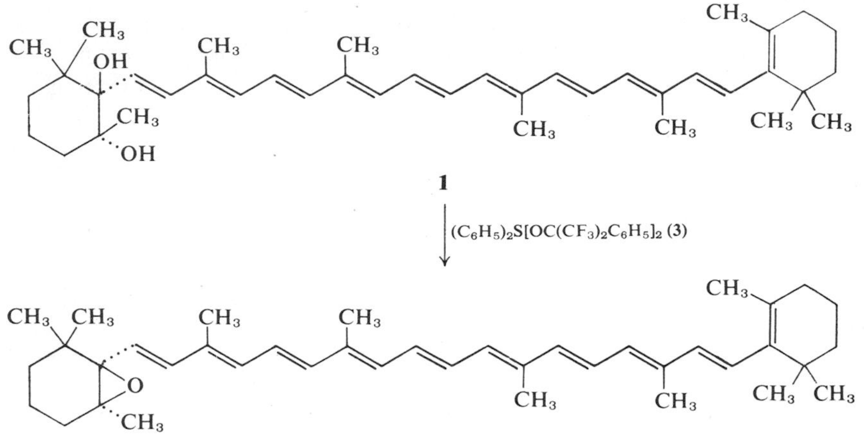

1

$(C_6H_5)_2S[OC(CF_3)_2C_6H_5]_2$ (3)

2

[1] W. Eschenmoser and C. H. Eugster, *Helv.*, **61**, 822 (1978).

Diphenyl diselenide–Bromine–Hexabutyldistannoxane,
$(C_6H_5Se)_2-Br_2-[(n-C_4H_9)_3Sn]_2O$ (**1**).

α-Phenylseleno ketones. This combination presumably forms $C_6H_5SeOSnBr_3$; in any event it adds to double bonds to form α-phenylseleno ketones, as formulated for a typical reaction with styrene (equation I). Usually in the reaction with

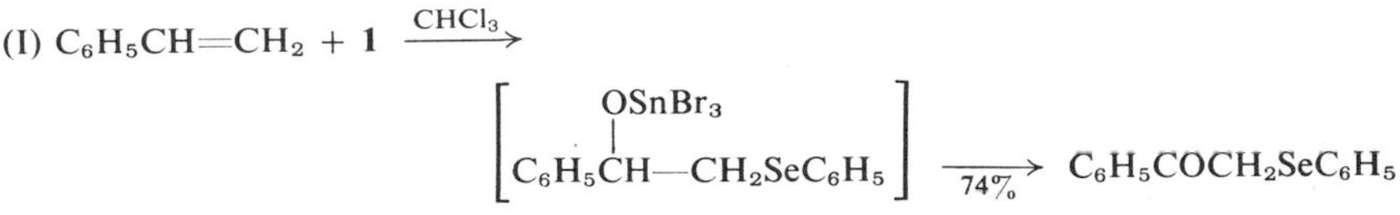

(I) $C_6H_5CH{=}CH_2$ + **1** $\xrightarrow{\text{CHCl}_3}$

$$\left[\underset{C_6H_5\overset{|}{C}H-CH_2SeC_6H_5}{\overset{OSnBr_3}{}} \right] \xrightarrow[74\%]{} C_6H_5COCH_2SeC_6H_5$$

terminal alkenes a mixture of the α-phenylseleno ketone and the α-phenylseleno aldehyde is formed in which the former product predominates. Internal alkenes react rather slowly and an excess of **1** is desirable for reasonable yields.[1]

Other examples:

$C_8H_{17}CH{=}CH_2 \xrightarrow[68\%]{1,\ CCl_4} C_8H_{17}COCH_2SeC_6H_5 \ + \ C_8H_{17}CH(SeC_6H_5)CHO$

68:32

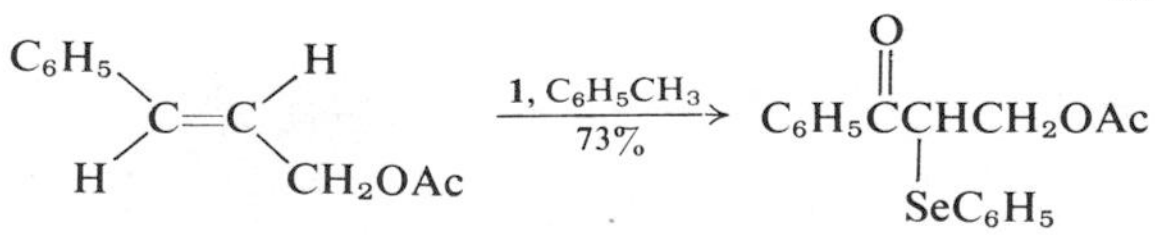

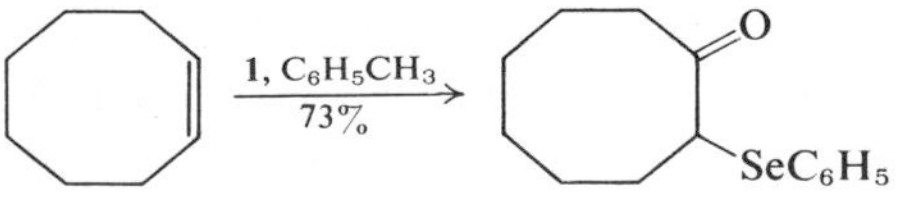

[1] I. Kuwajima and M. Shimizu, *Tetrahedron Letters*, 1277 (1978).

Diphenyl disulfide, 5, 276–277; **6,** 235–238; **7,** 137.

Aromatization of cyclohexanones.[1] Cyclohexanones undergo dehydrogenative sulfenylation when treated with diphenyl disulfide in methanolic sodium methoxide to give *o*-phenylthiophenols. One limitation is that only one alkyl group can be present at an α-position. This reaction appears to be unique to cyclohexanones; cycloheptanone is converted under these conditions to *o*-phenylthiocycloheptanone and *o,o'*-bisphenylthiocycloheptanone.

Examples:

[1] B. M. Trost and J. H. Rigby, *Tetrahedron Letters*, 1667 (1978).

N-(Diphenylmethylene)methylamine (1). Mol. wt. 195.25, b.p. 96–101°/0.01 mm.
Preparation:

$$(C_2H_5)_3Al + H_2NCH_3 \xrightarrow[\text{55} \to \text{110}°]{C_6H_5CH_3,} [(C_2H_5)_2AlNHCH_3] \xrightarrow[\text{90\%}]{(C_6H_5)_2C=O} (C_6H_5)_2C=NCH_3$$
$$\mathbf{1}$$

Aminomethylation.[1] The reagent is deprotonated by LDA in THF at −60° to give **2**, which reacts with alkyl bromides (also allyl bromides) to form imines **(3)**, which on acid hydrolysis give amines **(4)**.

$$\mathbf{1} \xrightarrow{LDA} \left[(C_6H_5)_2C\cdots\overset{Li^+}{\overset{..}{N}}\cdots CH_2 \right] \xrightarrow[\text{22-64\%}]{RBr} RCH_2N=C(C_6H_5)_2 \xrightarrow[\text{17-70\%}]{H_3O^+} RCH_2NH_2$$
$$\quad\quad\quad\quad\quad \mathbf{2} \quad\quad\quad\quad\quad\quad \mathbf{3} \quad\quad\quad\quad \mathbf{4}$$

The anion **(2)** also reacts with ketones to give, after hydrolysis, alkylideneamino alcohols **(5)**. On treatment with acid **5** is converted into β-amino alcohols **(6)**; on dehydration 2-aza-1,3-butadienes **(7)** are formed.

$$2 + R^1CR^2 \xrightarrow[\substack{2) H_2O \\ 32-52\%}]{1) THF} \underset{R^2}{\overset{R^1}{C}}\!\!\!\!-OH \atop C-CH_2N\!\!=\!\!C(C_6H_5)_2 \xrightarrow[48-80\%]{HCl} \underset{R^2}{\overset{R^1}{C}}\!\!\!\!-OH \atop C-CH_2NH_2$$

(with carbonyl O on R^1CR^2)

5   **6**

42–56% | SOCl₂, Py

$$\underset{R^2}{\overset{R^1}{C}}\!\!=\!\!CHN\!\!=\!\!C(C_6H_5)_2$$

7

[1] T. Kauffmann, H. Berg, E. Köppelmann, and D. Kuhlmann, *Ber.*, **110**, 2659 (1977).

Diphenyl phosphoroazidate, 4, 210–211; **5**, 280; **6**, 193; **7**, 138.

$OH \rightarrow N_3$. Alcohols are converted into azides by treatment with diphenyl phosphoroazidate, triphenylphosphine, and diethyl azodicarboxylate in 60–90% yield. The reaction involves inversion, even in the case of Δ^5-stenols. Hindered alcohols are unreactive.[1]

α-*Aryl carboxylic acids.*[2] Alkyl aryl ketones (**1**) can be converted into these acids (**4**) by the three-step process shown in equation (I). If the synthesis is carried out by a one-flask procedure overall yields are 90–100%.

(I)

$$\underset{R^2}{\overset{R^1}{C}}HCOAr \xrightarrow[55-80\%]{\substack{\text{N} \\ H}, BF_3\cdot(C_2H_5)_2O} \underset{R^2}{\overset{R^1}{C}}\!\!=\!\!\overset{Ar}{\underset{N}{C}} \xrightarrow[THF]{(C_6H_5O)_2PN_3,}$$

1   **2**

$$\left[\underset{R^2}{\overset{R^1}{}} \!\!\!\!\!\overset{Ar}{\underset{\substack{N=N \\ O}}{\underset{N-P(OC_6H_5)_2}{C-C-N}}} \right] \xrightarrow[65-85\%]{-N_2} R^1\!-\!\underset{R^2}{\overset{}{C}}\!-\!\underset{NP(OC_2H_5)_2 \atop O}{\overset{Ar\ \ N}{C}} \xrightarrow[90\%]{KOH} R^1\!-\!\underset{R^2}{\overset{Ar}{C}}\!-\!COOH$$

a   **3**   **4**

Thallium(III) nitrate has been used for this transformation (**4**, 496), but this reagent is toxic.

The nonsteroidal antiinflammatory agent naproxen (**6**)[3] has been prepared by this route.

[1] B. Lal, B. N. Pramanik, M. S. Manhas, and A. K. Bose, *Tetrahedron Letters*, 1977 (1977).
[2] T. Shioiri and N. Kawai, *J. Org.*, **43**, 2936 (1978).
[3] J. Riegel, M. L. Madox, and I. T. Harrison, *J. Med. Chem.*, **17**, 377 (1974).

5   **6**

Diphenylsulfonium cyclopropylide, 4, 211–214; **5**, 281; **6**, 242–243; **7**, 190.

Cyclobutanones. Trost's original cyclobutanone synthesis (**4**, 211–214; definitive papers[1]) by rearrangement of oxaspiropentanes with lithium salts results in selective formation of the cyclobutanone in which the new carbon to carbonyl bond is introduced on the more sterically hindered face of the ketone. He and Scudder[2] have now found that the isomeric cyclobutanone becomes the predominant product if the oxaspiropentane is rearranged through a selenoxide. An example is the rearrangement of **1**. When lithium perchlorate is used the cyclobutanone **2** is obtained. On treatment with sodium selenophenolate followed by oxidation, the cyclobutanone **4** becomes the major product. In some cases this new selenoxide route is stereospecific and results in essentially only one cyclo-butanone.

1   **3**   **4**

2

Double homologation of enones. Trost and Frazee[3] have reported a method of chain extension of an α,β-unsaturated carbonyl system both at the β-carbon and at the carbonyl group; the method is formulated for the case of methyl vinyl ketone (**1**). First **1** is converted into the cyclopropane **2**; the keto group is then

1 + BrCH[COOC(CH₃)₃]₂ →(KOC(CH₃)₃)→ **2** →(1) ▷=S(C₆H₅)₂ 2) LiBF₄ 82%)→

3 (two isomers) → **4** + **5**

4 →(NaOCH₃ 80%)→ **6**

5 →(84% NaOCH₃)→ **7**

allowed to react with diphenylsulfonium cyclopropylide to form **3**. The corresponding methyl esters are prepared and separated by VPC to give the pure isomers **4** and **5**. Treatment of **4** or **5** with base results in Grob fragmentation to the isomerically pure (E)- or (Z)-olefin (**6** and **7**). The overall process allows addition of a two-carbon and a three-carbon unit to the original enone with migration of the double bond to the position between the α-carbon and the carbonyl carbon of the original enone with stereo- and regioselectivity.

The same sequence was applied to 2-methacrolein (**8**). In this case a single cyclobutanone (**9**) was formed. This product was subjected to fragmentation of the corresponding cyclobutanol, which resulted in the unsaturated alcohol **10** in high yield.

CH₂=CCHO **8** → **9** →(1) NaBH₄, CH₃OH, Mg(OCH₃)₂ 2) Δ 85% overall)→ **10**

The driving force for this new reaction is the cleavage of the cyclobutanone ring; apparently the *gem*-carboalkoxy groups on the cyclopropane ring are necessary also for successful fragmentation.

[1] B. M. Trost and M. J. Bogdanowicz, *Am. Soc.*, **95**, 5311, 5321 (1973).
[2] B. M. Trost and P. H. Scudder, *ibid.*, **99**, 7601 (1977).
[3] B. M. Trost and W. J. Frazee, *ibid.*, **99**, 6124 (1977).

Diphosgene (Trichloromethyl chloroformate), $ClCOOCCl_3$, Mol. wt. 197.85, b.p. 128°, 49°/50 mm.; toxic, asphyxiating gas used in the first world war as a poison gas. This reagent is generally prepared by chlorination (*hv*) of methyl formate or methyl chloroformate.[1,2]

Isocyanides.[3] This reagent is more effective than the more dangerous phosgene for dehydration of N-monosubstituted formamides to form isocyanides (equation I).

(I) $2 \ RNHCHO + ClCOOCCl_3 + 4 \ N(C_2H_5)_3 \longrightarrow$

$$2 \ RN{\equiv}C + 2 \ CO_2 + 4 \ N(C_2H_5)_3 \cdot HCl$$
$$(50\text{–}98\%)$$

[1] V. Grignard *et al.*, *Compt. rend.*, **169**, 1075, 1143 (1919).
[2] H. C. Ramsperger and G. Waddington, *Am. Soc.*, **55**, 214 (1933).
[3] G. Skorna and I. Ugi, *Angew. Chem., Int. Ed.*, **16**, 259 (1977).

Dipotassium tetracarbonylferrate, $K_2Fe(CO)_4$. Mol. wt. 246.09, m.p. 270–273° dec. The reagent can be prepared in a one-flask operation by addition of $Fe(CO)_5$ to $K(s\text{-}C_4H_9)_3BH$ (K-Selectride); yield 95–100%. Unlike $Na_2Fe(CO)_4$, this ferrate is not spontaneously flammable in air. It probably can be used in similar reactions.[1]

$$K_2Fe(CO)_4 \xrightarrow[\substack{100\%}]{\substack{1) \ n\text{-}C_8H_{17}Br \\ 2) \ P(C_6H_5)_3; \ HOAc}} CH_3(CH_2)_7CHO$$

[1] J. A. Gladysz and W. Tam, *J. Org.*, **43**, 2279 (1978).

2,2′-Dipyridyl disulfide, 5, 285–286; **6,** 246–247.

Sulfenylation. Direct conversion of the cyclopentanone (**1**) into the natural product methyl dehydrojasmonate (**4**) by sulfenylation followed by sulfoxide elimination is not possible because electrophiles react preferentially at C_2 rather than C_5. The desired reaction was accomplished by activation of the C_5-position by formylation to give **2**. The anion of this derivative is not sulfenylated by the usual reagents, but is converted into the thioether **3** by the more reactive dipyridyl disulfide with loss of the formyl group. The cyclopentenone **4** is obtained in 73% yield on sulfoxide elimination from **3**.[1] Instead of dipyridyl disulfide, 2-pyridinesulfenyl bromide ($C_5H_4NSBr\text{-}o$) can be used equally well. This reagent is prepared immediately before use by reaction of 2,2'-dipyridyl disulfide with bromine in CHCl₃.

This sulfoxide elimination reaction was also used for conversion of jasmolactone (**5**) into tuberolactone (**7**).

¹ P. Dubs and R. Stüssi, *Helv.*, **61**, 998 (1978).

2,2'-Dipyridyl disulfide–Triphenylphosphine, 5, 285–286; **6**, 246–247; **7**, 141–142.

Lactonization. The disulfide **1** (7, 141) was used instead of 2,2'-dipyridyl disulfide by Corey and Bhattacharyya[1] in a total synthesis of macrocyclic lactone enterobactin (**2**) involved in the transport of iron in certain bacteria. The lactone ring was constructed from three L-serine units, with protection of the amino group as the CBZ group and of the hydroxyl group as the THP ether. The three serine units were condensed in steps using **1** and triphenylphosphine to give the hydroxy acid **3**. This product was lactonized to **4** in about 40% yield also with **1** and triphenylphosphine.

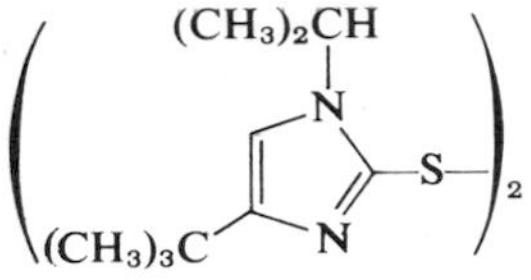

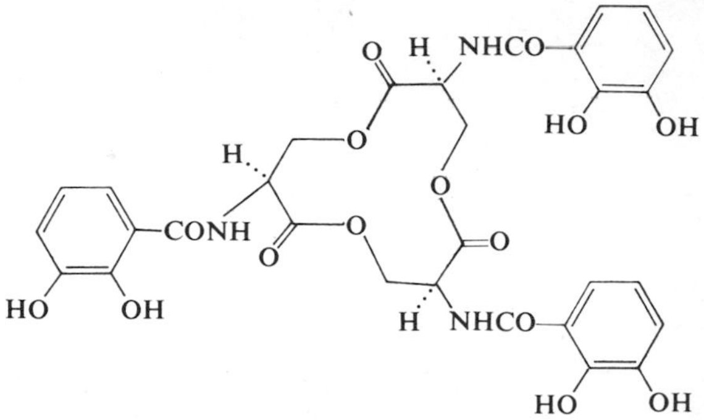

2

Remaining steps were removal of the CBZ protecting groups (H_2, Pd/C) and acylation of the amino groups with 2,3-hydroxybenzoyl chloride. The dihydroxybenzene units are involved in the transport of iron.

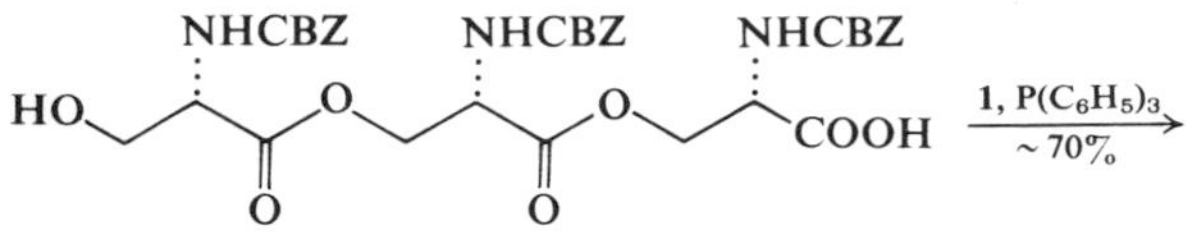

3

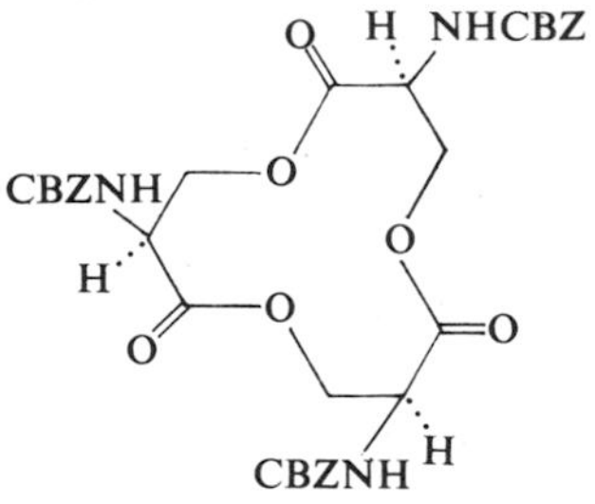

4

[1] E. J. Corey and S. Bhattacharyya, *Tetrahedron Letters*, 3919 (1977).

Disodium tetracarbonylferrate, 3, 267–268; **4**, 461–465; **5**, 624–625; **6**, 550–552; **7**, 341.

Hydroacylation of Michael acceptors. The organotetracarbonylferrates obtained by alkylation of disodium tetracarbonylferrate undergo insertion reactions with Michael-type acceptors to give eventually γ-keto esters, ketones, and nitriles. The last example shows an interesting synthesis of a cyclopentanone by an intramolecular insertion reaction.[1]

Examples:

$$n\text{-}C_6H_{13}Br + Na_2Fe(CO)_4 \longrightarrow n\text{-}C_6H_{13}\overset{-}{Fe}(CO)_4 \overset{Na^+}{\underset{93\%}{\xrightarrow{\begin{array}{l}1)\ CH_2{=}CHCOOC_2H_5\\2)\ CH_3COOH\end{array}}}}$$

$$n\text{-}C_6H_{13}\overset{O}{\overset{\|}{C}}CH_2CH_2COOC_2H_5$$

$$C_2H_5I + CH_2{=}CHCOC_2H_5 \xrightarrow{Na_2Fe(CO)_4} C_2H_5\overset{O}{\overset{\|}{C}}CH_2CH_2\overset{O}{\overset{\|}{C}}C_2H_5$$

$$I(CH_2)_3\overset{H}{\underset{H}{C}}{=}CCOOC_2H_5 \xrightarrow{Na_2Fe(CO)_4}$$

$$\xrightarrow[85\%]{H^+}$$

[1] M. P. Cooke, Jr., and R. M. Parlman, *Am. Soc.*, **99**, 5222 (1977).

Disodium tetrachloropalladate, 4, 369; **5**, 500–501; **6**, 447–449.

Allylic alkylation. Complete details for allylic alkylation of olefins via π-allylpalladium complexes have been published. The preferred reagent is Na$_2$-PdCl$_4$ in combination with NaOAc and CuCl$_2$. The latter salt increases the regioselectivity. The order of olefin reactivity varies with the substitution: tri- > di- > monosubstitution. The hydrogen that is abstracted is usually allylic to the more substituted carbon atom of the double bond and the relative reactivity of the various hydrogens is CH$_3$ > CH$_2$ ≫ CH.[1] The complexes normally do not react with nucleophiles in the absence of activating ligands, such as phosphines and phosphites.[2] Even so, only a limited number of nucleophiles are suitable for the alkylation step, such as the anions of malonate and of β-keto sulfones, sulfoxides, and sulfides. However, the products of alkylation with these nucleophiles are subject to further transformations such as decarbomethoxylation, desulfenylation, desulfonylation, and sulfoxide elimination. In these reactions stoichiometric amounts of Na$_2$PdCl$_4$ are required, but the palladium that results can be recycled.[3]

[1] B. M. Trost, P. E. Strege, L. Weber, T. J. Fullerton, and T. J. Dietsche, *Am. Soc.*, **100**, 3407 (1978).
[2] B. M. Trost, L. Weber, P. E. Strege, T. J. Fullerton, and T. J. Dietsche, *ibid.*, **100**, 3416 (1978).
[3] *Idem, ibid.*, **100**, 3426 (1978).

1,3-Dithiane, 2, 187; **3**, 135–136; **4**, 216–218; **5**, 287; **6**, 248; **7**, 142–143.

2-Hydroxycyclobutanone. A new efficient synthesis of 2-hydroxycyclobutanone involves as the first step reaction of 2-lithio-1,3-dithiane with 3-bromopropanal ethylene acetal to form **1**. Treatment of **1** with *n*-butyllithium at − 25 to − 15° for several hours effects cyclization to the alcohol **2**. This is converted into the THP ether **3**, and then the dithiane ring is hydrolyzed to a ketone group to give **4**.[1]

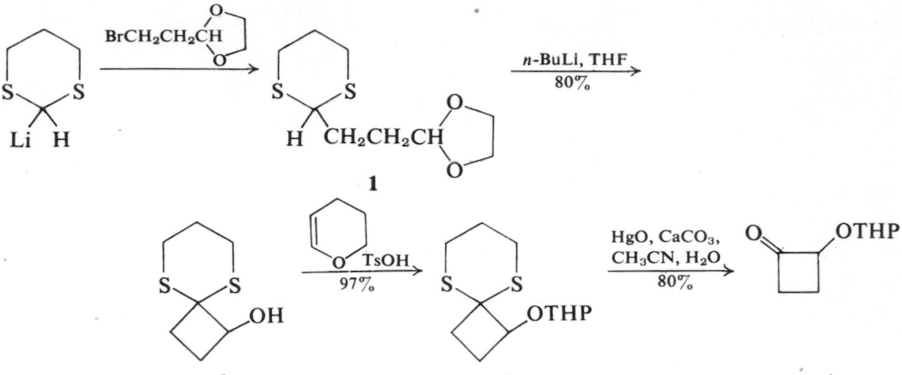

1

2   **3**   **4**

[1] A. Murai, M. Ono, and T. Masamune, *J.C.S. Chem. Comm.*, 573 (1977).

Dodecamethylcyclohexasilane,

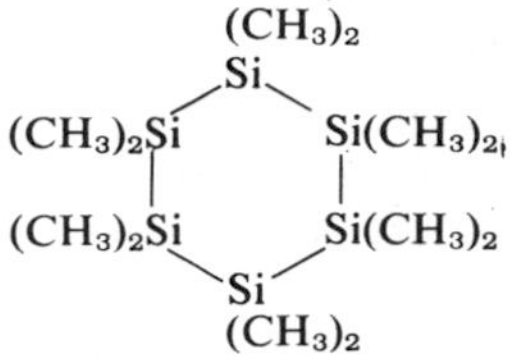

(**1**). Mol. wt. 318.93,

m.p. 250–252°. The reagent is prepared in 60% yield (pure) by controlled reaction of dichlorodimethylsilane in THF with lithium wire and a trace of triphenyl-silyllithium as catalyst.[1]

Silyl enol ethers.[2] Silyl enol ethers are obtained in isolated yields of 50–90% on irradiation (low-pressure mercury lamp) of an ether solution of **1** and a carbonyl compound. The actual reagent is dimethylsilylene, $(CH_3)_2Si$:, which is generated by photolysis of **1**.[3]

[1] H. Gilman and R. A. Tomasi, *J. Org.*, **28**, 1651 (1963).
[2] W. Ando and M. Ikeno, *Chem. Letters*, 609 (1978).
[3] M. Ishikawa and M. Kumada, *J. Organometal. Chem.*, **42**, 325 (1972).

E

2-Ethoxyallylidenetriphenylphosphorane.

$$
\begin{array}{c}
CH_2 \\
\parallel \\
HC\diagdown C \diagup OC_2H_5 \\
\parallel \\
P(C_6H_5)_3
\end{array}
$$

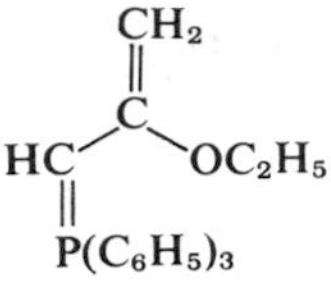

(1). Mol. wt. 370.41.

This reagent is prepared[1] *in situ* by treatment of (2-ethoxyl-1-propenyl)-triphenylphosphonium iodide[2] with *n*-butyllithium.

Cyclohexenone annelation.[1] Cyclohexenones can be prepared in fair to good yields by reaction of **1** with enones in THF at $-50°$. The annelation is an extension of the use of allylidenetriphenylphosphorane for the synthesis of 1,3-cyclohexa-dienes (**5**, 7–8). The ethoxyl function in **1** allows for conversion into cyclohexe-nones on acid hydrolysis.

Examples:

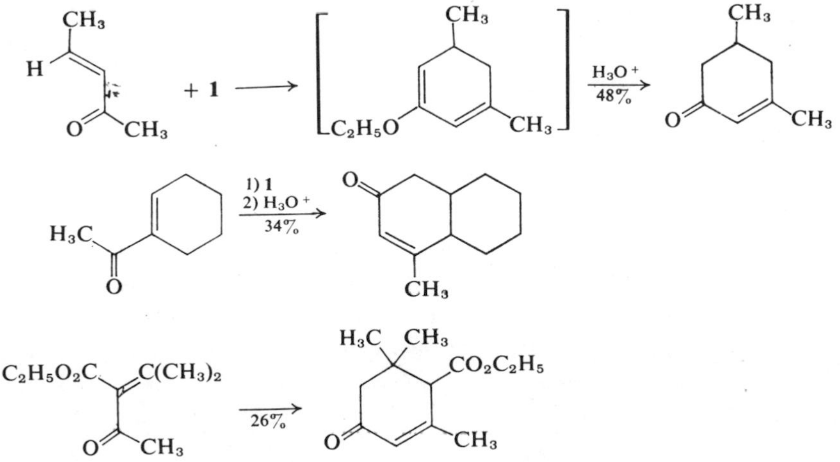

[1] S. F. Martin and S. R. Desai, *J. Org.*, **42**, 1664 (1977).
[2] F. Ramirez and S. Dershowitz, *ibid.*, **22**, 41 (1957).

1-Ethoxy-1-trimethylsiloxycyclopropane,

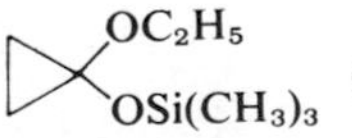

$$
\begin{array}{c}
\diagup OC_2H_5 \\
\triangleright \\
\diagdown OSi(CH_3)_3
\end{array}
$$

(1). Mol. wt. 174.31.

This reagent is prepared by treatment of ethyl 3-chloropropanoate with sodium and chlorotrimethylsilane (compare acyloin condensation, **2**, 435–436; **3**, 311–312).

Reaction with aldehydes.[1] This cyclopropanone ketal in the presence of TiCl$_4$ behaves as the equivalent of the homoenolate anion of ethyl propanoate,

$$\bar{C}H_2CH_2\overset{\overset{\displaystyle O}{\|}}{C}OC_2H_5.$$ Thus the reaction with *p*-nitrobenzaldehyde gives the adduct **2** in 88% yield. In the case of aliphatic aldehydes the initial products cyclize

to the γ-lactone, as in the second example. Ketals also react with **1**, but ketones themselves do not give adducts in isolable amounts.

Only TiCl$_4$ has been found to effect these reactions, probably because it can coordinate with cyclopropane rings.

[1] E. Nakamura and I. Kuwajima, *Am. Soc.*, **99**, 7360 (1978).

Ethyl *trans*-1,3-butadiene-1-carbamate (1), **7**, 147.

Diels-Alder reaction. The Diels-Alder reaction of **1** with *trans*-crotonaldehyde (**2**) was used in a short, stereospecific synthesis of (±)-pumiliotoxin C (**4**), a toxin from the frog *Dendrobates pumilio*, in about 45% overall yield. The cycloaddition is noteworthy in that three chiral centers are established in **3**.[1]

[1] L. E. Overman and P. J. Jessup, *Tetrahedron Letters*, 1253 (1977).

(Z)-2-Ethoxyvinyllithium, $\begin{array}{c} H \\ \diagdown \\ Li \end{array} C = C \begin{array}{c} H \\ \diagup \\ OC_2H_5 \end{array}$ **(1).**

Preparation from ethoxyacetylene (**1**, 357–360; **2**, 190; **5**, 290–291):[1,2]

$$HC \equiv COC_2H_5 \xrightarrow[94\%]{(C_4H_9)_3SnH} \begin{array}{c} H \\ \diagdown \\ \cdot(C_4H_9)_3Sn \end{array} C = C \begin{array}{c} H \\ \diagup \\ OC_2H_5 \end{array} \xrightarrow[\text{THF}, -78°]{C_4H_9Li,} \mathbf{1}$$

It can also be obtained by halide–lithium exchange by reaction of (Z)-$BrCH=CHOC_2H_5$ with *n*-butyllithium.

α,β-*Unsaturated aldehydes.* Two groups[1,2] have simultaneously reported the preparation of this anion (**1**) and its reaction with aldehydes and ketones to form enol ethers, which are converted readily into α,β-unsaturated aldehydes by acetic acid or chromatography on silica gel or Florisil. The reagent thus functions as the equivalent of the anion of acetaldehyde in an aldol condensation. It does not undergo conjugate addition.

Examples:

$$C_6H_5CHO + \mathbf{1} \xrightarrow[76\%]{} C_6H_5CH=CHCHO$$

The reagent also reacts with alkyl halides; an example is the reaction with 1-iodododecane to form dodecanal:

$$CH_3(CH_2)_8CH_2I + \mathbf{1} \xrightarrow[\text{HMPT}]{\text{THF,}} \begin{array}{c} H \\ \diagdown \\ CH_3(CH_2)_8CH_2 \end{array} C = C \begin{array}{c} \cdot \\ H \\ \diagup \\ OC_2H_5 \end{array} \xrightarrow[\text{THF, H}_2O]{\text{HOAc,}} CH_3(CH_2)_{10}CHO$$

A vinylcuprate reagent (**2**) derived from **1** has also been prepared by reaction with CuI and $(CH_3)_2S$. This reagent undergoes conjugate addition with α,β-unsaturated carbonyl compounds and also direct displacement with activated halides.[3]

Examples:

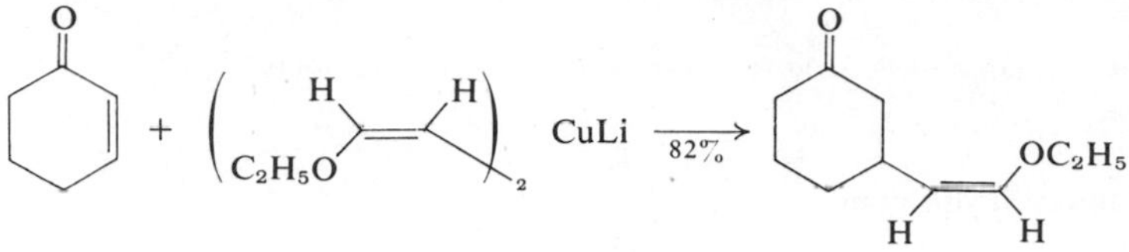

2

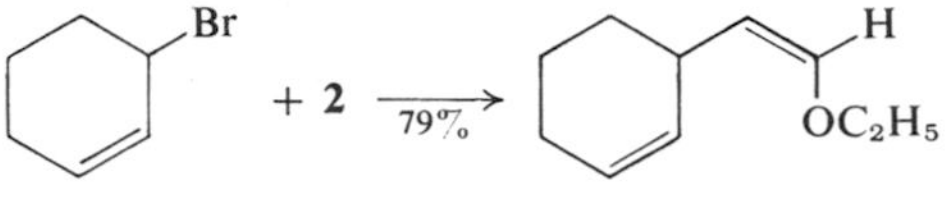

[1] R. H. Wollenberg, K. F. Albizati, and R. Peries, *Am. Soc.*, **99**, 7365 (1977).
[2] J. Ficini, S. Falou, A.-M. Touzin, and J. d'Angelo, *Tetrahedron Letters*, 3589 (1977).
[3] K. S. Y. Lau and M. Schlosser, *J. Org.*, **43**, 1595 (1978).

Ethyl diazoacetate, 1, 367–370; **2**, 193–195; **3**, 138–139; **4**, 228–230; **5**, 295–300; **6**, 252–253.

Ring expansion (**1**, 369–370; **6**, 252–253). The ring expansion of ketones to the next higher homolog with ethyl diazoacetate requires hydrolysis and decarboxylation of the intermediate β-keto ester, a step that is sometimes troublesome. Baldwin and Landmesser[1] have used benzyl diazoacetate[2] and allyl diazoacetate[3] as alternative reagents. The benzyl β-keto esters are cleaved and decarboxylated on hydrogenation; both benzyl and allyl keto esters are reduced by sodium in liquid ammonia to ketones.

Example:

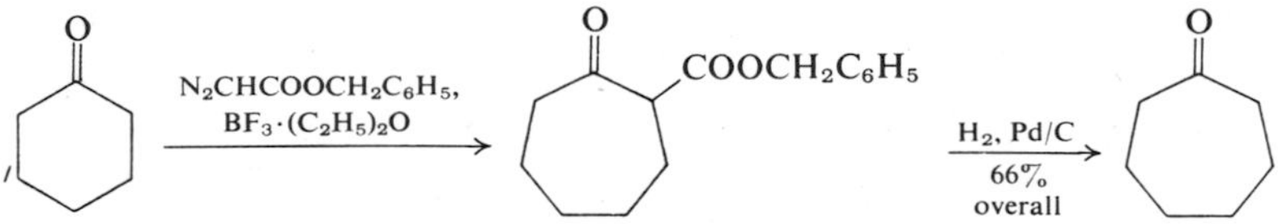

[1] S. W. Baldwin and N. G. Landmesser, *Syn. Comm.*, **8**, 413 (1978).
[2] T. DoMinh, O. P. Strausz, and H. E. Gunnina, *Tetrahedron Letters*, 5237 (1968).
[3] H. Ledon, G. Linstrumelle, and S. Julia, *Tetrahedron*, **29**, 3609 (1973).

Ethyl 4,4-dimethoxy-2-phenylthiobutyrate (1). Mol. wt. 284.37, b.p. 138–141°/1 mm.

Preparation:

$$C_6H_5S\underset{|}{\overset{Na}{C}}HCOOC_2H_5 + BrCH_2CH(OCH_3)_2 \xrightarrow[69\%]{DMF,\ 60^\circ} (CH_3O)_2CHCH_2CH(SC_6H_5)(COOC_2H_5)$$

1

3-*Substituted furanes*. The reaction of **1** with alkyl halides can be used for the synthesis of 3-substituted furanes, as shown for the synthesis of 3-benzylfurane.[1]

$$1 + C_6H_5CH_2Br \xrightarrow[82\%]{NaH,\ DMF} (CH_3O)_2CHCH_2\overset{\displaystyle CH_2C_6H_5}{\underset{\displaystyle COOC_2H_5}{C}}{-}SC_6H_5 \xrightarrow[98\%]{LiAlH_4}$$

$$(CH_3O)_2CHCH_2\overset{\displaystyle CH_2C_6H_5}{\underset{\displaystyle CH_2OH}{C}}{-}SC_6H_5 \xrightarrow[C_6H_6]{TsOH,} \left[\ \right] \xrightarrow[91\%]{\Delta}$$

[1] H. Kotaki, K. Inomata, S. Aoyama, and H. Kinoshita, *Chem. Letters*, 73 (1977).

2,2′-Ethylenebis(1,3-dithiane), 6, 493.

γ-*Keto esters*. A new route to these substances is shown in equation (I).[1]

(I) $\xrightarrow[\substack{1)\ n\text{-BuLi, THF} \\ 2)\ RX \\ 3)\ n\text{-BuLi} \\ 4)\ CH_3SSCH_3}]{}$ $\xrightarrow[25-85\%]{\substack{HgCl_2,\ HgO, \\ R'OH,\ H_2O}}$ $RCOCH_2CH_2CO_2R'$

[1] W. D. Woessner and P. S. Solera, *Syn. Comm.*, **8**, 279 (1978).

3-Ethyl-2-fluorobenzothiazolium tetrafluoroborate,   (**1**). Mol. wt. 269.05. This reagent is prepared by reaction of 2-fluorobenzothiazole (Fluka) with Meerwein's reagent.

Alkyl halides.[1] The reagent (**1**) reacts with alcohols at −78° to form a 2-alk-oxybenzothiazolium salt (**a**), which reacts with alkali metal halides (NaI, LiBr, LiCl) to form alkyl halides in 55–90% yield. In the case of optically active secondary alcohols the reaction occurs with inversion of configuration.

Example:

$$1 + HO-\underset{C_6H_{13}}{\overset{H}{\underset{|}{C}}}CH_3 \xrightarrow[CH_2Cl_2,\ -60°]{N(C_2H_5)_3,} \left[\text{benzothiazolium structure } \mathbf{a} \right]$$

a

$$\downarrow \text{NaI, } (CH_3)_2C=O, -45°$$

$$CH_3-\underset{C_6H_{13}}{\overset{H}{C}}-I \ + \ \text{(benzoxazolone structure)}$$

(87%, α_D −53°)

Optical interconversion of sec-alcohols.[2] This reaction can be accomplished readily by treatment of the salt (**a**, above) with trichloroacetic acid and triethylamine to give the ester **2**, with inverted configuration. Alkaline hydrolysis of **2** then affords the alcohol **3**. A typical example is the conversion of (R)-(−)-2-octanol into (S)-(+)-2-octanol, formulated in equation (I).

(I) $\mathbf{a} + CCl_3COOH \xrightarrow[CH_2Cl_2,\ -50°]{N(C_2H_5)_3,} CH_3-\underset{C_6H_{13}}{\overset{H}{C}}-OCOCCl_3 \ +$ (benzoxazolone)

2 (61%)

$$\downarrow \overset{\text{quant.}}{} \begin{array}{l} KOH,\ H_2O, \\ CH_3OH \end{array}$$

$$CH_3-\underset{C_6H_{13}}{\overset{H}{C}}-OH$$

3 (α_D +10.2°)

Biogenetic-like cyclization of polyenic alcohols.[3] The reaction of *trans,cis-*farnesol (**1**) with the salt **2** and tri-*n*-butylamine in CH_2Cl_2 at −40° for 2 hours

$$1 \qquad + \qquad 2 \xrightarrow[70-75\%]{\substack{N(C_4H_9)_3,\\ CH_2Cl_2,\ -40°}} 3\ (E/Z = 1:1)$$

affords α-bisabolene (**3**) in about 70–75% yield with only traces of γ-bisabolene (less than 5%); no other polycyclic products are detectable. Under the same conditions, **3** is obtained in only about 35% yield from *trans,trans*-farnesol or from geraniol. Nerolidol (**4**) is cyclized rather slowly by **2**, but α- and γ-bisabolene (**5**) can be obtained in about 70% yield.

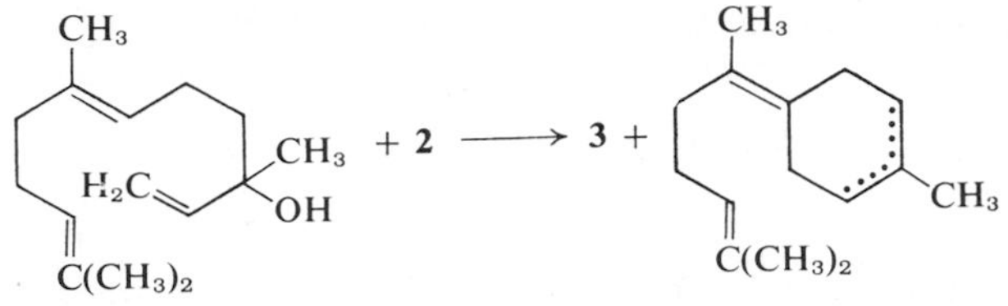

4

5 (70%)

[1] T. Mukaiyama and K. Hojo, *Chem. Letters*, 619 (1976).
[2] *Idem., ibid.*, 893 (1976).
[3] S. Kobayashi, M. Tsutsui, and T. Mukaiyama, *ibid.*, 1169 (1977).

Ethyl lithioacetate, $LiCH_2COOC_2H_5$. This reagent is prepared by metalation of ethyl acetate with LDA in THF at $-78°$.

   δ-*Substituted* δ-*lactones.* The lactones can be prepared by addition of $LiCH_2$-$COOC_2H_5$ to a cyclohexanone substituted with an *n*-butylthiomethylene group, which can be transformed into a formyl group. The complete steps are formulated in equation (I).[1]

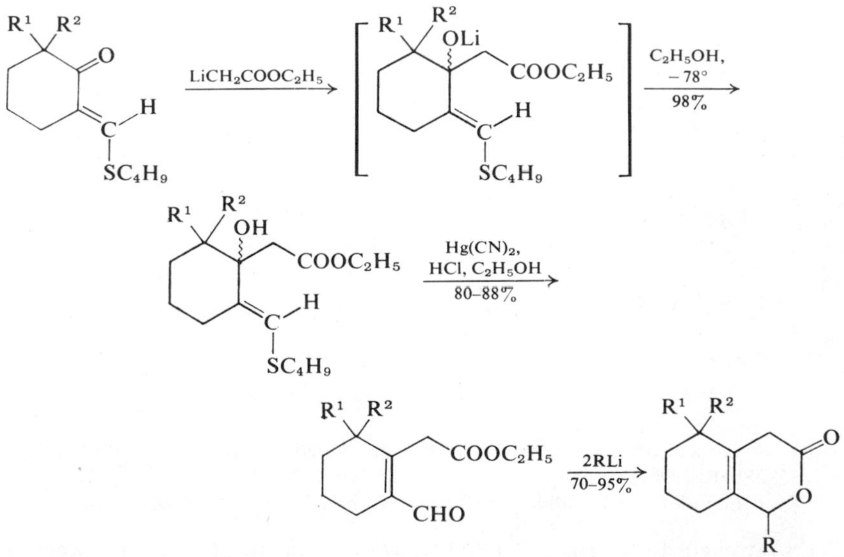

[1] A. J. G. M. Peterse, Ae. de Groot, P. M. van Leeuwen, N. H. G. Penners, and B. H. Koning, *Rec. trav.*, **97**, 124 (1978).

N-Ethyl-5-phenylisoxazolium 3′-sulfonate (Woodward's Reagent K), **1**, 384–385; **2**, 197–198; **5**, 306.

Peptide synthesis. Two examples of the use of this inner salt for peptide synthesis under somewhat different conditions have been published in *Org. Syn.*[1] Advantages and limitations of this method are discussed.

[1] R. B. Woodward and R. A. Olofson, *Org. Syn.*, **56**, 88 (1977).

Ethyl β-(1-pyrrolidinyl)acrylate, [structure] N—CH=CHCOOC$_2$H$_5$ **(1).** Mol. wt. 169.22, m.p. 41–43°. This reagent is prepared by the reaction of pyrrolidine with ethyl propiolate (40–60°); yield 56%.[1]

Cyclopentenones and butenolides. Treatment of **1** with *t*-butyllithium at temperatures below −100° gives the lithium derivative **2** in almost quantitative yield. This derivative reacts with α,β-unsaturated carbonyl compounds to form 2-cyclopentene-1-ones such as **3**. Compound **2** also undergoes [3 + 2] cyclo-additions with aldehydes and even esters to form butenolides such as **4**.[2]

[1] Y. Postowskü, E. I. Grünblat, L. Trifilova, *Zh. Obshch. Khim.*, **31**, 400 (1961); *C.A.*, **55**, 23541d (1961).
[2] R. R. Schmidt and J. Talbiersky, *Angew. Chem., Int. Ed.*, **17**, 204 (1978).

Ethyl trimethylsilylacetate, 7, 150–152. Detailed instructions are available for the preparation from ethyl bromoacetate and chlorotrimethylsilane and zinc (excess) initiated with CuCl; yield 63–74%.

Silylation. Details are available for conversion of 3-pentanone (**1**) into (Z)-3-trimethylsiloxy-2-pentene (**2**) in the presence of tetra-*n*-butylammonium fluoride (a preparation of this salt is included).[1]

$$
\underset{\textbf{1}}{C_2H_5\overset{\overset{O}{\|}}{C}C_2H_5} + (CH_3)_3SiCH_2COOC_2H_5 \xrightarrow[73-80\%]{\underset{THF}{(C_4H_9)_4NF,}} \underset{\textbf{2}}{C_2H_5\diagup\overset{\overset{OSi(CH_3)_3}{|}}{C}=\underset{\underset{H}{|}}{C}\diagdown CH_3}
$$

Other acyclic ketones also afford only (Z)-enol silyl ethers. The regioselectivity appears to differ from that observed with LDA (equation I).

$$
\text{(I)} \quad \underset{\underset{SC_6H_5}{|}}{CH_3\overset{\overset{OSi(CH_3)_3}{|}}{C}=CCH_3} \xleftarrow{(CH_3)_3SiCH_2COOC_2H_5} \underset{\underset{SC_6H_5}{|}}{CH_3\overset{\overset{O}{\|}}{C}CHCH_3} \xrightarrow[ClSi(CH_3)_3]{LDA,}
$$

$$
\underset{\underset{SC_6H_5}{|}}{CH_2=C\overset{\overset{OSi(CH_3)_3}{|}}{}CHCH_3} \underset{7:3}{+} \underset{\underset{SC_6H_5}{|}}{CH_3\overset{\overset{OSi(CH_3)_3}{|}}{C}=CCH_3}
$$

[1] E. Nakamura, K. Hashimoto, and I. Kuwajima, *Tetrahedron Letters*, 2079 (1978); *Org. Syn.*, submitted (1978).

F

Fenchone, (1). Mol. wt. 152.24, m.p. 5–7°, b.p. 63–65°/13 mm.,

α_{546} 75°. Supplier: Fluka.

Dehydrogenation. Reetz and Eibach[1] have developed a new method for dehydrogenation of dihydroarenes based on deprotonation–hydride elimination. Potassium fencholate (formed by reduction of fenchone with KH) serves as base and fenchone (1)[2] as hydride acceptor. Since potassium fencholate is regenerated in the aromatization, only catalytic quantities are required. Yields of arenes are 70–85% (isolated). The method is not useful for dehydrogenation to form alkenes. Fenchone is particularly suitable since it is reduced by hydride transfer because of steric reasons.

[1] M. T. Reetz and F. Eibach, *Angew. Chem., Int. Ed.*, **17**, 278 (1978); *Ann.*, 1598 (1978).
[2] M. T. Reetz and C. Weiss, *Synthesis*, 135 (1977).

Ferric chloride, 1, 390–392; **2**, 199; **3**, 145; **4**, 236; **5**, 307–308; **6**, 260; **7**; 153–155.

Carbohydrate acetonides. Carbohydrates can be converted into acetonides (O-isopropylidene derivatives) by reaction with dry acetone and anhydrous $FeCl_3$ as catalyst at 35°. Yields of pure derivatives are 65–75%.[1]

[1] P. P. Singh, M. M. Gharia, F. Dasgupta, and H. C. Srivastava, *Tetrahedron Letters*, 439 (1977).

Ferric chloride–Acetic anhydride, 6, 260.

Conversion of —C≡N *into* —$\overset{\overset{O}{\|}}{C}$NHCOCH$_3$. Watt et al.[1] treated the α-t-butoxyacrylonitrile **1** with this combination of reagents, expecting to cleave the ether group. Instead **2** was obtained, as a result of the formal addition of CH$_3$-COOH to the C≡N group, as well as the expected cleavage. Unfortunately this novel transformation of a nitrile to an imide is limited to α-t-butoxy- and α-acetoxyacrylonitriles.

[1] P. M. Diamond, S. E. Dinizio, R. W. Freerksen, R. C. Haltiwanger, and D. S. Watt, *J.C.S. Chem. Comm.*, 298 (1977).

Ferric chloride–Silica gel, $FeCl_3$–SiO_2.

Reactions.[1] $FeCl_3$ in solution has been used mainly as a mild oxidant. When it is adsorbed on silica gel, it is an effective reagent for rapid dehydration of allylic, tertiary, and sterically hindered secondary alcohols at room temperature. Selective dehydration of polyhydroxylic substrates is possible.

Examples:

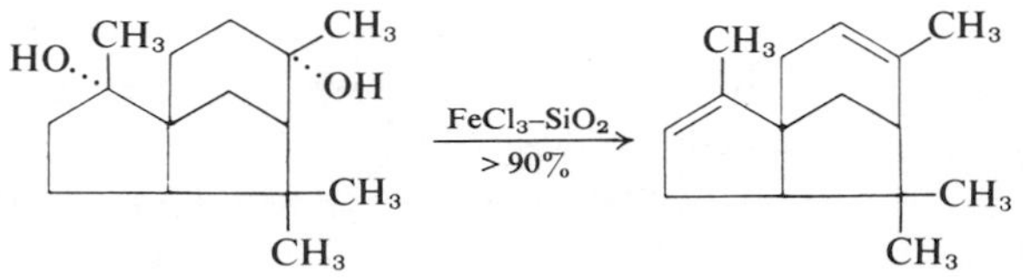

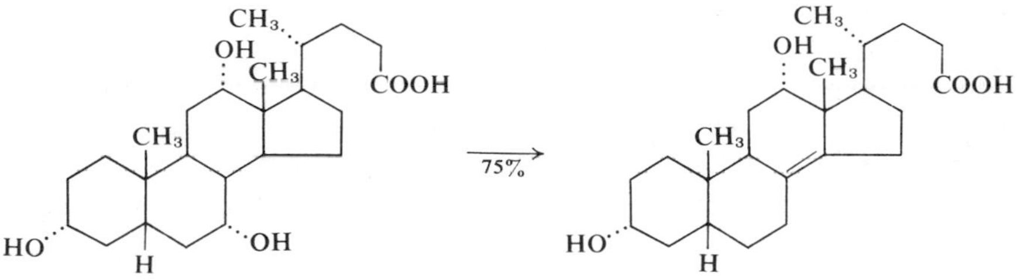

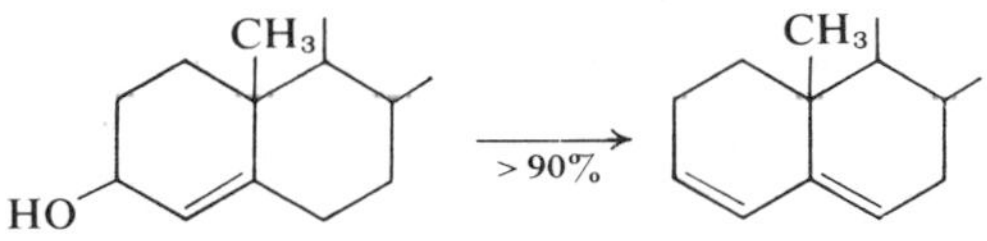

Reagent containing 2% water is less efficient for dehydration; this material (bright yellow) epimerizes tertiary alcohols efficiently to the equilibrium mixture of epimers. This wet reagent converts epoxides into diols.

Example:

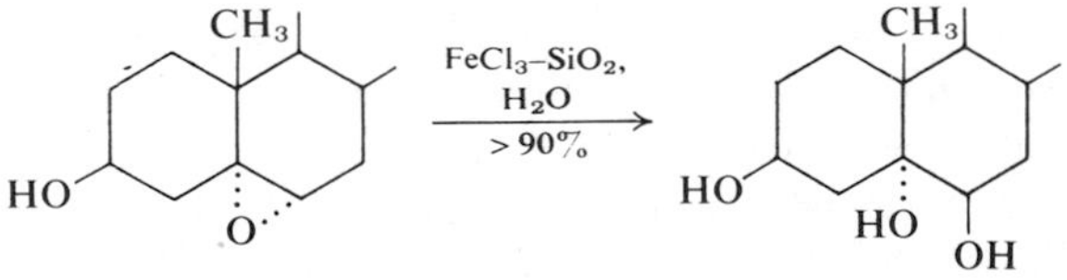

The $FeCl_3$–SiO_2 reagent is also useful for pinacol and acyloin rearrangements. Examples:

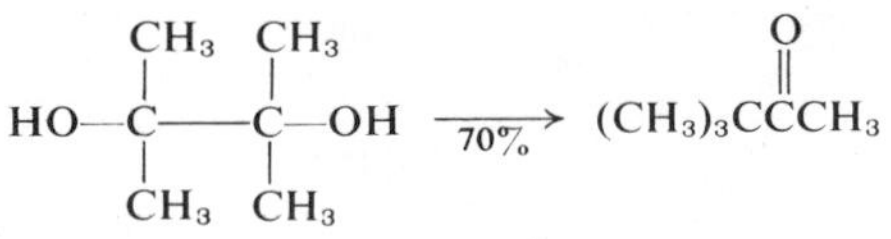

[1] E. Keinan and Y. Mazur, *J. Org.*, **43**, 1020 (1978).

9-Fluorenyl chloroformate, 3, 145–146; 4, 237.

Solid-phase peptide synthesis.[1] The combination of the base-labile N-α-fluorenylmethoxycarbonyl (Fmoc) amino acids and the acid-labile *t*-butyl protecting group is valuable for solid-phase peptide synthesis, particularly with polar resins. Intermediate Fmoc-peptide resins are deprotected with 20% piperidine or 5% piperazine in DMF. Six amino acid groups can be added per day without difficulty.[2] This new strategy was used for synthesis of human β-endorphin (31 residues), with 29 residues added as the anhydrides of Fmoc-amino acids. The last residue was the N-α-Boc derivative of O-*t*-butyltyrosine. The peptide resin was cleaved with anhydrous CF_3COOH. The overall yield of isolated polypeptide was 41%.[3] This method does not require vigorous acidic conditions.

[1] B. W. Erickson and R. B. Merrifield, *The Proteins*, Academic Press, New York 1976, p. 257.
[2] E. Atherton, H. Fox, D. Harkiss, C. J. Logan, R. C. Sheppard, and B. J. Williams, *J.C.S. Chem. Comm.*, 537 (1978).
[3] E. Atherton, H. Fox, D. Harkiss, and R. C. Sheppard, *ibid.*, 539 (1978).

2-Fluoro-1,3,5-trinitrobenzene,   (1). Mol. wt. 231.10, m.p. 127–128°. This reagent is prepared by nitration of 1-fluoro-2,4-dinitrobenzene with fuming HNO_3 and 25% fuming H_2SO_4.

Amides; esters. Treatment of a carboxylic acid with **1** and triethylamine in CH_3CN at room temperature gives an intermediate, probably **a**, that reacts with aniline to form anilides in high yield. The intermediate reacts with an alcohol to form an ester. Both reactions are subject to steric hindrance.[1]

[1] H. Kotake, K. Inomata, H. Kinoshita, K. Tanabe, and O. Miyano, *Chem. Letters*, 647 (1977); K. Inomata, H. Kinoshita, H. Fukuda, K. Tanabe, and H. Kotake, *Bull. Chem. Soc. Japan*, **51**, 1866 (1978).

Formaldehyde, 1, 397–402; **2**, 200–201; **4**, 238–239; **5**, 312–315; **6**, 264–267; **7**, 158–160.

Dimethylation of primary amines (**4**, 238). Methylation of primary amines in methanol with paraformaldehyde (large excess) followed by reduction with sodium borohydride can be used for dimethylation of primary amines. If $NaBH_4$ is replaced by $NaCNBH_3$, dimethylation of primary amines or methylation of secondary amines can be achieved without reduction of carbonyl groups in the substrate.[1]

$$RNH_2 \xrightarrow{CH_2O,\ CH_3OH} RN(CH_2OCH_3)_2 \xrightarrow{NaCNBH_3} RN(CH_3)_2$$

Branched-chain furanoses. Ho[2] has described an efficient one-step synthesis of branched carbohydrates by a crossed aldol reaction of a sugar aldehyde and formaldehyde in the presence of potassium carbonate (pH 10).

Example:

This reaction was used to synthesize D-hamamelose (**3**).

Mannich reaction with acetals. The Mannich reaction of formaldehyde with acetals of aliphatic aldehydes with more than three carbon atoms gives acrylaldehydes (**1**) in 50–80% yield (equation I). In the case of a chiral acetal, the reaction proceeds with an optical yield of ~80%.[3]

(I)
$$\overset{*}{R}CH_2CH(OR')_2 \xrightarrow[50-80\%]{HCHO,\ HN(CH_3)_2 \cdot HCl,\ 80°} \overset{*}{R}\underset{\underset{CH_2}{\|}}{C}CHO$$

1

[1] H. Kapnang, G. Charles, B. L. Sondergam, and J. H. Hemo, *Tetrahedron Letters*, 3469 (1977).
[2] P.-T. Ho, *ibid.*, 1623 (1978).
[3] R. Menicagli and M. L. Wis, *J. Chem. Res. (S)*, 262 (1978).

Formamidenesulfinic acid (Thiourea dioxide), 4, 506; **5,** 668–669.

Reduction of sulfoxides. Sulfoxides are reduced to sulfides by this reagent when iodine is present as catalyst. The reaction is carried out in refluxing acetonitrile. Yields are 90–95%.[1]

[1] J. Drabowicz and M. Mikołajczyk, *Synthesis*, 542 (1978).

Formic acid, 1, 404–407; **2,** 202–203; **3,** 147; **4,** 239–240; **5,** 316–319; **7,** 160.

Heterocyclization. Olefinic ω-ethoxylactams are cyclized by formic acid (20°); the actual intermediate is probably a cyclic acylimmonium species.[1]

Examples:

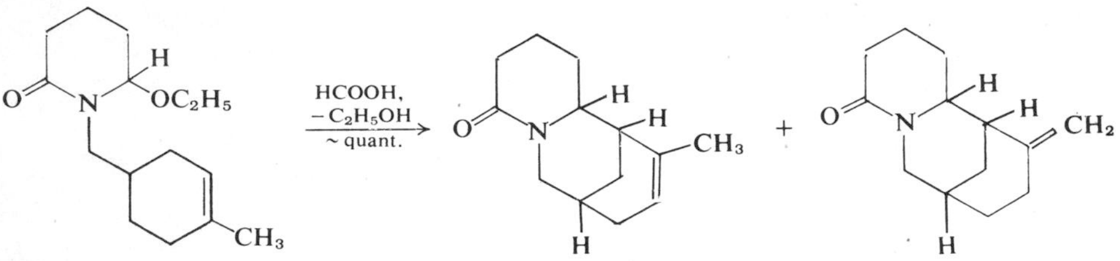

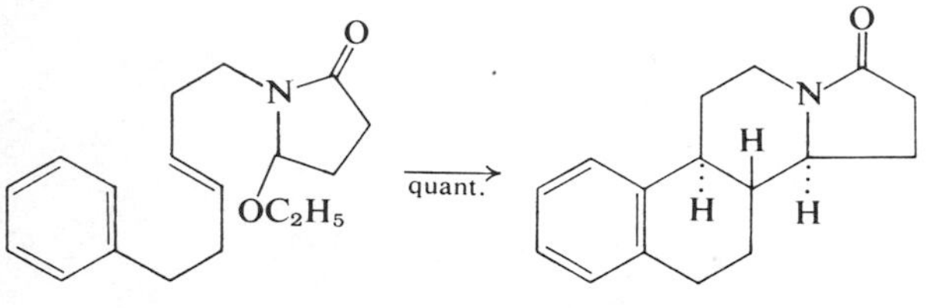

[1] H. E. Schoemaker, J. Dijkink, and W. N. Speckamp, *Tetrahedron*, **34**, 163 (1978);
J. Dijkink and W. N. Speckamp, *ibid.*, **34**, 173 (1978).

Formic acid–Formamide.

Reductive cleavage of indole alkaloids. Le Men *et al.* have reported the reductive cleavage of indolenine alkaloids (equation I)[1] and of oxindole alkaloids (equation II)[2] with formic acid and formamide.

(I)

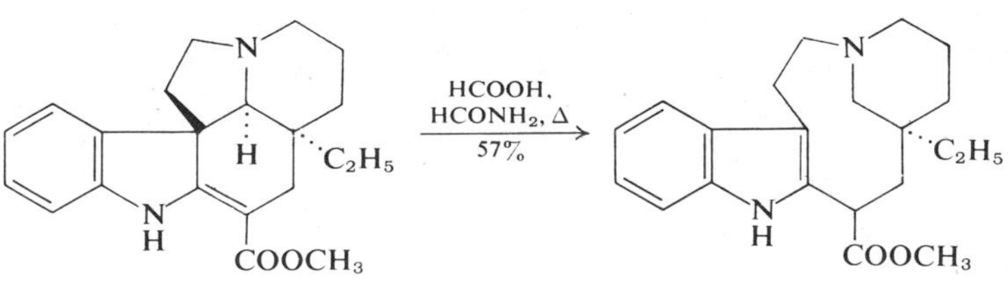

(II)

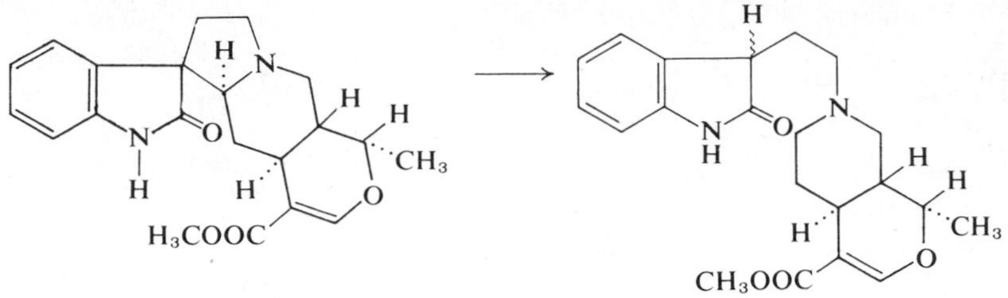

This cleavage has also been extended to indole alkaloids (reserpine) as shown in equation III.[3]

(III)

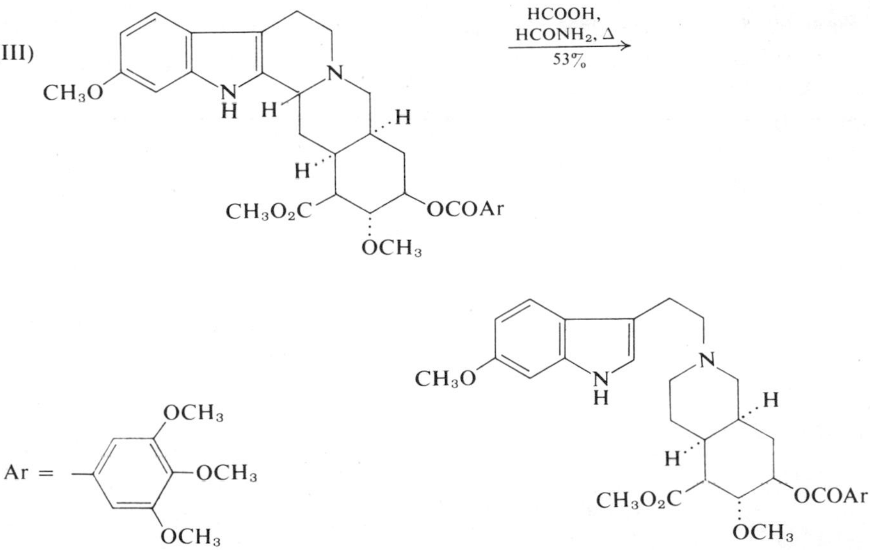

$$\xrightarrow[53\%]{\substack{HCOOH, \\ HCONH_2, \Delta}}$$

$$Ar = \underset{\underset{OCH_3}{|}}{\overset{OCH_3}{\overset{|}{\diagdown}}} OCH_3$$

[1] M. J. Hoizey, L. Olivier, J. Lévy, and J. Le Men, *Tetrahedron Letters*, 1011 (1971).
[2] F. Sigaut-Titeux, L. Le Men-Olivier, J. Lévy, and J. Le Men, *Heterocycles*, **6**, 1129 (1977).
[3] S. Sakai and M. Ogawa, *Chem. Pharm. Bull. Japan*, **26**, 678 (1978).

N-Formylimidazole, OHC—N **(1).** Mol. wt. 96.09. This reagent is prepared *in situ* from formic acid and N,N′-carbonyldiimidazole in THF at 0–5°.

Isoflavones. These compounds can be prepared by reaction of an *o*-hydroxy-desoxybenzoin with the reagent. Protection of hydroxyl groups other than the one participating in the cyclization is not necessary.[1]

Example:

[1] H. G. Krishnamurty and J. S. Prasad, *Tetrahedron Letters*, 3071 (1977).

Formylmethylenetriphenylphosphorane, 6, 267–268.

Conjugated (E,Z)-dienes.[1] These dienes can be prepared as shown in equation (I). Both Wittig reactions are stereoselective, and usually the final diene consists of the (E,Z)-isomer to the extent of 89–95%; the (Z,Z)-isomer is present to the extent of 3–6%, and the (E,E)-form and (Z,E)-form are only present in traces.

This synthesis was used to obtain bombykol (equation II) and various homologs.

(I) $R^1CHO + (C_6H_5)_3P{=}CHCH{=}O \xrightarrow[30-65\%]{}$

$$R^1\overset{H}{\underset{H}{C}}{=}CCH{=}O \xrightarrow[45-90\%]{R^2CH=P(C_6H_5)_3} R^1\overset{H}{C}{=}C\overset{H}{\underset{H}{C}}{=}CR^2$$

(II) $CH_3OCO(CH_2)_8CH{=}CH_2 \xrightarrow[66\%]{\substack{1)\,O_3 \\ 2)\,(C_6H_5)_3P}}$

$$CH_3OCO(CH_2)_8CH{=}O \xrightarrow[79\%]{\substack{1)\,(C_6H_5)_3P{=}CHCHO \\ 2)\,C_3H_7CH{=}P(C_6H_5)_3}}$$

$$C_3H_7\overset{H}{C}{=}\overset{H}{C}{-}\overset{H}{\underset{H}{C}}{=}C(CH_2)_8COOCH_3 \xrightarrow[92\%]{LiAlH_4} C_3H_7\overset{H}{C}{=}\overset{H}{C}{-}\overset{H}{\underset{H}{C}}{=}C(CH_2)_8CH_2OH$$

[1] H. J. Bestmann, O. Vostrowsky, H. Paulus, W. Billmann, and W. Stransky, *Tetrahedron Letters*, 121 (1977).

G

Grignard reagents, 1, 415–424; **2**, 205; **5**, 321; **6**, 269–270; **7**, 163–164.

Vinyl halides. The method of Normant *et al.* (**6**, 270) for preparation of vinylcopper compounds can be used to obtain vinyl halides.[1] Reaction of **1** with iodine gives vinyl iodides directly, but this reaction when extended to Br_2 or Cl_2 gives mainly dimers. The desired vinyl chlorides and bromides can be obtained with NCS or NBS in fair to good yields. The replacement occurs with retention of initial stereochemistry. The American group also stresses the importance of the purity of the copper salt and uses House's cuprous bromide complex with dimethyl sulfide (**6**, 270).

$$RMgBr + CuBr \cdot S(CH_3)_2 \longrightarrow RCu \cdot MgBr_2 \xrightarrow[\text{Ether}]{R^1C\equiv CH,}$$

(X = Cl, Br)

Chromens. Resorcinols when heated with an α,β-unsaturated aldehyde and pyridine are converted into chromens. Hydroquinones do not undergo this reaction. However, phenoxymagnesium bromides can be chromenylated, probably because of formation of a magnesium chelate. Thus treatment of the hydroquinone **1** with ethylmagnesium bromide and then with the dimethyl acetal of phytal in refluxing benzene gives **2**, dehydro-α-tocopherol. In the same way **3** can be converted into **4**.[2]

3 + CH₃CH=CHCH(OC₂H₅)₂ $\xrightarrow[\sim 40\%]{C_6H_6, \Delta}$ **4**

2-Substituted-1,3-benzodithioles.[3] These useful aldehyde derivatives (**2**) can be prepared easily by reaction of Grignard reagents with 2-alkoxy-1,3-benzo-

1 + R²MgBr $\xrightarrow[70-90\%]{Ether, \Delta}$ **2** + MgBrOR¹

dithioles, readily available by reaction of benzyne with CS_2 in the presence of alcohols.[4] In one example reported *n*-butyllithium was used.

1,3- and 1,4-Dienylic sulfoxides. 3-Sulfolenes react with 2 equiv. of Grignard reagents to give 1,3-dienylic sulfoxides in 20–66% yield. Usually the product is the (Z)-isomer, which can be converted into the (E)-isomer by irradiation in the presence of iodine. The reaction has been extended to bicyclic sulfolenes, obtainable by reduction of the dichlorocarbene adducts of sulfolenes.[5]

(R = H, CH₃)   (Z) $\xrightarrow[100\%]{I_2, h\nu}$ (E)

Synthesis of hindered ketones. Ketones can be prepared readily by condensation of an acid chloride with a Grignard reagent with catalysis by CuCl, which forms an organic cuprate of the Grignard reagent.[6] Subsequent work[7] has shown that the reaction is applicable to synthesis of highly hindered ketones, since highly hindered esters can be prepared by deprotonation of esters with LDA followed by alkylation (**5**, 402). Thus α,α,α-trisubstituted esters can be obtained by use of LDA and HMPT (which promotes enolization). However, the order of introduction of the alkyl groups is important. It is better to introduce the less bulky group in the early stages, an unexpected order. The synthesis of the most highly hindered ketone prepared in this way is shown in equation (I). The main

$$\text{(I)} \quad \underset{\underset{\text{CH}(CH_3)_2}{|}}{\overset{\overset{C(CH_3)_3}{|}}{C_2H_5C}}\text{—COCl} + (CH_3)_3CCH_2MgCl \xrightarrow[91\%]{CuCl} \underset{\underset{\text{CH}(CH_3)_2}{|}}{\overset{\overset{C(CH_3)_3}{|}}{C_2H_5C}}\text{—CO—CH}_2C(CH_3)_3$$

limit to the method is that highly hindered alkyl halides, the precursors to the Grignard reagent, are difficult to prepare.[8]

$R(CH_3)CuMgBr$. Mixed cuprates of this type were first prepared[9] by reaction of methylcopper (1 equiv.) and $RMgBr$ (1 equiv.), but these cuprates lack selectivity in transfer of the R group.[10] However, in the presence of excess $RMgBr$ the selectivity of transfer of the R group is increased, without decreasing the selectivity of 1,4-addition. In this case the reactive species is probably $(R)_2(CH_3)Cu(MgBr)_2$ or $(R)_3(CH_3)Cu(MgBr)_3$ or a mixture of the two.[11]

Examples:

$$\mathbf{1} + CH_3Cu + 5n\text{-}C_4H_9MgBr \xrightarrow[73\%]{\text{Ether}} \mathbf{2}\quad 97:3$$

$$\mathbf{1} + CH_3Cu + 3CH_2\!=\!CHMgBr \xrightarrow{78\%} \quad + \quad \mathbf{2}\quad 98:2$$

$$\mathbf{1} + CH_3Cu + 3(CH_3)_3SiC\!\equiv\!CCH_2CH_2MgBr \longrightarrow \quad (46\%) \quad + \quad \mathbf{2}\ (23\%)$$

A reagent of the stoichiometry $R_4(CH_3)Cu_3(MgBr)_2$ (**3**) has been prepared according to equation (I). It forms a yellow-orange powder suspended in a violet

$$\text{(I)} \quad CH_3Li + 3CuI + 4RMgBr \xrightarrow{\text{Ether}} \underset{\mathbf{3}}{R_4(CH_3)Cu_3(MgBr)_2} + 2MgBrI + LiI$$

ethereal phase. Both the powder and the ether-soluble reagent have essentially the same reactivity. These reagents all exhibit only 1,4-addition to enones with transfer of only the ligand R. They react even with hindered enones and are able to transfer even *t*-butyl groups. (The cuprates discussed above cannot transfer this group.) In addition the presence of excess RMgBr is not required for satisfactory yields.[12]

[1] A. B. Levy, P. Talley, and J. A. Dunford, *Tetrahedron Letters*, 3545 (1977).

[2] J. O. Asgill, L. Crombie, and D. A. Whiting, *J.C.S. Chem. Comm.*, 59 (1978).

[3] S. Ncube, A. Pelter, and K. Smith, *Tetrahedron Letters*, 255 (1977).

[4] J. Nakayama, *Synthesis*, **38**, 170 (1975).

[5] Y. Gaoni, *Tetrahedron Letters*, 4521 (1977).

[6] J.-E. Dubois and M. Boussu, *Tetrahedron*, **29**, 3943 (1973).

[7] J.-E. Dubois and J. A. McPhee, *J.C.S. Perkin I*, 694 (1977).

[8] C. Lion, J.-E. Dubois, and Y. Bonzougou, *J. Chem. Res. (M)*, 826 (1978).

[9] D. E. Bergbreiter and J. M. Killough, *J. Org.*, **41**, 2750 (1976).

[10] F. Leyendecker, J. Drouin, J. J. Debesse, and J.-M. Conia, *Tetrahedron Letters*, 1591 (1977).

[11] J. Drouin, F. Leyendecker, and J.-M. Conia, *Nouv. J. Chem.*, **2**, 267 (1978).

[12] F. Leyendecker, J. Drouin, and J.-M. Conia, *ibid.*, **2**, 271 (1978).

H

Hexachloroacetone, CCl_3COCCl_3. Mol. wt. 264.75, b.p. 66–70°/6 mm.

Chlorination of enamines. Hexachloroacetone can be used as a source of positive chlorine in a reaction of enamines to give, after acid hydrolysis, α-chloro ketones.[1]

One advantage observed with this reagent is that dichlorinated ketones are not observed. Moreover, the stereochemistry of the reaction can be controlled by the experimental conditions as shown in the chlorination of 6-methyl-1-pyrrolidino-1-cyclohexene (1), which can lead selectively to either *cis-* or *trans*-6-chloro-2-methylcyclohexanone.[2]

2 (86%)	:	3 (4%)	:	4 (10%)
2 (4%)	:	3 (94%)	:	4 (2%)

[1] F. M. Laskovics and E. M. Schulman, *Tetrahedron Letters*, 759 (1977).
[2] *Idem, Am. Soc.*, **99**, 6672 (1977).

Hexafluoroantimonic acid, 5, 309–310; **6**, 272–273; **7**, 166–167.

Tricyclic spiroenones. 1,3-Diarylpropanes of type **1** cyclize to spiroenones (**2**) in this super acid. The *p*-methoxy group in one aryl group is essential for the

reaction; the presence of a methoxy or acetylamino in the other phenyl group improves the reactivity.[1]

4,4-Dialkyl-1-tetralones. Alkyl phenyl ketones of type **1** are cyclized to tetralones (**2**) in this super acid through a carbonium ion intermediate.[2]

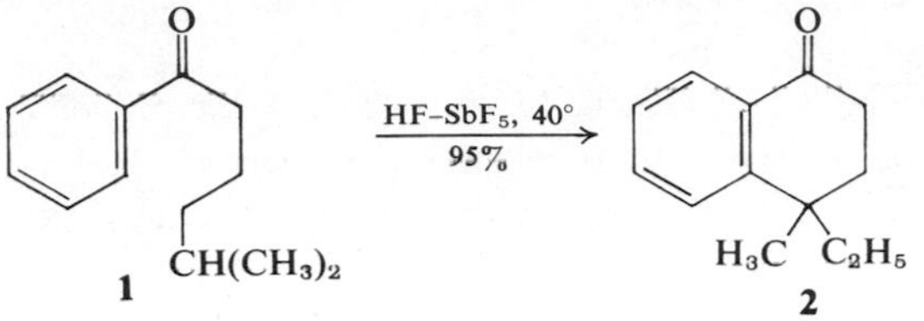

Ozonization of 3-ketosteroids. Ozonization of 3-ketosteroids in $HF-SbF_5$ results in oxidation to 3,6- and 3,7-diketosteroids.[3]

Example:

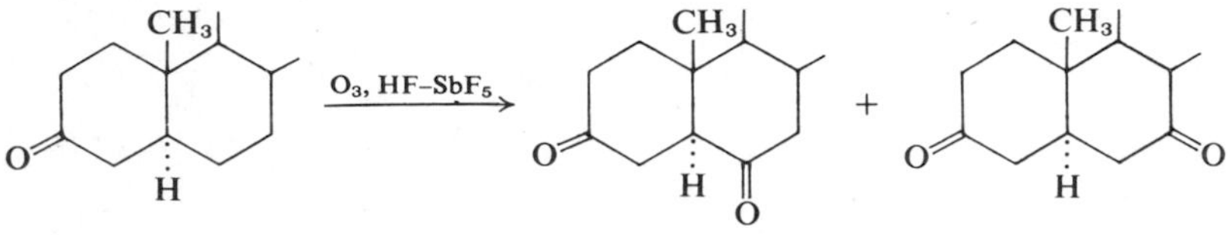

[1] J.-C. Jacquesy and M. P. Jouannetaud, *Bull. soc.*, II-202 (1978).
[2] N. Yoneda, Y. Takahashi, and A. Suzuki, *Chem. Letters*, 231 (1978).
[3] J.-C. Jacquesy and J.-F. Patoiseau, *Tetrahedron Letters*, 1499 (1977).

Hexamethyldisiloxane, $[(CH_3)_3Si]_2O$. Mol. wt. 162.38, b.p. 101. Supplier: Aldrich.

Aromatic acid chlorides. Trichloromethylarenes are converted into benzoyl chlorides by reaction with hexamethyldisiloxane in the presence of $FeCl_3$ at room temperature. Yields are in the range 75–85%.[1]

$$ArCCl_3 + [(CH_3)_3Si]_2O \xrightarrow[-(CH_3)_3SiCl]{FeCl_3,} [ArCCl_2OSi(CH_3)_3] \xrightarrow[75-85\%]{-(CH_3)_3SiCl} ArCOCl$$

[1] T. Nakano, K. Ohkawa, H. Matsumoto, and Y. Nagai, *J.C.S. Chem. Comm.*, 808 (1977).

Hexamethyldisilthiane, $[(CH_3)_3Si]_2S$. Mol. wt, 178.44, b.p. 163°. The disilthiane is prepared by reaction of chlorotrimethylsilane with sodium sulfide (96% yield).[1]

Reduction of sulfoxides. This reagent and the related hexamethylcyclotrisilthiane reduce sulfoxides to sulfides at 25–60° in 70–90% yield.[2]

[1] E. W. Abel, *J. Chem. Soc.*, 4933 (1961).
[2] H. S. D. Soysa and W. P. Weber, *Tetrahedron Letters*, 235 (1978).

Hexamethylphosphoric triamide (HMPT), 1, 430–431; **2,** 208–210; **3,** 149–153; **4,** 244–247; **5,** 323–325; **6,** 273–274; **7,** 168–170.

Reaction of methyl 1-lithio-1-selenophenyl acetate (**1**) *with enones.* The reagent

1 undergoes 1,2-addition to enones in THF at $-78°$, but in the presence of HMPT conjugate addition is the major reaction (equation I).[1] Similar results

obtain with methyl 1-lithio-1-selenomethyl acetate. The same solvent effect has been reported for reactions of $(C_6H_5CHCN)^-Li^+$.[2]

Alkylation of sulfoxides. The last steps in a recent synthesis of biotin (**4**) are shown in equation (I). The sulfide **1** was converted into the sulfoxide **2** since alkylation of sulfoxides is highly stereoselective, leading to substitution *trans* to the $S \rightarrow O$ bond. Alkylation of the carbanion of **2**, generated with CH_3Li in THF,

with *t*-butyl ω-iodovalerate at $-25°$ proceeded in only 42% yield. However, addition of HMPT increased the yield dramatically to 80%. Remaining steps to

4 involved reduction of the sulfoxide group, hydrolysis of the ester group, and debenzylation.[3]

Deprotonation of **2H-*thiopyrane*.** Deprotonation of 2*H*-thiopyrane (**1**) with *n*-butyllithium (and a trace of TMEDA) in THF at −60° gives 6-lithio-2*H*-thiopyrane (**2**), as shown by subsequent methylation. If pure, dry HMPT is added to the solution of **2**, the mesomeric anion **3** is formed within an hour. The transformation may involve protonation and deprotonation. The direct conversion of **1** to **3** occurs on reaction with either *n*-butyllithium or LDA at −60° in a 4:1 mixture of THF and HMPT. The tendency for formation of charge-

delocalized anions increases as the polarity of the solvent increases and is probably a rather common phenomenon.[4]

Reaction of C_6H_5OMgBr *with HCHO.* Phenoxymagnesium bromide (**1**) reacts with formaldehyde in benzene solution to form **2** and **3** through an intermediate *o*-quinone methide (**a**). In the presence of stoichiometric amounts of

HMPT, 2-hydroxybenzaldehyde (**4**) is obtained in yields as high as 85%. This oxidation–reduction process is an attractive synthetic route to 2-hydroxybenz-aldehydes.[5]

Claisen rearrangement of allyl esters. Ireland and Wilcox[6] have extended the studies on the Claisen rearrangement of enolates of allyl esters to γ,δ-un-saturated acids (**4**, 307–308; **6**, 276–279) to a stereoselective route to these unsaturated acids containing a β-methoxy group (aldol-type systems). As in previous work, stereoselective formation of the enolate can be achieved by use of THF as solvent to favor the (E)-type enolate *a* and of 23% HMPT–THF to favor the (Z)-type enolate *b*. These enolates rearrange to give diastereomeric products. Thus when the allyl ester **1** is converted into the silyl enol ether in THF alone and allowed to rearrange, a mixture of **2** and **3** in the ratio 82:18 (total yield 80%) is obtained. When these reactions are conducted in THF–HMPT, **2** and **3** are again obtained, but in the ratio 20:80 (total yield 75%). It is thus possible to obtain either product diastereomer irrespective of thermodynamic stability.

Smiles rearrangement. The Smiles rearrangement is ordinarily limited to substrates in which the aromatic ring is activated by an electron-withdrawing substituent.[7] However, if HMPT is used as solvent, 2-aryloxy-2-methylpropana-mides (**1**) are rearranged by sodium hydride to N-aryl-2-hydroxy-2-methyl-propanamides (**2**) in 50–85% yield. The products are hydrolyzed by acid to anilines (**3**).[8]

Modified Wittig reaction. The reaction of methylenetriphenylphosphorane (prepared *in situ* from methyltriphenylphosphonium bromide and *n*-butyllithium) with the hydroxy ketone **1** in ether is markedly improved by addition of 3 equiv. of HMPT. As the reaction proceeds, a complex of LiBr and HMPT separates from the ether solution. Probably the Wittig reaction becomes salt free, with enhancement of the rate and increase in yield from 24 to 66%. The product (**2**) is converted by hydrogenation into *erythro*-3,7-dimethylpentadecane-2-ol, esters of which are sex attractants of pine saw flies.[9]

Dehydrobromination. King and Paquette[10] have reported that HMPT is an excellent solvent for dehydrobrominations with lithium carbonate and lithium fluoride.[11] Powdered glass is added to facilitate CO_2 evolution.[12] These conditions were used for preparation of 1,3-cycloalkadienes by a bromination–dehydrobromination sequence.

Examples:

[1] J. Luchetti and A. Krief, *Tetrahedron Letters*, 2697 (1978).

[2] R. Sauvetre and J. Seyden-Penne, *ibid.*, 3949 (1976).

[3] S. Lavielle, S. Bory, B. Moreau, M. J. Luche, and A. Marquet, *Am. Soc.*, **100**, 1558 (1978).

[4] R. Gräfing and L. Brandsma, *Rec. trav.*, **97**, 208 (1978).

[5] G. Casiraghi, G. Casnati, M. Cornia, A. Pochini, G. Puglia, G. Sartori, and R. Ungaro, *J.C.S. Perkin I*, 318 (1978).

[6] R. E. Ireland and C. S. Wilcox, *Tetrahedron Letters*, 2839 (1977).

[7] J. F. Bunnett and R. E. Zahler, *Chem. Rev.*, **49**, 362 (1951); W. E. Truce, E. M. Kreider, and W. W. Brand, *Org. React.*, **18**, 99 (1970).

[8] R. Bayles, M. C. Johnson, R. F. Maisey, and R. W. Turner, *Synthesis*, 33 (1977).

[9] G. Magnusson, *Tetrahedron Letters*, 2713 (1977).

[10] M. R. Detty and L. A. Paquette, *ibid.*, 347 (1977); P. F. King and L. A. Paquette, *Synthesis*, 279 (1977).

[11] R. G. Pearson and J. Songstad, *Am. Soc.*, **89**, 1827 (1967).

[12] M. S. Newman and M. C. VanderZwan, *J. Org.*, **38**, 319 (1973).

Hydrazine, 1, 434–445; **2**, 211; **3**, 153; **4**, 248; **5**, 327–329; **6**, 280–281; **7**, 170–171.

Reaction with α,β-epoxy ketones. Swiss chemists[1] have reported a case in which the Wharton reaction (**1**, 439–440) led not only to the expected allylic alcohol but also to a product of cyclization. Stork and Williard[2] have investigated this reaction in detail and have observed several more examples of the formation of cyclized allylic alcohols in the Wharton reaction as shown in the examples. On the other hand, several related systems were converted only into the products expected from a Wharton reaction. At the present time, the subtle factors that lead to cyclization are not well understood.

Examples:

[1] G. Ohloff and G. Unde, *Helv.*, **53**, 531 (1970); K. N. Schulte-Elte, V. Rautenstrauch, and G. Ohloff, *ibid.*, **54**, 1805 (1971).

[2] G. Stork and P. G. Williard, *Am. Soc.*, **99**, 7067 (1977).

Hydriodic acid, 1, 449–450; **2**, 213–214.

α-Trimethylsilyl ketones. Treatment of a silyloxirane such as **1** with magnesium iodide results in rearrangement to a trimethylsilyl ketone (**3**).[1] However, yields are generally improved if **1** is first converted into the iodohydrin **2** by reaction with HI. The product **2** results from exclusive cleavage at the silylated carbon atom. On treatment with *n*-butyllithium **2** is converted into the desired ketone **3**. The yields in both steps are essentially quantitative.[2]

[1] M. Obayashi, K. Utimoto, and H. Nozaki, *Tetrahedron Letters*, 1807 (1977).
[2] *Idem, ibid.*, 1383 (1978).

Hydrogen chloride, 2, 215; **4**, 252; **5**, 335–336; **6**, 285; **7**, 172–173.

α,β-Dehydro amino acid esters. α-Imino amino acid esters (**2**), prepared by N-chlorination and dehydrochlorination of amino acid esters (**1**), rearrange to hydrochlorides of α,β-dehydro amino acid esters when treated with hydrogen chloride in absolute ether at −70°. The esters (**3**) are liberated from the salts by ammonia.[1]

The dehydro amino acid esters react with various aldehydes, including pyridoxal, to form imines. These imines (**4**) are believed to be involved in several reactions of pyridoxal-containing enzymes.[2]

[1] U. Schmidt and E. Öhler, *Angew Chem., Int. Ed.*, **16**, 327 (1977).
[2] U. Schmidt and E. Prantz, *ibid.*, **16**, 328 (1977).

Hydrogen peroxide, 1, 457–471; **2**, 216–217; **3**, 154–155; **4**, 253–255; **5**, 337–339; **6**, 286; **7**, 174.

Allylic hydroperoxides. Allylic chlorides can be converted into allylic hydroperoxides by reaction with excess 98% H_2O_2 (FMC Corp.) and either silver nitrate or silver triflate in the dark under argon.[1]

Examples:

Oxidation of stabilized phosphorus ylides. β-Carotene and other symmetrical carotenoids can be prepared conveniently by oxidation of phosphorus ylides with hydrogen peroxide (30%) and sodium carbonate in aqueous solution.[2]

Cleavage of α-ketols.[3] α-Ketols of type **1** are oxidatively cleaved by alkaline hydrogen peroxide to carboxylic acids and ketones. Acyloins (**2**) are also cleaved to carboxylic acids and to aldehydes, which are oxidized further to carboxylic acids. This oxidation has possible use for cleavage of α-ketols containing groups susceptible to oxidation with the usual reagents used for this purpose: HIO_4, $Pb(OAc)_4$, RCOOOH.

[1] A. A. Frimer, *J. Org.*, **42**, 3194 (1977).
[2] A. Nürrenbach, J. Paust, H. Pommer, J. Schneider, and B. Schulz, *Ann.*, 1146 (1977).
[3] Y. Ogata, Y. Sawaki, and M. Shiroyama, *J. Org.*, **42**, 4061 (1977).

Hydrogen peroxide–Diisopropylcarbodiimide.

Arene oxides. The reaction of this carbodiimide (supplier: Aldrich) with hydrogen peroxide generates a peroxycarboximidic acid (1) (*cf.* Payne's reagent **6**, 455–456). In combination with acetic acid, 1 converts phenanthrene into the 9,10-oxide (28% yield) and pyrene into the 4,5-oxide (27% yield). Some other carbodiimides can be used in this reaction.[1]

$$(CH_3)_2CHN{=}C{=}NCH(CH_3)_2 \xrightarrow{H_2O_2} (CH_3)_2CHNHC\overset{\displaystyle NCH(CH_3)_2}{\underset{\displaystyle OOH}{\diagup\!\!\diagdown}}$$

1

[1] S. Krishnan, D. G. Kuhn, and G. A. Hamilton, *Tetrahedron Letters*, 1369 (1977).

Hydrogen peroxide–Formic acid.

Penicillin and cephalosporin sulfoxides. A number of reagents have been used to prepare these sulfoxides. Actually hydrogen peroxide (32%), one of the cheapest reagents, in CH_2Cl_2 in combination with excess formic acid (or acetic acid) appears to be the best reagent. Yields of 85–100% of sulfoxides are obtainable in this way.[1] Excess acid minimizes sulfone formation and decreases the reaction time.[1]

[1] A. Mangia, *Synthesis*, 361 (1978).

Hydrogen peroxide–Hydrochloric acid, 7, 174–175.

vic-Dihalides.[1] The reaction of alkenes with conc. HCl or HBr and 30% H_2O_2 results in *vic*-dichlorides or dibromides. Use of a cosolvent (CCl_4) and catalysis with benzyltriethylammonium chloride improve the yield by suppression of formation of chlorohydrins. Yields are in the range 55–95%, being higher in the case of *vic*-dibromides. This halogenation is an extension of the method used to convert phenol into chloranil (**7**, 174–175).

[1] T.-L. Ho, B. G. B. Gupta, and G. A. Olah, *Synthesis*, **676** (1977).

Hydrogen peroxide–Mercury(II) trifluoroacetate, H_2O_2–$Hg(OCOCF_3)_2$.

Bicyclic peroxides. Reaction of 1,5-cyclooctadiene (1) with H_2O_2 (85–95%) and $Hg(OCOCF_3)_2$ (**6**, 360) in CH_2Cl_2 at 25° gives equal amounts (~40% yields) of the bicyclic peroxide **2** and the corresponding ether **3**. Reduction of **2** in CH_2Cl_2

with alkaline sodium borohydride gives the previously unknown cyclic peroxide **4** in about 25% yield together with **5**.[1]

The strain-free endoperoxide **7** has been obtained in the same way from 1,4-cyclooctadiene (**6**).[2]

[1] W. Adam, A. J. Bloodworth, H. J. Eggelte, and M. E. Loveitt, *Angew. Chem., Int. Ed.*, **17**, 209 (1978).
[2] A. J. Bloodworth and J. A. Khan, *Tetrahedron Letters*, 3075 (1978).

Hydrogen peroxide–Silver trifluoroacetate, 7, 175.

Prostaglandin endoperoxide.[1] Reaction of $9\beta,11\beta$-dibromo-9,11-dideoxy prostaglandin $F_{2\alpha}$ (**1**) with H_2O_2 (98%)[2] and silver trifluoroacetate (preparation: **5**, 522, 523) in ether gives prostaglandin H_2 methyl ester (**2**) in 20–25% yield. The product has been isolated by high-pressure liquid chromatography.

[1] N. A. Porter, J. D. Byers, R. C. Mebane, D. W. Gilmore, and J. R. Nixon, *J. Org.*, **43**, 2088 (1978).
[2] *Caution:* Prepared from 90% H_2O_2 by crystallization or by drying ethereal H_2O_2 solutions.

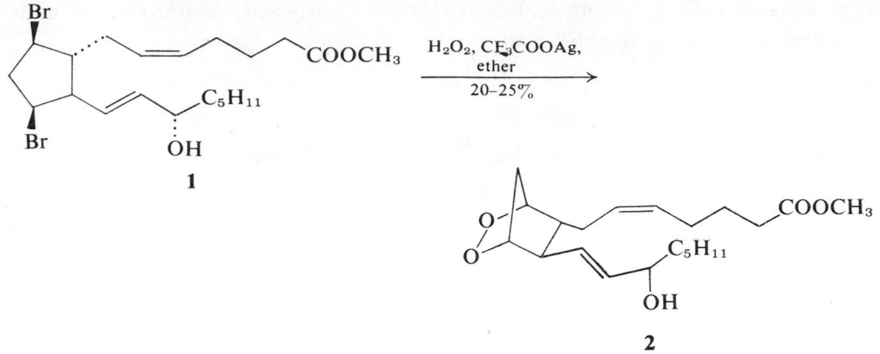

1

2

Hydrogen peroxide–Sodium peroxide, H_2O_2–Na_2O_2.

Alkyl hydroperoxides. N-alkyl-N'-tosylhydrazines are oxidized by H_2O_2 and Na_2O_2 in THF at 20° to give hydroperoxides in 90–95% yield. Tosylhydrazones are not intermediates, since these compounds are oxidized in low yields to dihydroperoxides $\left[\triangleright C(OOH)_2 \right]$.[1]

Examples:

$$(CH_3)_3CCH_2NHNHTs \xrightarrow[87\%]{H_2O_2,\ Na_2O_2} (CH_3)_3CCH_2OOH$$

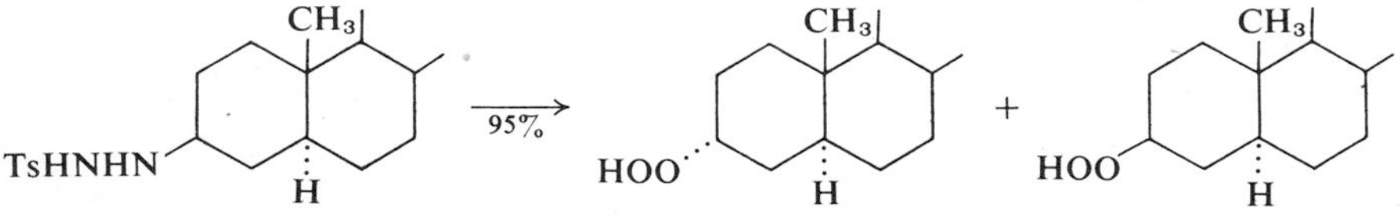

[1] L. Cagliotti, F. Gasparrini, D. Misiti, and G. Palmieri, *Tetrahedron*, **34**, 135 (1978).

Hydroxylamine-O-sulfonic acid, 1, 481–484; **2,** 217–219; **3,** 156–157; **4,** 256; **5,** 343–344; **6,** 290.

Hydrodeamination.[1] Primary amino groups can be replaced with hydrogen by treatment with excess hydroxylamine-O-sulfonic acid in an alkaline medium at 0°. The reaction is believed to involve N-amination to a hydrazine (**1,** 482) followed by reaction with nitrene (HN:), formed by reaction of hydroxylamine-O-sulfonic acid with base. Evidence for this mechanism is the conversion of benzyl-hydrazine to toluene (73% yield) under the same conditions.

$$RNH_2 \xrightarrow[OH^-,\ 0°]{NH_2OSO_3H,} \left[RNHNH_2 \xrightarrow{HN:} RN{=}NH \right] \xrightarrow[25-70\%]{-N_2} RH$$

The reaction is applicable to a wide range of substrates, in particular amino acids. Thus **1** can be converted into **2** in 31% yield.

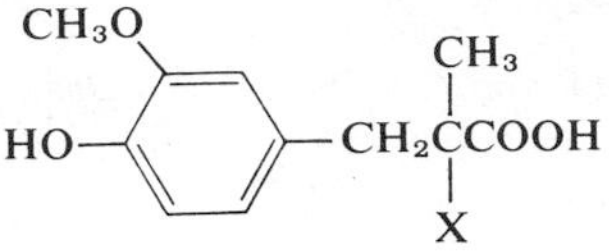

1 (X = NH$_2$)
2 (X = D)

[1] G. A. Doldouras and J. Kollonitsch, *Am. Soc.*, **100**, 341 (1978).

3-Hydroxy-3-methylbutylidenetriphenylphosphorane, lithium salt (1). Mol. wt. 354.35.

Preparation:

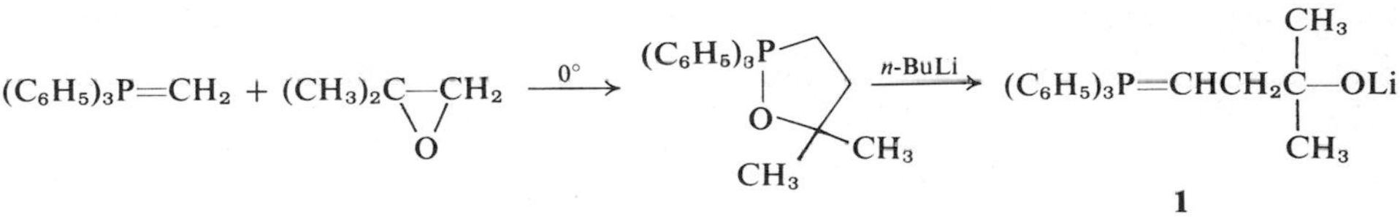

Stereoselective Wittig reaction.[1] The reagent has been used in a novel synthesis of 25-hydroxycholesterol (**5**). Thus the *i*-steroid aldehyde **2** reacts with **1** to form **3**. Two features of the reaction are noteworthy. The 20-methyl group does not epimerize and the Δ^{22}-double bond of the major product has the E-con-

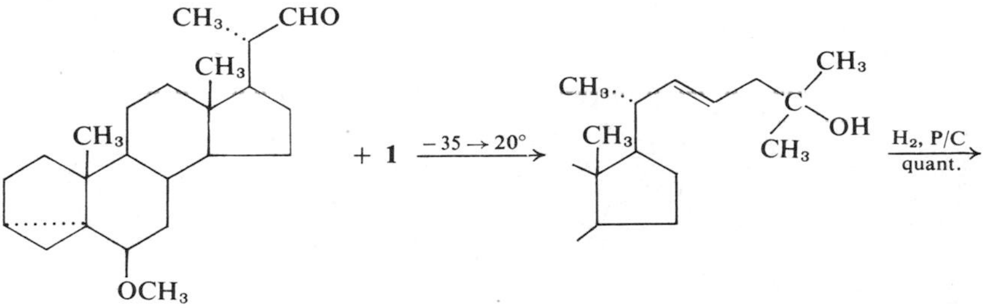

2

3 (E/Z = 85:15)

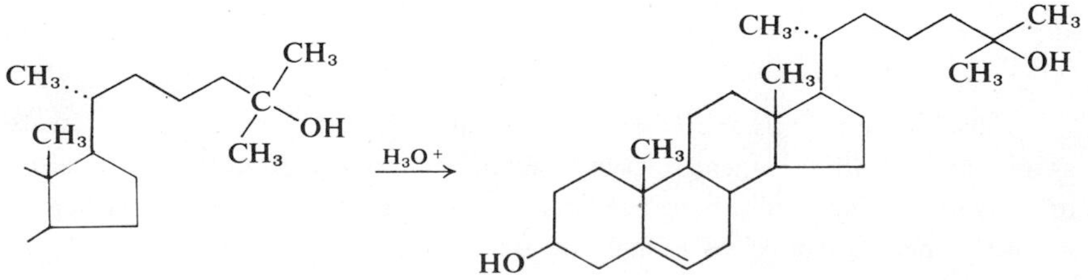

4

5

figuration. Normally, Z-olefins are the major products. The 25-alkoxide function is apparently responsible for this stereoselectivity, because replacement by $OSi(CH_3)_3$ results in Z-selectivity.

[1] W. G. Salmond, M. A. Barta, and J. L. Havens, *J. Org.*, **43**, 790 (1978).

2-Hydroxymethylanthraquinone, **(1)**. Mol.

wt. 238.24, m.p. 198–201°. This quinone is prepared by bromination of 2-methyl-anthraquinone followed by hydrolysis.

Protection of carboxyl groups. This reagent forms esters by reaction with a carboxylic acid in the presence of DCC and hydroxybenzotriazole. These esters have been used in syntheses of small peptides; they are cleaved by catalytic hydrogenation. A more unusual and very selective deblocking is achieved quantitatively by reduction to the hydroquinone by sodium hydrosulfite.[1]

[1] D. S. Kemp and J. Reczek, *Tetrahedron Letters*, 1031 (1977).

(−)-3-Hydroxy-5-methylhydantoin, **(1)**.   Mol.   wt.   130.11,

m.p. 163–164°, α_D −36.0°.
   Preparation:

Asymmetric peptide synthesis. This chiral hydroxyhydantoin has been used as an acyl activating reagent for synthesis of asymmetric peptides. Optical yields of 50–80% for dipeptides have been reported; even higher induction is obtained when the methyl group in **1** is replaced with isobutyl.[1]

[1] T. Teramoto and T. Kurosaki, *Tetrahedron Letters*, 1523 (1977).

(2S,2′S)-2-Hydroxymethyl-1-[(1-methylpyrrolidin-2-yl)methyl] pyrrolidine (1). Mol. wt. 198.31, b.p. 112°/4.5 mm., α_D −130°.

Preparation[1]:

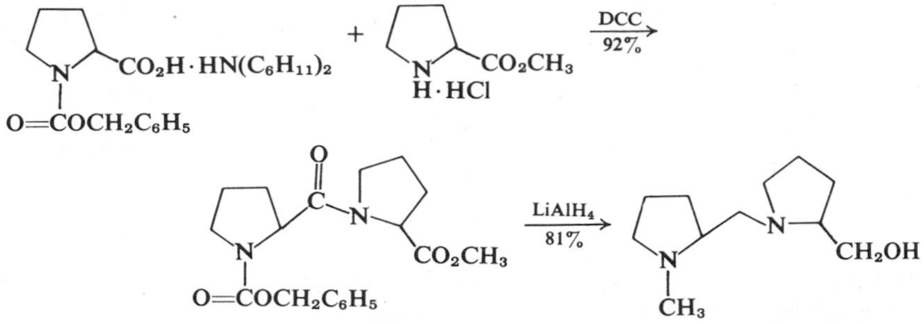

Asymmetric synthesis of alcohols. This ligand is effective for asymmetric addition of an alkyllithium to an aldehyde.[1] In more recent work[2] it has been found that the lithium anion (2) of **1**, prepared with *n*-butyllithium, is particularly effective for asymmetric addition of dialkylmagnesium compounds to aldehydes. All the alcohols have the (R)-configuration. Toluene was found to be a superior solvent for this reaction; lower temperatures increase optical yields. Only one example of use of an aliphatic aldehyde was reported; only a low optical yield was obtained.[2]

Examples:

$$C_6H_5CHO + (C_2H_5)_2Mg \xrightarrow[74\%]{2} C_6H_5\overset{*}{C}HC_2H_5$$
$$(92\% \text{ ee})$$

$$(CH_3)_2CHCHO + (n\text{-}C_4H_9)_2Mg \xrightarrow[70\%]{2} (CH_3)_2CH\overset{*}{C}HC_4H_9\text{-}n$$
$$(22\% \text{ ee})$$

[1] T. Mukaiyama, K. Soai, and S. Kobayashi, *Chem. Letters*, 219 (1978).
[2] T. Sato, K. Soai, K. Suzuki, and T. Mukaiyama, *ibid.*, 601 (1978).

N-Hydroxymethylphthalimide, **(1), 1, 484.** Additional suppliers: Aldrich, Eastman.

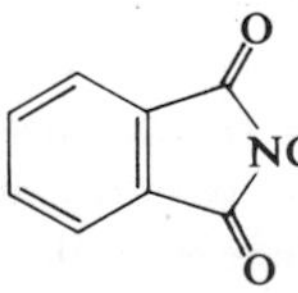

Sulfinic acids. A useful route to these intermediates in synthesis is shown in equation (I).[1]

(I) 1 $\xrightarrow{\text{HBr}}$ [structure] N—CH$_2$Br $\xrightarrow{\text{RSH}}$ [structure] N—CH$_2$SR $\xrightarrow{\text{H}_2\text{O}_2}$

[structure] N—CH$_2$SO$_2$R $\xrightarrow[\substack{80-100\% \\ \text{overall}}]{\text{C}_2\text{H}_5\text{ONa}}$ RSO$_2$Na + [structure] NCH$_2$OC$_2$H$_5$

[1] M. Uchino, K. Suzuki, and M. Sekiya, *Synthesis*, 794 (1977); *idem, Chem. Pharm. Bull. Japan*, **26**, 1837 (1978); *Org. Syn.*, submitted (1978).

3-Hydroxy-2-pyrone, 6, 291–292.

Diels-Alder reactions (**6**, 291–292). Under usual conditions, the Diels-Alder reactions of this pyrone are accompanied by loss of CO_2 from the adducts. If the reaction is conducted under high pressure (20–40 kbar), room temperatures suffice and loss of CO_2 is not observed. The products consist of *endo-* and *exo-* isomers, with the former predominating.[1]

Examples:

[reaction scheme: vinyl methyl ketone + pyrone → adduct, 100%]

[reaction scheme: chloroacrylonitrile + pyrone → adduct, 100%]

[1] J. A. Gladysz, S. J. Lee, J. A. V. Tomasello, and Y. S. Yu, *J. Org.*, **42**, 4170 (1977).

Hypofluorous acid, HOF. Mol. wt. 35.01; unstable above 77°K. The material is prepared in small quantities (*caution*) by the reaction of fluorine with ice at $-40°$ in a special apparatus.[1,2]

Hydroxylation of arenes.[2] The reagent behaves as though it is polarized in the sense $HO^{\delta+}—F^{\delta-}$ in contrast to HOCl, which is polarized $HO^{\delta-}—Cl^{\delta+}$. Thus HOCl chlorinates arenes, whereas HOF hydroxylates arenes. Benzene is converted into phenol (18.7%) and *o*-catechol (3.9%). *p*-Xylene gives 35.2% of 2,5-xylenol. Nitrobenzene reacts sluggishly. HOF is somewhat less reactive than Fenton's reagent (**1**, 472–474; **5**, 340), a radical reagent. The hydroxylation reaction with HOF resembles microsomal hydroxylation: both occur with a shift of hydrogen from the position of hydroxylation to the *o*-position.

A porphyrin N-oxide.[3] The first known oxide of this type has been obtained by reaction of octaethylporphyrin with HOF ($\sim 64\%$ yield). A more polar product may be an N,N'-dioxide. The N-oxygen atom is lost rather readily.

[1] E. H. Appelman, *Accts. Chem. Res.*, **6**, 113 (1973).
[2] E. H. Appelman, R. Bonnett, and B. Mateen, *Tetrahedron*, **33**, 2119 (1977).
[3] R. Bonnett, R. J. Ridge, and E. H. Appelman, *J.C.S. Chem. Comm.*, 310 (1978).

Hypophosphorus acid, H_3PO_2, 1, 489–491.

Reduction of arenediazonium tetrafluoroborates. H_3PO_2 in $CHCl_3$ in the presence of a trace of CuO reduces these salts to arenes in yields generally in the range of 90–99%. In one case, *p*-methoxybenzenediazonium tetrafluoroborate, the yield of *p*-methoxybenzene was only 67%; but when 18-crown-6 was added the yield improved to 88%.[1]

[1] S. H. Korzeniowski, L. Blum, and G. W. Gokel, *J. Org.*, **42**, 1469 (1977).

I

N-Imino-N,N-dimethyl-2-hydroxypropanaminium ylide (1). Mol. wt. 104.16.
Preparation:

$$H_2N-N(CH_3)_2 + \overset{O}{\triangle} \xrightarrow{CH_3CH(OH)_2, 50°} H\bar{N}-\overset{\overset{\displaystyle CH_3}{|}}{\underset{\overset{\displaystyle |}{CH_3}}{N^+}}-CH_2CH_2OH$$

1

Nitriles.[1] The reagent converts aldehydes into nitriles at room temperature in 70–90% yield (equation I).

$$(I) \quad RCHO + 1 \longrightarrow [RCH{=}N{-}\overset{+}{N}(CH_3)_2CH_2CH_2OH \ \bar{O}H] \longrightarrow$$

$$RC{\equiv}N + H_2O + (CH_3)_2NCH_2CH_2OH$$

[1] I. Ikeda, Y. Machii, and M. Okahara, *Synthesis*, 301 (1978).

Iodine, 1, 495–500; **2,** 220–222; **3,** 159–160; **4,** 258–260; **5,** 346–347; **6,** 293–295; **7,** 179–181.

Prostacyclin (PGI$_2$). This unstable intermediate in the biosynthesis of prostaglandins is a potent vasodilator and inhibitor of aggregation of human blood platelets. The synthesis was achieved in five laboratories in 1977.[1–5] The syntheses all involve conversion of prostaglandin F$_{2\alpha}$ methyl ester (**1**) into a cyclic halo ether (**2**) followed by dehydrohalogenation to **3**, the methyl ester of PGI$_2$. A number of positive halogen reagents can be used in the first step, but iodine or NBS[1] seems to be most useful. Both reagents favor formation of **2a**, which undergoes dehydrohalogenation more readily than **2b**. This elimination step can be carried out with either DBN or DBU. Sodium ethoxide can also be used, in which case prostacyclin is obtained as the stable sodium salt.

Stereoselective iodolactonization; stereoselective epoxidation. Iodolactonization of γ,δ- and δ,ε-unsaturated acids with iodine in the presence of $NaHCO_3$ exhibits only slight stereoselectivity (kinetic control). In the absence of base, equilibration occurs to give the more stable *trans*-isomers. Epoxides are formed in quantitative yield on methanolysis ($CH_3OH + Na_2CO_3$) of the iodolactones.[6]

Examples:

γ-*Methylenebutyrolactones.* These lactones can be prepared by orthoester Claisen rearrangement–hydrolysis of allylic alcohols (1) followed by iodolactonization of the products (3) either with iodine (alone) or a KI_3 solution in the presence of sodium bicarbonate. The final step is dehydroiodination with DBU in benzene at 60–80°. One advantage of this route is that the products are virtually free from the endocyclic double bond isomers.[7]

$$HOCH_2CH=C\underset{R^2}{\overset{R^1}{\diagdown}} \xrightarrow[145°]{CH_3C(OR)_3,} \left[CH_2=CH\underset{R^2}{\overset{R^1}{\underset{|}{C}}}CH_2COOR \right] \xrightarrow{OH^-}$$

1 (R = H, CH₃, C₂H₅)   **2**

$$CH_2=CH\underset{R^2}{\overset{R^1}{\underset{|}{C}}}CH_2COOH \xrightarrow[H_2O]{I_2, NaHCO_3,} \quad \mathbf{4} \xrightarrow[\substack{48-56\% \\ overall}]{DBU, \\ C_6H_6, 60°} \quad \mathbf{5}$$

3   **4**   **5**

trans,vic-*Iodohydrins.* Iodine with water in 1:1 tetramethylene sulfone–chloroform converts alkenes into *trans,vic*-iodohydrins, often in high yields. Use of sodium acetate in place of water results in *trans*-iodo acetates.[8]

$$\text{cyclohexene} \xrightarrow[81\%]{I_2, H_2O} \text{(2-iodocyclohexanol)}$$

$$C_6H_5CH=CH_2 \xrightarrow{87\%} C_6H_5\underset{OH}{\overset{}{\underset{|}{CH}}}-CH_2I$$

The same solvent system was shown to be useful in other electrophilic reactions of iodine. Thus the reaction of cyclohexene with sodium azide and iodine in this medium results in the *trans*-iodo azide (95% yield).

α,β-*Epoxy esters.*[9] An alternative to the Darzens route to these compounds is the oxidation of dianions of β-hydroxy esters, readily available by reaction of ethyl lithioacetate with carbonyl compounds (3, 172). The dianion of the hydroxy ester on oxidation with iodine at −78° is converted into the less hindered oxide; that is, the smaller of the R groups is *cis* to the carboethoxy group. Overall yields from the carbonyl compound are in the range 30–50%.

$$R^1-\underset{R^2}{\overset{OH}{\underset{|}{\underset{|}{C}}}}-CH_2COOC_2H_5 \xrightarrow{2LDA} \left[R^1-\underset{R^2}{\overset{OLi}{\underset{|}{\underset{|}{C}}}}-CH=\overset{OLi}{\underset{|}{C}}OC_2H_5 \right] \xrightarrow{I_2, -78°} R^1\underset{R^2}{\overset{O}{\diagup\diagdown}}\underset{COOC_2H_5}{H}$$

(30–50%)

2-Alkynoates; 1-alkynyl aryl ketones. The synthesis of acetylenes by reaction of iodine with lithium 1-alkynyltrialkylborates (**5**, 346) has been modified to provide a synthesis of acetylenic esters and ketones. Lithium diisopropylamide is used as base rather than *n*-butyllithium as in the original method for preparation of the ate complex.[10]

Examples:

$$HC{\equiv}CCOOC_2H_5 \xrightarrow[\text{quant.}]{\text{LDA, ether, } -76°} LiC{\equiv}CCOOC_2H_5 \xrightarrow{R_3B, THF}$$

$$Li[R_3BC{\equiv}CCOOC_2H_5] \xrightarrow[70–80\%]{I_2, 76°} RC{\equiv}CCOOC_2H_5$$

$$LiC{\equiv}CCOC_6H_5 \longrightarrow Li[R_3BC{\equiv}CCOC_6H_5] \xrightarrow[70–80\%]{I_2} RC{\equiv}CCOC_6H_5$$

Iodocyclization. Iodine reacts slowly with 1,6-heptadiene to form a mixture of acyclic mono and bis addition products. In contrast, the *gem*-dimethyl derivative **1** is converted in 85% yield into a mixture of two *cis* and *trans* cyclized diiodo compounds (**2**), whose structures were established by conversion to **3** and **4**.

Only acyclic addition compounds are obtained from monosubstituted 1,6-heptadienes. Use of bromine leads only to addition products. Iodocylization is also observed when transannular interactions are possible: 1,5-cyclodecadiene → 1,4-diodo-*cis*-decalin.[11]

[1] E. J. Corey, G. E. Keck, and I. Székely, *Am. Soc.*, **99**, 2006 (1977); E. J. Corey, H. L. Pearce, I. Székely, and M. Ishiguro, *Tetrahedron Letters*, 1023 (1978).

[2] R. A. Johnson, F. H. Lincoln, J. L. Thompson, E. G. Nidy, S. A. Mizsak, and U. Axen, *Am. Soc.*, **99**, 4182 (1977).

[3] K. C. Nicolaou, W. E. Barnette, G. P. Gasic, R. L. Magolda, and W. J. Sipio, *J.C.S. Chem. Comm.*, 630 (1977); K. C. Nicolaou, W. E. Barnette, and R. L. Magolda, *Org. Syn.*, submitted (1978).

[4] I. Tömösközi, G. Gambos, V. Simonidesz. and G. Kovaćs, *Tetrahedron Letters*, 2627 (1977).

[5] N. Whittaker, *ibid.*, 2805 (1977).

[6] P. A. Bartlett and J. Myerson, *Am. Soc.*, **100**, 3950 (1978).

[7] V. Jäger and H. J. Günther, *Tetrahedron Letters*, 2543 (1977).

[8] R. C. Cambie, W. I. Noall, G. J. Potter, P. S. Rutledge, and P. D. Woodgate, *J.C.S. Perkin I*, 226 (1977).

[9] G. A. Kraus and M. J. Taschner, *Tetrahedron Letters*, 4575 (1977).

[10] K. Yamada, M. Miyaura, M. Itoh, and A. Suzuki, *Synthesis*, 679 (1977).

[11] H. J. Günther, V. Jäger, and P. S. Skell, *Tetrahedron Letters*, 2539 (1977).

Iodine–Copper(II) acetate.

vic-*Alkoxyiodoalkanes.* Cyclohexene reacts with iodine in methanol in the presence of $Cu(OAc)_2 \cdot H_2O$ to give *trans*-1-iodo-2 methoxycyclohexane in 93% yield (equation I). The reaction is widely applicable, even to α,β-unsaturated (allylic) alcohols. Yields in general are in the range 30–95%.[1]

(I)

[1] C. Georgoulis and J. M. Valery, *Synthesis*, 402 (1978).

Iodine–Thallium(I) acetate.

α-*Iodo ketones.* α-Iodo ketones can be prepared by reaction of enol acetates with iodine (1 equiv.) and thallium(I) acetate in CH_2Cl_2 or $CHCl_3$ at 20°. Yields vary considerably from 50 to 90%.[1]

Examples:

[1] R. C. Cambie, R. C. Hayward, J. L. Jurlina, P. S. Rutledge, and P. D. Woodgate, *J.C.S. Perkin I*, 126 (1978).

Iodine azide, 1, 500–501; 2, 222–223; 3, 160–161; 4, 262; 5, 350–351; 6, 297.

2H-*Azirines* (2, 223). An example of the use of this reagent for synthesis of a 2*H*-azirine has been published in *Org. Syn.*[1]

$$C_6H_5\diagdown C=C \diagup^H_{CH(OCH_3)_2} \xrightarrow[97-98\%]{\substack{ICl,\ NaN_3,\\ CH_3CN}} C_6H_5\overset{\overset{\displaystyle I}{|}}{C}H\overset{\overset{\displaystyle }{|}}{C}H CH(OCH_3)_2 \xrightarrow[68-76\%]{\substack{KOC(CH_3)_3,\\ ether}}$$

(with N_3 below the central carbon)

$$C_6H_5\diagdown C=C \diagup^H_{CH(OCH_3)_2} \xrightarrow[78-93\%]{CHCl_3,\ \Delta} \underset{CH(OCH_3)_2}{C_6H_5{-}\!\!\!\diagdown\!\!=\!\!N} \xrightarrow[30\%]{H_3O^+} \underset{CHO}{C_6H_5{-}\!\!\!\diagdown\!\!=\!\!N}$$

(with N_3 below left carbon)

[1] A. Padwa, T. Blacklock, and A. Tremper, *Org. Syn.*, **57**, 83 (1977).

Iodine chloride, 1, 502.

Vinyl iodides; vinyl chlorides. Iodine chloride adds to *trans*-vinylsilanes to give mainly the adducts **a**, which on treatment with KF in DMSO are converted mainly into vinyl iodides. The reaction of *cis*-vinylsilanes is more dependent on the size of the R group. When R is bulky the intermediate **b** is favored; this is converted by KF into a vinyl chloride.[1]

(I) $$\underset{H}{\overset{R}{\diagdown}}C=C\underset{Si(CH_3)_3}{\overset{H}{\diagup}} \xrightarrow{ICl} \left[R{-}\overset{\overset{\displaystyle Cl}{|}}{C}{-}\overset{\overset{\displaystyle I}{|}}{C}{-}H\right] \xrightarrow[\sim 50-60\%]{KF} \underset{H}{\overset{R}{\diagdown}}C=C\underset{H}{\overset{I}{\diagup}}$$
(with H and Si(CH₃)₃ below the two central carbons)

R = *n*-Bu or C(CH₃)₃   **a**   (main product)

(II) $$\underset{H}{\overset{(CH_3)_3C}{\diagdown}}C=C\underset{H}{\overset{Si(CH_3)_3}{\diagup}} \xrightarrow{ICl}$$

$$\left[(CH_3)_3C{-}\overset{\overset{\displaystyle I}{|}}{\underset{\underset{\displaystyle H}{|}}{C}}{-}\overset{\overset{\displaystyle Cl}{|}}{\underset{\underset{\displaystyle H}{|}}{C}}{-}Si(CH_3)_3\right] \xrightarrow[\sim 70\%]{KF} \underset{H}{\overset{(CH_3)_3C}{\diagdown}}C=C\underset{H}{\overset{Cl}{\diagup}}$$

b   (main product)

[1] R. B. Miller and G. McGarvey, *Syn. Comm.*, **8**, 291 (1978).

Iodotrimethylsilane $(CH_3)_3SiI$. Mol. wt. 200.10, b.p. 106°, stable in the dark under N_2; reacts vigorously with H_2O. Supplier: Aldrich. The reagent is prepared most conveniently by reaction of hexamethyldisiloxane, aluminum powder, and iodine (equation I).[1]

(I) $(CH_3)_3SiCl + H_2O \xrightarrow[93\%]{C_6H_5N(CH_3)_2} (CH_3)_3SiOSi(CH_3)_3 \xrightarrow[87.5\%]{I_2,\ Al,\ 60°} (CH_3)_3SiI$

It has also been obtained in $\sim 55\%$ yield by reaction of trimethylphenylsilane with iodine.[2] This latter reaction has been used by Ho and Olah[3] for *in situ* generation of the reagent in some of the reactions mentioned below.

Ester hydrolysis. Two laboratories[3,4] have reported that esters are hydrolyzed in 80–95% yield by this reagent (equation II). Benzyl and *t*-butyl esters are hydrolyzed more readily than methyl or ethyl esters.

(II) $\qquad$ $RCOOR' + (CH_3)_3SiI \xrightarrow{\text{CCl}_4,\ 50°} RCOOSi(CH_3)_3 + R'I$

$$\downarrow{\scriptstyle 80-95\%\ |\ H_2O}$$

$$RCOOH + [(CH_3)_3Si]_2O$$

Ether cleavage.[5] The reagent is also useful for cleavage of aryl alkyl ethers to phenols (80–90% yield). It also cleaves dialkyl ethers to alkyl trimethylsilyl ethers which are then hydrolyzed to alcohols. This reaction is particularly useful for hydrolysis of methyl ethers (equation III).

(III)

Deoxygenation of sulfoxides.[6] Alkyl sulfoxides are reduced to sulfides on treatment with either bromo- or iodotrimethylsilane in CCl_4 at room temperature for 30 minutes (yields generally around 80%). Some halogenated products are also formed in the case of diaryl or dibenzyl sulfoxides.

Alkyl iodides. Alcohols (primary, secondary, and tertiary) can be converted into iodides by reaction with this reagent. The reaction can be carried out on the free alcohol or on the trimethylsilyl ether of the alcohol. Yields of alkyl iodides are usually about the same for both methods. The conversion proceeds with inversion.[7]

(IV) $\qquad$ $ROH \xrightarrow{(CH_3)_3SiI,\ CH_2Cl_2,\ 25°} RI + [(CH_3)_3Si]_2O$

$$ROH \xrightarrow[\quad]{(CH_3)_3SiCl,\ Py} ROSi(CH_3)_3 \Big] \Bigg\} \xrightarrow{(CH_3)_3SiI}$$

Cleavage of ketals.[8] Dimethyl and diethyl ketals are converted into ketones in 85–95% yield by treatment with iodotrimethylsilane in chloroform saturated with propene, added to eliminate traces of HI in the reagent. Ethylene ketals also react, but give a complex mixture of products. Thioketals are completely stable. The probable mechanism is shown in equation (I).

$$(I) \quad \underset{R^1}{\overset{R}{>}} \underset{OCH_3}{\overset{OCH_3}{C}} \xrightarrow{ISi(CH_3)_3} \underset{R^1}{\overset{R}{>}} \underset{OCH_3}{\overset{\overset{CH_3}{|}}{C}} \overset{O^+ - Si(CH_3)_3}{\underset{}{I^-}} \longrightarrow$$

$$\underset{R^1}{\overset{R}{>}} C \overset{+}{=} \overset{I^-}{O} - CH_3 + CH_3OSi(CH_3)_3$$

$$\downarrow$$

$$\underset{R^1}{\overset{R}{>}} C = O + CH_3I$$

Amines from alkyl carbamates. Iodotrimethylsilane reacts with alkyl carbamates to form trimethylsilyl carbamates, which on methanolysis afford amines (equation I).[9]

$$(I) \quad \underset{R^2}{\overset{R^1}{>}} NCOOR^3 \xrightarrow[-R^3I]{(CH_3)_3SiI} \underset{R^2}{\overset{R^1}{>}} NC \overset{OSi(CH_3)_3}{\underset{O}{<}} \xrightarrow[-CH_3OSi(CH_3)_3]{CH_3OH}$$

$$(95-100\%)$$

$$\left[\underset{R^2}{\overset{R^1}{>}} NC \overset{OH}{\underset{O}{<}} \right] \xrightarrow{-CO_2} \underset{R^2}{\overset{R^1}{>}} NH$$

$$(70-95\%)$$

[1] M. E. Jung and M. A. Lyster, *Org. Syn.*, submitted (1977).
[2] B. O. Pray, L. H. Sommer, G. M. Goldberg, G. T. Kerr, P. A. D. Giorgio, and F. C. Whitmore, *Am. Soc.*, **70**, 433 (1948).
[3] T.-L. Ho and G. A. Olah, *Angew. Chem.*, *Int. Ed.*, 744 (1976); *idem*, *Synthesis*, 917 (1977).
[4] M. E. Jung and M. A. Lyster, *Am. Soc.*, **99**, 968 (1977).
[5] *Idem*, *J. Org.*, **42**, 3761 (1977).
[6] G. A. Olah, B. G. B. Gupta, and S. C. Narang, *Synthesis*, 583 (1977).
[7] M. E. Jung and P. L. Ornstein, *Tetrahedron Letters*, 2659 (1977).
[8] M. E. Jung, W. A. Andrus, and P. L. Ornstein, *ibid.*, 4175 (1977).
[9] M. E. Jung and M. A. Lyster, *J.C.S. Chem. Comm.*, 315 (1978).

Ion-exchange resins, 1, 511–517; **2**, 227–228; **4**, 266–267; **5**, 355–356; **6**, 302–304; **7**, 182.

Nucleophilic substitutions.[1] Nucleophilic substitutions can be conducted in the solid phase by means of quaternary ammonium exchange resins. For example, alkylation of β-dicarbonyl compounds can be conducted by removal of the chloride ion from Amberlite IRA 900(Cl^-) and then treatment of the resulting resin with the β-dicarbonyl compound. The Amberlite IRA 900 β-diketonate is then treated with an alkyl halide in ethanol, hexane, or toluene.

Examples:

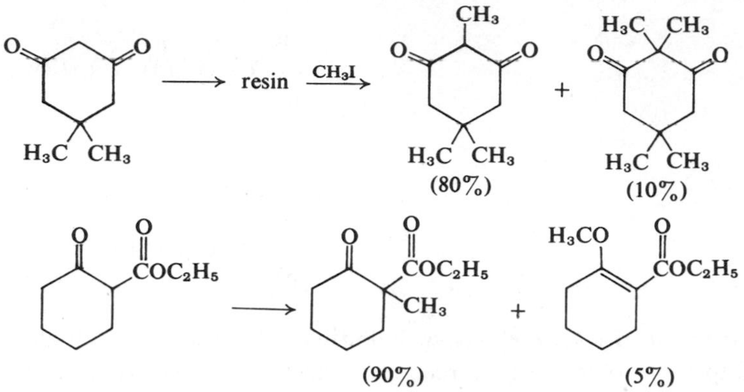

The same method was used to alkylate phenols with alkyl halides or mesylates and to effect halogen–NO$_2$ exchange. Yields compare favorably with other procedures.

F⁻-Anion exchange resins.[1] Fluoride ion immobilized on strongly basic anion exchange resins, particularly Amberlyst A27 and Dowex MSA-1, promote the various reactions that have been found to be promoted by alkali metal or tetraalkylammonium fluorides, such as C- and O-alkylation, sulfenylation, and Michael additions.

Examples:

$$CH_2(COCH_3)_2 + CH_3I \xrightarrow[70\%]{F^-} CH_3CH(COCH_3)_2$$

$$C_6H_5OH + CH_3I \xrightarrow[50\%]{} C_6H_5OCH_3$$

$$CH_3COCH{=}CH_2 + C_6H_5SH \xrightarrow[80\%]{} CH_3COCH_2CH_2SC_6H_5$$

[1] G. Gelbard and S. Colonna, *Synthesis*, 113 (1977).
[2] J. M. Miller, K.-H. So, and J. H. Clark, *J.C.S. Chem. Comm.*, 466 (1978).

Iron–Acetic acid.

Reductions of nitroarenes.[1] Nitroarenes were first reduced with iron and an acid by Bechamp in 1854. The use of iron in glacial acetic acid was reported in 1952 by Spring *et al.*[2] Owsley and Bloomfield have now made a systematic study of this method and conclude that it is a useful alternative to catalytic hydrogenation and is simpler to carry out than reductions with tin or tin(II) chloride with hydrochloric acid. Acetanilides are obtained as the product unless ethanol is used as cosolvent. Yields are generally satisfactory.

[1] D. C. Owsley and J. J. Bloomfield, *Synthesis*, 118 (1977).
[2] H. Berrie, G. T. Neubold, and F. S. Spring, *J. Chem. Soc.*, 2092 (1952).

Iron carbonyl [Fe(CO)$_5$], **1**, 519; **3**, 167; **5**, 357–358; **6**, 304–305; **7**, 183.

Deoxygenation of epoxides.[1] Iron carbonyl (1 equiv.) in N,N-dimethylacetamide or tetramethylurea deoxygenates epoxides (2 hours, 145°). No reaction occurs in refluxing THF. The reaction is not stereospecific: *trans*-stilbene oxide is converted into both *trans*-stilbene (56%) and *cis*-stilbene (22%). Epoxides of 1-alkenes are converted mainly into mixtures of internal alkenes.

Examples:

Carboxylic esters. Grignard reagents can be converted into esters with an alcohol and iron carbonyl as the supplier of CO (equation I).[2]

Ketones. Grignard reagents can be converted into ketones by reaction with Fe(CO)$_5$ in THF (25°, 1 hour) and then with an alkyl iodide (25°). Yields are somewhat improved by addition of 1-methylpyrrolidine-2-one to the solution.[3]

α-Diketones. A new synthesis of α-diketones involves the reaction of an ethylene dithioacetal of an aldehyde (**1**) with *n*-butyllithium followed by Fe(CO)$_5$ and then an alkyl iodide (equation I). Ethylene acetals are less useful in this reaction.[4]

Formylation and acylation of pyridine. Direct formylation and acylation at the β-position of pyridine is possible by reaction with phenyllithium and then with Fe(CO)₅. The reactions are formulated in scheme (I). 2-Phenylpyridine is also obtained in these reactions. Halide exchange of 3-bromopyridine with n-butyl-lithium to give 3-pyridinyllithium followed by reaction with Fe(CO)₅ gives only a 5% yield of 3-pyridinecarboxaldehyde.[5]

Scheme (I)

Lactones from dienes. Several 1,3-dienes have been converted into lactones by monoepoxidation and conversion to iron complexes with Fe(CO)₅ with insertion of CO. The lactone is obtained on oxidation with CAN.[6]
Examples:

Desulfurization. Iron carbonyl in the presence of potassium hydroxide forms potassium hydridotetracarbonyl ferrate, $K^+HFe(CO)_4^-$ (1). This hydride converts thioketones and thioamides to hydrocarbons and amines, respectively, in 50–80% yields.[7]

[1] H. Alper and D. D. Roches, *Tetrahedron Letters*, 4155 (1977).
[2] M. Yamashita and R. Suemitsu, *ibid.*, 1477 (1978).
[3] *Idem, ibid.*, 761 (1978).

⁴ *Idem, J.C.S. Chem. Comm.*, 691 (1977).
⁵ C.-S. Giam and K. Ueno, *Am. Soc.*, **99**, 3166 (1977).
⁶ G. D. Annis and S. V. Ley, *J.C.S. Chem. Comm.*, 581 (1977).
⁷ H. Alper and H.-N. Paik, *J. Org.*, **42**, 3522 (1977).

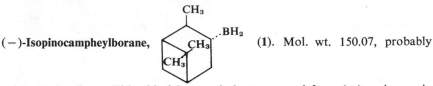

(−)-**Isopinocampheylborane,**   (1). Mol. wt. 150.07, probably

exists as the dimer. This chiral borane is best prepared from (+)-α-pinene via diisopinocampheylborane.¹

*Asymmetric hydroboration.*² Diisopinocampheylborane (**1**, 262–263; **4**, 161) is a useful reagent for asymmetric hydroboration of disubstituted olefins, but reacts only slowly with hindered trisubstituted olefins. For such substrates, the less hindered borane **1** is a useful reagent. Thus it reacts with 1-methylcyclopentene at −25°; oxidation leads to *trans*-2-methylcyclopentanol with an optical purity of 55.4%, with the new asymmetric center having the (S)-configuration. The paper reports asymmetric hydroboration of two other alkenes.

¹ H. C. Brown, J. R. Schevier, and B. Singaram, *J. Org.*, **43**, 4395 (1978).
² H. C. Brown and N. M. Yoon, *Am. Soc.*, **99**, 5514 (1977).

K

Ketene dimethyl thioacetal monoxide, $\begin{array}{c} CH_3S \\ \\ CH_3S \end{array}$ $C{=}CH_2$ (**1**). Mol. wt. 136.23.

(with S→O below)

Preparation (**5**, 302).

2-Alkyl-4-hydroxycyclopentenones. This system is a useful intermediate to prostaglandins. Schlessinger *et al.*[1] have developed a useful synthesis of **5** that uses **1** as a key reagent (equation I). The cyclization of **4** to **5** was carried out

$$(I) \quad 1 + CH_3\overset{O}{\overset{\|}{C}}CH(CH_2)_6COOCH_3 \xrightarrow[\substack{\text{quant.}\\(\text{crude})}]{\substack{KOC(CH_3)_3,\\HOC(CH_3)_3}} \overset{\substack{O\\\|\\CH_3C}}{\underset{\underset{\underset{\|}{O}}{\overset{CH_3S}{\underset{CH_3S}{\diagdown}}CH}}{\overset{SCH_3}{\overset{|}{\underset{}{C}}-(CH_2)_6COOCH_3}}} \xrightarrow[\substack{\text{quant.}\\(\text{crude})}]{\substack{HBF_4,\\CH_3CN}}$$

$$\underset{SCH_3}{|}$$

2   **3**

$$\overset{\substack{O\\\|}}{CH_3C}\overset{SCH_3}{\underset{\underset{\substack{|\\CH_2}}{C}}{\overset{|}{}}}(CH_2)_6COOCH_3 \xrightarrow[\substack{55\%\\\text{from 2}}]{\substack{\text{Adogen 464,}\\LiOH, H_2O, C_6H_6}} \text{(cyclopentenone with } (CH_2)_6COOCH_3, HO)$$

$$OHC$$

4   **5**

under phase-transfer conditions. Two other 2-alkyl-4-hydroxycyclopentenones were obtained in the same way in comparable yields.

[1] G. R. Kieczykowski, C. S. Pogonowski, J. E. Richman, and R. H. Schlessinger, *J. Org.*, **42**, 176 (1977).

L

Lead tetraacetate, 1, 537–563; 2, 234–238; 3, 168–171; 4, 278–282; 5, 365–370; 6, 313–317; 7, 185–188.

Ketone synthesis. Oxidation of N-acylamino acids with lead tetraacetate in HMPT–THF at 0–20° results in ketones in 60–80% yield.[1] This reaction was first reported in 1967,[2] but yields were lower because DMF was used as solvent.

Examples:

$$C_6H_5CH_2\underset{\underset{NHCOC_6H_5}{|}}{\overset{\overset{CH_2CH=C(CH_3)_2}{|}}{C}}-COOH \xrightarrow[82\%]{\underset{HMPT-THF}{Pb(OAc)_4,}} C_6H_5CH_2\overset{O}{\overset{||}{C}}CH_2CH=C(CH_3)_2$$

$$(CH_3)_2CH\underset{\underset{NHCOC_6H_5}{|}}{\overset{\overset{CH_2C_6H_4-Cl(p)}{|}}{C}}-COOH \xrightarrow[79\%]{} (CH_3)_2CH\overset{O}{\overset{||}{C}}CH_2C_6H_4-Cl(p)$$

The substrates are prepared by alkaline hydrolysis of 2-oxazoline-5-ones, available from amino acids by several routes. One route is formulated in equation (I).

$$(I) \quad R^1\underset{\underset{NHCOR^2}{|}}{CH}COOH \xrightarrow{Ac_2O} \underset{R^2}{\overset{R^1 \quad O}{\boxed{}}} \xrightarrow[2)\ R^3X]{1)\ Base}$$

$$\underset{R^2}{\overset{R^3 \quad O}{R^1-\boxed{}}} \xrightarrow[85-95\%]{OH^-} R^1-\underset{\underset{NHCOR^2}{|}}{\overset{\overset{R^3}{|}}{C}}-COOH$$

Cleavage of 1-trimethylsilyloxybicyclo[n.1.0]alkanes. These cyclopropane derivatives (1) are cleaved by LTA in glacial acetic acid to acyclic ω-unsaturated carboxylic acids (equation I). An intermediate cyclopropanol (*a*) is formed initially and then two bonds of the cyclopropane ring are cleaved by oxidation.[3]

(I)

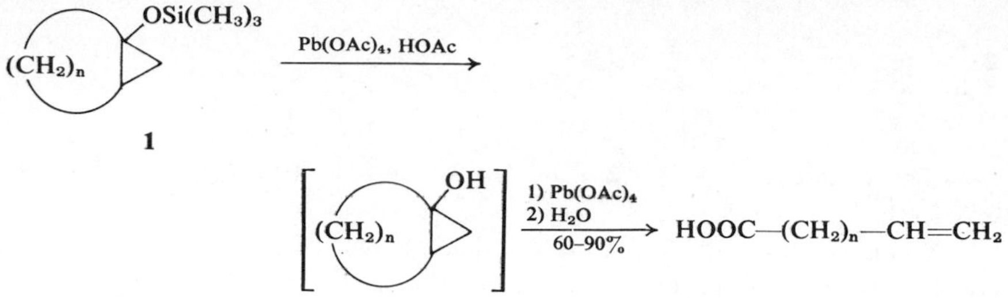

1

a   **2**

This cleavage reaction can also be applied to dichloronorcarenols such as **3** and **5** to give the carboxylic acids **4** and **6**. These can be converted by known methods

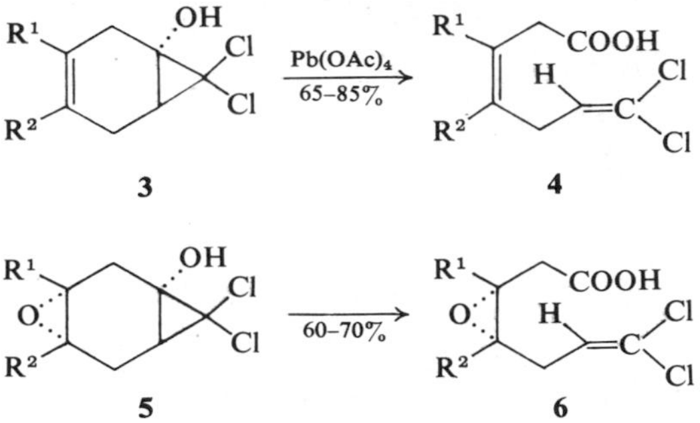

3   **4**

5   **6**

into 1,4-enynes. For example, crepenynic acid **7** was prepared from **4** ($R^1 = R^2 = H$) in about 25% overall yield.

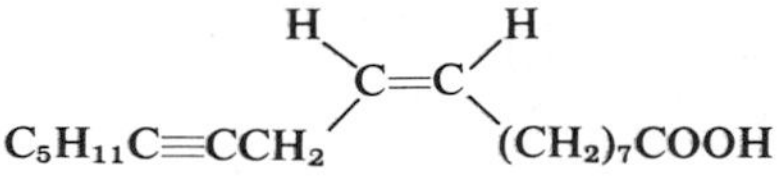

7

The starting materials (**3**) are prepared according to the sequence shown in equation (II).[4]

(II)

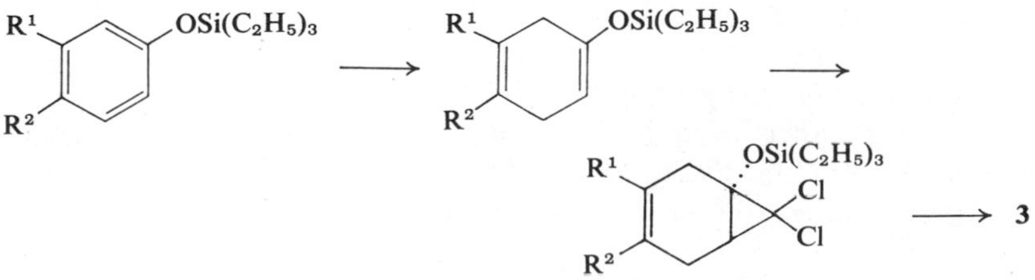

Oxidative decarboxylation (**1**, 554–557; **2**, 235–237; **3**, 168–169; **4**, 280; **6**, 313). This reaction is a key step in a new synthesis of barrelene [bicyclo[2.2.2]octa-2,5,7-triene (**4**)]. The overall yield is rather low, but the reactions are easy to carry out and involve inexpensive chemicals.[5]

Oxidation of o-*quinones*. *cis,cis*-Muconic acid dimethyl esters are obtained (60–90% yield) by oxidation of *o*-quinones with lead tetraacetate in C_6H_6–CH_3OH (1:1).[6]

[R¹–R⁴ = H, CH₃, C(CH₃)₃]

$[R^1–R^4 = H, CH_3, C(CH_3)_3]$

α-Diazo ketones. Oxidation of α-keto semicarbazones with Pb(OAc)₄ (2 equiv.) in CH_2Cl_2 followed by acidic work-up gives α-diazo ketones in 30–50% yield (equation I).[7]

Cleavage of a 1,2-diketone. Büchi and co-workers[8] have published an interesting synthesis of the dimethyl ester of betalamic acid (2), a unit in the red and yellow pigments of plants of the order Centrospermae. The acid is a dihydropyridine derivative and is very sensitive to oxidation. Consequently a synthesis was devised in which the dihydropyridine structure was obtained only in the final step. This step was the oxidative cleavage of the nonenolizable diketone 1 with lead tetraacetate in benzene–methanol (1:1, 0°) to give the dimethyl ester of betalamic acid.

1   2

[1] R. Lohmar and W. Steglich, *Angew. Chem., Int. Ed.*, **17**, 450 (1978).
[2] H. L. Needles and K. Ivanetich, *Chem. Ind.*, 581 (1967).
[3] G. M. Rubottom, R. Marrero, D. S. Drueger, and J. L. Schreiner, *Tetrahedron Letters*, 4013 (1977).
[4] T. L. Macdonald, *ibid.*, 4201 (1978).
[5] C. W. Jefford, T. W. Wallace, and M. Acar, *J. Org.*, **42**, 1654 (1977).
[6] M. Wiessler, *Tetrahedron Letters*, 233 (1977).
[7] D. Daniil, U. Merkle, and H. Meier, *Synthesis*, 535 (1978).
[8] G. Büchi, H. Fliri, and R. Shapiro, *J. Org.*, **42**, 2192 (1977).

L-*t*-Leucine *t*-butyl ester, $(CH_3)_3C\overset{*}{C}HCOOC(CH_3)_3$. Mol. wt. 187.28, α_D +54.6°.
$\underset{NH_2}{|}$

This ester is prepared from DL-*t*-leucine.[1]

Asymmetric alkylation of cyclohexanone. Hashimoto and Koga[2] have reported an asymmetric synthesis of α-alkylated cyclohexanones by conversion of cyclohexanone to the chiral imine 1 by reaction with the *t*-butyl ester of *t*-leucine. The imine is treated with LDA in THF at −78°, and after 30 minutes the alkylating agent is added to the lithioenamine (a). α-Alkylated cyclohexanones (2) are obtained in chemical yields of 60–75% and in optical yields of 84–98% (four examples).

This method was also applied to the asymmetric synthesis of α-methyl-α-phenylcyclopentanone and α-methyl-α-phenylcyclohexanone.

[1] E. Abderhalden, W. Faust, and E. Haase, *Z. physiol. Chem.*, **228**, 187 (1934).
[2] S. Hashimoto and K. Koga, *Tetrahedron Letters*, 573 (1978).

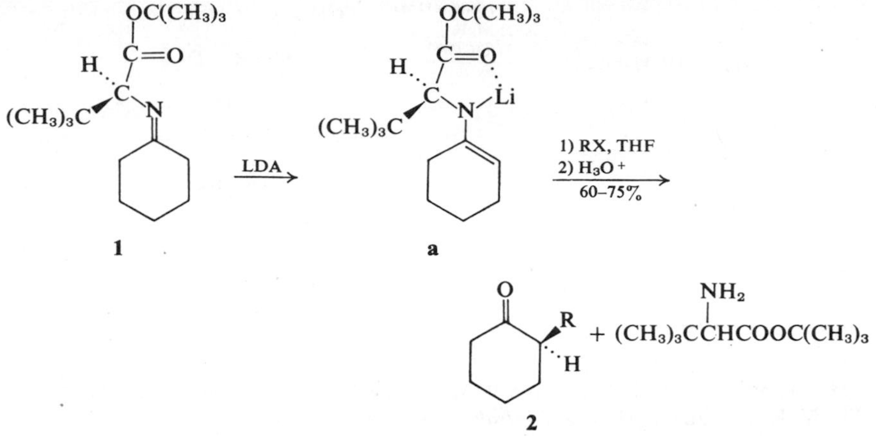

1) RX, THF
2) H_3O^+
60–75%

1   **a**

2

Lithioallyltrimethylsilane (1). Mol. wt. 120.19. This carbanion is prepared from allyltrimethylsilane (Petrarch Systems) by treatment with *n*-butyllithium in THF at -40 to $-78°$:

$$(CH_3)_3SiCH_2CH{=}CH_2 \xrightarrow{\textit{n-}\text{BuLi, THF}} (CH_3)_3Si\overset{\alpha}{CH}\cdots\overset{Li^+}{\underset{-}{CH}}\cdots\overset{\gamma}{CH_2}$$

1

γ-Butyrolactones.[1] This carbanion reacts with carbonyl compounds at the γ-position exclusively to form δ-hydroxyvinylsilanes with the (E)-configuration. These products are convertible into γ-lactols or γ-lactones. The sequence is illustrated for the reaction of **1** with cyclohexanone (equation I).

(I)

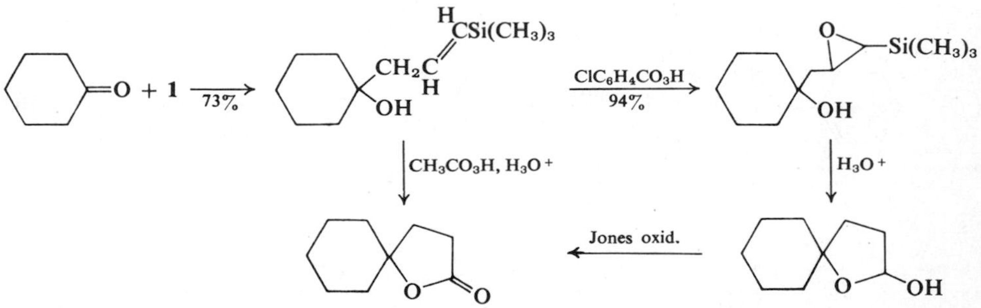

$$\text{O} + 1 \xrightarrow{73\%}$$

$\xrightarrow[94\%]{ClC_6H_4CO_3H}$

CH_3CO_3H, H_3O^+

H_3O^+

Jones oxid.

1,3-Dienes.[2] Aldehydes and ketones can be converted into 1,3-dienes (**2**) by reaction with the anion of allyltrimethylsilane (**1**) in the presence of 2 equiv. of freshly prepared $MgBr_2$ (**1**, 629–630). $MgBr_2$ is essential for control of the regioselectivity. In its absence γ-substitution predominates over the desired α-substitution.

$$CH_2=CHCH_2Si(CH_3)_3 \quad \xrightarrow{\begin{array}{l}1)\ t\text{-BuLi, HMPT}\\2)\ MgBr_2\\3)\ R^1COR^2\end{array}} \quad \begin{array}{l}CH_2=CHCHSi(CH_3)_3\\ \quad\quad\quad\quad |\\ R^1-C-OLi\\ \quad\quad |\\ \quad\quad R^2\end{array}$$

1   **a**

$$\xrightarrow[\text{40-55\%}]{SOCl_2}$$

$$CH_2=CHCH=C\begin{array}{l}R^1\\ \diagdown R^2\end{array}$$

2

[1] D. Ayalon-Chass, E. Ehlinger, and P. Magnus, *J.C.S. Chem. Comm.*, 772 (1977).
[2] P. W. K. Lau and T. H. Chan, *Tetrahedron Letters*, 2383 (1978).

2-Lithiobenzothiazole, [benzothiazole structure]—Li| (**1**). Mol. wt. 141.12. This reagent is generated by reaction of benzothiazole with *n*-butyllithium in ether at $-78°$.

Synthetic uses. Corey and Boger[1] have confirmed early observations of Gilman and Beel[2] that this reagent (denoted as BT-Li) adds to carbonyl compounds to form carbinols in good yield and have extended this reaction to a broad range of synthetic reactions. Thus the adduct of cyclohexanone with **1** can be used for the transformations shown in scheme (I). Hydrolysis of **4** or **6** to **5**

Scheme (I)

and **7**, respectively, can be accomplished by treatment with $AgNO_3$ in CH_3CN or by methylation with CH_3OSO_2F followed by treatment with aqueous K_2CO_3.

Unlike α,β-unsaturated aldehydes, which usually undergo 1,2-addition with carbon nucleophiles, vinyl BT's (enal equivalents) undergo 1,4-conjugate addition with various organolithiums in yields usually $>90\%$. A typical example is shown in scheme (II) using the same model compound (**2**). The allylic anion of **2** can be alkylated at the α-position to give a β,γ-unsaturated BT (**11**), convertible into **12**

Scheme (II)

by the usual conditions. The vinyl BT can also be hydrogenated in high yield to give **13**, which can be converted into the saturated aldehyde **14** or into the α-alkylated aldehyde **16** (scheme III)[3].

Scheme (III)

Conjugate addition to the vinyl BT derivative **2** can be used for annulation with either a five- or a six-membered ring (scheme IV). An angular alkyl group can also be introduced in both sequences.

Scheme (IV)

Spiroannulation has also been effected by alkylation of vinyl BT derivatives. An example is shown in scheme (V).

Scheme (V)

The same paper reports the synthesis of several other spiro systems, including **26** and **27**.[4]

26   **27**

[1] E. J. Corey and D. L. Boger, *Tetrahedron Letters*, 5 (1978).
[2] H. Gilman and J. A. Beel, *Am. Soc.*, **71**, 2328 (1949).
[3] E. J. Corey and D. L. Boger, *Tetrahedron Letters*, 9 (1978).
[4] *Idem, ibid.*, 13 (1978).

$$\overset{\displaystyle Li}{\underset{\displaystyle |}{}}$$

α-Lithio-α-chloromethyltrimethylsilane, $(CH_3)_3SiCHCl$ **(1); α-lithio-α-chloroethyltri-**

$$\overset{\displaystyle Li}{\underset{\displaystyle |}{}}$$

methylsilane, $(CH_3)_3SiC(Cl)CH_3$ **(2).** Both of these carbanions are obtained by treatment of the corresponding α-chloroalkyl(trimethyl)silane with *sec*-butyllithium in THF at $-80°$.

α,β-*Epoxysilanes*. Both of these reagents react with aldehydes and ketones to form α,β-epoxysilanes, which are useful precursors to carbonyl compounds. The α,β-epoxysilanes obtained in this way from **1** are converted into the homologous aldehyde by treatment with acid.[1] The α,β-epoxysilanes obtained from **2** are converted into homologous methyl ketones.[2]

Examples:

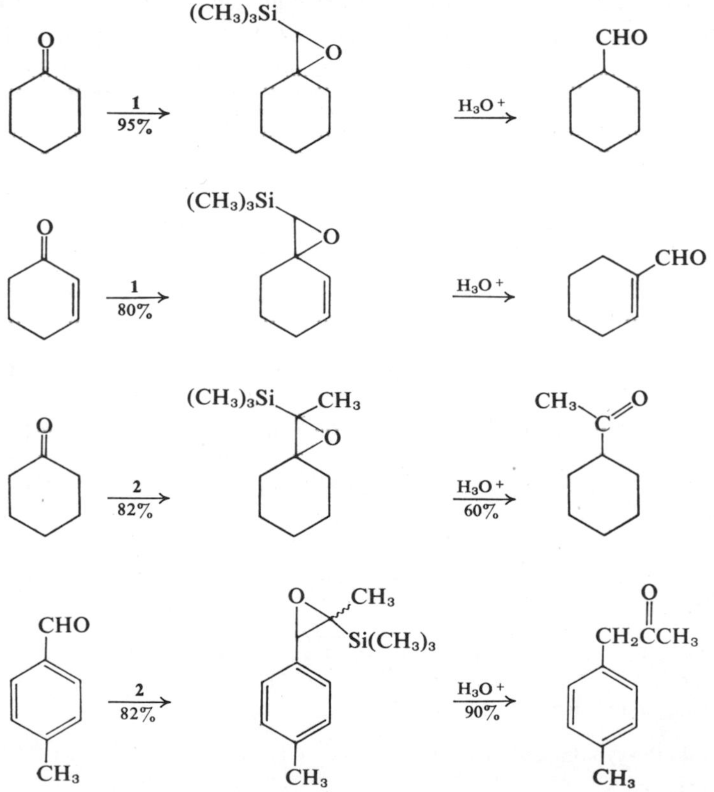

The reagent **1** has been used to convert a methyl ketone into a homologous enol formate. Thus dihydro-β-ionone **(3)** was transformed into *Latia* luciferin **(5)** by way of the intermediate α,β-epoxysilane **(4)**.[3]

The reagent **2** has been used to convert (−)-linalool (**6**) into (R)-(+)frontalin (**10**), the aggregation pheromone of the pine beetle *Dendroctonus frontalis*.[3] The overall yield is 23–29%, which is remarkably high for a chiral synthesis.

[1] C. Burford, F. Cooke, E. Ehlinger, and P. Magnus, *Am. Soc.*, **99**, 4536 (1977).
[2] F. Cooke and P. Magnus, *J.C.S. Chem. Comm.*, 513 (1977).
[3] P. Magnus and G. Roy, *ibid.*, 297 (1978).

1-Lithio-4-ethoxybutadiene, $LiCH=CHCH=CHOC_2H_5$ (**1**). Mol. wt. 104.08.
   Preparation:

$$HC\equiv CCH=CHOC_2H_5 + (n\text{-Bu})_3SnH \xrightarrow{AIBN;\ 90°}$$

$$(n\text{-Bu})_3SnCH=CHCH=CHOC_2H_5 \xrightarrow{n\text{-BuLi, THF, } -78°} 1$$

Conjugated dienals. This anion reacts with aldehydes and ketones to form the enol ethers (2), which rearrange to conjugated dienals (3) on chromatography or on brief treatment with TsOH.[1]

Examples:

$$C_6H_5CHO \xrightarrow{90\%} C_6H_5 \diagup\!\!\!\diagdown\!\!\!\diagup\!\!\!\diagdown CHO$$

(E/Z = ~5:1)

[1] R. H. Wollenberg, *Tetrahedron Letters*, 717 (1978).

2-(2-Lithio-4-methoxyphenyl)-4,4-dimethyl-2-oxazoline (1). Mol. wt. 211.18. The reagent is prepared by lithiation[1] of 2-(4-methoxyphenyl)-4,4-dimethyl-2-oxazoline.[2]

7,12-Dimethylbenz[a]*anthracene synthesis.*[3] The key step in a new synthesis of this system is the reaction of **1** with methyl 2-naphthyl ketone (**2**) to give, after acid hydrolysis, the phthalide **3** in 62% yield. This product is reduced to **4** with zinc in alkali. Remaining steps involve introduction of the 12-methyl group and cyclization. Similarly, methyl 1-naphthyl ketone affords 10-methoxy-7,12-dimethylbenz[a]anthracene in 27% overall yield.

[1] H. W. Gschwend and A. Hamdan, *J. Org.*, **40**, 2008 (1975).
[2] A. I. Meyers, D. L. Temple, D. Haidukewych, and E. D. Mihelich, *ibid.*, **39**, 2787 (1974).
[3] M. S. Newman and S. Kumar, *ibid.*, **43**, 370 (1978).

1 + [structure **2**] $\xrightarrow[\text{62\%}]{\text{Ether, 0}\rightarrow 35°\quad \text{H}_3\text{O}^+}$ [structure **3**] $\xrightarrow{\text{Zn, KOH}}$

[structure **4**] $\xrightarrow[\substack{33\% \\ \text{from 1}}]{\substack{1)\ \text{CH}_3\text{Li} \\ 2)\ \text{PPA}}}$ [structure **5**]

Lithiomethyl(diphenyl)arsine oxide, $(C_6H_5)_2\overset{\text{O}}{\overset{\|}{\text{As}}}CH_2Li$ (**1**). Mol. wt. 266.09. The reagent is obtained[1] in almost quantitative yield by reaction of methyl(diphenyl)-arsane oxide[2] with LDA in THF at $-40°$.

Homologation of alkyl halides. This reagent can be used for stepwise homologation of alkyl halides (equation I).

(I) $1 + CH_3(CH_2)_3Br \xrightarrow[\text{72\%}]{\text{THF, 65}°} (C_6H_5)_2\overset{\text{O}}{\overset{\|}{\text{As}}}CH_2(CH_2)_3CH_3 \xrightarrow[\text{98\%}]{\text{LiAlH}_4}$

$(C_6H_5)_2As(CH_2)_4CH_3 \xrightarrow{\text{Br}_2,\ 90°} Br(CH_2)_4CH_3 + BrAs(C_6H_5)_2$
$(79\%)(92.5\%)$

[1] T. Kauffmann, H. Fischer, and A. Woltermann, *Angew. Chem., Int. Ed.*, **16**, 53 (1977).
[2] G. J. Burrows and E. E. Turner, *J. Chem. Soc.*, **117**, 1381 (1920); *idem, ibid.*, **119**, 428 (1921); A. Merijanian and R. A. Zingaro, *Inorg. Chem.*, **5**, 187 (1966).

2-Lithio-3,3,6,6-tetramethoxy-1,4-cyclohexadiene, 1, 191–192.

Anthracyclinones. The use of latent quinone reagents such as 2-lithio-3,3,6,6-tetramethoxy-1,4-cyclohexadiene has been extended to a synthesis of an anthracyclinone. The starting material **1** was converted into the quinone bisketal **2** by anodic oxidation. The corresponding lithio compound was then condensed with dimethyl 3-methoxyphthalate (**3**). The reaction fortunately was stereoselective and resulted in **4** in satisfactory yield. The conversion of **4** to the anthracyclinone **5** was conducted in three steps without isolation of intermediates: reductive hydrolysis to the hydroquinone, saponification, and finally cyclization with hydrogen fluoride. The overall yield of **5** from 3-bromo-2,5-dimethoxybenzaldehyde, the precursor of **1**, was 8%.[1]

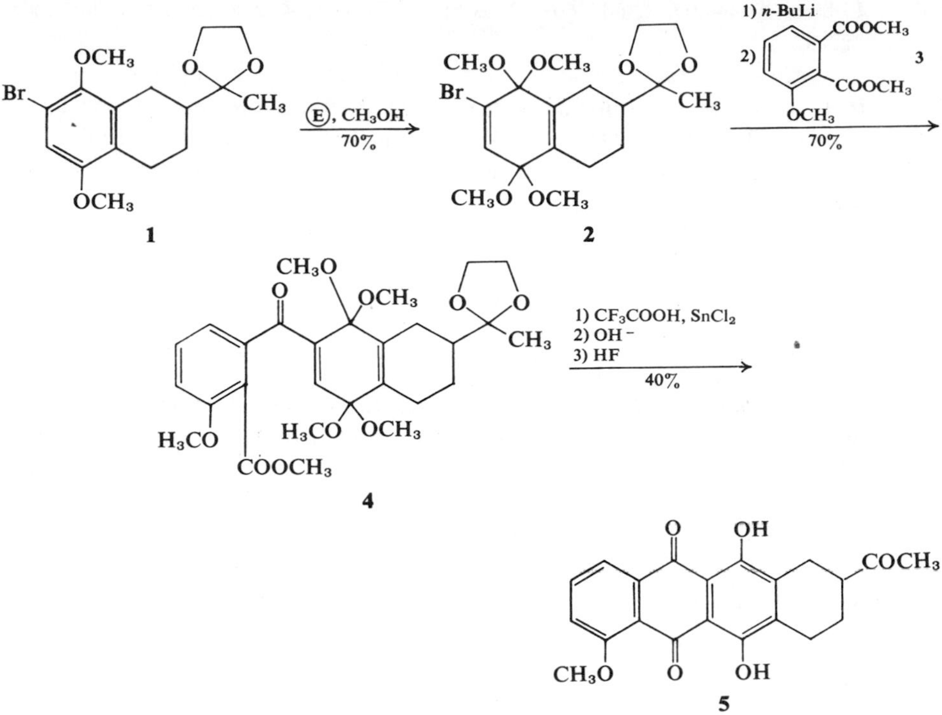

[1] P. W. Raynolds, M. J. Manning, and J. S. Swenton, *Tetrahedron Letters*, 2383 (1977); J. S. Swenton and P. W. Raynolds, *Am. Soc.*, **100**, 6188 (1978).

α-Lithiovinyl phenyl sulfide, $C_6H_5SC(Li)=CH_2$ **(1).** Mol. wt. 142.14. The reagent is prepared by metalation of phenyl vinyl sulfide[1] with LDA in THF–HMPT.

Ketene thioacetals. α-Lithiovinyl phenyl sulfide is useful for preparation of ketene thioacetals and related compounds.[2]

Examples:

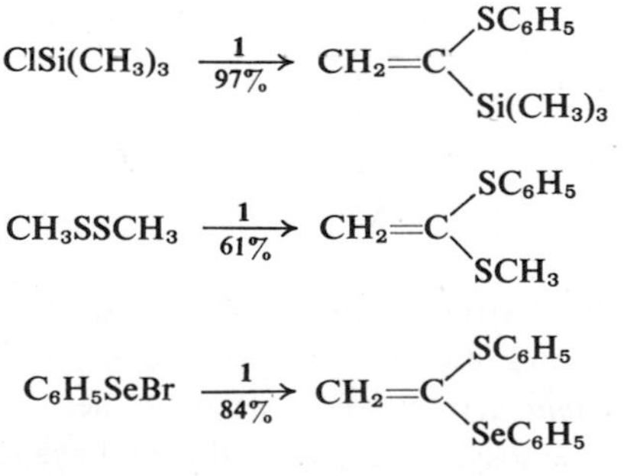

[1] H. Böhme and H. Bentler, *Ber.*, **89**, 1464 (1956).
[2] B. Harirchian and P. Magnus, *J.C.S. Chem. Comm.*, 522 (1977).

Lithium–Ammonia, 1, 601–603; **2**, 205; **3**, 179–182; **4**, 288–290; **5**, 379–381; **6**, 322–323; **7**, 195.

(E,E)-1,4-Dienes. Reduction of the air-sensitive 1,4-diynol **1** with lithium (6 equiv.) in liquid ammonia with *t*-butanol as the proton source and in the presence of $(NH_4)_2SO_4$ (10 equiv.) to prevent base-catalyzed isomerization and overreduction produces the (E,E)-dienes **2** in quantitative yield. The method is probably generally applicable.[1]

$$CH_3C{\equiv}C{-}CH_2{-}C{\equiv}C(CH_2)_3OH \xrightarrow[\text{quant.}]{\text{Li–NH}_3}$$

1   **2**

Reduction of aryl silyl ethers; nonconjugated cycloenones.[2] The isopropyldimethyl and *t*-butyldimethyl silyl ethers **1** are reduced by lithium–ammonia to dihydroaryl silyl ethers (**2**). These can be converted to lithium enolates (**a**). One

1 [R = $CH(CH_3)_2$
    or $C(CH_3)_3$]   **2**

a   **3** (70%)   **4** (25%)

useful transformation of **a** is acylation with acetic anhydride to give **3** and **4**. Unfortunately alkylation of **a** results in a plethora of products.

5 (96%)   **6** (<1%)

One particularly useful transformation of **2** is hydrolysis mediated by tetrabutylammonium fluoride to the nonconjugated enone (**5**).

Furanoid and pyranoid glycals. The classical method for preparation of glycals is reduction of pyranosyl halides with zinc and acetic acid. A more recent method is lithium–ammonia reduction of furanosyl and pyranosyl halides with a blocking acetonide group.[3]

Examples:

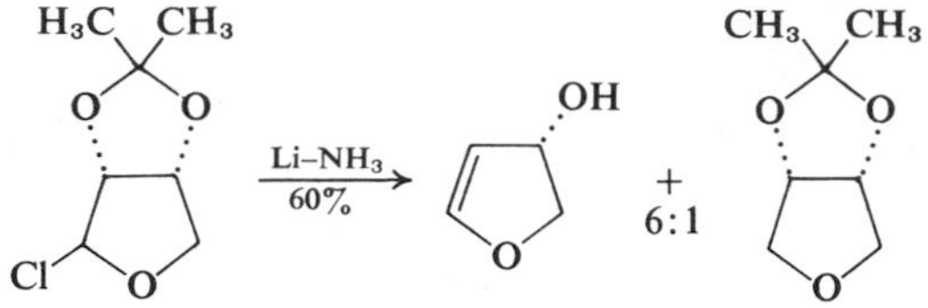

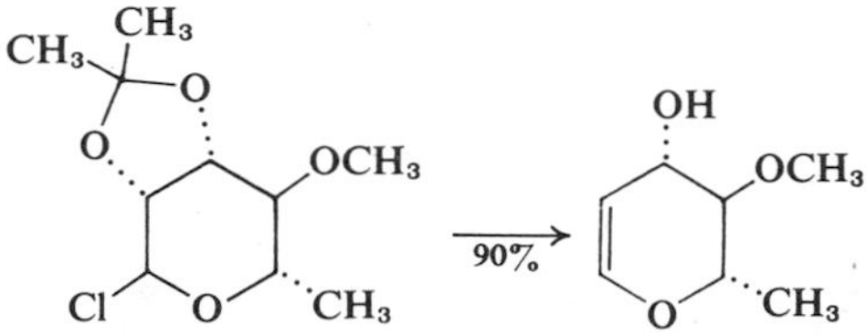

Reductive alkylation of α,β-unsaturated ketones (3, 179–181). An example of this useful method of Stork for alkylation of α,β-unsaturated ketones via the less stable lithium enolate has been described in *Org. Syn.*[4] Direct base-catalyzed alkylation of 3-methylcyclohexanone leads mainly to 2-alkyl-5-methylcyclohexanones. The Stork reaction requires an added proton source as a buffer; in the example shown water is used so that lithium hydroxide (a weak base) is present.

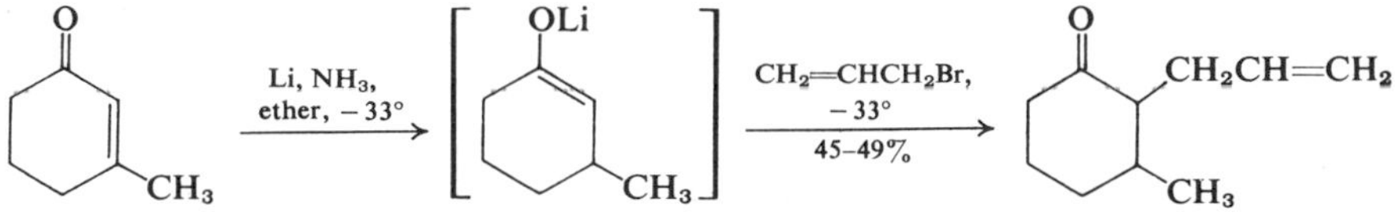

Reductive elimination–alkylation. One problem in the synthesis of the diterpene resin acids of the podocarpane type is the establishment of the proper stereochemistry (axial) of the carboxyl group at C_1. A useful procedure based on a new approach to the problem has been used in a recent total synthesis of (±)-callitrisic acid and (±)-podocarpic acid.[5] Synthesis of the latter acid (**5**) involved preparation of the tricyclic enone **1** by standard methods. Application of Stork's reductive carbomethoxylation method (**1**, 601–602) led to the keto ester **2**. Alkylation of the sodium enolate of **2** with chloromethyl methyl ether furnished the vinyl ether ester **3**. Reduction of **3** with lithium in liquid ammonia and DME gave the exocyclic ester enolate ion, which was alkylated exclusively from the less hindered equatorial direction to give **4**. Reduction, deoxygenation, and stereospecific methylation are thus accomplished in a single step. The ester group of **4** has been cleaved quantitatively by lithium *n*-propylmercaptide (3, 188), and the ether group has been cleaved with HI in acetic acid.

2,5-Dihydrofurane-2-carboxylic acid (2). Reduction of 2-furoic acid (**1**) in methanol and dry liquid ammonia with lithium (3 equiv.) gives this acid (**2**) in 90% yield.[6]

[1] R. K. Boeckman, Jr. and E. W. Thomas, *Am. Soc.*, **99**, 2805 (1977).

[2] R. E. Donaldson and P. L. Fuchs, *J. Org.*, **42**, 2032 (1977).

[3] R. E. Ireland, C. S. Wilcox, and S. Thaisrivongs, *ibid.*, **43**, 786 (1978).

[4] D. Caine, S. T. Chao, and H. A. Smith, *Org. Syn.*, **56**, 52 (1977).

[5] S. C. Welch, C. P. Hagan, J. H. Kim, and P. S. Chu, *J. Org.*, **42**, 2879 (1977).

[6] H. R. Divanfard and M. M. Joullié, *Org. Prep. Proc. Int.*, **10**, 94 (1978).

Lithium–Ethylamine, 1, 574–581; **2,** 241–242; **3,** 175; **4,** 287–288; **5,** 377–379; **6,** 322; **7,** 194–195.

Deoxygenation of alcohols. Acetates (or other esters) of sterically hindered secondary and tertiary alcohols are reduced to alkanes by lithium in ethylamine or by potassium in *t*-butylamine in the presence of 18-crown-6.[1]

Examples:

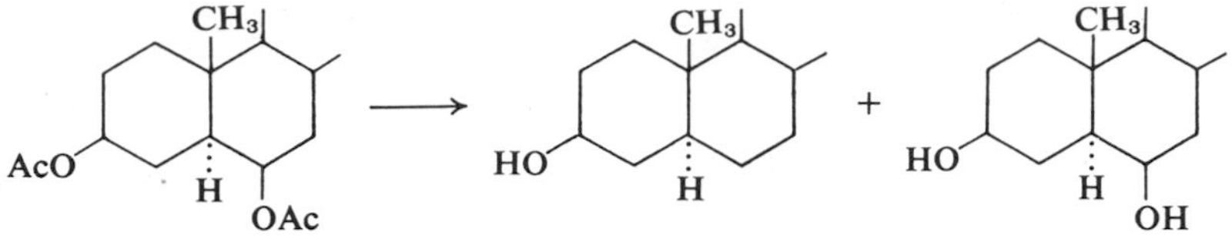

(60%)   (29%)

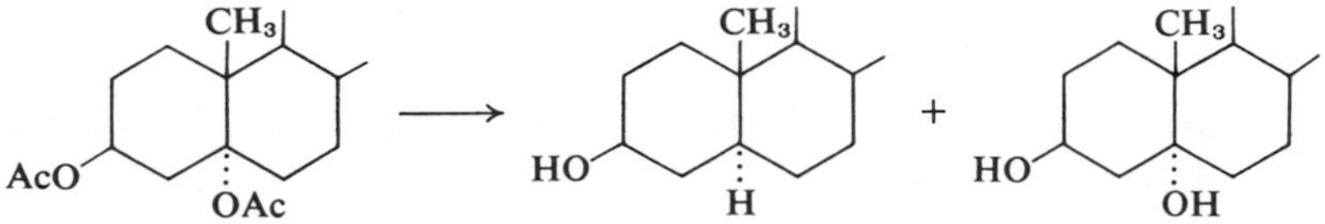

(66%)   (8%)

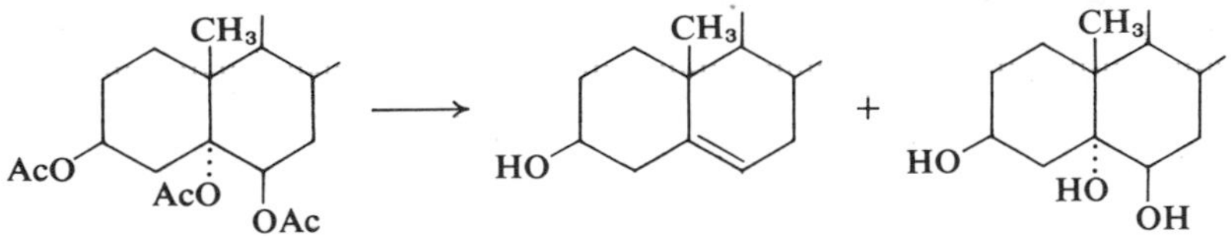

(81%)   (8%)

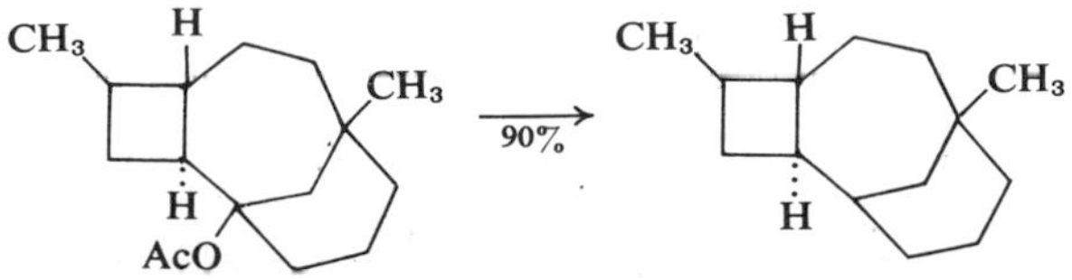

90%

[1] R. B. Boar, L. Joukhadar, J. F. McGhie, S. C. Misra, A. G. M. Barrett, D. H. R. Barton, and P. A. Prokopiou, *J.C.S. Chem. Comm.*, 68 (1978).

Lithium acetylide, 1, 573–574; **5,** 382; **6,** 324–325.

Allenes, acetylenes. Midland[1] has devised an allene synthesis by the overall transformation shown in equation (I). The actual steps involve condensation of an

$$(I) \qquad R_3B + LiC{\equiv}CH + O{=}C{\diagdown}^{R^1}_{R^2} \longrightarrow RCH{=}C{=}C{\diagdown}^{R^1}_{R^2}$$

aldehyde or ketone with lithium acetylide to form an ethynylcarbinol (**6,** 324), which is then acetylated to give **1.** This product is treated in sequence with *n*-butyllithium and a trialkylborane at $-120°$ and then allowed to warm to room temperature to form an allenic borane (*b*). Protonolysis results in the allene **2**; addition of water gives an acetylene (**3**).

$$HC\equiv CC-R^1 \xrightarrow[\text{2) } R_3B]{\text{1) } n\text{-BuLi}} Li^+ \left[R_2B^- - C\equiv C - C - R^1 \right] \longrightarrow \left[\begin{array}{c} R \\ R_2B \end{array} C=C=C \begin{array}{c} R^1 \\ R^2 \end{array} \right]$$

with OAc and R², R on the structures; intermediate **a**, **b**, starting material **1**

70–90% HOAc | H₂O 85–95%

$$\begin{array}{c} R \\ H \end{array} C=C=C \begin{array}{c} R^1 \\ R^2 \end{array} \qquad RC\equiv CC \begin{array}{c} R^1 \\ R^2 \end{array}$$
with H below

2   **3**

trans-2-*Ethynylcycloalkanols*. These products can be prepared by the reaction of cycloalkene epoxides with lithium acetylide conducted in DMSO.[2]

$$\text{(cyclohexene oxide)} + LiC\equiv CH \xrightarrow[92\%]{DMSO} \text{product with } C\equiv CH,\ H,\ OH$$

$$\text{(cyclopentene oxide)} \xrightarrow{64\%} \text{product with } C\equiv CH,\ H,\ OH$$

[1] M. M. Midland, *J. Org.*, **42**, 2650 (1977).
[2] M. Hanack, E. Kunzmann, and W. Schumacher, *ibid.*, **43**, 26 (1978).

Lithium aluminum hydride, 1, 581–595; **2**, 242; **3**, 176–177; **4**, 291–293; **5**, 382–389; **6**, 325–326; **7**, 196.

Vinyl sulfides. Lithium aluminum hydride adds to arylthioalkynes stereo-specifically in a *trans* manner to give vinyl sulfides. The reaction was developed for preparation of labeled sulfides (equation I).[1]

(I) $$C_6H_5SC\equiv CH \xrightarrow[93\%]{\substack{\text{1) LiAlD}_4 \\ \text{2) D}_2O}} \begin{array}{c} C_6H_5S \\ D \end{array} C=C \begin{array}{c} D \\ H \end{array}$$

$$\sim 100\% \left| \substack{\text{1) LiAlH}_4 \\ \text{2) D}_2O} \right.$$

$$C_6H_5SCD=CH_2$$

Retrosulfolene reaction. The formation of 1,3-dienes by extrusion of SO_2 from sulfolenes when heated is a useful reaction (**2**, 390) and of theoretical interest.

The reaction is essentially a reverse Diels-Alder and is usually conducted at 80–150°. Gaoni[2] has now found that the reaction is induced by lithium aluminum hydride in refluxing ether. He considers that this fragmentation is also a thermal cleavage and that lithium aluminum hydride serves as a catalyst and also as reducing agent for the liberated SO_2.

Examples:

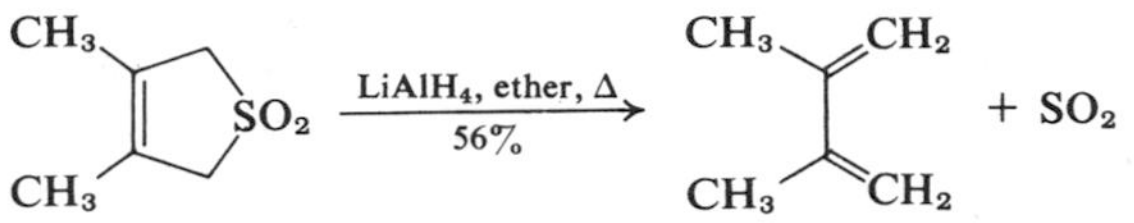

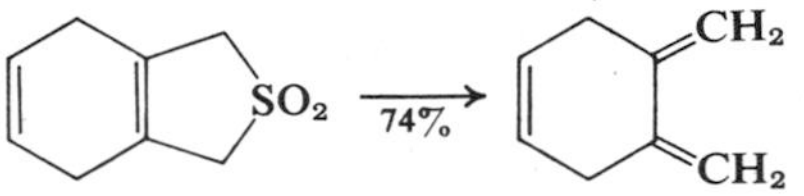

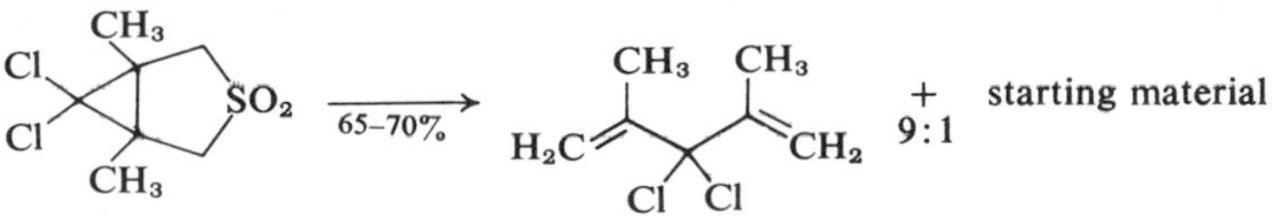

Contraction of sulfolanes to cyclobutenes (**5**, 385–386). Details have been published for the preparation of 7,9-dimethyl-*cis*-8-thiabicyclo[4.3.0]nonane 8,8-dioxide (**1**) and the reductive ring contraction to *cis*-7,8-dimethylbicyclo-[4.2.0]octene-7 (**2**). The procedure cites a number of other examples of this reaction (20–62% yield).[3]

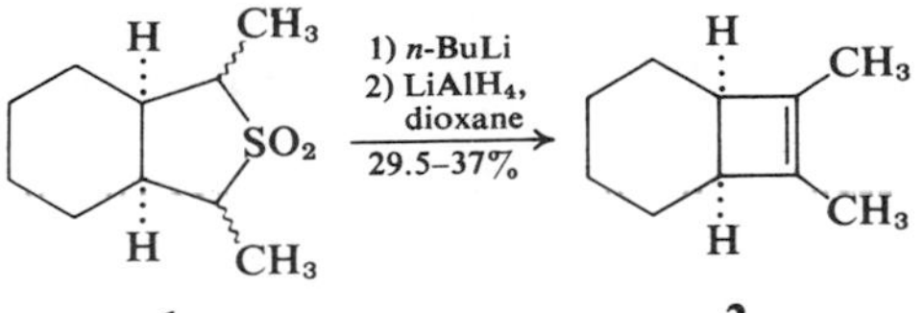

Primary alkyl amines. α-Alkoxy azides are reduced to primary amines by lithium aluminum hydride in ether. A side reaction noted occasionally, reduction to an ether, can be suppressed by addition of TMEDA.[4] Sodium bis(2-methoxy-ethoxy)aluminum hydride cannot replace lithium aluminum hydride in this reductive amination. The α-alkoxy azides are available by reaction of aldehydes with azidotrimethylsilane, by addition of hydrazoic acid to an enol ether, or by reaction of ketals with the 18-crown-6 complex of sodium azide.

$$R^1CR^2 \xrightarrow[\text{methods}]{\text{Several}} \underset{R^1 \quad R^2}{\overset{RO \quad N_3}{C}} \xrightarrow[80-95\%]{LiAlH_4} \underset{R^1 \quad R^2}{\overset{NH_2}{CH}}$$

Hydroalumination.[5] Titanium(IV) chloride is a very effective catalyst for addition of $LiAlH_4$ to olefinic double bonds. Relative rates for various olefins are $RCH=CH_2 > R_2C=CH_2 > RCH=CHR$. Alane is somewhat less reactive. Selective addition to the least hindered $C=C$ bond of nonconjugated dienes is possible. Ether is not an effective solvent; THF, ethylene glycol, and diglyme are satisfactory solvents.

Terminal acetylenes.[6] The hydroalumination of 1-alkenes followed by treatment with bromopropadiene with catalysis by CuCl results in terminal acetylenes.

Examples:

$$CH_2=CH_2 \xrightarrow[TiCl_4]{LiAlH_4,} Al(CH_2CH_3)_3 \xrightarrow[43\%]{\overset{CH_2=C=CHBr,}{CuCl}} CH_3CH_2CH_2C{\equiv}CH$$

Terminal allenes.[7] Use of 3-bromo-1-propyne in the preceding reaction leads to terminal allenes.

Examples:

$$CH_2=CH_2 \xrightarrow[TiCl_4]{LiAlH_4,} \xrightarrow[88\%]{\overset{HC{\equiv}CCH_2Br,}{CuCl}} CH_3CH_2CH=C=CH_2$$

Reduction of a furanose ditosylate. Joullié et al.[8] have reported a facile synthesis of a muscarine derivative (**4**) from the furanose **1**, obtainable from D-glucose. Reduction of **1** with lithium aluminum hydride gives a mixture of **2** and **3**, considered to be formed via an epoxide (**a**). Indeed reduction of **a**, prepared by a standard method, gives the same mixture of **2** and **3**. The former product was converted into D-epiallomuscarine (**4**).

1

a

2 $+$ 3:2   **3**

Several steps

4

Reduction of acetylenic primary alcohols to (E)-alkenols. This reaction is often one step in the synthesis of insect pheromones. Sodium in liquid ammonia is generally preferred for this reduction (**5**, 590). An alternative, general method is reduction with a large excess of lithium aluminum hydride in diglyme–THF.[9]

$$RC\equiv C-(CH_2)_n-OH \xrightarrow[85-95\%]{\substack{LiAlH_4, 140°, \\ diglyme, THF}} \underset{\underset{H}{\overset{\overset{H}{|}}{\parallel}}{RC=C}-(CH_2)_n-OH$$

$$n = 2-7$$

[1] M. Hojo, R. Masuda, and S. Takagi, *Synthesis*, 284 (1978).

[2] Y. Gaoni, *Tetrahedron Letters*, 947 (1977).

[3] J. M. Photis and L. A. Paquette, *Org. Syn.*, **57**, 53 (1977).

[4] E. P. Kyba and A. M. John, *Tetrahedron Letters*, 2737 (1977).

[5] F. Sato, S. Sato, and M. Sato, *J. Organometal. Chem.*, **131**, C26 (1977); F. Sato, S. Sato, H. Kodama, and M. Sato, *ibid.*, **142**, 71 (1977).

[6] F. Sato, H. Kodama, and M. Sato, *Chem. Letters*, 789 (1978).

[7] F. Sato, K. Ogura, and M. Sato, *ibid.*, 805 (1978).

[8] P.-C. Wang, Z. Lysenko, and M. M. Joullié, *Tetrahedron Letters*, 1657 (1978).

[9] R. Rossi and A. Carpita, *Synthesis*, 561 (1977).

Lithium aluminum hydride–Aluminum chloride, 1, 595–599; **2**, 243; **3**, 176–177; **4**, 293–294; **5**, 389–391.

1,2-Transposition of ketones.[1] A new method for effecting this reaction involves conversion of the ketone **1** into a vinylsilane (**2**) as indicated (*see* Chlorotrimethylsilane, this volume). The vinyl group is then epoxidized and reduced to a β-silanol (4) with LiAlH$_4$. This reduction is known to generally involve preferential hydride attack at the carbon bearing silicon.[2] Oxidation of the hydroxyl group then gives the transposed ketone 5. A typical example is formu-

(I)   CH$_3$(CH$_2$)$_2$$\overset{O}{\overset{\|}{C}}$(CH$_2$)$_2CH_3$ $\xrightarrow[\text{96\%}]{\text{1) TsNHNH}_2 \atop \text{2) RLi, ClSi(CH}_3)_3}$

1

CH$_3$(CH$_2$)$_2$$\overset{\text{Si(CH}_3)_3}{\overset{|}{C}}$=CHCH$_2CH_3$ $\xrightarrow[\text{97.5\%}]{\text{ClC}_6\text{H}_4\text{CO}_3\text{H}}$ (CH$_3$)$_3$Si$\underset{\text{CH}_3(\text{CH}_2)_2 \qquad \text{CH}_2\text{CH}_3}{\triangle}$O $\xrightarrow[\text{88\%}]{\text{LiAlH}_4}$

2   3

CH$_3$(CH$_2$)$_2$$\overset{\text{Si(CH}_3)_3}{\overset{|}{\text{CH}}}$—$\underset{\text{OH}}{\overset{|}{\text{CH}}}CH_2CH_3$ $\xrightarrow[\text{89\%}]{\text{H}_2\text{CrO}_4}$ CH$_3$(CH$_2$)$_3$$\overset{O}{\overset{\|}{C}}CH_2CH_3$

4   5

lated in equation (I). In the case of the activated vinylsilane 6, epoxidation leads directly to β-tetralone (7) without need for the reduction–oxidation step (equation II).*

(II) $\xrightarrow[\text{83\%}]{\text{ClC}_6\text{H}_4\text{CO}_3\text{H}}$

6   7

In some cases the effect of silicon is insufficient to prevent the usual *trans* diaxial ring cleavage in rigid systems. For example, LiAlH$_4$ reduction of 8 results in two products, formed by both α- and β-attack of hydride. Use of AlH$_2$Cl results in almost exclusive α-attack (equation III).

(III) $\xrightarrow[\text{95\%}]{\text{AlH}_2\text{Cl}}$ $\xrightarrow{\text{83\%}}$

8   9   10

[1] W. E. Fristad, T. R. Baily, and L. A. Paquette, *J. Org.*, **43**, 1620 (1978).
[2] J. J. Eisch and J. T. Trainor, *ibid.*, **28**, 487, 2870 (1963); C. M. Robbins and G. H. Whitham, *J.C.S. Chem. Comm.*, 697 (1976).

Lithium aluminum hydride–Cobalt(II) chloride, nickel chloride.

Reductions. Lithium aluminum hydride in combination with these metal chlorides reduces alkenes, alkynes, and organic halides in high yield. Terminal alkenes are reduced efficiently by $LiAlH_4$ and $NiCl_2$ or $CoCl_2$ in the ratio 1.0:0.1, but internal alkenes are reduced effectively only by stoichiometric amounts of the transition metal halide. The combination of $LiAlH_4$ and $NiCl_2$ (1.0:0.1) is highly effective for reduction of terminal alkynes to 1-alkenes (94–96% yield). Internal alkynes are reduced by this combination to *cis*-alkenes. Both $LiAlH_4$–$CoCl_2$ and $LiAlH_4$–$NiCl_2$ are about equally effective for reduction of halides. Primary halides require only catalytic amounts of the metal halide; secondary, tertiary, alicyclic, and aryl bromides are reduced in high yield by $LiAlH_4$ with stoichiometric amounts of $NiCl_2$ or $CoCl_2$.[1]

[1] E. C. Ashby and J. J. Lin, *Tetrahedron Letters*, 4481 (1977); *idem, J. Org.*, **43**, 1263, 2567 (1978).

Lithium 1,3-butadiene-1-olate, CH_2⎯$CHOLi$ (1). Mol. wt. 76.02. The reagent

is obtained[1] by reaction of 1-trimethylsilyloxy-1,3-butadiene (this volume) with *n*-butyllithium in THF (20°).

Annelation.[1] The reagent does not react with saturated ketones, but does undergo conjugate addition to α,β-unsaturated ketones and esters at $-78°$ in THF to give, after quenching with chlorotrimethylsilane, cyclohexene derivatives. An example is the reaction of ethyl acrylate (equation I). This reaction is an alternative to Diels-Alder reactions with alkoxybutadienes, which require much higher temperatures.

(I) CH_2=$CHCOOC_2H_5$ + 1 $\xrightarrow[\substack{60\%}]{\substack{1)\ THF,\ -78° \\ 2)\ (CH_3)_3SiCl}}$

[1] G. A. Kraus and H. Sugimoto, *Tetrahedron Letters*, 3929 (1977).

Lithium di-*n*-butyl cuprate, 2, 152; **5**, 187–188; **7**, 92–93.

Alkylation of cyclopropyl bromides.[1] Reaction of a bromocyclopropane with this cuprate (4–5 equiv.) in THF at -48 to $0°$ and then with an alkyl bromide (or iodide) at the same temperature results in stereospecific replacement of the bromo group by the alkyl group with retention of configuration. A mixed cuprate such as **a** is postulated as an intermediate. An example is the conversion of *trans*-1-bromo-2-phenylcyclopropane to *trans*-1-allyl-2-phenylcyclopropane (equation I). The yield in this reaction is higher than that obtained by bromine–lithium exchange followed by treatment with allyl bromide [23% yield of **2**].

Other examples:

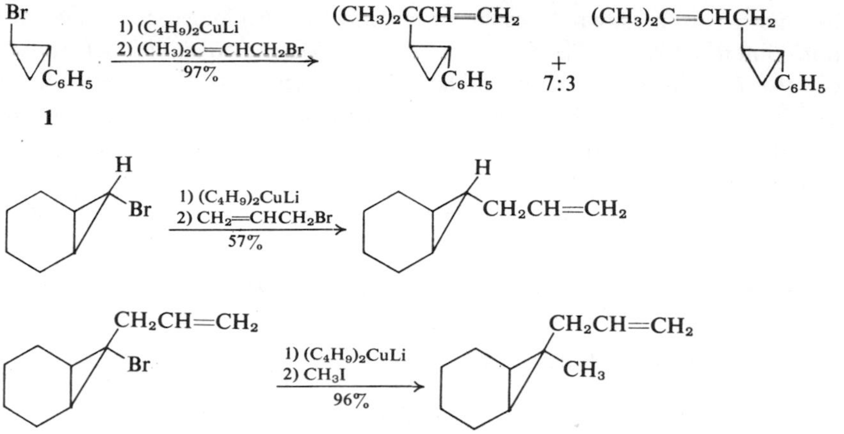

This alkylation may be applicable also to bromoarenes. Treatment of bromo-benzene according to the procedure shown in equation (I) affords allylbenzene in 64% yield.

[1] H. Yamamoto, K. Kitatani, T. Hiyama, and H. Nozaki, *Am. Soc.*, **99**, 5816 (1977).

Lithium diisobutylmethylaluminum hydride, 2, 248–249.

(*E*)-*Enynes. trans*-Monohydroalumination of 1,3-diynes (**1**) with this reagent (**2**) provides a simple route to *trans*-enynes (**3**). The reagent does not discriminate in addition to unsymmetrically substituted diynes; hence two isomeric *trans*-enynes are obtained in this case.[1]

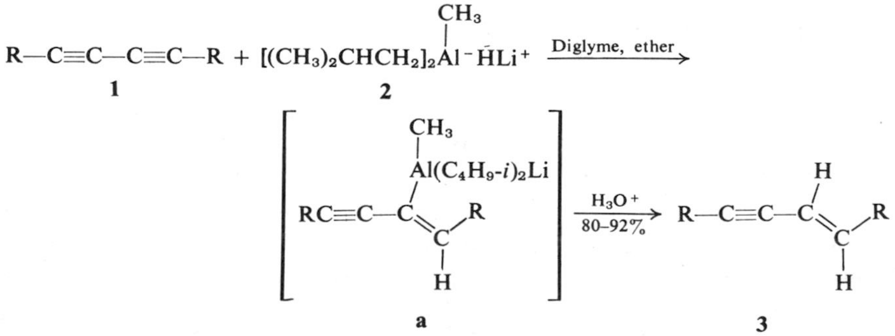

[1] G. Zweifel, R. A. Lynd, and R. E. Murray, *Synthesis*, 52 (1977).

Lithium diisopropylamide (LDA), 1, 611; **2**, 249; **3**, 184–185; **4**, 298–302; **5**, 400–406; **6**, 334–339; **7**, 204–207.

Preparation in hexane. Stock solutions of this base cannot be prepared in solvents such as ether, THF, DME, or HMPT, because they are deprotonated by LDA at temperatures above 0°. However, House *et al.*[1] have found that stable solutions of LDA in hexane or hexane–pentane can be prepared by dilution of *n*-BuLi in hexane (Foote Mineral Co.) with additional hexane or pentane and then treatment with diisopropylamine. These solutions (0.5–0.6 *M*) are stable for weeks at 25°, provided they are not cooled or concentrated. They can be diluted with cold ethereal solvents for use in reactions.

Lithium enolates of methyl ω-bromoalkyl ketones. House *et al.*[1] used solutions of LDA in ether–hexane to generate the lithium enolates (**a**) of ω-bromo ketones. On addition of an activating ligand, particularly HMPT, the enolates undergo intramolecular cyclization to six-membered cyclic ketones. An example is shown in equation (I). Replacement of LDA by potassium *t*-butoxide results mainly in formation of internal enolates, and **2** is a minor product (14% yield).

With limitations, this method is applicable to synthesis of cycloheptanones.

Indole synthesis. *o*-Tolyl isocyanide (**1**) is lithiated by treatment with LDA (2 equiv.) in diglyme at −78°; on warming to 20°, indole (**2**) is obtained in almost quantitative yield. The intermediate **a** can also be used for preparation of 3-alkyl-indoles (**4**).[2]

Metalation of phenyl vinyl selenide (**1**). The reaction of this selenide and *n*-butyllithium follows mainly two courses: Se—C cleavage and addition to the double bond. The most satisfactory reagent for metalation is LDA; however, depending on the solvent, elimination can occur as well. For example, reaction of **1** with 1.5 equiv. of LDA at $-78°$ in THF–HMPT (20:1) and then with decyl bromide gives 2-phenylseleno-1-dodecene (**2**) in 70% yield. Elimination to **3** predominates in THF–HMPT (3:1).[3]

$$CH_2{=}CHSeC_6H_5 \xrightarrow{\text{LDA}} \left[CH_2{=}\underset{\underset{Li}{|}}{C}SeC_6H_5 \right] \xrightarrow[70\%]{\substack{CH_3(CH_2)_9Br,\\ \text{THF–HMPT (20:1)}}} CH_2{=}\underset{\underset{(CH_2)_9CH_3}{|}}{C}SeC_6H_5$$

$$\mathbf{1} \qquad\qquad \mathbf{a} \qquad\qquad\qquad\qquad \mathbf{2}$$

$$65\% \downarrow CH_3(CH_2)_9Br, \text{THF–HMPT (3:1)}$$

$$C_6H_5Se(CH_2)_9CH_3 + HC{\equiv}CH$$
$$\mathbf{3}$$

α-*Lithiation of phenoxyacetic acid.* Phenoxyacetic acid (**1**) can be converted into the lithium α-lithiocarboxylate (**a**) by reaction with LDA (excess) in THF at $-78°$. (Ordinarily a carbanion is destabilized by an adjacent oxygen atom.) The reaction of **a** with common electrophiles proceeds normally (equation I).[4]

$$(\text{I}) \qquad C_6H_5OCH_2COOH \xrightarrow[>90\%]{\text{LDA}} C_6H_5O\underset{\underset{Li}{|}}{C}H\overset{\overset{O}{\|}}{C}OLi \xrightarrow[50-70\%]{RX} C_6H_5O\underset{\underset{R}{|}}{C}HCOOH$$

$$\mathbf{1} \qquad\qquad\qquad \mathbf{a} \qquad\qquad\qquad \mathbf{2}$$

Stereoselective aldol condensation. Aldol condensation has been shown to be subject to kinetic stereoselection, with (Z)-enolates giving mainly the *erythro*-aldol and (E)-enolates giving mainly the *threo*-aldol.[5] This observation has been extended to preformed (Z)- and (E)-lithium enolates generated under kinetic control with LDA in THF or ether at $-72°$.[6] When one of the alkyl groups is sterically demanding, complete kinetic stereospecificity can be obtained. As the bulk of the alkyl group decreases, stereoselection also decreases. Thus the kinetic enolate of ethyl *t*-butyl ketone (**1**) is the (Z)-isomer **a**, and it reacts with benzaldehyde to form the *erythro*-aldol (**2**) with no detectable amounts of the *threo*-aldol.

$$CH_3CH_2\overset{\overset{O}{\|}}{C}C(CH_3)_3 \xrightarrow[\text{THF, }-72°]{\text{LDA,}} \left[\underset{CH_3}{\overset{H}{\diagup}}C{=}C\underset{OLi}{\overset{C(CH_3)_3}{\diagdown}} \right] \xrightarrow{C_6H_5CHO} C_6H_5\overset{\overset{OH}{|}}{\underset{\underset{CH_3}{|}}{C}H}\overset{\overset{O}{\|}}{C}C(CH_3)_3$$

$$\mathbf{1} \qquad\qquad\qquad \mathbf{a\ (Z)} \qquad\qquad\qquad \mathbf{2\ (\textit{erythro})}$$

(E)-Lithium enolates of acyclic ketones of the type required are not readily available, but the kinetic enolate of ethyl mesityl ketone (3) consists of the (E)- and (Z)-isomers in the ratio 92:8. Reaction of this mixture affords the *threo*-aldol (4) and the *erythro*-aldol in a 92:8 ratio.

$$C_2H_5\overset{\overset{\displaystyle O}{\|}}{C}C_6H_2(CH_3)_3 \xrightarrow[\text{THF, } -72°]{\text{LDA,}} \left[\underset{CH_3}{\overset{H}{}}C=C\underset{C_6H_2(CH_3)_3}{\overset{OLi}{}} \right] \xrightarrow{C_6H_5CHO}$$

      3   **b (E)**

$$C_6H_5\overset{\overset{\displaystyle OH}{\cdot\cdot}}{\underset{\underset{\displaystyle CH_3}{|}}{C}}\overset{\overset{\displaystyle O}{\|}}{C}C_6H_2(CH_3)_3$$

4 (*threo*)

In some cases, a *threo*-aldol can be obtained indirectly from the (Z)-silyl enol ether by use of a tetraalkylammonium fluoride (6, 44). Thus the silyl enol ether **5** of the ketone **1** reacts with benzaldehyde in the presence of catalytic amounts of benzyltrimethylammonium fluoride to give the *threo*-aldol **6**.

$$\underset{CH_3}{\overset{H}{}}C=C\underset{OSi(CH_3)_3}{\overset{C(CH_3)_3}{}} + C_6H_5CHO \xrightarrow[\text{THF, 25°}]{C_6H_5CH_2N(CH_3)_3F,} C_6H_5\overset{\overset{\displaystyle (CH_3)_3SiO}{\cdot\cdot}}{\underset{\underset{\displaystyle CH_3}{|}}{C}}\overset{\overset{\displaystyle O}{\|}}{C}C(CH_3)_3$$

           5   **6 (*threo*)**

This approach has been extended to preparation of chiral β-hydroxy carboxylic acids by use of a ketone in which the R group is convertible into a hydroxyl group. To this end the enolate formed under kinetic control from **7** was allowed to

$$CH_3CH_2\overset{\overset{\displaystyle O}{\|}}{C}\overset{\overset{\displaystyle CH_3}{|}}{\underset{\underset{\displaystyle CH_3}{|}}{C}}OSi(CH_3)_3 \xrightarrow[\text{2) } C_6H_5CHO]{\text{1) LDA, THF, } -70°}$$

      7

$$C_6H_5\underset{\underset{\displaystyle OH}{|}}{\overset{\overset{\displaystyle CH_3}{|}}{C}}\overset{\overset{\displaystyle CH_3}{|}}{\underset{\underset{\displaystyle O \ CH_3}{}}{C}}OSi(CH_3)_3 \xrightarrow{HIO_4} C_6H_5\underset{\underset{\displaystyle OH}{|}}{\overset{\overset{\displaystyle CH_3}{|}}{C}}COOH$$

           8   **9**

react with benzaldehyde to form the *erythro*-aldol **8**, which was cleaved by periodic acid to the β-hydroxy carboxylic acid **9**.

The ultimate objective of the investigation is synthesis of polyhydroxy carboxylic acids, which in nature are formed by condensation of acetyl and propionyl units. And indeed condensation of ketone enolates with aldehydes having a chiral center adjacent to the carbonyl group gives only two of the four possible diastereomers.[7]

Cyclopropanes from α,β-unsaturated carbonyl compounds. This new transformation is illustrated for conversion of cinnamaldehyde into phenylcyclopropane (equation I). The sequence involves Michael addition of sodium thiophenoxide

(I) $C_6H_5CH=CHCHO$ $\xrightarrow[\substack{100\%}]{\substack{1)\ C_6H_5SNa \\ 2)\ NaBH_4}}$ $\underset{C_6H_5S}{\overset{C_6H_5}{\diagdown}}CHCH_2CH_2OH$ $\xrightarrow[\substack{88\%}]{H_2O_2,\ HOAc}$

$\underset{C_6H_5SO_2}{\overset{C_6H_5}{\diagdown}}CHCH_2CH_2OH$ $\xrightarrow[\substack{95\%}]{CH_3SO_2Cl,\ Py}$ $\underset{C_6H_5SO_2}{\overset{C_6H_5}{\diagdown}}CHCH_2CH_2OSO_2CH_3$ $\xrightarrow[\substack{100\%}]{\substack{LDA, \\ -78 \to 20°}}$

$\xrightarrow[\substack{83\%}]{Na(Hg)}$

to the enal, reduction of the carbonyl group, oxidation to the phenyl sulfone, conversion of the hydroxyl group to a tosylate or mesylate, and finally deprotonation followed by cyclization. The sulfone group can be eliminated by treatment with 6% sodium amalgam. The yield of cyclopropanes is markedly lower if the sulfide group is not converted into the sulfone before treatment with base.[8]

Stobbe condensation.[9] One limitation of the Stobbe condensation is that α-keto esters cannot be used because of their sensitivity to potassium t-butoxide. This disadvantage can be overcome by use of diethyl lithiosuccinate (1), generated with LDA in THF at −78°. This anion reacts with α-keto esters (2) to form γ-butyrolactones (3). These products are converted into the Stobbe products (4) by potassium t-butoxide in t-butyl alcohol (equation I).[10]

(I) $\underset{\overset{|}{CH_2COOC_2H_5}}{LiCHCOOC_2H_5}$ + $R^1\overset{\overset{O}{\|}}{C}COOR^2$ $\xrightarrow[\substack{25-80\%}]{THF,\ -78°}$

      1   **2**

$\xrightarrow[\substack{70-75\%}]{\substack{KOC(CH_3)_3, \\ HOC(CH_3)_3}}$

        3   **4**

Bis(trimethylsiloxy)cyclohexadienes. LDA and related lithium dialkyl-amides appear to be specific for generation of the anion of keto trimethylsilyl enol ethers; reaction of these anions with $(CH_3)_3SiCl$ gives the disiloxycyclohexa-dienes **1** and **2**, which cannot be prepared directly from the 1,2- and 1,3-dike-

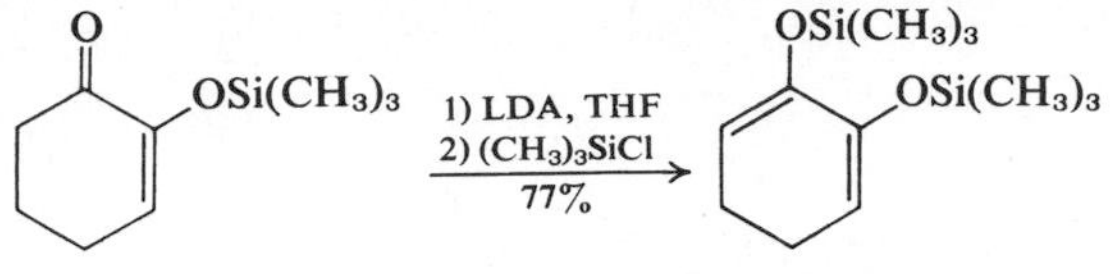

1

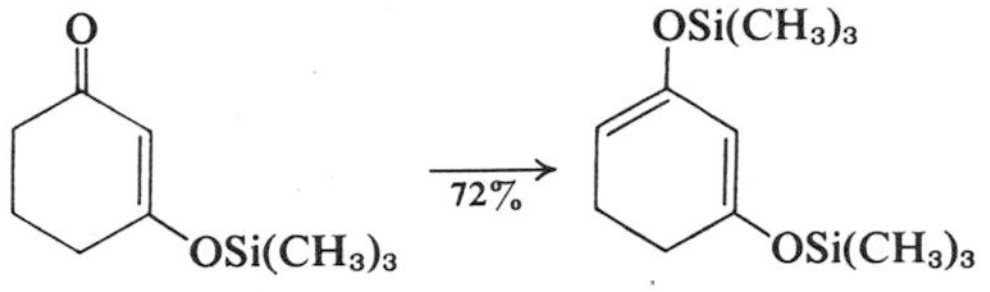

2

tones. The disilyl enol ethers (**3** and **4**) of cyclohexanedione-1,4 can be prepared as shown.[11]

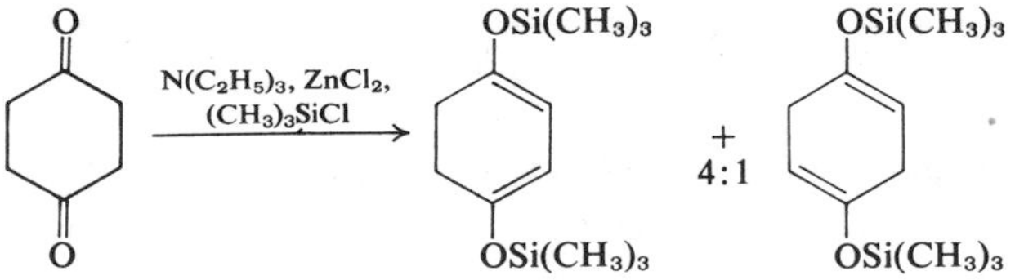

3   **4**

Cyclization of 1,5-diketones; 1-alkyl-3-oxocyclohexenes. Cyclization of 1,5-diketones (**1**) by base or acid in a protic medium results in formation predominantly of the cyclohexenone (**2**). French chemists[12] report that cyclization with LDA in ether or HMPT results, after dehydration (p-TsOH), in the isomeric 1-

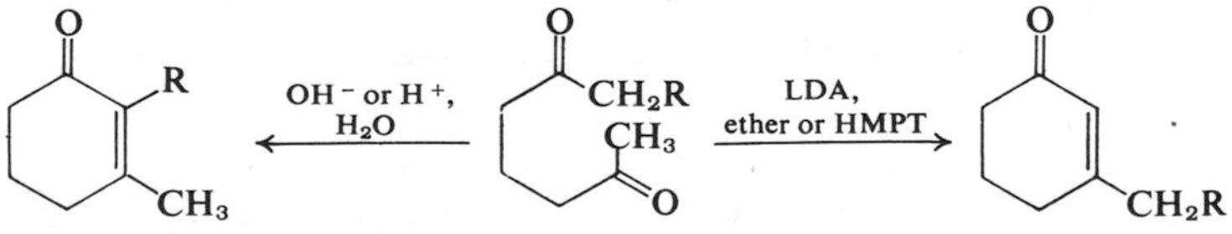

2   **1**   **3**

alkyl-3-oxocyclohexene (**3**). The former cyclization is under thermodynamic control, the latter under kinetic control.

Example:

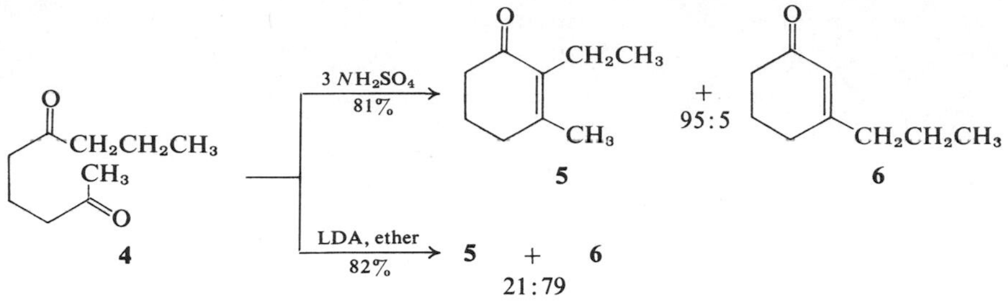

5 + **6**
21:79

Ketene thioacetals. A recent preparation of these useful intermediates is shown in equation (I).[13]

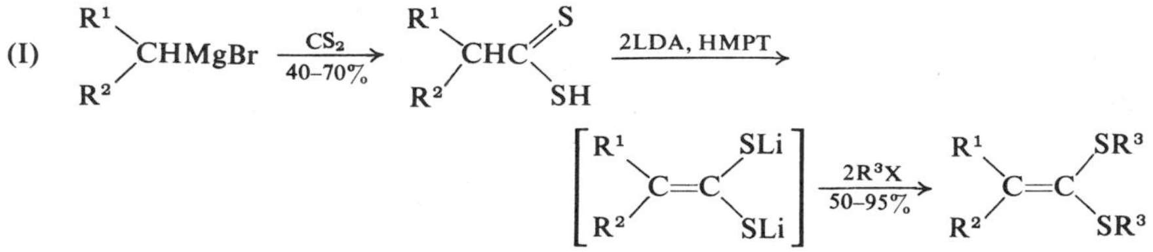

Alkoxy enediolates.[14] Ethyl mandelate (**1**) can be converted by LDA (excess) into the alkoxy enediolate **a**, which reacts with a number of electrophiles to give useful products. This deprotonation is also successful with ethyl lactate, but fails with ethyl glycolate ($CH_2OHCOOC_2H_5$). Tertiary and aryl halides do not

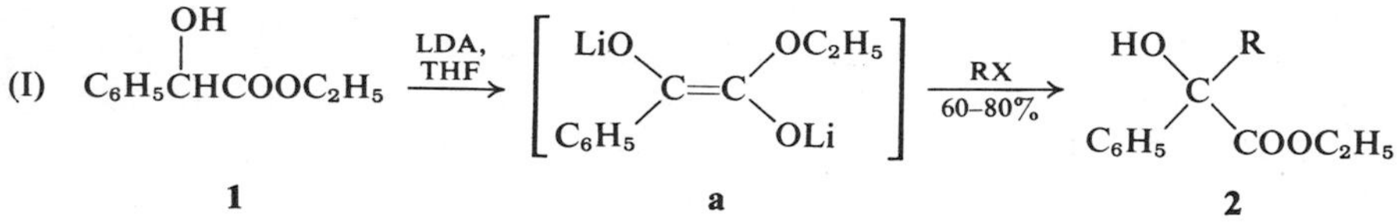

react with **a**, but a wide range of alkyl halides can be used: CH_3I (80%), C_6H_5-CH_2Br (85%), $(CH_3)_2CHI$ (57%), $CH_2=CHCH_2Br$ (82%). Aldehydes, ketones, and epoxides can also be used as shown in equations (II) and (III).

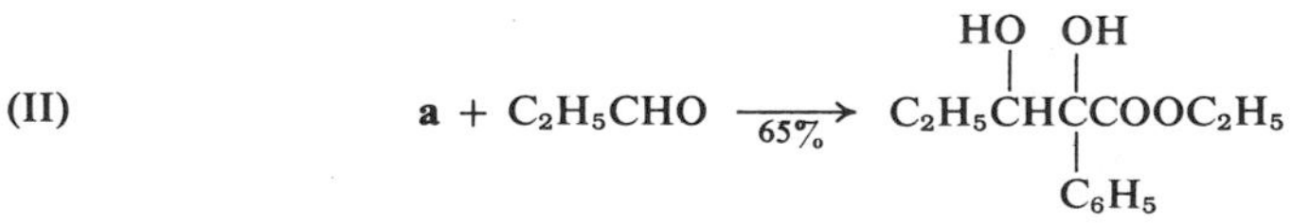

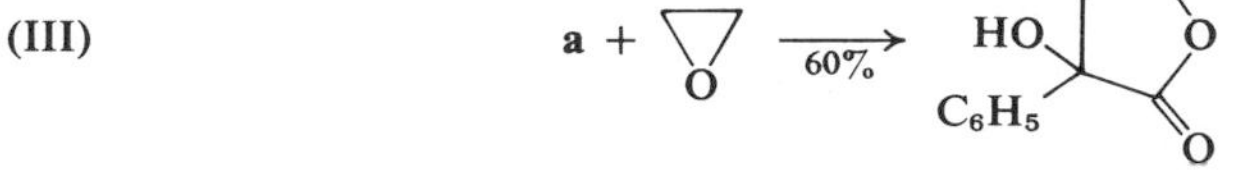

(II) $a + C_2H_5CHO \xrightarrow[65\%]{}$

(III) $a +$

Epoxide isomerization.[15] The alcohol **3** (decahydro-2,4,7-metheno-1*H*-cyclopenta[*a*]pentalene-3-ol) can be prepared by epoxidation of **1** (the Diels-Alder adduct of norbornene and cyclopentadiene) followed by isomerization of the resulting oxide (**2**) with LDA (0–35°, 48 hours). This reaction involves a transannular carbenoid insertion.

$ClC_6H_4CO_3H$, CH_2Cl_2
98% crude

LDA, ether
82%

1   2   3

Reduction of ketones. There have been scattered reports of reduction of nonenolizable ketones by lithium amides, but this reaction has attracted little attention. For example, benzophenone is reduced by LDA in THF even at $-78°$ to benzhydrol and products of higher molecular weight, including tetraphenylethylene oxide. Lithium 2,2,6,6-tetramethylpiperidide, which has no β-hydrogens, does not reduce benzophenone to benzhydrol but only to higher molecular weight products. Therefore the carbinol hydrogen of benzhydrol comes from LDA either by hydride transfer or by electron transfer followed by transfer of hydrogen atom.[16]

LDA is also able to reduce some enolizable α-halo and α-methoxy ketones, such as $C_6H_5COCH_2Br$ and 2,2-dimethoxycyclobutanone, to the corresponding *sec*-alcohols in surprisingly high yields.[17]

[1] H. O. House, W. V. Phillips, T. S. B. Sayer, and C.-C. Yan, *J. Org.*, **43**, 700 (1978).
[2] Y. Ito, K. Kobayashi, and T. Saegusa, *Am. Soc.*, **99**, 3532 (1977).
[3] M. Sevrin, J. N. Denis, and A. Krief, *Angew. Chem., Int. Ed.*, **17**, 526 (1978).
[4] W. Adam and H.-H. Fick, *J. Org.*, **43**, 772 (1978).
[5] J.-E. Dubois and M. Dubois, *Tetrahedron Letters*, 4215 (1967); J.-E. Dubois and P. Fellman, *ibid.*, 1225 (1975).
[6] W. A. Klesick, C. T. Buse, and C. L. Heathcock, *Am. Soc.*, **99**, 247 (1977).
[7] C. T. Buse and C. H. Heathcock, *ibid.*, **99**, 8109 (1977).
[8] Y.-H. Chang and H. W. Pinnick, *J. Org.*, **43**, 373 (1978); *see also* Y.-H. Chang, D. E. Campbell, and H. W. Pinnick, *Tetrahedron Letters*, 3337 (1977).
[9] W. S. Johnson and G. H. Daub, *Org. React.*, **6**, 1 (1951).
[10] V. Reutrakul, K. Kusamran, and S. Wattanasin, *Heterocycles*, **6**, 715 (1977).
[11] S. Torkelson and C. Ainsworth, *Synthesis*, 431 (1977).
[12] M. Larchevêque, G. Valette, and T. Cuvigny, *Synthesis*, 424 (1977).
[13] F. E. Ziegler and C. M. Chan, *J. Org.*, **43**, 3065 (1978).
[14] L. J. Ciochetto, D. E. Bergbreiter, and M. Newcomb, *ibid.*, **42**, 2948 (1977).
[15] J. R. Neff and J. E. Nordlander, *Tetrahedron Letters*, 499 (1977).
[16] L. T. Scott, K. J. Carlin, and T. H. Schultz, *ibid.*, 4637 (1978).
[17] C. Kowalski, X. Creary, A. J. Rollin, and M. C. Burke, *J. Org.*, **43**, 2601 (1978).

Lithium (3,3-dimethyl-1-butynyl)-1,1-diethoxy-2-propenyl cuprate.

$$Li^+\left[(CH_3)_3CC\equiv C-\overset{-}{Cu}-\underset{\underset{CH_2}{\|}}{C}CH(OC_2H_5)_2\right]$$

Mol. wt. 268.78.

This mixed cuprate is prepared from 3,3-dimethyl-1-butynlcopper and 2-lithio-3,3-diethoxypropene.

The diorganocuprate derived from 2-lithio-3,3-diethoxypropene (**6**, 330–331) is difficult to handle because of solubility problems.[1] However, the mixed cuprate **1** is soluble in ether and THF. Moreover, this cuprate also undergoes 1,4-addition to enones, although the reaction is sluggish with enones substituted with bulky groups at the β-position. The adducts have also been shown to be suitable precursors to α-methylene-γ-lactones. The reagent undergoes coupling with allylic halides but not with vinyl halides.

Examples:

[1] R. K. Boeckman, Jr., and M. Ramaiah, *J. Org.*, **42**, 1581 (1977).

Lithium dimethyl cuprate, 2, 151–153; **3,** 106–113; **4,** 177–183; **5,** 234–244; **6,** 209–215; **7,** 120–122.

Solvent effects on chemoselectivity. Use of donor solvents (THF or HMPT) enhances the rate of displacement reactions of lithium organocuprates with alkyl halides, but retards conjugate addition to enones. House and Lee[1] have examined the effect of solvents on the reaction of $(CH_3)_2CuLi$ with the bromo enone **1.** Reaction of **1** with the cuprate in ether–$(CH_3)_2S$ gives the product of

$$Br(CH_2)_3\overset{\overset{\textstyle CH_3}{|}}{\underset{\underset{\textstyle CH_3}{|}}{C}}-CH=CH\overset{\overset{\textstyle O}{\|}}{C}C(CH_3)_3$$

1

$(CH_3)_2CuLi, (C_2H_5)_2O,$
$S(CH_3)_2, H_2O$
83–92%

$(CH_3)_2CuLi, (C_2H_5)_2O,$
$S(CH_3)_2, HMPT, H_2O$
72–97%

$$Br(CH_2)_3\overset{\overset{\textstyle H_3C \quad CH_3}{| \quad |}}{\underset{\underset{\textstyle CH_3}{|}}{C}}-CHCH_2\overset{\overset{\textstyle O}{\|}}{C}C(CH_3)_3$$

2

$$CH_3(CH_2)_3\overset{\overset{\textstyle CH_3}{|}}{\underset{\underset{\textstyle CH_3}{|}}{C}}-CH=CH\overset{\overset{\textstyle O}{\|}}{C}C(CH_3)_3$$

3

conjugate addition (**2**) as expected because conjugate additions with cuprates are typically faster than displacement reactions. However, when the reaction is carried out with added HMPT, only **3** is formed in a somewhat slower reaction. This chemoselectivity is probably general for cuprates that are sufficiently stable to survive the rather slow coupling reaction.

Reaction with conjugated allenic ketones and esters. Lithium dimethyl cuprate adds 1,4- to these unsaturated systems, and the intermediate lithium enolate **a** can be alkylated, as in related reactions (equation I). This reaction thus permits

(I) $CH_2=C=CHCOCH_3$ $\xrightarrow[\text{ether, } -90°]{(CH_3)_2CuLi,}$

$$\left[\overset{\overset{\textstyle CH_3}{\diagdown}}{\underset{\underset{\textstyle CH_2}{\diagup}}{C}}-CH=C\overset{\overset{\textstyle OLi}{\diagup}}{\underset{\underset{\textstyle CH_3}{\diagdown}}{}} \right] \xrightarrow[\text{RX, DME, } -30°]{} \overset{\overset{\textstyle CH_3}{\diagdown}}{\underset{\underset{\textstyle CH_2}{\diagup}}{C}}-\overset{\overset{\textstyle R}{|}}{C}HCOCH_3$$

a

the addition of two substituents to a carbon skeleton and was used successfully for a synthesis of lavandulol (**3**), a nonisoprenoid terpene in oil of lavender (equation II).[2]

(II) $CH_2{=}C{=}CHCOOC_2H_5$ $\xrightarrow{\substack{1)\ (CH_3)_2CuLi \\ 2)\ (CH_3)_2C=CHCH_2Br}}$

$\xrightarrow[\substack{50\% \\ \text{overall}}]{LiAlH_4}$

2   **3**

Deblocking of allyl esters. Allyl esters are cleaved in 75–85% yield by reaction with lithium dimethyl cuprate.[3]

$$RCOOCH_2CH{=}CH_2 + (CH_3)_2CuLi \longrightarrow RCOOH + CH_3Cu + CH_2{=}CHCH_2CH_3$$

[1] H. O. House and J. M. Wilkins, *J. Org.*, **43**, 2443 (1978); H. O. House and T. V. Lee, *ibid.*, **43**, 4369 (1978).
[2] M. Bertrand, G. Gil, and J. Viala, *Tetrahedron Letters*, 1785 (1977).
[3] T.-L. Ho, *Syn. Comm.*, **8**, 15 (1978).

Lithium diphenylphosphide, 4, 303–304; **5,** 408–410; **6,** 340–341. The reagent can also be prepared by reaction of lithium foil (excess) with chlorodiphenylphosphine in THF. The excess Li is removed by filtration. This material can be used directly for preparation of alkyldiphenylphosphines (70–95% yield).[1]

Demethylation of methyl aryl ethers (**5,** 409). The cleavage of 4-ethoxy-3-methoxybenzaldehyde ethylene acetal to 4-ethoxy-3-hydroxybenzaldehyde with $(C_6H_5)_2PLi$ has now been published.[2]

[1] P. W. Clark, *J. Organometal. Chem.*, **139**, 385 (1977); *Org. Syn.*, submitted (1978).
[2] R. E. Ireland and D. M. Walba, *ibid.*, **56**, 44 (1977).

Lithium di-(E)-propenyl cuprate, 7, 141.

γ,δ-Unsaturated aldehydes. The reagent reacts with the α,β-unsaturated aldehyde **1** to give a mixture of the *trans-* and *cis*-adducts **2** and **3**. Methylenation of **2** gives aucentene (**4**), a constituent of brown algae.[1]

1   **2** $\xleftarrow{K_2CO_3,\ CH_3OH}$ **3**

$74\% \Big| (C_6H_5)_3P{=}CH_2$

4

[1] H.-J. Lin and E. N. C. Browne, *Canad. J. Chem.*, **56**, 306 (1978).

Lithium B-isopinocampheyl-9-borabicyclo[3.3.1]nonyl hydride (1). Mol. wt. 266.21. This trialkylborohydride is prepared by hydroboration of (+)-α-pinene with 9-BBN followed by reaction with *t*-butyllithium in THF (−78°).

1

Asymmetric reduction of ketones. This borohydride reduces even hindered ketones and with high stereoselectivity, as noted with other hindered trialkyl-borohydrides. Thus 2-methylcyclohexanone is reduced to *cis*-2-methylcyclohexanol (99% isomeric purity). Of even greater interest, the alcohols formed on reduction are consistently enriched in the (R)-enantiomer to the extent of ∼10–40%.[1]

[1] S. Krishnamurthy, F. Vogel, and H. C. Brown, *J. Org.*, **42**, 2534 (1977).

Lithium β-lithiopropionate, LiCH₂CH₂COOLi (1). Mol. wt. 85.94. This carbanion can be prepared by reaction of β-bromopropionic acid with 1 equiv. of *n*-butyllithium in THF followed by reaction in the same solvent with lithium naphthalenide (**2**, 288–289; **3**, 208; **4**, 348–349; **5**, 468; **6**, 415). The reagent is formed only in low yield by reaction of β-bromopropionic acid with 2.2 equiv. of *n*-butyllithium.

γ-*Butyrolactones.* The reagent reacts with aldehydes or ketones at −70 to 20° to give γ-hydroxy carboxylic acids, which are lactonized on treatment with a trace of acid (*p*-TsOH).[1]

Examples:

$$C_6H_5CHO + LiCH_2CH_2COOLi \xrightarrow{\text{THF}} \left[\begin{array}{c} C_6H_5CHOH \\ | \\ CH_2CH_2COOH \end{array} \right] \xrightarrow[56\%]{\text{TsOH}}$$

[1] D. Caine and A. S. Frobese, *Tetrahedron Letters*, 883 (1978).

Lithium methyl mercaptide, CH₃SLi. Mol. wt. 54.04, stable off-white solid. The material is prepared by the reaction of methyllithium with methyl mercaptan in anhydrous ether at 0°(N₂). A slurry of CH₃SLi is formed, from which ether and

CH_3SH are removed by evaporation. Store under N_2. LiH in HMPT can also be used for the preparation.

Cleavage of esters, ethers, and lactones.[1] The reagent resembles C_3H_7SLi (**3**, 188; **5**, 415) and C_2H_5SNa (**4**, 465) in reactivity for cleavage of hindered esters and methyl aryl ethers. In addition it cleaves lactones to methylthio carboxylic acids; an example is the conversion of butyrolactone to $CH_3SCH_2CH_2CH_2COOH$ in 69% yield. A direct comparison indicates that this reagent is superior to lithium thiophenoxide for cleavage of the lactone **1** to **2**.

The final step in a total synthesis of methylenomycin A (**4**), an antibiotic isolated from *Streptomyces violaceoruber*, involved selective opening of the lactone ring of **3**. A wide variety of methods were unsuccessful, but reaction of **2** with this reagent in HMPT led directly to **4** in 68% yield. The lactone ring is cleaved and then CH_3SH is eliminated.[2]

Azo compounds.[3] A mild new synthesis of azo compounds from dicarbamates (prepared by Diels-Alder reactions with dimethyl azodicarboxylate) involves cleavage with $LiSCH_3$ to form a dilithium dicarbamate followed by oxidative didecarboxylation with $K_3Fe(CN)_6$ at 0°.

Example:

[1] T. R. Kelly, B. B. Dali, and W.-G. Tsang, *Tetrahedron Letters*, 3859 (1977).
[2] R. M. Scarborough, Jr., and A. B. Smith III, *Am. Soc.*, **99**, 7085 (1977).
[3] R. D. Little and M. G. Venegas, *J. Org.*, **43**, 2921 (1978).

Lithium methylthioformaldine (4,5-Dihydro-5-methyl-1,3,5-dithiazin-2-yllithium) (1). Mol. wt. 141.18. Fluka supplies the immediate precursor, 4,5-dihydro-5-methyl-1,3,5-dithiazine.

Preparation[1]:

$$CH_2O + CH_3NH_2 + H_2S \longrightarrow$$

(20 g.)

(32 g.) **1**

Synthesis of aldehydes.[2] The reagent is readily alkylated, particularly by primary alkyl halides. The main advantage of this reagent over lithio-1,3-dithiane is that the adducts are easily cleaved to aldehydes in high yield at room temperature.

$$1 + RX \xrightarrow[50-100\%]{-78 \to 0°} \qquad \xrightarrow[\sim 90\%]{HgCl_2, HgO, CH_2Cl_2, H_2O, 20°} RCHO$$

Unfortunately this approach cannot be extended to a synthesis of ketones because the intermediate adduct cannot be metalated under any known conditions.

[1] J. Graymore, *J. Chem. Soc.*, 865 (1935).
[2] R. D. Balanson, V. M. Kobal, and R. R. Schumaker, *J. Org.*, **42**, 393 (1977).

Lithium naphthalenide, 2, 288–289; **3,** 208; **4,** 348–349; **5,** 468; **6,** 415.

Alkyllithium reagents. 1-Alkenes can be converted into these reagents in two steps: anti-Markownikoff addition of thiophenol followed by cleavage of the resulting alkyl phenyl sulfides with a form of lithium metal (equation I). The

$$\text{(I)} \quad RCH{=}CH_2 + C_6H_5SH \xrightarrow[70-100\%]{AIBN} RCH_2CH_2SC_6H_5 \xrightarrow{2Li} RCH_2CH_2Li + C_6H_5SLi$$

cleavage with lithium metal in THF is slow, but yields of organolithiums in the range 55–95% can be achieved. The cleavage is more facile with lithium naphthalenide in THF and yields of products are in the range 60–100%; but the presence of naphthalene can be inconvenient. This method should be useful when the alkyl halides for the more usual exchange method are not readily available.[1]

Lithium naphthalenide (2 equiv.) is a superior reagent for reductive lithiation of cyclopropanone dithioketals. The reaction is complete in THF at $-70°$ in less than 15 minutes.[2]

Examples:

(two isomers)

Retro-Baeyer-Villiger oxidation. Benzhydryl benzoates (2) are reduced to benzhydryl aryl ketones (3) by lithium naphthalenide (1) in THF at 20°. Aromatic carboxylic acids (4) are converted under these conditions into α-diketones (5).[3]

2, (R = H, CH$_3$, C$_6$H$_5$)   **3**

4, (R = H, OCH$_3$)   **5**

[1] C. G. Screttas and M. Micha-Screttas, *J. Org.*, **43**, 1064 (1978).
[2] T. Cohen, W. M. Daniewski, and R. B. Weisenfeld, *Tetrahedron Letters*, 4665 (1978).
[3] W. Tochtermann and R. G. H. Kirrstetter, *Ber.*, **111**, 1228 (1978).

Lithium selenophenolate, LiSeC$_6$H$_5$. This reagent is prepared by reduction of diphenyl diselenide with aqueous hypophosphorous acid to selenophenol, which is converted into lithium selenophenolate by *n*-BuLi.

Cleavage of cyclopropyl ketones.[1] Cyclopropyl ketones are cleaved in high yield by reaction with lithium selenophenolate in refluxing benzene containing 12-crown-4.

Examples:

Cyclopropyl nitriles are cleaved more efficiently by sodium selenophenolate. Example:

Cyclopropyl esters and lactones are stable to both reagents.

[1] A. B. Smith III and R. M. Scarborough, Jr., *Tetrahedron Letters*, 1649 (1978).

Lithium 2,2,6,6-tetramethylpiperidide (LiTMP), 4, 310–311; **5**, 417; **6**, 345–348; **7**, 213–215.

Enol carbonates.[1] An efficient route to these substances involves deprotonation of a ketone with LiTMP (1.1 equiv.) in THF at $-78°$, dilution of the enolate solution after it is warmed to 20° with HMPT, and addition of ethyl chloroformate (1.1 equiv.). Yields are generally 75–90% and no C-acyl products are formed. The success with LiTMP results from the lack of reactivity of HMPT with chloroformates.

Trimethylsilylylcarbene; trimethylstannylcarbene. This base converts chloromethyltrimethylsilane (1) into trimethylsilylcarbene, $(CH_3)_3SiĊH$, evidenced by formation of silylcyclopropanes (2) from alkenes. Trimethylstannylcarbene,

$(CH_3)_3Sn\ddot{C}H$, can be generated in the same way from chloromethyltrimethyltin (3).[2]

$$(CH_3)_3SnCH_2Cl + \bigcirc \xrightarrow[21\%]{LiTMP} (CH_3)_3Sn-\triangleleft\text{(bicyclic)}$$

3   **4**

5,6-Dihydro-2H-pyrane-2-ones. The dianion of 2-butynoic acid (**6**, 345–346) reacts with aldehydes to form both α- and γ-adducts. The latter have been used for synthesis of several naturally occurring 5,6-dihydro-2*H*-pyrane-2-ones, such as massoialactone (equation I).[3]

(I) $LiCH_2C{\equiv}C-C\overset{O}{\underset{OLi}{\diagup}} + C_5H_{11}CHO \longrightarrow \xrightarrow{CH_2N_2}$

$$CH_2{=}C{=}C\overset{COOCH_3}{\underset{\underset{C_5H_{11}}{\overset{|}{CHOH}}}{\diagup}} + \underset{C_5H_{11}}{\overset{HO}{\diagup}}CHCH_2C{\equiv}CCOOCH_3 \xrightarrow[85\%]{H_2, Cat.}$$

(35%)   (35%)

$$\xrightarrow[87\%]{HCl, THF} $$

massoialactone ring

¹ R. A. Olofson, J. Cuomo, and B. A. Bauman, *J. Org.*, **43**, 2073 (1978).
² R. A. Olofson, D. H. Hoskin, and K. D. Lotts, *Tetrahedron Letters*, 1677 (1978).
³ H. H. Meyer, *Ann.*, 337 (1978).

Lithium tri-*sec*-butylborohydride, 4, 312–313; **6**, 348; **7**, 307.

Reduction-alkylation. One step in a synthesis of (±)-7,9-deoxydaunomycinone (**4**) involved conversion of the Knoevenagel product **1** to **3**. Hydrogenation was undesirable because of loss of the bromine. The conversion can be conducted in two steps: sodium borohydride reduction and alkylation. A more efficient method is *in situ* alkylation of the product of reduction of **1** with lithium tri-*sec*-butyl-borohydride (overall yield 89%).[1]

¹ J. S. Swenton and P. W. Raynolds, *Am. Soc.*, **100**, 6188 (1978).

1

1) LiB(sec-Bu)$_3$H
2) BrCH$_2$COOCH$_3$
89%

87% | 1) NaH
2) BrCH$_2$COOCH$_3$

2

4

3

Lithium triethylborohydride (Super-Hydride), 4, 313–314; **6,** 348–349; **7,** 215–216.

Hydroboration of styrenes.[1] This borohydride adds to styrene to form a tetraalkylborate (**1**), which can be hydrolyzed to ethylbenzene (**2**) or converted into the Markownikoff mixed trialkylborane **3** by strong acids. Alkaline oxidation of **3** yields 1-phenylethanol (**4**).

$$C_6H_5CH{=}CH_2 + Li(C_2H_5)_3BH \xrightarrow{20°} C_6H_5CHCH_3Li^+ \xrightarrow{H_2O, THF, 65°}$$

$$^-B(C_2H_5)_3$$

1

$$C_6H_5CH_2CH_3 + B(C_2H_5)_3$$

2 (92–99%)   (100%)

$$\downarrow CH_3SO_3H$$

$$C_6H_5CHCH_3 + C_6H_5CH_2CH_3 \xleftarrow{H_2O_2, OH^-} \begin{array}{c} C_6H_5CHCH_3 \\ | \\ B(C_2H_5)_2 \end{array} + C_2H_6 + CH_3SO_3Li$$

OH

4 (90–96%)   4–5%   **3**

Selective reduction of neopinone (**1**). The ketone **1** is reduced by this bulky borohydride or by lithium tri-*sec*-butylborohydride almost exclusively to neopine (**2**) because of steric factors. Reduction with sodium borohydride in alcoholic solvents gives about equal amounts of **2** and the epimer **3** (isoneopine). In this case alkoxyborohydrides are the actual reducing agents. If **1** is reduced in an aqueous medium containing potassium hydroxide with sodium borohydride, then **2** and

1

2 (R^1 = OH, R^2 = H)
3 (R^1 = H, R^2 = OH)

3 are obtained in the ratio 11:89. After chromatography of the mixture, isoneo-pine can be obtained in 88% yield. In this reduction torsional strain predominates over steric effects.[2]

Reduction of **tert-*amides***. Reduction of tertiary amides with this borohydride follows an unusual course: fission of the C—N bond to give an alcohol. The reaction may proceed through the aldehyde.[3]

$$R^1\overset{\overset{\displaystyle O}{\|}}{C}-N\overset{R^2}{\underset{R^3}{\diagdown}} \quad \xrightarrow[50-100\%]{2Li(C_2H_5)_3BH,\ THF,\ 0-25°} \quad R^1CH_2OH$$

Primary and secondary amides are not reduced by the reagent.

Dialkyl selenides; dialkyl diselenides. Reaction of selenium with 2.1 equiv. of this hydride in THF gives a suspension of milky white lithium selenide. On addition of alkyl halides dialkyl selenides are formed in 60–95% yield (equation I).

(I) $Se + 2Li(C_2H_5)_3BH \xrightarrow[-H_2]{} 2(C_2H_5)_3B + Li_2Se \xrightarrow[60-95\%]{2RX} R_2Se$

(II) $2Se + 2Li(C_2H_5)_3BH \xrightarrow[-H_2]{} 2(C_2H_5)_3B + Li_2Se_2 \xrightarrow[50-75\%]{2RX} R_2Se_2$

Reduction with only 1 equiv. leads to the dark brownish-red lithium diselenide, which reacts with alkyl halides to form dialkyl diselenides (equation II).[4]

For large-scale preparations use of sodium borohydride in water or ethanol[5] in place of lithium triethylborohydride may be more economical.

[1] H. C. Brown and S. C. Kim, *J. Org.*, **42**, 1482 (1977).
[2] S. W. Wunderly and E. Brochmann-Hanssen, *ibid.*, **42**, 4277 (1977).
[3] H. C. Brown and S. C. Kim, *Synthesis*, 635 (1977).
[4] J. A. Gladysz, J. L. Hornby, and J. E. Garbe, *J. Org.*, **43**, 1204 (1978).
[5] D. L. Klayman and T. S. Griffin, *Am. Soc.*, **95**, 197 (1973).

M

Magnesium ethyl malonate, (structure) **(1)**. Mol. wt. 152.42, stable under an inert gas. The material is prepared by reaction of ethyl malonate (**6**, 255) with magnesium turnings in ethanol (dibromoethane catalysis).[1]

Conjugate addition of an ethyl acetate group. The reagent undergoes Michael addition to enones; the adducts are decarboxylated when warmed in acetic acid. The reaction is limited to disubstituted enones.[2]

Examples:

$$CH_3(CH_2)_4CH=CHCOCH_3 \xrightarrow[98\%]{} CH_3(CH_2)_4\underset{CH_2COOC_2H_5}{CHCH_2COCH_3}$$

γ,δ-Unsaturated β-keto esters. The magnesium complex (**1**) reacts with α,β-unsaturated acid chlorides to form γ,δ-unsaturated β-keto esters in 55–75% yield.[3]

[1] G. Bram and M. Vilkas, *Bull. soc.*, 945 (1964).
[2] J. E. McMurry, W. A. Andrus, and J. H. Musser, *Syn. Comm.*, **8**, 53 (1978).
[3] P. Pollet and S. Gelin, *Synthesis*, 142 (1978).

Magnesium hydride–Copper(I) iodide, MgH_2–CuI.

Reduction of alkynes. This reagent reduces alkynes stereoselectively to the corresponding *cis*-alkenes; alkanes are not formed. Typical yields are 80–95%. Cuprous iodide can be replaced by copper(I) *t*-butoxide with equally satisfactory results.[1]

[1] E. C. Ashby, J. J. Lin, and A. B. Goel, *J. Org.*, **43**, 757 (1978).

Manganese dioxide, 1, 637–643; **2**, 257–263; **3**, 191–194; **4**, 317–318; **5**, 422–424; **6**, 357.

endo-**Disulfides.** The *endo*-disulfide analog **5** of PGH$_2$ has been prepared from **1** by the steps indicated in equation (I). The product has interesting biological activity and is more stable than PGH$_2$.[1]

cis-*Azobenzenes*. These substances can be obtained in high yield ($\sim$ 90–95%) by oxidation of several hydrazobenzenes with MnO$_2$ (Attenburrow or Alfa) in CHCl$_3$ at 25°. At higher temperatures the *cis*-products rearrange to the *trans*-isomers. However, the *trans*-isomers are obtained directly if Ar or Ar' has an electron-releasing group (OAc, NHAc) in the *para*-position.[2]

[1] H. Miyake, S. Iguchi, H. Itoh, and M. Hayashi, *Am. Soc.*, **99**, 3536 (1977).
[2] J. A. Hyatt, *Tetrahedron Letters*, 141 (1977).

Manganese(II) iodide, MnI$_2$, **7**, 222.

Organomanganese(II) iodides, RMnI. These reagents are prepared by reaction of an organolithium or a Grignard reagent (RLi, RMgX) with MnI$_2$ in ether. The resulting reagents, RMnI, react with carboxylic acid chlorides or bromides to form ketones in good to high yield. One limitation is that α,β-unsaturated acid chlorides form products of polymerization.[1]

Examples:

$$n\text{-}C_7H_{15}MnI + (CH_3)_3CCOCl \xrightarrow[80\%]{} n\text{-}C_7H_{15}\overset{\overset{\displaystyle O}{\|}}{C}C(CH_3)_3$$

$$(CH_3)_2C{=}CHMnI + C_6H_5COCl \xrightarrow[85\%]{} (CH_3)_2C{=}CH\overset{\overset{\displaystyle O}{\|}}{C}C_6H_5$$

$$n\text{-}C_4H_9C{\equiv}CMnI + (CH_3)_3CCOCl \xrightarrow[73\%]{} n\text{-}C_4H_9C{\equiv}C\overset{\overset{\displaystyle O}{\|}}{C}C(CH_3)_3$$

Similar results can be obtained with R_2CuLi, but in this reaction, an excess of the reagent is necessary for satisfactory yields.

[1] G. Cahiez, D. Bernard, and J. F. Normant, *Synthesis*, 130 (1977).

Meldrum's acid,

(1). Mol. wt. 144.12, m.p. 94–95° dec., pK_a 4.83.

The chemical is prepared by condensation of malonic acid with acetone in the presence of acetic anhydride.[1] It was shown to have the structure 2,2-dimethyl-1, 3-dioxane-4,6-dione (1) by Davidson and Bernhard.[2] Most of the synthetic uses of 1 involve the Knoevenagel condensation with aldehydes or ketones to form 5-methylene derivatives (an example is cited in **6**, 309–310). A review (90 references) has been published.[3]

β-Keto esters. Meldrum's acid (1) reacts with acyl chlorides in CH_2Cl_2 in the presence of pyridine to form an acyl Meldrum's acid (2) in almost quantitative yield. These derivatives undergo alcoholysis readily to form β-keto esters (3).[4]

Ethyl indolepropionates. The substances can be prepared in a one-pot synthesis by reaction of 1, an aldehyde (2), and indole (3) in acetonitrile (30°) followed by ethanolysis and concomitant decarboxylation in ethanol–pyridine containing copper powder.[5]

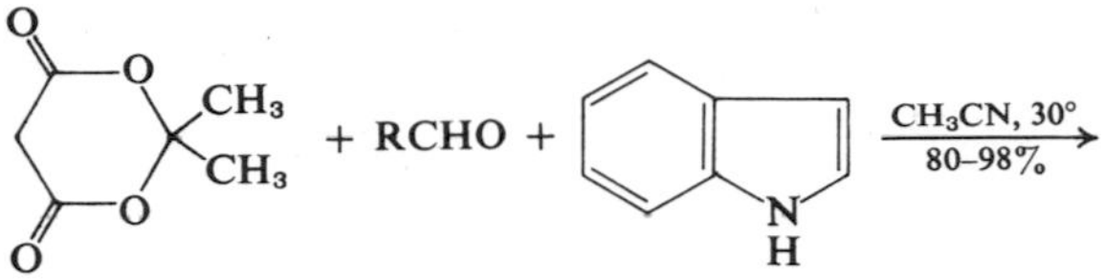

1　　　**2**　　　**3**

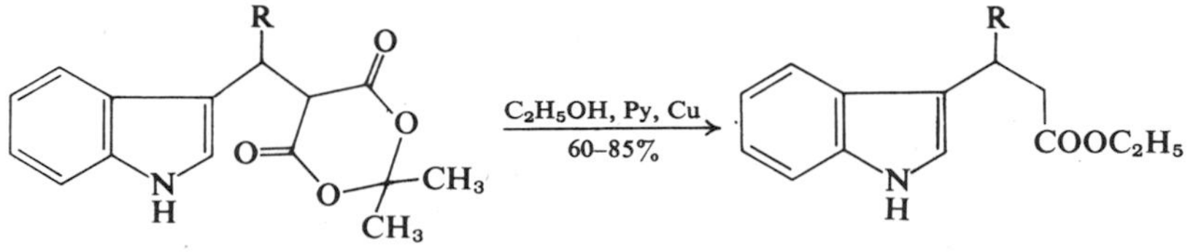

4　　　　　　　　　　　　　**5**

[1] A. N. Meldrum, *J. Chem. Soc.*, **93**, 598 (1908).
[2] D. Davidson and S. A. Bernhard, *Am. Soc.*, **70**, 3426 (1948).
[3] H. McNab, *Chem. Soc. Rev.*, 345 (1978).
[4] Y. Oikawa, K. Sugano, and O. Yonemitsu, *J. Org.*, **43**, 2087 (1978).
[5] Y. Oikawa, H. Hirasawa, and O. Yonemitsu, *Tetrahedron Letters*, 1759 (1978).

α-Mercapto-γ-butyrolactone, **(1).** Mol. wt. 118.1, b.p. 118–120°/9 mm.

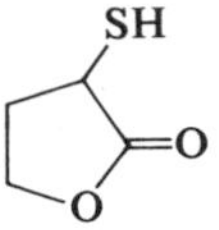

Preparation[1]:

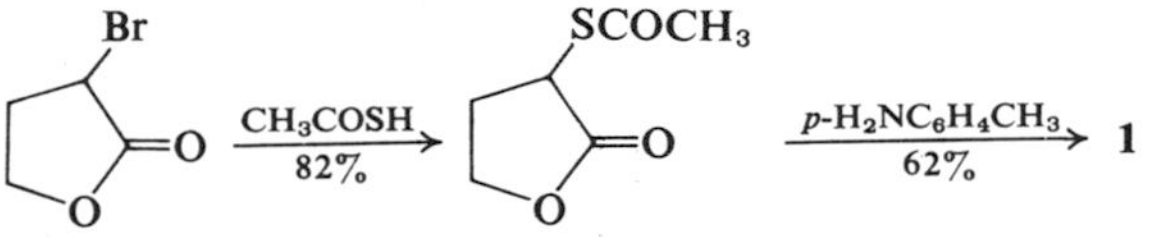

α-Alkylidene-γ-butyrolactones.[2] The lactone (**1**) forms a dianion (**a**) when treated with LDA (2 equiv.) and TMEDA in THF at −78°. This dianion is a highly reactive nucleophile as shown by the reaction with carbonyl compounds to form adducts (**2**), which are converted into α-alkylidene-γ-butyrolactones (**3**) when treated with ethyl chloroformate. The double bond of **3** has the (E)-configuration either exclusively or predominantly. Overall yields are 45–70%.

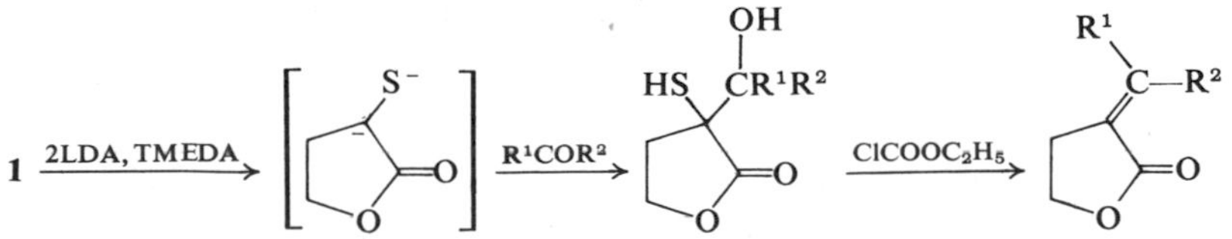

a　　　　　　　**2**　　　　　　**3**

[1] G. Fuchs, *Ark. Kemi*, **26**, 111 (1966).
[2] K. Tanaka, H. Uneme, N. Yamagishi, N. Ono, and A. Kaji, *Chem. Letters*, 653 (1978).

Mercury(II) acetate, 1, 644–652; **2**, 264–267; **3**, 194–196; **4**, 319–323; **5**, 424–427; **6**, 358–359; **7**, 222–223.

γ-*Methylene-γ-butyrolactone* (**3**). An efficient route to this lactone (also known as α′-angelicalactone) involves oxidation of 4-pentyne-1-ol (**1**, Chem. Samples) to 4-pentynoic acid (**2**) followed by cyclization with mercuric acetate.[1]

$$HC\equiv C(CH_2)_2CH_2OH \xrightarrow[]{CrO_3, \ H_2SO_4} HC\equiv C(CH_2)_2COOH \xrightarrow[74\%]{Hg(OAc)_2, \ CH_2Cl_2}$$

1   **2**   **3**

Hydrolysis of vinyl chlorides. Two laboratories[2,3] have reported use of mercury(II) acetate for this reaction. Formic acid or trifluoroacetic acid is used as the medium; glacial acetic acid with BF_3 etherate is also suitable.

Examples:

$$C_6H_5CH_2CH_2CH=CHCl \xrightarrow[80\%]{} C_6H_5CH_2(CH_2)_2CHO$$

Cleavage of cinnamyl esters. Cinnamyl esters can be cleaved by mercuration in CH_3OH catalyzed with nitric acid; after neutralization and removal of solvent, demercuration is effected with excess potassium thiocyanate in cyclohexane–water; overall yields are about 90%.[4]

[1] R. A. Amos and J. A. Katzenellenbogen, *J. Org.*, **43**, 560 (1978).
[2] M. Julia and C. Blasioli, *Bull. soc.*, 1941 (1976).
[3] S. F. Martin and T. Chow, *Tetrahedron Letters*, 1943 (1978); *J. Org.*, **43**, 1027 (1978).
[4] E. J. Corey and M. A. Tius, *Tetrahedron Letters*, 2081 (1977).

Mercury(II) oxide, 1, 655–658; **2**, 267–268; **4**, 323–324; **5**, 428; **6**, 360; **7**, 224.

γ-Methylenebutyrolactones.[1] γ,δ-Acetylenic carboxylic acids (**1**) cyclize to γ-methylenebutyrolactones (**2**) in the presence of HgO when heated neat or in

various solvents at 60° or above if necessary. Yields are generally high, 80–100%. A similar reaction applied to **3** gives **4** in 88% yield.

Oxidation of phenols and hydroquinones.[2] Mercuric oxide in methanol oxidizes these substances to quinones. Mercury(II) trifluoroacetate can also be used, but then an acid scavenger is necessary to neutralize the trifluoroacetic acid formed. The oxidation is analogous to that with thallium(III) salts. Yields in the oxidation of *p*-hydroquinones with HgO are in the range 70–95%.

[1] M. Yamamoto, *J.C.S. Chem. Comm.*, 649 (1978).
[2] A. McKillop and D. W. Young, *Syn. Comm.*, **7**, 467 (1977).

Mercury(II) trifluoroacetate, 6, 360–361.

Cyclomercuration.[1] The reaction of the allylic hydroperoxide **1** with mercury (II) trifluoroacetate at $-40°$ results in cyclomercuration with formation of the mercury-substituted 1,2-dioxetane (**a**) and allylic mercuration to give **b**. The formation of these very sensitive products was demonstrated by bromination and reduction to give **2**, **3**, and **4**.

Scheme (I)

[3,3]Sigmatropic rearrangement. Mercury(II) trifluoroacetate is an effective catalyst for equilibration of allylic carbamates. Isolated yields after equilibration are greater than 90%. No products from other reactions have been detected. Thus this method is useful in cases where the acid-catalyzed rearrangement of the alcohols themselves leads to other rearrangements, cyclization, or dehydration. An example is the rearrangement of the sterol **1** to **2** in 96% yield in 1 hour

at room temperature. In some cases, contrathermodynamic allylic isomerization can be achieved with use of mercury(II) trifluoroacetate in molar excess.[2]

[1] W. Adam and K. Sakanishi, *Am. Soc.*, **100**, 3935 (1978).
[2] L. E. Overman, C. B. Campbell, and F. M. Knoll, *ibid.*, **100**, 4822 (1978).

Mesitylenelithium,   (1). Mol. wt. 126.12. This hindered reagent is obtained by reaction of 1-bromo-2,4,6-trimethylbenzene with 2 equiv. of *t*-butyllithium.[1]

1,3-Dicarbonyl compounds.[2] In a new preparation of 1,3-diketones by reaction of lithium enolates with acid chlorides, Seebach *et al.* prepared the kinetic lithium enolates of ketones directly with this aryllithium at −78°. The mesitylene (b.p. 165°) that results is readily separated by filtration through silica gel. LDA is not a useful base in this case because the free amine formed presents problems in the acylation step.

Examples:

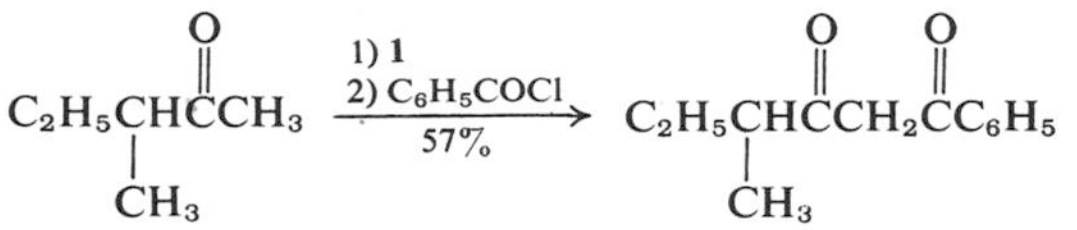

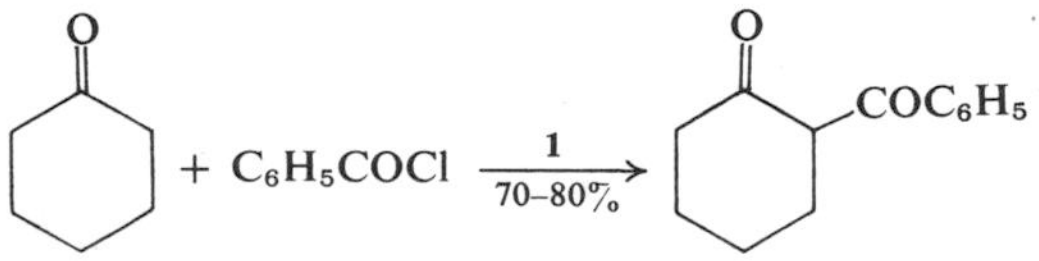

Yields are comparable to those obtained with kinetic lithium enolates generated by the usual method (generation of the kinetic silyl enol ether and cleavage with CH_3Li).

[1] D. Seebach and H. Neumann, *Tetrahedron Letters*, 4839 (1976).
[2] A. K. Beck, M. S. Hoekstra, and D. Seebach, *ibid.*, 1187 (1977).

2-Mesitylenesulfonyl chloride, 4, 661; 6, 625.

Lactamization. A key step in the first total synthesis of a maytansenoid, macrolactams with antitumor activity, involved cyclization of the amino acid **1**. This reaction was effected by conversion of the acid into the soluble tetra-*n*-butylammonium salt and addition of this salt to a solution of mesitylenesulfonyl chloride and diisopropylethylamine in benzene (3 hours, 35°). The lactam **2** was

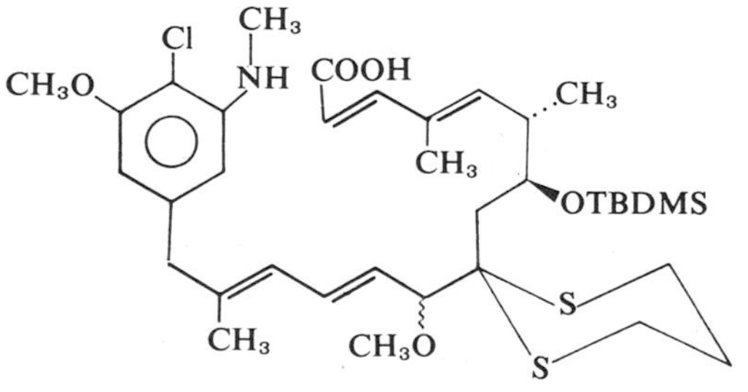

1 [TBDMS = $SiC(CH_3)_3$]

2

3

obtained together with the C_{10}-epimer in ~65% yield. The lactam (**2**) was transformed into (+)-N-methylmaysenine (**3**) in three steps.[1]

Protection of the guanidino group. The mesitylene-2-sulfonyl group has been used to protect the guanidino group of arginine. The Mts group is stable to TFA, $1N$ NaOH, and hydrazine hydrate. It is somewhat resistant to Na in liquid ammonia. It is cleaved quantitatively by methanesulfonic acid, trifluoromethane-sulfonic acid, or hydrogen fluoride. This new protective group was used success-fully for the synthesis of an arginine-containing undecapeptide amide, hypothala-mic substance P.[2]

[1] E. J. Corey, L. O. Weigel, D. Floyd, and M. G. Bock, *Am. Soc.*, **100**, 2916 (1978).
[2] H. Yajima, M. Takeyama, J. Kanaki, and K. Mitani, *J.C.S. Chem. Comm.*, 482 (1978).

O-Mesitylenesulfonylhydroxylamine (MSH), 5, 430–433; 6, 320.

Review.[1] The synthesis and synthetic applications of this reagent for amina-tion have been reviewed (107 references).

[1] Y. Tamura, J. Minamikawa, and M. Ikeda, *Synthesis*, 1 (1977).

Methaneselenenic acid, CH_3SeOH. This unstable reagent can be prepared *in situ*[1] from methaneseleninic acid, CH_3SeO_2H, m.p. 122°,[2] by treatment with aqueous 50% hypophosphorous acid at 20°.

β-Hydroxy selenides. The reagent reacts with alkenes at room temperature to form β-hydroxy selenides by *trans*-addition of CH_3Se and OH. The CH_3Se group is attached to the less substituted carbon atom in the case of trisubstituted alkenes.[1]

Examples:

$$CH_3\!-\!C\!=\!C\overset{H}{\underset{(CH_2)_2CHC_2H_5}{\vert}} \xrightarrow{86\%} CH_3\!-\!C\!-\!C\cdots H$$

(structure showing CH_3, CH_3, $(CH_2)_2CHC_2H_5$, CH_3 on left; and CH_3, $SeCH_3$, OH, $(CH_2)_2CHC_2H_5$, CH_3 on right)

Benzeneselenenic acid[2] can be prepared and used in the same way as methaneselenenic acid; it reacts somewhat more readily.

[1] D. Labar, A. Krief, and L. Hevesi, *Tetrahedron Letters*, 3967 (1978).
[2] F. Wohler and J. Dean, *Ann.*, **97**, 1 (1856).

(S)-N-Methanesulfonylphenylalanyl chloride, $CH_3SO_2NH\!-\!\overset{COCl}{\underset{CH_2C_6H_5}{C}}\!-\!H$

(1, RCOCl)

This reagent was prepared from L-phenylalanine by reaction first with methanesulfonyl chloride (Schotten-Baumann conditions) and then with PCl_5 in benzene.

Chemical resolution of a **meso-diol.**[1] Acylation of the symmetrical *meso*-diol *cis*-2-cyclopentene-1,4-diol (**2**) with 1 equiv. of the chiral reagent **1** in pyridine affords a mixture of the monoesters **3** and **4** in 51% yield, together with some of the diester. The mono esters were separated readily by fractional crystallization to give pure, optically active **3** and **4**. These esters were converted into the (+)- and (−)-**5** hydroxy ethers, respectively, and then into the optically active lactones (+)- and (−)-**6** by Claisen rearrangement and lactonization (**6**, 608–609). These lactones have been converted into both natural and unnatural prostaglandins. See scheme (I) at top of page 321.

[1] S. Terashima, S. Yamada, and M. Nara, *Tetrahedron Letters*, 1001 (1977).

Methoxyallene, 7, 225–226.

(E)-Enones.[1] A simple new synthesis of (E)-α,β-enones (**4**), involves conversion of methoxyallene (**1**) into the allenic ether (**2**).[2] This product is metalated and

(I) $CH_2\!=\!C\!=\!CHOCH_3 \xrightarrow[\text{2) R}^1\text{X}]{\text{1) } n\text{-BuLi, THF}}$

1

$$CH_2\!=\!C\!=\!C\overset{OCH_3}{\underset{R^1}{\diagdown}} \xrightarrow[80-90\%]{\text{1) } n\text{-BuLi}\;\;\text{2) R}^2\text{X}} \overset{R^2}{\underset{H}{\diagdown}}C\!=\!C\!=\!C\overset{OCH_3}{\underset{R^1}{\diagup}}$$

2   **3**

$$\downarrow 80-90\%\;\;H_3O^+$$

$$\overset{R^2}{\underset{H}{\diagdown}}C\!=\!C\overset{H}{\underset{COR^1}{\diagup}}$$

4

Scheme (I)

alkylated to form **3**. The final step involves mild acid hydrolysis. The sequence can be conducted in "one pot."

(*E*)-Enediones can be obtained by substitution of R^2X in the above equation by an alkyl dimethylamide (equation II).

(*E*)-α,β-Unsaturated aldehydes can be prepared by another variation. Terminal metalation of *t*-butoxyallene (**6**) with the hindered base lithium dicyclohexylamide followed by alkylation and hydrolysis gives the aldehyde **7** with traces of the ketone **8** (equation III).

$$(III) \quad CH_2=C=C\begin{smallmatrix}OC(CH_3)_3\\H\end{smallmatrix} \xrightarrow[\text{2) RI; } H_3O^+]{\text{1)} \left(\langle\;\rangle\right)_2 NLi} \begin{smallmatrix}R\\H\end{smallmatrix}C=C\begin{smallmatrix}H\\CHO\end{smallmatrix} + CH_2=CHCR$$

6   **7** (86–73%)   **8** (5–6%)

2-Substituted-3-methoxy-1,3-dienes; α-alkylidene ketones. A route to these compounds from methoxyallene is shown in equation (I). The organoheterocuprate reagent is prepared from RMgX and CuBr and is activated with LiBr.[3]

$$(I) \quad H_2C=C=C\begin{smallmatrix}OCH_3\\H\end{smallmatrix} \xrightarrow[\text{2) } R^1COR^2]{\text{1) } n\text{-BuLi}} H_2C=C=C\begin{smallmatrix}OCH_3\\R^1\\ \quad C\\LiO \quad R^2\end{smallmatrix} \xrightarrow{CH_3SCl}$$

$$H_2C=C=C\begin{smallmatrix}OCH_3\\R^1\\ \quad C\\CH_3SO \quad R^2\\ \quad\quad O\end{smallmatrix} \xrightarrow[70-95\%]{[RCuBr]MgX\cdot LiBr, THF} H_2C=C-C=C\begin{smallmatrix}OCH_3 \quad R^1\\ \quad \quad R^2\\R\end{smallmatrix} \xrightarrow[70-90\%]{H_3O^+, DMSO}$$

(E and Z)

$$H_2C=C-C-CH\begin{smallmatrix}O \quad R^1\\ \quad \quad R^2\\R\end{smallmatrix}$$

[1] J. C. Clinet and G. Linstrumelle, *Tetrahedron Letters*, 1137 (1978).
[2] S. Hoff, L. Brandsma, and J. F. Arens, *Rec. trav.*, **87**, 916 (1968).
[3] H. Kleijn, H. Westmijze, and P. Vermeer, *Tetrahedron Letters*, 1133 (1978).

2-Methoxyallyl bromide (1), 4, 327–328.

Acetonylation.[1] The reagent can be used for acetonylation of anions of esters, ketones, enamines, and imines. The methoxyallyl product is hydrolyzed by weak acids to 2-acetonyl derivatives. This process can be used to prepare substituted cyclopentenones (first example).

Examples:

$$CH_3CH_2CH_2COOCH_3 \xrightarrow[79\%]{} \begin{array}{c} CH_3CH_2CHCOOCH_3 \\ | \\ CH_2COCH_3 \end{array}$$

[1] R. M. Jacobson, R. A. Raths, and J. H. McDonald III, *J. Org.*, **42**, 2545 (1977).

3-Methoxyallylidenetriphenylphosphorane, $(C_6H_5)_3P{=}CHCH{=}CHOCH_3$ **(1)**.
Mol. wt. 332.38.
    Preparation:

$$CH_2{=}C{=}CHOCH_3 \xrightarrow[68\%]{P(C_6H_5)_3 \cdot HBr} (C_6H_5)_3\overset{+}{P}CH_2CH{=}CHOCH_3\ Br^- \xrightarrow[\text{THF, } -50°]{n\text{-}C_4H_9Li,} 1$$

    α,β-*Unsaturated aldehydes.* The reagent has been used in a three-carbon
homologation of ketones to (E)-α,β-unsaturated aldehydes. The intermediate
1-methoxy-1,3-butadienes can be isolated if desired.[1]
    Examples:

$$\begin{array}{c} O \\ \parallel \\ (CH_3)_2C{=}CHCCH_3 \end{array} + 1 \xrightarrow[41\%]{} (CH_3)_2C{=}CH\overset{\overset{\displaystyle CH_3}{|}}{C}{=}CHCH{=}CHOCH_3$$

[1] S. F. Martin and P. J. Garrison, *Tetrahedron Letters*, 3875 (1977).

π-(2-Methoxyallyl)nickel bromide (1), 5, 437.

Isocoumarins; dihydroisocoumarins.[1] Synthesis of these compounds by acetonylation of 2-halobenzoic esters is formulated in equation (I).

[1] D. E. Korte, L. S. Hegedus, and R. K. Wirth, *J. Org.*, **42**, 1329 (1977).

3-Methoxy-3-methylbutynylcopper, $CH_3O-\overset{\overset{\displaystyle CH_3}{|}}{\underset{\underset{\displaystyle CH_3}{|}}{C}}-C\equiv CCu$ **(1).** Mol. wt. 160.67, red oil, soluble THF, insoluble hexane.

Preparation *in situ*[1]:

Mixed cuprates.[2] Mixed cuprates from this copper reagent are useful because of high solubility in THF and because $(CH_3)_3CC\equiv CCu$ is an expensive reagent. They are prepared by reaction of **1** and RLi in THF at $-78°$.

Examples:

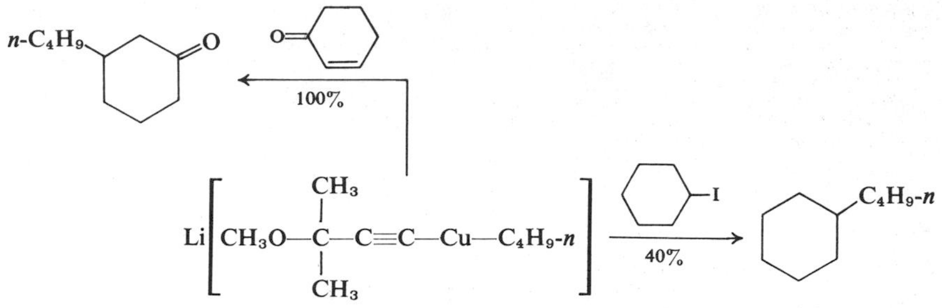

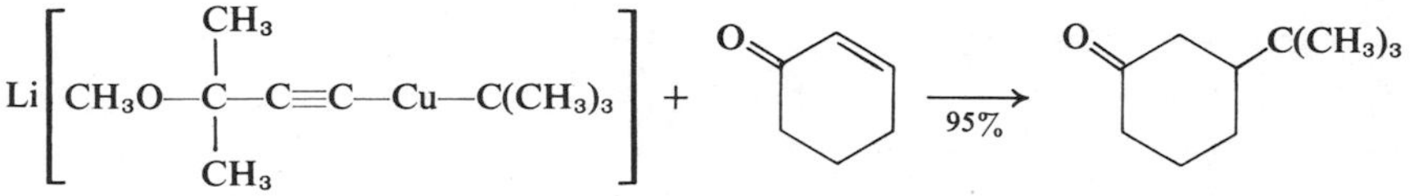

The reagent was developed for the synthesis of a key intermediate (**2**) in the synthesis of maytansine, an antitumor agent.

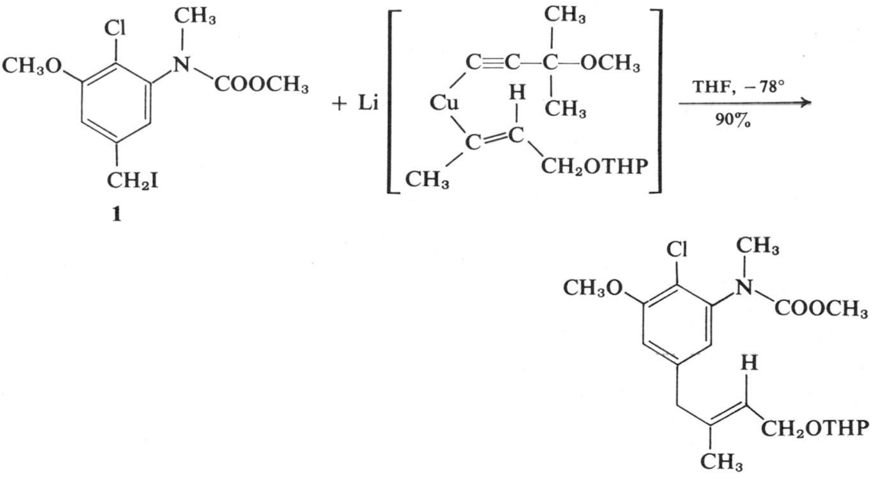

[1] E. J. Corey, D. Floyd, and B. H. Lipshutz, *J. Org.*, **43**, 3418 (1978).
[2] E. J. Corey, M. G. Bock, A. P. Kozikowski, A. V. Rama Rao, D. Floyd, and B. H. Lipshutz, *Tetrahedron Letters*, 1051 (1978).

6-Methoxy-4-methyl-2-pyrone (**1**). Mol. wt. 126.11, m.p. 54–55°.

Preparation:

Diels-Alder reactions (*cf.* 3-Hydroxy-2-pyrone, **6**, 291–293). This reagent undergoes regioselective cycloaddition reactions with naphthoquinones. Thus Diels-Alder addition of **1** to naphthoquinone followed by oxidation and demethylation gives the natural anthraquinone pachybasin in 64% yield. The same regioselectivity is shown in the reaction with juglone to give crysophanol (62% yield). This work was carried out as a model for synthesis of rhodomycinones.[1]

Examples:

[1] M. E. Jung and J. A. Lowe, *J.C.S. Chem. Comm.*, 95 (1978).

p-**Methoxyphenylthionophosphine sulfide dimer,**

$$CH_3O-\langle\bigcirc\rangle-\overset{\overset{S}{\|}}{P}\overset{S}{\underset{S}{\diagdown}}\overset{S}{\underset{\|}{P}}-\langle\bigcirc\rangle-OCH_3 \quad (1)$$

Mol. wt. 404.46, m.p. 229°. This reagent is prepared by reaction of anisole with P_4S_{10}; yield 71%.[1]

Thiation.[2] The reagent is one of the most efficient substances known to date for thiation of ketones, amides, and esters; the yields of thiocarbonyl compounds are usually 90–100%. The reaction is conducted in toluene or xylene at 110–140° with 0.6 eq. of **1**.

Examples:

$$(C_6H_5)_2C=O \xrightarrow[98\%]{1,\,110°} (C_6H_5)_2C=S$$

$$CH_3(CH_2)_5\overset{\overset{O}{\|}}{C}OC_2H_5 \xrightarrow[91\%]{140°} CH_3(CH_2)_5\overset{\overset{S}{\|}}{C}OC_2H_5$$

$$CH_3\overset{\overset{O}{\|}}{C}-SC_6H_5 \xrightarrow[90\%]{} CH_3\overset{\overset{S}{\|}}{C}-SC_6H_5$$

[1] B. S. Pedersen, S. Scheibye, N. H. Nilsson, and S.-O. Lawesson, *Bull. Soc. Chim. Belg.*, **87**, 223 (1978).

[2] B. S. Pedersen, S. Scheibye, K. Clausen, and S.-O. Lawesson, *ibid.*, **87**, 293 (1978).

3-Methoxy-1-phenylthio-1-propene, $C_6H_5SCH=CHCH_2OCH_3$ (**1**). Mol. wt. 180.27, b.p. 107/5 mm.

Preparation:

$$C_6H_5SH + \overset{ClCH_2}{\underset{O}{\triangle}} \xrightarrow[85\%]{NaOH} \overset{C_6H_5SCH_2}{\underset{O}{\triangle}} \xrightarrow{NaH,\,THF}$$

$$C_6H_5SCH=CHCH_2ONa \xrightarrow[82\%]{CH_3I} \mathbf{1},\ E/Z = 1:1$$

α,β-Unsaturated aldehydes. The reagent can be used for synthesis of α,β-unsaturated aldehydes as shown in equation (I).

$$(I)\ \mathbf{1} \xrightarrow[\substack{2)\ RX \\ 60-80\%}]{1)\ LDA,\ THF} C_6H_5S\overset{\overset{R}{|}}{C}HCH=CHOCH_3 \xrightarrow[55-95\%]{HgCl_2,\ HCl,\ CH_3CN,\ 20°} RCH=CHCHO$$

[(E)- and (Z)-isomers, but (E) predominates]

[1] M. Wada, H. Nakamura, T. Taguchi, and H. Takei, *Chem. Letters*, 345 (1977).

trans-**1-Methoxy-3-trimethylsilyloxy-1,3-butadiene, 6, 370–372; 7, 233.**

cis-Δ^1-*3-Octalones* (**6**, 371–372). A key step in Danishefsky's synthesis of *dl*-vernolepin (**6**) and *dl*-vernomenin (**7**)[1] involved the Diels-Alder reaction of this butadiene derivative (**1**) with the dienophile **2**, prepared as shown. The desired

cis-octalone (**3**) was obtained in about 60% yield. This was converted as shown into the dienone lactone (**5**), which is suitably substituted for conversion to the

Vernolepin (**6**)   Vernomenin (**7**)

ring systems of both **6** and **7**. The final step in the synthesis involved bis-α-methylenation, accomplished with dimethyl(methylene)ammonium iodide (**7**, 130–132).

Cyclohexadienone synthesis. Danishefsky et al.[2] have developed a cyclohexadienone synthesis based on Diels-Alder reactions of this diene (**1**) with dienophiles substituted with an α-phenylsulfinyl group, $C_6H_5S(O)$—, which is readily eliminated as C_6H_5SOH from the adducts with introduction of a double bond. An example is shown in equation (I). This cycloaddition is obviously also applicable to a synthesis of phenols as shown by the synthesis of the tetralone in equation (II).

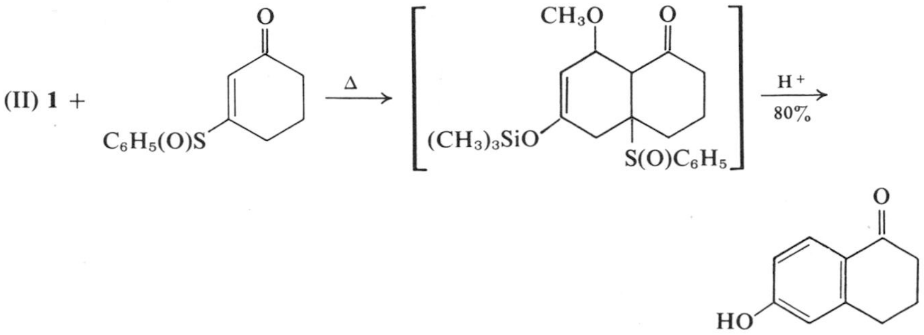

(II) **1** +

This new route to cyclohexadienones has been used in a total synthesis of the disodium salt of prephenic acid (**2**),[3] an unstable intermediate in biosynthesis of various phenols. The first attempt used the dienophile **3** in which the potential

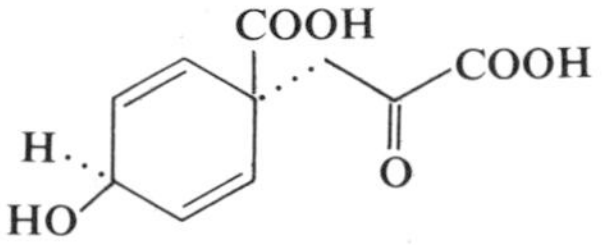

2

carbonyl group of **2** is protected as the dimethyl ketal. The product obtained was the desired dienone (**4**). On reduction with 9-BBN, two epimeric dienols were obtained. On treatment of **5** with aqueous NaOH in THF, the disodium salt of the dimethyl ketal of prephenic acid (**6**) was obtained. Unfortunately even mild acid treatment of **6** or the epimer resulted in dehydration and decarboxylation to give the dimethyl ketal of phenylpyruvic acid.

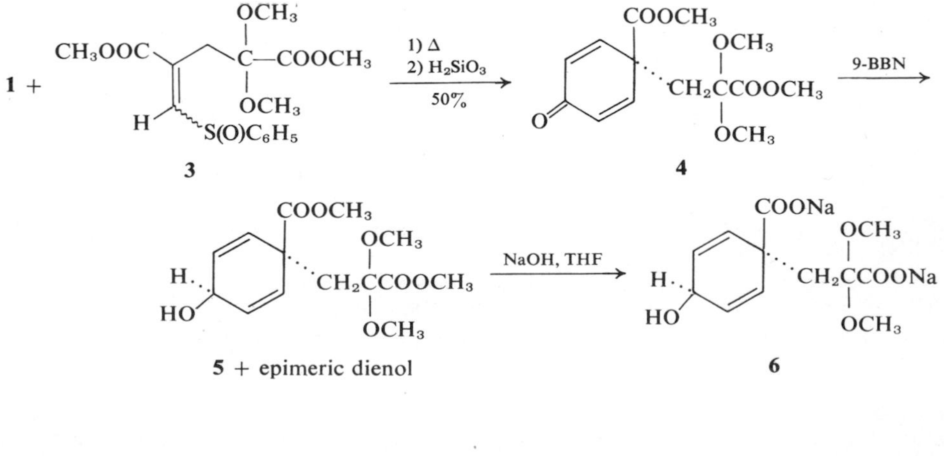

1 +

3

4

5 + epimeric dienol

6

Success was then achieved by blocking the carboxyl group and the ketone group as a methoxylactone, which could be unmasked by mild base. Condensation of **1** with **7** gave the desired dienone **8**, but in low yield. This was reduced as before to the epimeric dienols **9** and **10**. The latter on treatment with sodium

hydroxide in aqueous methanol at room temperature was converted into the disodium salt of prephenic acid. The disodium salt of epiprephenic acid was obtained from **9** in the same way.[3]

An alternate approach to **8** has been reported.[4] In this case the starting material was **11**, which was converted into the dienone (**8**) by two successive reactions with C_6H_5SeCl followed by selenoxide elimination.

[1] S. Danishefsky, P. F. Schuda, T. Kitahara, and S. J. Etheridge, *Am. Soc.*, **99**, 6066 (1977).
[2] S. Danishefsky, R. K. Singh, and T. Harayama, *ibid.*, **99**, 5810 (1977).
[3] S. Danishefsky and M. Hirana, *ibid.*, **99**, 7740 (1977).
[4] W. Gramlich and H. Plieninger, *Tetrahedron Letters*, 475 (1978).

α-Methoxyvinyllithium, 6, 372–373; **7**, 233–234.

Branched-chain sugars. The reaction of 1,2;5,6-di-O-isopropylidene-α-D-hexofuranose-3-ulose (**2**) with this reagent (**1**) in THF at −65° gives the adduct **3**, which can be converted as shown into branched chain sugars as formulated in scheme (I),[1] shown on page 332.

This method has been used to synthesize methyl α-pillaroside (**10**), a derivative of pillarose, a sugar component of an anthracycline antibiotic from *S. flavovirens*. The starting material was **8**, readily available from L-rhamnose. Reaction with **1** led to **9**, which was oxidized to **10**.[2] Fraser-Reid and Walker have synthesized **10** by a different route.[3]

[1] J. S. Brimacombe and A. M. Mather, *Tetrahedron Letters*, 1167 (1978).
[2] J. S. Brimacombe, A. M. Mather, and R. Hanna, *ibid.*, 1171 (1978).
[3] D. L. Walker and B. Fraser-Reid, *Am. Soc.*, **97**, 6251 (1975).

N-Methylanilinium trifluoroacetate, $CF_3COO^- H_2\overset{+}{N}\overset{CH_3}{\underset{C_6H_5}{\diagdown}}$ (**1**). Mol. wt. 221.18,

m.p. 66.5°. The salt is obtained by reaction of 1 equiv. each of trifluoroacetic acid and N-methylaniline at 0° (90% yield).

α-Methylene ketones. The reaction of a ketone with *s*-trioxane and a catalytic amount (0.1 equiv.) of **1** in THF or dioxane (reflux) results in direct formation of α-methylene ketones in yields of 85–95%.[1]

Examples:

$$C_6H_5CCH_2CH_3 \xrightarrow[85\%]{} C_6H_5CCCH_3$$

Scheme (I)

[1] J.-L. Gras, *Tetrahedron Letters*, 2111 (1978).

Methyl   2,3-bis-O-diphenylphosphino-4,6-O-benzylidene-α-D-glucopyranoside   (1).
Mol. wt. 486.42, m.p. 130–131°, α_D − 9°. This substance is prepared by reaction of methyl 4,6-benzylidene-α-D-glucopyranoside with chlorodiphenylphosphine.

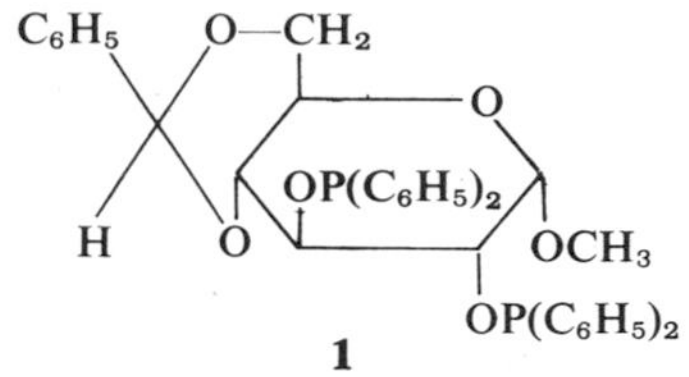

1

Asymmetric hydrogenation.[1] This diphosphinite has been incorporated into a cationic rhodium catalyst (2) by reaction of 1 with chloronorbornadienerhodium(I) dimer, [(NBD)RhCl]$_2$,[2] and AgPF$_6$ in acetone. Use of 2 in the hydrogenation of α-acetamidoacrylic acids and esters results in amino acids with the natural (S)-configuration in optical yields of 50–80% (equation I). The acetamido group is essential for the high optical yields.

(I)

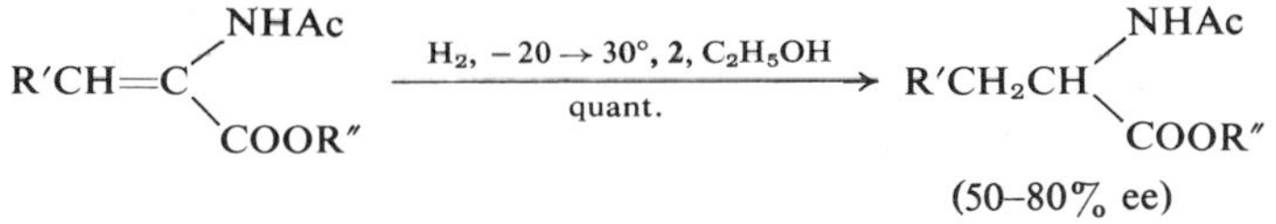

(50–80% ee)

[1] W. R. Cullen and Y. Sugi, *Tetrahedron Letters*, 1635 (1978).
[2] R. R. Schrock and J. A. Osborn, *Am. Soc.*, **93**, 3089 (1971).

3-Methyl-1,3-butadienyl phenyl sulfoxide (1). Mol. wt. 192.28.
   Preparation:

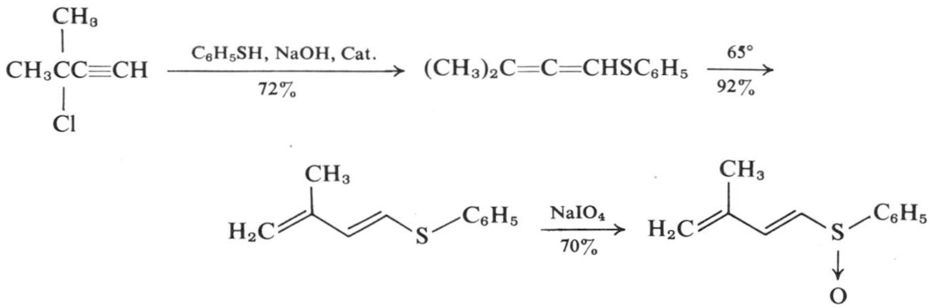

1

Conjugated dienones. When **1** reacts with the anion **2** in THF at − 78 to 20°, the Michael adduct (3) is formed as a mixture of isomers. The adduct is deprotected by acid, and the cyanohydrin **4** is treated with 0.5 N NaOH for 1 hour. The carbonyl group is liberated, and sulfoxide elimination occurs to give a mixture of (E)- and (Z)-tagetone (5).[1]

$$1 + (CH_3)_2CHCH_2\overset{\overset{\displaystyle CN}{|}}{\underset{\underset{\displaystyle CH_3}{|}}{\underset{\displaystyle OCHOC_2H_5}{|}}}{C}\text{—Li} \xrightarrow{\text{THF, } -78 \to 20°}$$

2

$$(CH_3)_2CHCH_2\overset{\overset{\displaystyle CN}{|}}{\underset{\underset{\underset{\displaystyle CH_3}{|}}{\displaystyle O}}{\underset{\displaystyle C_2H_5OHC—O}{|}}}{C}\text{—}\overset{\overset{\displaystyle CH_3}{|}}{CH_2C}\text{=}CHCH_2\overset{\overset{\displaystyle O}{\uparrow}}{S}C_6H_5 \xrightarrow{H_3O^+}$$

3

$$(CH_3)_2CHCH_2\overset{\overset{\displaystyle CN}{|}}{\underset{\underset{\displaystyle OH}{|}}{C}}\overset{\overset{\displaystyle CH_3}{|}}{CH_2CH}\text{=}CHCH_2\overset{\overset{\displaystyle O}{\uparrow}}{S}C_6H_5 \xrightarrow[\substack{50\% \\ \text{from 1}}]{NaOH, H_2O} (CH_3)_2CHCH_2\overset{\overset{\displaystyle O}{\|}}{C}CH\text{=}\overset{\overset{\displaystyle CH_3}{|}}{C}\text{—}CH\text{=}CH_2$$

4   **5** (E/Z = 55:45)

[1] E. Guittet and S. Julia, *Tetrahedron Letters*, 1155 (1978).

B-(3-Methyl-2-butyl)-9-borabicyclo[3.3.1]nonane (B-Siamyl-9-BBN),

$$\text{BCHCH}(CH_3)_2 \quad (1). \text{ Mol. wt. 192.15. Preparation.}^{[1]}$$
$$\underset{\displaystyle CH_3}{|}$$

Reduction of aldehydes.[2] This borane reduces aldehydes and α,β-enals to the corresponding alcohols in 55–90% yield (isolated). The reaction is chemoselective in the presence of keto groups.

[1] H. C. Brown, E. F. Knights, and C. G. Scouten, *Am. Soc.*, **96**, 7765 (1974).
[2] M. M. Midland and A. Tramontano, *J. Org.*, **43**, 1470 (1978).

Methyl (carboxysulfamoyl)triethylammonium hydroxide inner salt, 5, 442. The preparation of this reagent has been published in *Org. Syn.*[1] together with an example of the use for preparation of a urethane of a primary alcohol. Secondary and tertiary alcohols are dehydrated to olefins under these conditions.

$$n\text{-}C_6H_{13}OH + (C_2H_5)_3\overset{+}{N}SO_2\overset{-}{N}COOCH_3 \xrightarrow[51-52\%]{95°} n\text{-}C_6H_{13}NHCOOCH_3 + SO_3 + N(C_2H_5)_3$$

[1] E. M. Burgess, H. R. Penton, Jr., E. A. Taylor, and W. M. Williams, *Org. Syn.*, **56**, 40 (1977).

Methylcopper–Boron trifluoride complex, $CH_3Cu \cdot BF_3$.

γ-*Methylation of allylic halides.* Japanese chemists[1] have reported methylation of allylic halides with complete allylic rearrangement. After noting that methylcopper alone reacted with cinnamyl chloride to give predominantly the product of γ-substitution, they examined the behavior of various boron complexes. Of these the complex with boron trifluoride was found to give exclusively the product of γ-methylation (equation I). The complex is formed *in situ* from CH_3Cu and BF_3 etherate; the reaction is conducted in THF at -70 to $20°$.

(I) $\qquad C_6H_5CH{=}CHCH_2Cl \xrightarrow[90\%]{CH_3Cu \cdot BF_3,\ THF} C_6H_5\underset{\underset{CH_3}{|}}{C}HCH{=}CH_2$

The reagent is probably an ate complex, $CH_3B^-F_3Cu^+$. This reaction is general for allylic halides. Moreover, various alkylcopper compounds show the same behavior. About 10 examples have been reported in which γ-alkylation was effected to the extent of 90% or more.

Examples:

$$CH_3CH{=}CHCH_2Cl + n{-}C_4H_9Cu \cdot BF_3 \xrightarrow[99\%]{} CH_3\underset{\underset{C_4H_9{-}n}{|}}{C}HCH{=}CH_2$$

$$(CH_3)_2C{=}CHCH_2Br + CH_3Cu \cdot BF_3 \xrightarrow[92\%]{} (CH_3)_3CCH{=}CH_2$$

Allylic alcohols also can be alkylated with this new reagent and again with almost complete allylic rearrangement. Best results are obtained with 3 equiv. of the reagent. Regioselectivity is considerably lower with allylic acetates. Although

(I) $C_6H_5CH{=}CHCH_2OH \xrightarrow[96\%]{CH_3Cu \cdot BF_3} C_6H_5\underset{\underset{CH_3}{|}}{C}HCH{=}CH_2 + C_6H_5CH{=}CHCH_2CH_3$

$$19:1$$

the regioselectivity in the reaction of allylic alcohols is somewhat less than that observed with allylic chlorides, the latter substrates are less readily obtained.[2]

[1] K. Maruyama and Y. Yamamoto, *Am. Soc.*, **99**, 8068 (1977).
[2] Y. Yamamoto and K. Maruyama, *J. Organometal. Chem.*, **156**, C9 (1978).

Methyl cyclobutenecarboxylate (**1**). Mol. wt. 112.12, b.p. $45°/1.3$ mm.
    Preparation[1,2]:

1,5-Dienes.[2] Under irradiation **1** undergoes cycloaddition with cycloalkenes to form tricyclic products. The products have been converted to 1,5-dienes by pyrrolysis. Three examples are formulated.

[1] W. G. Dauben and J. R. Wiseman, *Am. Soc.*, **89**, 3545 (1967).
[2] P. A. Wender and J. C. Lechleiter, *ibid.*, **99**, 267 (1977).

Methyl 4-diethylphosphonocrotonate, (**1**). Mol. wt. 236.21.

The reagent (**1**) is prepared[1] by dropwise addition of methyl 4-bromocrotonate (1.0 equiv.) to hot (120°) triethyl phosphite (1.05 equiv.). Ethyl bromide is distilled from the reaction mixture as it is formed, and the residue is distilled to afford 87% of product, bp 112–122° (0.35 mm).[2] Triethyl phosphonocrotonate is similarly prepared in 86% yield.[3] The latter reagent is currently available from Aldrich Chemical Co.

The reagents can be utilized in a modified Wadsworth-Emmons synthesis of (E,E)-diene esters.[2–4] This method has recently been applied by Roush in the initial stages of a total synthesis of dendrobine.[5] The lithium anion of **1**, formed by treatment of **1** at −78° with either lithium hexamethyldisilazide or lithium diisopropylamide[6] (LDA) in THF, was treated at −40° with isobutyraldehyde.

The solution was warmed to 23°, extracted, and distilled to afford **2**, 93% isomerically pure, in 84–86% yield. By comparison, application of the NaH method[3] to **1** and isobutyraldehyde afforded only 30–40% of **2** that was 60–70% isomerically pure.

Diene **2** was converted by five steps into **3**, which (as a trimethylsilyl ether) underwent an intramolecular Diels-Alder reaction in refluxing toluene. Desilylation and chromatography afforded **4** as the major cyclization products (*endo/exo* = 5:1; total yield 84%). Alcohols **4** were elaborated by a lengthy (16 steps) sequence into the natural product.

[1] W. R. Roush, personal communication.
[2] R. S. Burden and L. Crombie, *J. Chem. Soc.*, 2477 (1969).
[3] K. Sato, S. Mizuno, and M. Hirayama, *J. Org.*, **32**, 177 (1967).
[4] Review: W. S. Wadsworth, Jr., *Org. React.*, **25**, 73 (1977).
[5] W. R. Roush, *Am. Soc.*, **100**, 3599 (1978).
[6] E. J. Corey and B. W. Erickson, *J. Org.*, **30**, 821 (1974).

Methyl diformylacetate, (1). Mol. wt. 130.11, b.p. 54–55°/3.5 mm.

The reagent is prepared by acylation of methyl 3,3-dimethoxypropionate with methyl formate in the presence of sodium.[1]

Iridoids. Iridoids are cyclopentanoid monoterpenes that occur widely as glucosides. The first synthesis of an iridoid glucoside, loganin (**4**), was achieved by Büchi *et al.*[1] The key step involved photoannelation of **1** to the tetrahydropyranyl ether of 3-cyclopentenol (**2**), which resulted mainly in **3**. Remaining steps to **4**

are shown in **4**, 64. The method has been adapted to an asymmetric synthesis of **4** (**4**, 161–162).

The [2 + 2] photoannelation of **1** to 1,4-cyclohexadiene (**5**) is the key step in a recent synthesis of secologanin aglycone O-methyl ether (**7**). In both cases the photoproducts possess the *cis*-ring junction.[2]

[1] G. Büchi, J. A. Carlson, J. E. Powell, Jr., and L.-F. Tietze, *Am. Soc.*, **92**, 2165 (1970); **95**, 540 (1973).
[2] R. C. Hutchinson, K. C. Maltes, M. Nakane, J. J. Partridge, and M. R. Uskoković, *Helv.*, **61**, 1221 (1978).

Methylene iodide–Zinc–Trimethylaluminum, CH_2I_2–Zn–$(CH_3)_3Al$.

Methylenation. Methylenation of aldehydes (but not ketones) has been reported with CH_2I_2 and zinc (large excess), but yields are not satisfactory. However, addition of trimethylaluminum gives a reagent that converts both aldehydes and ketones into terminal alkenes in 60–85% yield. No Simmons-Smith products are formed from unsaturated carbonyl compounds.

$$RCOR' + CH_2I_2 \xrightarrow[\text{60–85\%}]{\text{Zn, Al(CH}_3)_3} RR'C{=}CH_2$$

The system CH_2Br_2–Zn–$TiCl_4$ can also be used for methylenation of ketones.

[1] K. Takai, Y. Hotta, K. Oshima, and H. Nozaki, *Tetrahedron Letters*, 2417 (1978).

Methylenetriphenylphosphorane, 1, 678; **6,** 380–381.

α-Methylene ketones.[1] These compounds can be obtained very simply and in high yield by reaction of α-keto ketals with the Wittig reagent, preferably prepared from $CH_3P(C_6H_5)_3Br$ and sodium 2-methyl-2-butoxide as base, followed by deketalization with wet siiica gel.

Methyltriphenylphosphonium bromide can be replaced by ethyl- or isopropyl-triphenylphosphonium bromide to obtain alkylidene ketones.

α-Cyclopropylidene ketones[2] can be obtained in the same way by the reaction of α-keto ketals with cyclopropylidenetriphenylphosphorane (**2,** 95–96).[3]

1,2-Dimethylenecycloalkanes.[4] These substances (**4**) can be prepared from cycloalkanones (**1**) by a Mannich reaction followed by a Wittig reaction (conditions of Greenwald, Chaykovsky, and Corey, **1,** 310); the final step is a Hofmann degradation to give **4**. Overall yields are 10–50%. The order of these reactions is important. The compounds **4** cannot be obtained by reaction of methylenetriphenylphosphorane with a 2-methylenecycloalkanone because the reagent undergoes Michael addition rather than attack of the carbonyl group.

[1] F. Huet, M. Pellet, and J. M. Conia, *Tetrahedron Letters*, 3505 (1977).
[2] F. Huet, A. Lechevallier and J. M. Conia, *ibid.*, 2521 (1977).
[3] K. Utimoto, M. Tamura, and K. Sisido, *Tetrahedron*, **29**, 1169 (1973).
[4] J. W. van Straten, J. J. van Norden, T. A. M. van Schaik, G. Th. Franke, W. H. de Wolf, and F. Bickelhaupt, *Rec. trav.*, **97**, 105 (1978).

N-Methylephedrinium chloride, supported, $C_6H_5CH-\overset{\overset{\displaystyle CH_3}{|}}{CH}-\overset{\overset{\displaystyle CH_3}{|}}{\underset{|}{\overset{+}{N}}}HCl^-$ (1). This

with OH and $CH_2-(P)$

catalyst is prepared by heating a polystyrene-divinylbenzene resin with N-methylephedrine. The treated resin is then washed with dilute hydrochloric acid.

Optically active α,β-epoxy sulfones.[1] The Darzens reaction of ethyl methyl ketone with chloromethyl *p*-tolyl sulfone in a two-phase system in the presence of chiral ammonium salts such as N-ethylephedrinium bromide results in α,β-epoxy sulfones with 0–2.5% optical yields. However, if the supported catalyst (**1**) is used, optical yields of up to 23% can be obtained as in the example formulated in equation (I). On the other hand, the reaction is slower when the catalyst is supported. The presence of a hydroxy group β to the nitrogen atom of the catalyst is essential for asymmetric induction.

(I) $CH_3COC_2H_5 + p\text{-}CH_3C_6H_4SO_2CH_2Cl \xrightarrow[40\%]{1,\ NaOH,\ H_2O,\ CH_3CN}$

$$\underset{C_2H_5}{\overset{CH_3}{\diagdown}} \overset{SO_2C_6H_4CH_3\text{-}p}{\diagup}$$

2 (23% ee)

[1] S. Colonna, R. Fornasier, and W. Pfeiffer, *J.C.S. Perkin I*, 8 (1978).

Methyl fluorosulfonate, 3, 202; 4, 339–340; 5, 445–446; 6, 381–383; 7, 240–241.

Methylation. Kevill and Lin[1] have studied the nucleophilic substitution reactions of the two most generally used methyl esters of superacids, methyl trifluoromethanesulfonate and methyl fluorosulfonate, and of methyl perchlorate. This last ester is a highly explosive material that has been prepared only in solution.[2] Trimethyloxonium tetrafluoroborate was also included in the study. In several kinetic studies the order of reactivity of these four alkylating reagents was found to be: $(CH_3)_3OBF_4 > CH_3OSO_2CF_3 > CH_3OSO_2F > CH_3OClO_3$. In terms of commercial availability and expense, these chemists conclude that methyl fluorosulfonate may well be the reagent of choice for a relatively difficult methylation on a synthetic scale.

Methylation of protomeric tautomers.[3] Generally protomeric tautomers are methylated by this reagent at the heteroatom remote from the mobile proton. Examples:

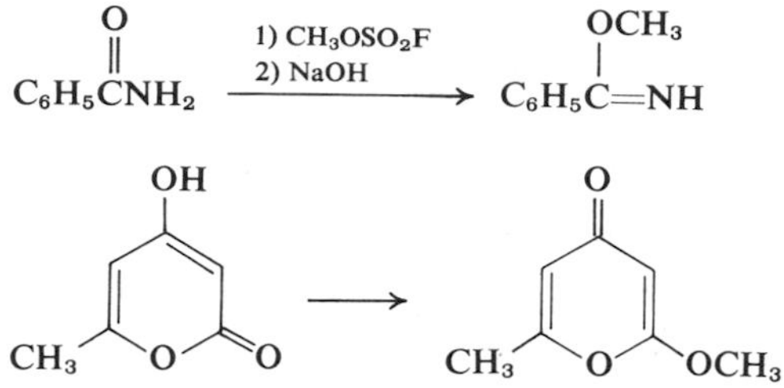

[1] D. N. Kevill and G. M. L. Lin, *Tetrahedron Letters*, 949 (1978).
[2] D. N. Kevill and A. Wang, *J.C.S. Chem. Comm.*, 618 (1976).
[3] P. Beak, J. Lee, and B. G. McKinnie, *J. Org.*, **43**, 1367 (1978).

2-(N-Methyl-N-formyl)aminopyridine (1). Mol. wt. 136.15, b.p. 71–72°/0.05 mm.
Preparation:

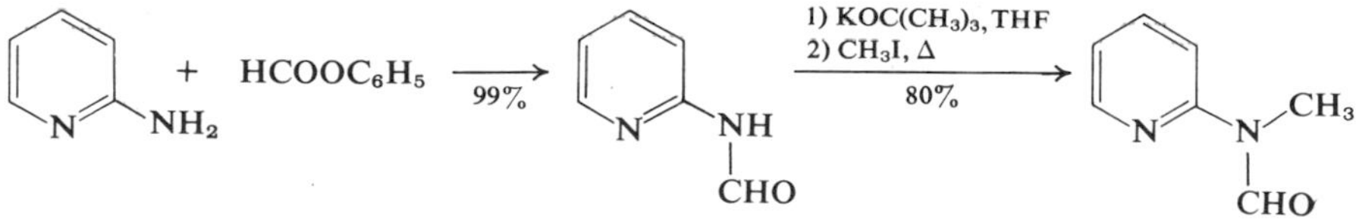

Formylation of Grignard reagents.[1] Grignard reagents are converted into aldehydes in high yield by reaction with **1** followed by quenching with dilute aqueous acid (equation I).

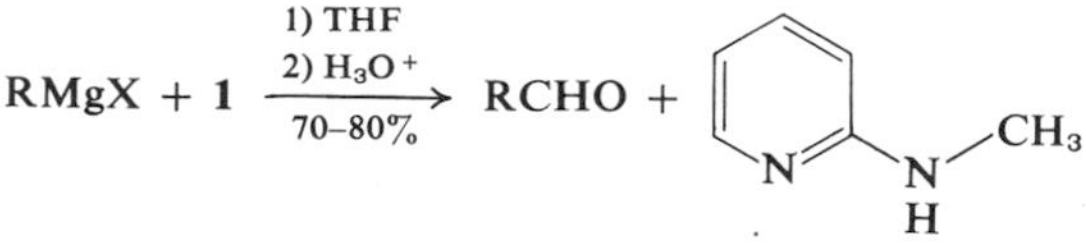

[1] D. Comins and A. I. Meyers, *Synthesis*, 403 (1978).

N-Methylhydroxylamine-O-sulfonic acid, CH_3NHOSO_3H. Mol. wt. 127.1, hygroscopic.
Preparation[1]:

$$CH_3NHOH + ClSO_3H \xrightarrow[88\%]{} CH_3NHSO_3H$$

Nitrones.[2] Nitrones can be prepared by the reaction of this reagent with imines of aldehydes or ketones.

Examples:

$$C_6H_5CH{=}NC_6H_5 \xrightarrow[89\%]{CH_3NHOSO_3H,\ CH_3OH,\ 0-10°}$$

$+ C_6H_5NHSO_3H$

[1] E. Schmitz, R. Ohme, and D. Murawski, *Ber.*, **98**, 2516 (1965).
[2] M. Abou-Gharbia and M. M. Joullié, *Synthesis*, 318 (1977); *Org. Syn.*, submitted (1978).

Methyl iodide, 1, 682–685; **2**, 274; **3**, 206; **4**, 341; **5**, 447–448; **6**, 384.

   trans-*Iodopropenylation of alkyl halides* (**5**, 448). Details for this reaction have been published in *Org. Syn.*[1]

[1] K. Hirai and Y. Kishida, *Org. Syn.*, **56**, 77 (1977).

Methyllithium, 1, 686–689; **2**, 274–278; **3**, 202–204; **5**, 448–454; **6**, 384–385; **7**, 242–243.

   Halide-free reagent. Commercial preparations (Alfa, Lithium Corp. of America) contain a full molar equivalent of lithium bromide formed in the reaction of methyl bromide and lithium. For several applications, halide-free methyllithium is desirable. Complete details for preparation of such material have been submitted to *Org. Syn.*[1] Four different assemblies are required, and of course the usual precautions for handling lithium are mandatory.

   2-Adamantyl methyl ketone (**2**) can be obtained in high yield by reaction of 2-cyanoadamantane (this volume) with methyllithium followed by acidic hydrolysis. The yield of **2** by reaction of 2-adamantanecarboxylic acid with methyl-lithium is only about 5%. Other alkyllithium and aryllithium reagents can be used.[2]

Acetylenic ketones. Coke *et al.*[3] have reported an alternative method to fragmentation of α,β-epoxy ketones (**2**, 419–422; **3**, 293; **6**, 232–233). The substrate is a β-halo-α,β-unsaturated ketone. The reaction involves addition of an alkyl-lithium to form an alkoxide, which on pyrolysis cleaves to an acetylenic ketone.

Examples:

$$CH_3CCH_2CCH_2C{\equiv}CH$$

$$CH_3C(CH_2)_2C{\equiv}CCH_3$$

Intramolecular cyclization of carbenoids.[4] The most efficient route to the tricyclo[4.1.0.0^{2,7}]heptene-3 ring system (*e.g.*, **3**) is the methyllithium-promoted cyclization of a 7,7-dibromobicyclo[4.1.0]heptene-3 such as **2**. The precursors (**2**) are available by dibromocarbene addition to 1,4-cyclohexadienes (**1**). The cyclization involves generation of a carbenoid and intramolecular C—H insertion.

In the case of 1,4-cyclohexadiene itself, the yield of the tricyclic compound is only 1–5%. A single methyl group at a ring position increases the yield to about 50% of the two possible products. Possibly the reactivity of the intermediate carbenoid or the stability of the product is enhanced by steric factors.

[1] M. J. Lusch, W. V. Phillips, R. F. Sieloff, and H. O. House, *Org. Syn.*, submitted (1977).
[2] A. M. van Leusen and D. van Leusen, *Syn. Comm.*, **8**, 397 (1978).

[3] J. L. Coke, H. J. Williams, and S. Natarajan, *J. Org.*, **42**, 2380 (1977).
[4] L. A. Paquette and R. T. Taylor, *Am. Soc.*, **99**, 5708 (1977); *Org. Syn.*, submitted (1977).

Methyl methylthiomethyl sulfoxide, 4, 341–342; **5**, 456–457; **6**, 390–392.

Conjugate addition. The lithio derivative (**1**) of this reagent reacts with some enones to give the 1,4-adduct in addition to the 1,2-adduct. The ratio of these two products depends on substituent and temperature effects.[1]

Examples:

The paper cites a new use of the 1,2-adducts (equation I).

(I)

α-Keto esters from nitriles. Ogura *et al.*[2] have modified the synthesis of α-amino acids from nitriles using this reagent (**5**, 457) to a synthesis of α-keto esters. The first steps to give **1** and **2** are the same as before. Treatment of **1** with CuCl$_2$ in ethanol gives the ethyl ester of an α-keto acid (**3**), possibly by hydrolysis of the enamine followed by Pummerer rearrangement. Reaction of **1** with $CuCl_2$ in methylene chloride affords the methanethiol ester of an α-keto acid (**4**). This product can also be obtained from **2** by peracid oxidation. α-Keto amides (**5**) can be prepared from **2** by treatment with ammonia in isopropanol followed by peracid oxidation.

Aldehyde synthesis. Grignard reagents, RMgX, in THF react with this sulfoxide to form dimethyl dithioacetals of RCHO. These are deacetalated by sulfuryl chloride in the presence of wet silica gel. The method is particularly useful for synthesis of aromatic aldehydes.[3]

Aliphatic aldehydes are more conveniently prepared by reaction of alkyl halides with the carbanion of the sulfoxide (**4**, 341).

[1] K. Ogura, M. Yamashita, and G. Tsuchihashi, *Tetrahedron Letters*, 1303 (1978).
[2] K. Ogura, N. Katoh, I. Yoshimura, and G. Tsuchihashi, *ibid.*, 375 (1978).
[3] M. Hojo, R. Mazuda, T. Saeki, K. Fujimori, and S. Tsutsumi, *ibid.*, 3883 (1977).

N,N-Methylphenylaminotri-*n*-butylphosphonium iodide, $(n\text{-}C_4H_9)_3\overset{+}{P}N(CH_3)C_6H_5I^-$ (**1**). Mol. wt. 435.37, m.p. 120.5°. The reagent is prepared by reaction of tri-*n*-butylphosphine with phenyl azide (ether, reflux) and then with excess methyl iodide; yield 90%.

γ-*Alkylation and arylation of allylic alcohols.*[1] Organolithium compounds in conjunction with **1** alkylate allylic alcohols selectively at the γ-position with formation of rearranged olefins (equation I). The reaction proceeds satisfactorily with primary, secondary, and tertiary allylic alcohols. In some cases only the

(E)-olefins are formed; in other cases the (É)-isomer predominates.
Examples:

[1] Y. Tanigawa, H. Ohta, A. Sonoda, S.-I. Murahashi, *Am. Soc.*, **100**, 4610 (1978).

N,N-Methylphenylaminotriphenylphosphonium iodide, (1), $(C_6H_5)_3\overset{+}{P}N(CH_3)C_6H_5I^-$
6, 392. Mol. wt. 495.34, m.p. 238.5°. The reagent is prepared by reaction of phenyliminotriphenylphosphorane with methyl iodide (reflux, 2 hours).

Alkene synthesis.[1] Alcohols can be alkylated or arylated by treatment with organolithium compounds and **1** (equation I). The main interest in this synthesis is the reaction of allylic alcohols to form alkenes. The reaction is regioselective and stereoselective (inversion of configuration).

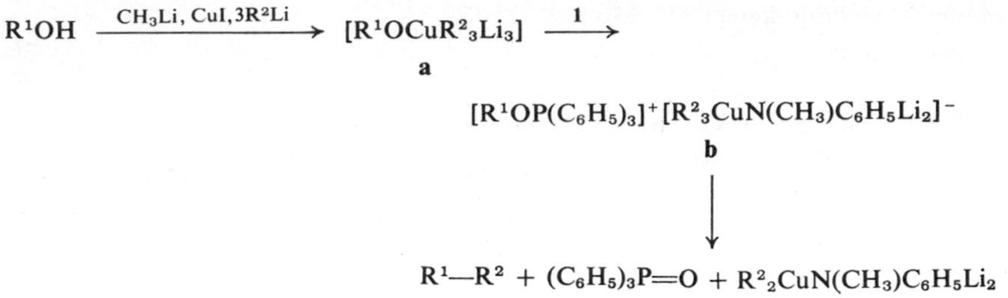

$$R^1OH \xrightarrow{\text{CH}_3\text{Li, CuI, 3R}^2\text{Li}} [R^1OCuR^2_3Li_3] \xrightarrow{\mathbf{1}}$$

a

$$[R^1OP(C_6H_5)_3]^+ [R^2_3CuN(CH_3)C_6H_5Li_2]^-$$

b

$$R^1\!\!-\!\!R^2 + (C_6H_5)_3P\!\!=\!\!O + R^2_2CuN(CH_3)C_6H_5Li_2$$

Examples:

(91%)   (9%)

(96%)   (4%)

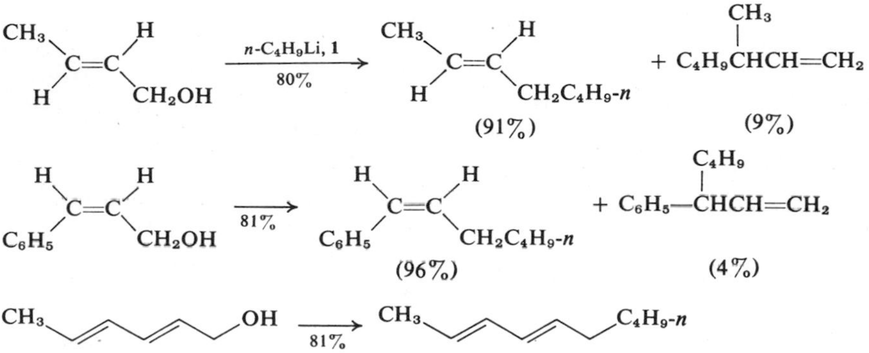

[1] Y. Tanigawa, H. Kanamaru, A. Sonoda, and S.-I. Murahashi, *Am. Soc.*, **99**, 2361 (1977).

Methyl phenylsulfinylacetate, $C_6H_5S(O)CH_2COOCH_3$. Mol. wt. 198.24.

5-Substituted resorcinols.[1] A typical synthesis involves the base-catalyzed conjugate addition of **1** to **2** followed by cyclization to form **3** as a mixture of isomers. This product loses benzenesulfenic acid when heated to give 5-phenyl-resorcinol (**4**).

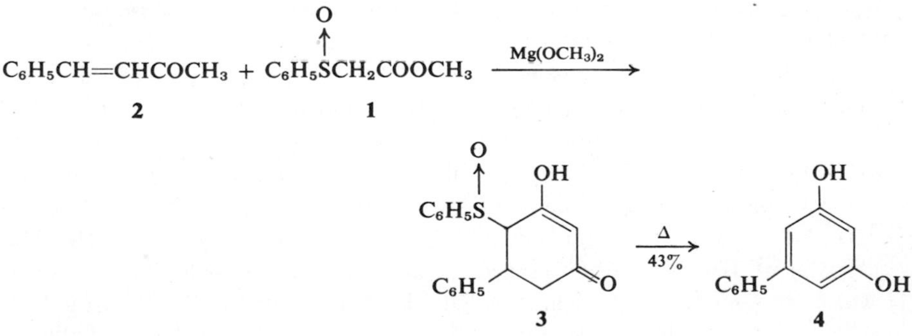

[1] A. A. Jaxa-Chamiec, P. G. Sammes, and P. D. Kennewell, *J.C.S. Chem. Comm.*, 118 (1978).

Methyl(phenylthio)ketene (1). Mol. wt. 254.22.

Preparation:

$$CH_3CHClCOOC_2H_5 + C_6H_5SNa \longrightarrow \xrightarrow{OH^-}$$

$$CH_3CHCOOH \xrightarrow[\text{2) N(C}_2\text{H}_5)_3]{\text{1) SOCl}_2} \begin{array}{c} CH_3 \\ \diagdown \\ C_6H_5S \end{array} C{=}C{=}O$$
$$\underset{SC_6H_5}{|}$$

1

[2 + 2] *Cycloadditions.* The reagent can be used as the equivalent of methyl-eneketene in cycloadditions with alkenes and imines.[1]

Examples:

$$C_6H_5N{=}CHC_6H_5 \xrightarrow[58\%]{1}$$

[1] T. Minami, M. Ishida, and T. Agawa, *J.C.S. Chem. Comm.*, 12 (1978).

4-(4′-Methyl-1′-piperazinyl)-3-butyne-2-one, $CH_3N \overbrace{} N{-}C{\equiv}CCH_3$ (1). Mol.

wt. 166.23, m.p. 18–19°. The ynone is prepared by bromination of the corresponding enone followed by dehydrobromination with potassium *t*-butoxide in THF (59% yield).[1]

Synthesis of amides[2] and peptides.[3] The simplest compound of this type, dimethylaminopropynal $(CH_3)_2NC{\equiv}CCHO$, is known, but it polymerizes within minutes at room temperature. The thermal stability is improved by introduction of bulky constituents. A whole series of these acetylenes have been prepared; of these **1** and a few related compounds have been found useful for synthesis of CO—NH bonds by a "push–pull" mechanism shown in equation (I). The reaction of **1** with a carboxylic acid proceeds by Michael addition to give *a*, which rearranges by a cyclic intermediate (**b**) to the enol ester (**2**). Reaction of **2** with an amine yields an amide (**3**) and the water adduct (**4**) of **1**. Yields of amides are in the range 85–95%. This sequence cannot be used for esterification of carboxylic acids.

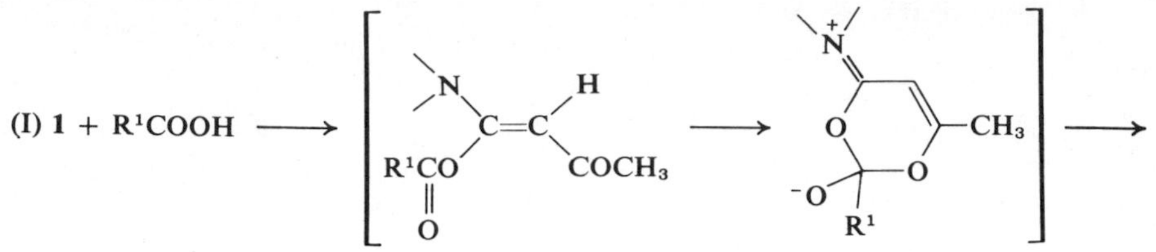

$$(I) \; 1 + R^1COOH \longrightarrow$$

a

b

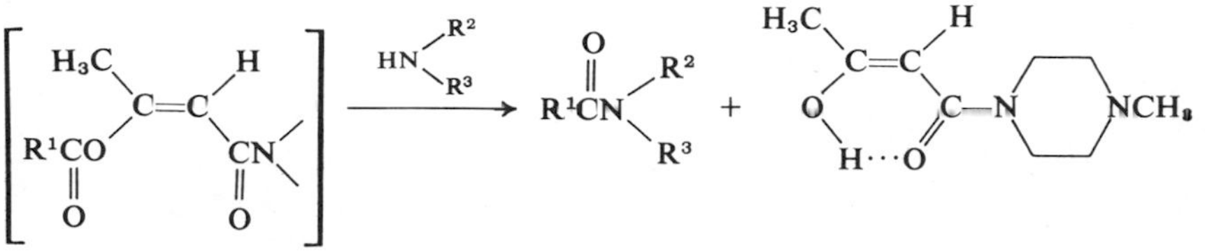

2   3   4

Acetylenes of the type of **1** are useful for peptide synthesis with CBZ-protected amino acids (equation II). Since the intermediate enol ester (**2**) does not react

$$(II) \; CBZ\text{—}NHCHCOOH + H_2NCHCOOR \xrightarrow[75\text{–}90\%]{1} CBZ\text{—}NHCHCONHCHCOOR$$

with —OH and —SH groups, serine, tyrosine, and hydroxyproline can be used as such without protection.

The related ynone **5** is also useful for peptide synthesis; yields are somewhat higher and the water adduct corresponding to **4** is more easily separable from the peptide.

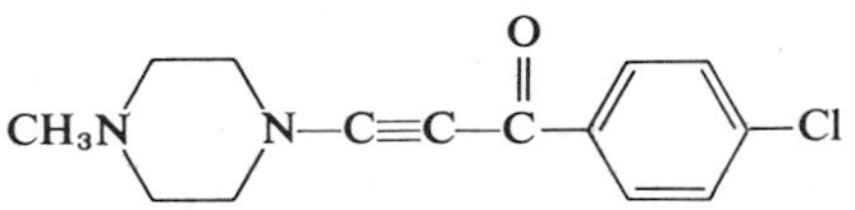

5

[1] U. Lienhard, H.-P. Fahrni, and M. Neuenschwander, *Helv.*, **61**, 1609 (1978).
[2] M. Neuenschwander, U. Lienhard, H.-P. Fahrni, and B. Hurni, *Chimia*, **32**, 212 (1978).
[3] M. Neuenschwander, H.-P. Fahrni, and U. Lienhard, *ibid.*, **32**, 214 (1978).

Methylthioacetonitrile, CH_3SCH_2CN **(1).** Mol. wt. 87.15, b.p. 72°/30 mm. The reagent is prepared by reaction of chloroacetonitrile and CH_3SNa in methanol, yield 83%.[1]

α,β- and β,γ-Unsaturated nitriles. Typical reactions with this reagent are formulated in equations (I) and (II).[2]

(I) $ArCHO + 1$ $\xrightarrow[\text{75–95\%}]{\text{Triton B, THF}}$

$$\underset{Ar}{\overset{H}{C}}=\underset{CN}{\overset{SCH_3}{C}} \quad + \quad \underset{Ar}{\overset{H}{C}}=\underset{SCH_3}{\overset{CN}{C}}$$

$\xrightarrow[\text{90–95\%}]{\substack{\text{NaBH}_4 \\ \text{CH}_3\text{OH}}}$ $\underset{Ar}{\overset{H}{C}}=\underset{SCH_3}{\overset{CN}{C}}$ $\xrightarrow[\text{65–95\%}]{\text{Raney Ni}}$ $\underset{Ar}{\overset{H}{C}}=\underset{CN}{\overset{H}{C}}$

$ArCH_2CCH_3$ (with =O) $\xrightarrow[\text{2) NBS}]{\text{1) OH}^-}$ $ArCH_2\underset{CH_3}{\overset{SCH_3}{C}}CN$ $\xrightarrow{\text{LDA, CH}_3\text{I, DMSO}}$ $ArCH_2\underset{CN}{\overset{SCH_3}{CH}}$ $\xrightarrow[\text{85–90\%}]{\substack{\text{1) ClC}_6\text{H}_4\text{CO}_3\text{H} \\ \text{2) }\Delta}}$ $\underset{Ar}{\overset{H}{C}}=\underset{H}{\overset{CN}{C}}$

(II) cyclohexanone $+ 1$ $\xrightarrow[\text{60–95\%}]{\text{Triton B, THF}}$ cyclohexylidene$\overset{SCH_3}{\underset{CN}{C}}$ $\xrightarrow[\text{60–85\%}]{\substack{\text{1) LDA} \\ \text{2) RI}}}$ cyclohexenyl$\overset{R}{\underset{CN}{C}}SCH_3$

[1] R. Dijkstra and H. J. Backer, *Rec. trav.*, **73**, 569 (1954).
[2] S. Kano, T. Yokommatsu, T. Ono, S. Hibino, and S. Shibuya, *Chem. Pharm. Bull. Japan*, **26**, 1874 (1978).

3-Methylthio-1,4-diphenyl-5-triazolium iodide, $\cdot HI^-$ **(1), 6,** 396.

RCH$_2$COOH → RCHO.[1] This transformation can be conducted as shown in equation (I). The key step is acetoxylation effected with iodobenzene diacetate

(I) RCH$_2$COCl + 1 $\xrightarrow{\text{NaH, DMF}}$ RCH$_2$— ... I$^-$ $\xrightarrow[]{\text{1) I}_2\text{, C}_6\text{H}_5\text{I(OAc)}_2 \\ \text{2) Na}_2\text{S}_2\text{O}_4 + \text{KI, H}_2\text{O}}$

RCH— ... I$^-$ $\xrightarrow[\text{20–55\%, overall}]{\text{1) NaOCH}_3\text{, CH}_3\text{OH} \\ \text{2) H}_3\text{O}^+\text{, KI}}$ RCHO + CH$_3$COOCH$_3$ + 1

in the presence of iodine. Tl(OAc)$_3$ or Pb(OAc)$_4$ is ineffective for this reaction. In addition, only the iodide salts are susceptible to this acetoxylation, the chloride and tetrafluoroborate being unreactive. Only tars are formed when R = C$_6$H$_5$.

[1] G. Doleschall, *Tetrahedron Letters*, 381 (1977).

2-Methylthiomaleic anhydride,   **(1)**. Mol. wt. 144.15, m.p. 36–37°.

The anhydride is prepared in 68–77% yield by addition of methanethiol to acetylenedicarboxylic acid to give a mixture of 2-methylthiofumaric and 2-methylthiomaleic acid, followed by cyclization with thionyl chloride.

β-Keto esters; α-methylene ketones.[1] The anhydride (1) undergoes Diels-Alder reactions with dienes at 0–80° to give only *endo*-adducts; usually only one of the two possible *endo*-adducts is formed. The double bond in the adducts can be reduced without desulfurization with H$_2$ and Pd(OH)$_2$ on carbon[2]. The adducts are convertible by oxidative decarboxylation with NCS (6, 116–117) to the ketal and/or the enol thioether of a β-keto ester, both of which are hydrolyzable to

β-keto esters. α-Methylene ketones are obtained by reduction of the ester group before hydrolysis of the vinyl sulfide grouping.[3]

Examples: See page 353.

[1] B. M. Trost and G. Lunn, *Am. Soc.*, **99**, 7079 (1977).
[2] M. Pearlman, *Tetrahedron Letters*, 1663 (1967).
[3] W. F. Erman, *J. Org.*, **32**, 765 (1967).

Methylthiotrimethylsilane, 6, 399. Supplier: Petrach Systems.

Thioketalization. Contrary to the earlier report cited in **6**, 399, the reaction of aldehydes and ketones with this reagent is initiated by traces of Lewis acids. The most satisfactory conditions use diethyl ether as solvent and zinc iodide as catalyst. The initial reaction results in O-trimethylsilyl hemithioketals, which in the presence of a second equivalent of the reagent are converted into dimethyl thioketals:

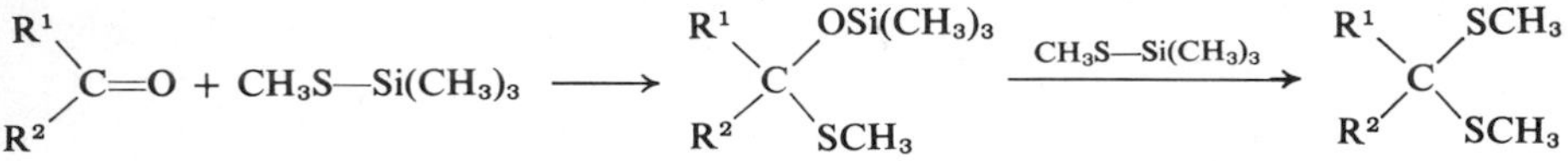

The yields of derivatives are usually $> 95\%$. Moreover this method is useful for selective monothioketalization. Thus Δ^4-androstene-3,17-dione can be converted into the 3-dimethyl thioketal in 92% yield.[1]

[1] D. A. Evans, L. K. Truesdale, K. G. Grimm, and S. L. Nesbitt, *Am. Soc.*, **99**, 5009 (1977).

Methyl tributylstannyl sulfide, $(n\text{-}C_4H_9)_3SnSCH_3$ (**1**). Mol. wt. 337.13, b.p. 104°/ 0.2 mm. The sulfide is prepared by the reaction of bis(tri-n-butyltin) oxide with methyl mercaptan (2 equiv.).[1]

1-Thioglycosides.[2] These glycosides are prepared in high yield by reaction of an alkyl tributylstannyl sulfide with peracetyl pyranosyl or furanosyl halides in the presence of stannic chloride. An anomeric mixture is usually formed.

Examples:

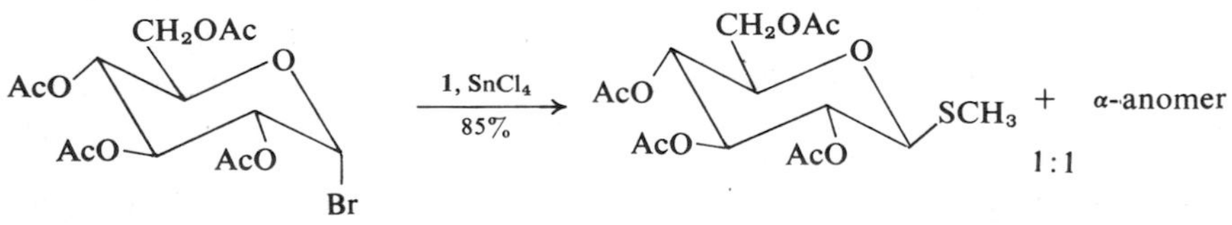

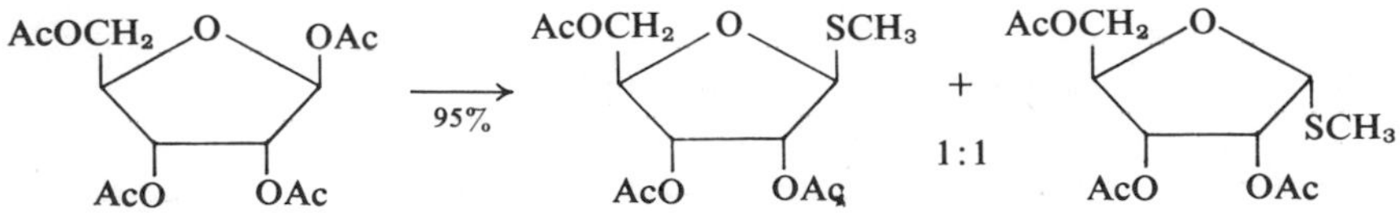

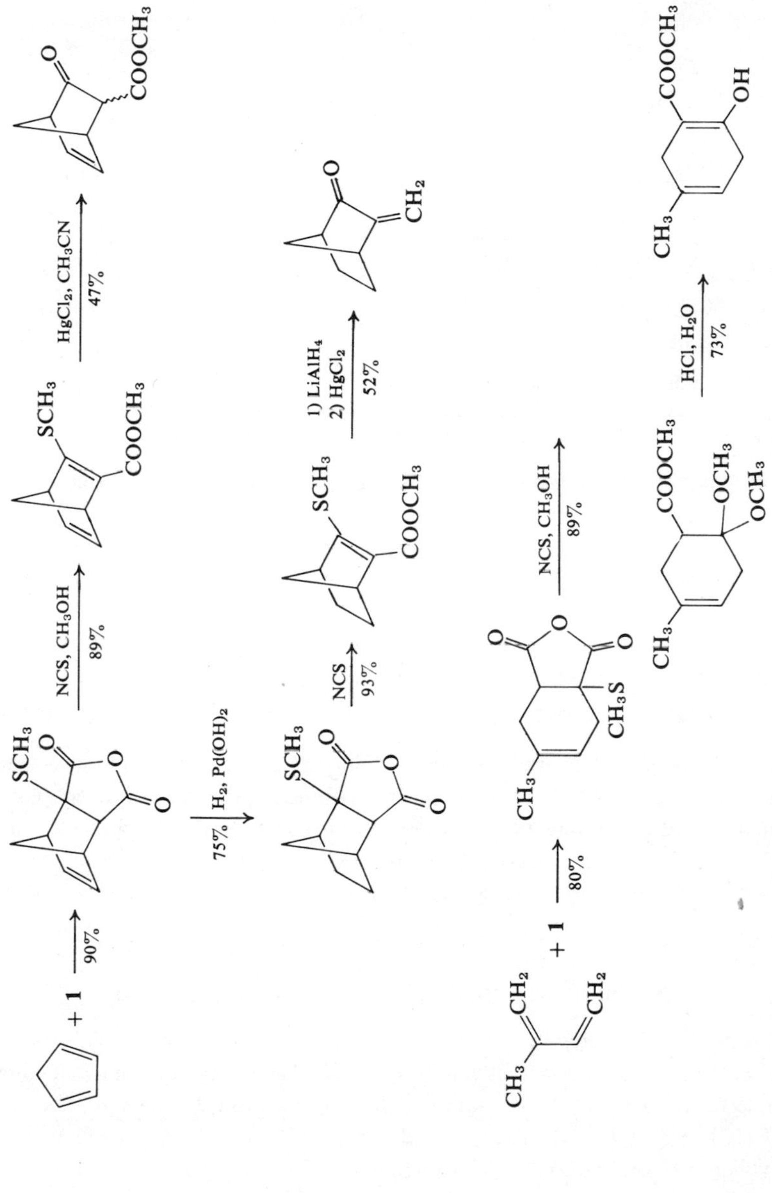
+ 1
90%
SCH3
NCS, CH3OH
89%
SCH3
COOCH3
HgCl2, CH3CN
47%
COOCH3
75%
H2, Pd(OH)2
SCH3
NCS
93%
SCH3
COOCH3
1) LiAlH4
2) HgCl2
52%
CH2
CH3
CH2
CH2
+ 1
80%
CH3
CH3S
NCS, CH3OH
89%
CH3
COOCH3
OCH3
OCH3
HCl, H2O
73%
CH3
COOCH3
OH

This reaction was used in a synthesis of $(-)$-*cis*-rose oxide (**5**) from D-glucose (**1**).[3]

1

2

3 $(6\alpha/6\beta = 1:1)$

4

5

[1] M. E. Peach, *Canad. J. Chem.*, **46**, 211 (1968).
[2] T. Ogawa and M. Matsui, *Carbohydrate Res.*, **54**, C17 (1977).
[3] T. Ogawa, N. Takasaka, and M. Matsui, *ibid.*, **60**, C4 (1978).

Methyltri(phenoxy)phosphonium iodide (Triphenylphosphite methiodide), **1**, 1249; **2**, 446; **4**, 448; 557–559; **6**, 649.

Deoxygenation of epoxides. This reagent in combination with BF_3 etherate converts epoxides into olefins in 75–99% yield. A 1:1 mixture of $CH_3CN-C_6H_6$ is used as solvent; the reaction is conducted at room temperature. The iodide is used in large excess (10 molar equiv.), and generally an excess of BF_3 etherate (3 molar equiv.) is also required for satisfactory yields.[1]

[1] K. Yamada, S. Goto, H. Nagase, Y. Kyotani, and Y. Hirata, *J. Org.*, **43**, 2076 (1978).

Monochloroborane–Dimethyl sulfide, $H_2BCl \cdot S(CH_3)_2$; **Dichloroborane–Dimethyl sulfide**, $HBCl_2 \cdot S(CH_3)_2$. These stable hydroborating agents are prepared[1] from borane–dimethyl sulfide (**4**, 124, 191; **5**, 47) and boron trichloride–dimethyl sulfide[2] (equations I and II). They possess the advantage over chloroborane etherates of indefinite stability. They show regiospecificity comparable to the

(I) $2BH_3 \cdot S(CH_3)_2 + BCl_3 \cdot S(CH_3)_2 \xrightarrow{25°} 3BH_2Cl \cdot S(CH_3)_2$

(II) $BH_3 \cdot S(CH_3)_2 + 2BCl_3 \cdot S(CH_3)_2 \xrightarrow{25°} 3BHCl_2 \cdot S(CH_3)_2$

chloroborane etherates in hydroboration–oxidation of alkenes. The product of the reaction of olefins with $BH_2Cl \cdot S(CH_3)_2$ is a dialkylchloroborane–dimethyl sulfide from which R_2BCl is obtained free by vacuum distillation. The reaction of $BHCl_2 \cdot S(CH_3)_2$ with alkenes in the presence of BCl_3 (*cf.* **4**, 130) yields an alkyldichloroborane directly with formation of $BCl_3 \cdot S(CH_3)_2$.

[1] H. C. Brown and N. Ravindran, *J. Org.*, **42**, 2533 (1977).
[2] *Idem, Inorg. Chem.*, **16**, 2938 (1977).

N

$$\text{H}_2\text{N}-\underset{\underset{\text{H}}{|}}{\overset{\overset{\text{CH}_3}{|}}{\text{C}}}$$

(S)-1-α-Naphthylethylamine, 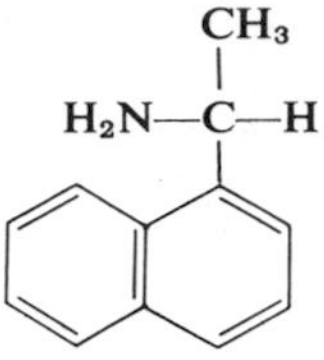 Mol. wt. 161.23, α_D +50.5. Supplier:
Norse Chem. Co.

Convergent resolution of enantiomers. Trost *et al.*[1] have devised an enantio-convergent prostanoid synthesis from **1**, in which a 1,3-hydroxy shift interconverts the two enantiomers. The interconversion was conducted on the urethane of **1** with (S)-1-α-naphthylethylamine catalyzed with mercury(II) trifluoroacetate. The equilibrium ratio of **2** to **3** is about 2:1. By repetition of this equilibration optically active **1** can be obtained without loss of one enantiomer. Optically active **1** was used for synthesis of the optically pure prostanoid **4**.

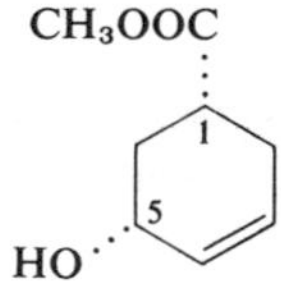

1a (1R, 5R) 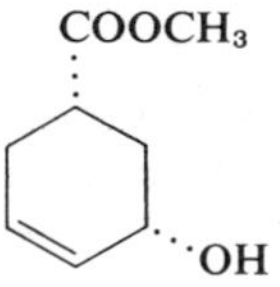   **1b (1S, 5S)**

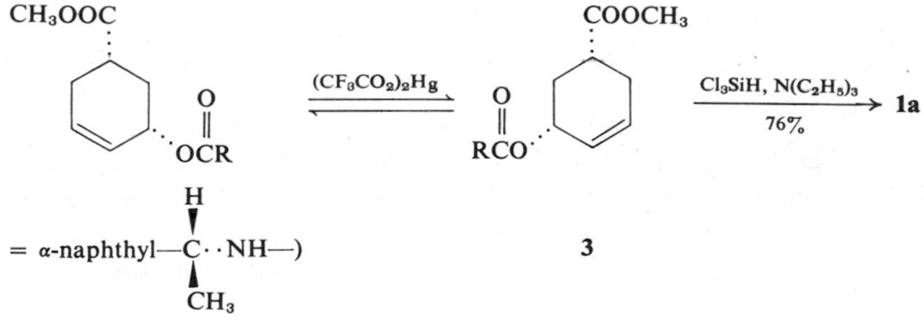

$(CF_3CO_2)_2Hg$

$Cl_3SiH, N(C_2H_5)_3$
76% → **1a**

2 (R = α-naphthyl—$\overset{\overset{\text{H}}{\blacktriangledown}}{\underset{\underset{\text{CH}_3}{\blacktriangle}}{\text{C}}}$··NH—)   **3**

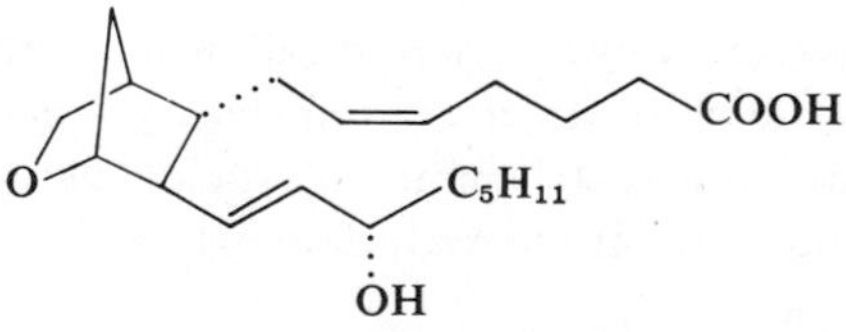

4 (α_D − 8.5°)

[1] B. M. Trost, J. M. Timko, and J. L. Stanton, *J.C.S. Chem. Comm.*, 436 (1978).

(R)-(−)-1-(Naphthyl)ethyl isocyanate (1). Mol. wt. 197.23, b.p. 106–108°/0.16 mm., α_D −50.5°. This isocyanate can be prepared from (R)-(+)-1-(naphthyl)-ethylamine (Norse Chem. Co.) by two methods.

Resolution of hydroxylic compounds by chromatography.[1,2] The chiral isocyanate is useful for large scale chromatographic separation of diastereomeric carbamates. The elution order is correlated with the structure and the stereochemistry.

The reagent was used to resolve 5-cyanopentane-2-ol (1), a precursor to the four possible isomeric lactones (3). One of the *cis*-enantiomers is the carpenter bee pheromone.[3]

Chiral 1,3-dialkylallenes. These allenes can be obtained by reaction of lithium dialkylcuprates with one of the diastereomeric carbamates obtained from racemic secondary propargylic alcohols and the reagent (1). An example is the synthesis of chiral pheromone of the male bean weevil, (R)-(−)-6, as outlined in scheme (I).[4]

Thus reaction of the alcohol 2 with 1 leads to the two diastereomeric cyano-carbamates 3 formed as a 1:1 mixture. They are converted into the diastereomeric carbamates 4, which are separated by liquid chromatography on silica gel. The reaction of one carbamate (4a) with lithium di-*n*-octylcuprate affords the chiral allene 5. The final step to 6 involves introduction of the *trans*-double bond.[5]

[1] W. H. Pirkle and M. S. Hoekstra, *J. Org.*, **39**, 3904 (1974).
[2] W. H. Pirkle and J. R. Hauske, *ibid.*, **42**, 1839 (1977).
[3] W. H. Pirkle and P. E. Adams, *ibid.*, **43**, 378 (1978).

Scheme (I)

[4] W. H. Pirkle and C. W. Boeder, *ibid.*, **43**, 2091 (1978).
[5] P. J. Kocienski, G. Cernigliaro, and G. Feldstein, *ibid.*, **42**, 353 (1977).

Nickel peroxide, 1, 731–732; **2,** 294–295; **5,** 474; **7,** 250–251.

Dehydrogenation of dihydro derivatives of thiazoles, imidazoles, furanes. In a study directed toward synthesis of a natural product containing a bithiazole group, Hecht *et al.*[1] examined various reagents for oxidation of the thiazoline **1** to the corresponding thiazole **2.** Nickel peroxide was found to be superior to activated MnO_2, which effected this conversion in a yield of 65%. No other

reagents were found to give significant conversions. This reagent was found to be generally useful for dehydrogenation of thiazolines, imidazolines, dihydrofuranes, and related heterocycles.

Examples:

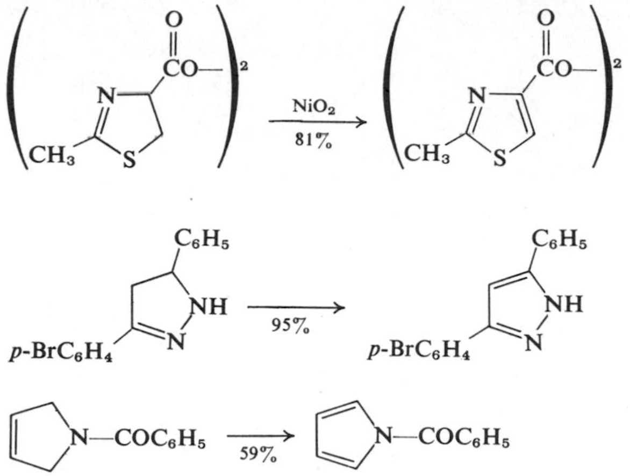

ArCHO → ArCOOH. This oxidation can be carried out with NiO_2 in aqueous sodium hydroxide. Yields are generally in the range 94–100%. α-Furoic acid is obtained in this way from furfural in 93.5% yield. p-Tolualdehyde is oxidized in part to terephthalic acid.[2]

[1] D. A. McGowan, U. Jordis, D. K. Minster, and S. M. Hecht, *Am. Soc.*, **99**, 8078 (1977); D. K. Minster, U. Jordis, D. L. Evans, and S. M. Hecht, *J. Org.*, **43**, 1624 (1978).

[2] K. Nakagawa, S. Mineo, and S. Kawamura, *Chem. Pharm. Bull. Japan*, **26**, 299 (1978).

Nitrobenzene, **2**, 295–296; **7**, 251.

Aromatic hydroxylation.[1] Some isoquinoline alkaloids can be hydroxylated by a method based on a reaction first noted by Buck and Köbrich.[2] The method is illustrated for hydroxylation of laudanosine (**1**) (equation I). When nitrobenzene

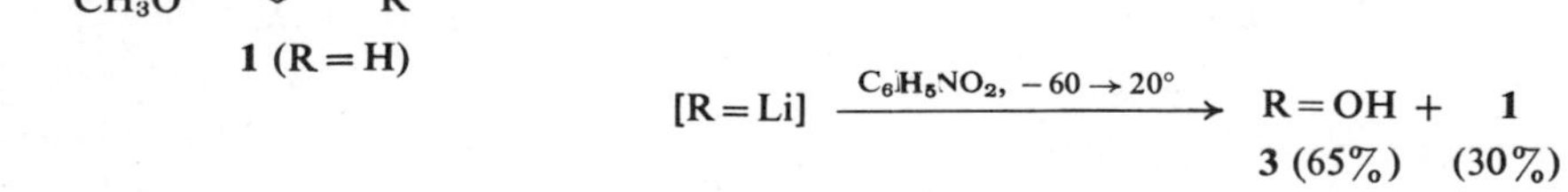

is replaced by t-butyl perbenzoate the yield of **3** is only 13%. Other nitroarenes are also less effective than nitrobenzene.

Benzylic hydroxylation.[3] Irradiation of the estrane **1** with nitrobenzene in CH_3CN with a medium pressure mercury lamp through a Pyrex filter gives a mixture of **3a** and **3b** as the major product; the dehydration product **2** is also formed.

[1] P. Wiriyachitra and M. P. Cava, *J. Org.*, **42**, 2274 (1977).
[2] G. Köbrich and P. Buck, *Ber.*, **103**, 1412 (1970).
[3] J. Libman and E. Berman, *Tetrahedron Letters*, 2192 (1977).

2-Nitrobenzeneseleninic acid, $2\text{-}NO_2C_6H_4Se$; **2,4-Dinitrobenzeneseleninic**

acid, $2,4\text{-}(NO_2)_2C_6H_3Se$. These reagents are generated *in situ* by hydrogen peroxide oxidation of bis(*o*-nitrophenyl) diselenide and of bis(2,4-dinitrophenyl) diselenide, respectively.

Epoxidation of alkenes.[1] Benzeneseleninic acid in stoichiometric amounts together with H_2O_2 has been used for epoxidation (this volume). This reagent is not satisfactory in catalytic amounts because it is degraded to SeO_2. Hori and Sharpless[1] have tested nine areneseleninic acids as catalysts for epoxidation and find that the nitro and dinitro compounds are the best catalysts. H_2O_2 (30%) is used, and $MgSO_4$ is added to scavenge excess water. The most satisfactory solvent is CH_2Cl_2. On a preparative scale yields of epoxides can be as high as 90–95%. However, this method does not lead to the stereoselective *syn*-epoxidation shown by carboxylic peracids.

[1] T. Hori and K. B. Sharpless, *J. Org.*, **43**, 1689 (1978).

o-Nitrobenzenesulfenyl chloride, 1, 745.

Protection of amino group. The *o*-nitrophenylsulfenyl group has been used for protection of side-chain amino groups in peptide synthesis. The classical method for removal is treatment with acid, even acetic acid, but this method is not selective. The selective removal is now possible by cleavage with 2-mercapto-pyridine (2-thiopyridone) and 1 equiv. of HOAc in CH_3OH, DMF, or CH_2Cl_2.[1]

A polymeric form of the reagent has also been used.[2]

[1] A. Tun-Kyi, *Helv.*, **61**, 1086 (1978).
[2] H. Ito, K. Ogawa, and I. Ichikizaki, *Chem. Letters*, 7 (1978).

p-Nitrobenzyl bromide, 1, 736.

Protection of carboxylic acids. The *p*-nitrobenzyl group is not useful for protection of carboxyl groups of penicillins and cephalosporins because hydrogenolysis can present a problem with sulfur-containing substrates. Lilly chemists[1] have found that the esters can be reduced with sodium sulfide ($Na_2S \cdot 9H_2O$) in aqueous THF, DMF, or acetone at $0°$ (25–35 minutes). In the case of 3-cephem esters, the reduction is accompanied by isomerization of the double bond to the Δ^2-position (last example).

Examples:

[1] S. R. Lammert, A. I. Ellis, R. R. Chauvette, and S. Kukolja, *J. Org.*, **43**, 1243 (1978).

N-Nitrocollidinium tetrafluoroborate,   **(1)**. The reagent **(1)**

$$CH_3$$

is prepared[1] *in situ* by the reaction of nitronium tetrafluoroborate and 2,4,6-trimethylpyridine. The reagent is less reactive than nitronium tetrafluoroborate but safer to use. It converts alcohols into nitrate esters, $ROH \rightarrow RONO_2$, in acetonitrile at $0°$ (yields 40–100%). Diol and triol nitrates can be prepared, but are not isolated for reasons of safety.

[1] G. A. Olah, S. C. Narang, R. L. Pearson, and C. A. Cupas, *Synthesis*, 452 (1978).

Nitrogen tetroxide, 1, 737–738.

Deamination of primary and secondary amines. Reaction of cyclohexylamine, nitrogen tetroxide, and DBU (or DBN) in ether at $-78°$ furnishes cyclohexyl nitrate in 75% yield. In the absence of the base, a mixture of the nitrate, cyclohexanol, and another product is obtained in about 50% yield. The new conditions are also applicable to primary alkyl amines: *n*-octylamine $\rightarrow$, *n*-octyl nitrate (81% yield). The reaction proceeds with retention of configuration in the case of 3α- and 3β-aminocholesterol.

[1] D. H. R. Barton and S. C. Narang, *J.C.S. Perkin I,* 1114 (1977).

Nitronium tetrafluoroborate, 1, 742–743; **4,** 358; **5,** 476–477. Suppliers: Aldrich, Cationics.

Oxidative cleavage of alkyl methyl ethers.[1] Methyl ethers are cleaved to carbonyl compounds by treatment with $NO_2^+BF_4^-$ in CH_2Cl_2.

Examples:

$$C_6H_5CH_2OCH_3 \xrightarrow[89\%]{NO_2^+BF_4^-} C_6H_5CHO$$

$$\underset{\underset{OCH_3}{|}}{CH_3(CH_2)_5CHCH_3} \xrightarrow{63\%} CH_3(CH_2)_5COCH_3$$

$$\xrightarrow{69\%} \underset{\underset{CH=NOH}{|}}{\overset{\overset{COOCH_3}{|}}{(CH_2)_4}}$$

[1] T.-L. Ho and G. A. Olah, *J. Org.*, **42**, 3097 (1977).

4-Nitrophenyl phenyl phosphorochloridate, C_6H_5O

—NO_2 **(1).** Mol.

wt. 297.03, m.p. 78–80°. The reagent can be prepared in about 65% yield by the reaction of phenyl phosphorodichloridate and 4-nitrophenol in the presence of 5-chloro-1-ethyl-2-methylimidazole (**2**) as catalyst.

Phosphorylation.[1] This reagent is useful for the first step in the phosphotriester approach to oligonucleotide synthesis,[2,3] which involves conversion of a free 3′-hydroxy group of a protected nucleoside into a phosphotriester of type **3**. Treatment of **3** with *p*-thiocresol and triethylamine at 20° leads to the triethyl-

$$1 + ROH \xrightarrow[\text{80–90\%}]{2,\ CH_3CN}$$

ammonium salts of phosphodiesters (**4**), which are the actual intermediates in this route to oligonucleotides. The paper includes syntheses of two dinucleoside phosphates, shown to have only 3′ → 5′ internucleotide linkages.

[1] C. B. Reese and Y. T. Y. Kui, *J.C.S. Chem. Comm.*, 802 (1977).
[2] J. H. van Boom, P. M. J. Burgers, G. R. Owen, C. B. Reese, and R. Saffhill, *ibid.*, 869 (1971).
[3] C. B. Reese, *Phosphorus and Sulfur*, **1**, 245 (1976).

o-**Nitrophenyl selenocyanate, 6,** 420–422; **7,** 252–253.

Cyanoselenenylation of aldehydes.[1] Aldehydes react with aryl selenocyanates in the presence of tri-*n*-butylphosphine in THF at 20° to form cyano selenides in 75–99% yield (equation I). Ketones do not undergo this reaction. The products can be used for a variety of transformations, as formulated in equations (II) and (III).

[1] P. A. Grieco and Y. Yokoyama, *Am. Soc.*, **99**, 5210 (1977).

(II)

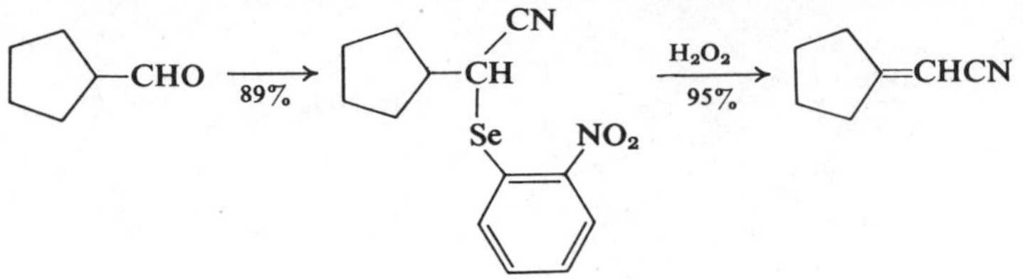

(III) $CH_3(CH_2)_5CH$ with CN and SeC_6H_5 $\xrightarrow[87\%]{\text{1) LDA} \\ \text{2) } CH_3I}$ $CH_3(CH_2)_5C-CH_3$ with CN and SeC_6H_5

91% $\downarrow$ 1) LDA 2)

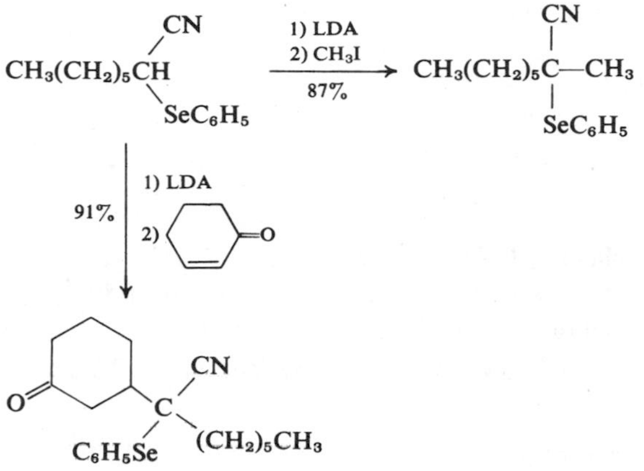

β-Nitropropionyl chloride, $NO_2CH_2CH_2COCl$. Mol. wt. 89.53, b.p. 78°/4 mm.

4-Hydroxy-2-cyclopentene-1-ones.[1] Compounds of this type can be obtained by reaction of the lithium enolate of a ketone with β-nitropropionyl chloride to form 5-nitro-1,3-diketones (2). The products cyclize in an alkaline medium to hydroxycyclopentenones (3) with loss of HNO_2. These products can be isolated or rearranged during a prolonged reaction period to 4-hydroxy-2-cyclopentene-1-ones (4).

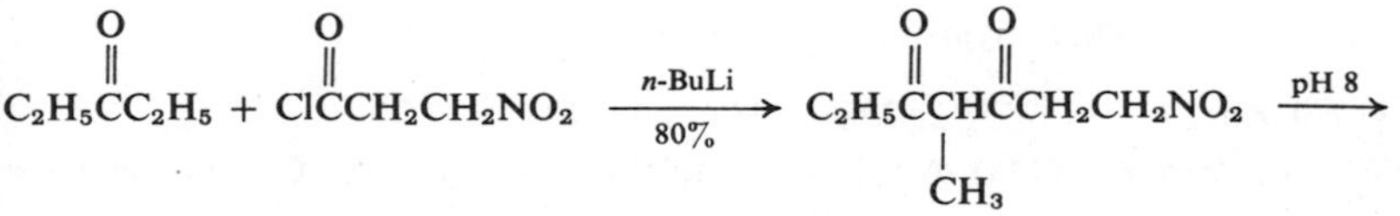

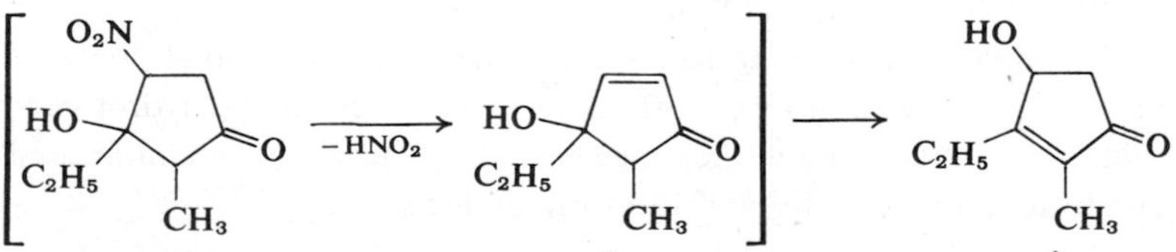

The method can be used for cyclopentenone annelation of cycloalkanones, as shown in the second example.

[1] D. Seebach, M. S. Hoekstra, and G. Protschuk, *Angew. Chem., Int. Ed.*, **16**, 321 (1977).

Nitrosonium tetrafluoroborate, 1, 747–748; **7**, 253–254.

Cleavage of benzylic esters. Benzylic esters are cleaved in 80–98% yield by nitrosonium tetrafluoroborate. The reaction involves a hydride transfer (equation I). Benzylic esters have usually been cleaved by hydrogenation.[1]

[1] T.-L. Ho and G. A. Olah, *Synthesis*, 418 (1977).

Nitrosyl chloride, 1, 748–755; **2**, 298–299; **6**, 422–423; **7**, 254.

Nitrosolysis (**6**, 422–423). A full report of this method for C–C bond cleavage of ketones and ketals is available. The reaction is a useful supplement to the Beckmann fragmentation of α-oximino ketones.[1]

Oxidative cleavage of oximes. Ketones are regenerated from oximes by reaction with NOCl in Py to form nitrimines, which can be hydrolyzed by either acid or base. Yields are in the range 80–95%. Under the conditions used, olefinic bonds do not react with NOCl to form adducts.[2]

[1] M. M. Rogić, J. Vitrone, and M. D. Swerdloff, *Am. Soc.*, **99**, 1156 (1977).
[2] C. R. Narayanan, P. S. Ramaswamy, and M. S. Wadia, *Chem. Ind.*, 454 (1977).

O

Osmium tetroxide–*t*-Butyl-N-chloro-N-argentocarbamate, OsO_4–

$(CH_3)_3CO\overset{\overset{\displaystyle O}{\|}}{C}\overset{-}{N}ClAg^+$. The salt is generated[1] *in situ* by reaction of *t*-butyl N-chlorosodiocarbamate[2] with silver nitrate in acetonitrile.

Caution: N-Chlorosodiocarbamates are unstable; do not prepare on a large scale.

Oxyamination. Vicinal oxyamination of olefins can be carried out with OsO_4 in catalytic amounts with this N-chloro-N-argentocarbamate (*cf.* **Osmium tetroxide–Chloramine-T, 7,** 256). A major improvement over chloroamine-T is that the —$COOC(CH_3)_3$ group is readily eliminated.

$$(CH_3)_3CO\overset{\overset{\displaystyle O}{\|}}{C}\overset{-}{N}ClAg^+ + RCH{=}CHR \xrightarrow[\text{60–70\%}]{OsO_4, CH_3CN} (CH_3)_3CO\overset{\overset{\displaystyle O}{\|}}{C}\overset{H}{N}{-}CHR$$
$$HO{-}CHR$$

[1] E. Herranz, S. A. Biller, and K. B. Sharpless, *Am. Soc.*, **100**, 3597 (1978).
[2] J. Johnson, Ph.D. Thesis, Heriot-Watt University, Edinburgh, Scotland, 1975.

Oxalyl chloride, 1, 767–772; **2,** 301–302; **3,** 216–217; **4,** 361; **5,** 481–482; **6,** 424; **7,** 257–258.

Angeloyl chloride, $\overset{H}{\underset{CH_3}{>}}C{=}C\overset{CH_3}{\underset{COCl}{<}}$. The preparation of this acid chloride is of interest because angelate esters are known natural products.[1] The reaction of angelic acid with either PCl_3 or thionyl chloride results in the isomeric tigloyl chloride. Use of oxalyl chloride under standard conditions gives a 2:1 mixture of angeloyl chloride and tigloyl chloride.[2] This reaction, however, does result only in the former chloride if conducted with the potassium salt of the acid in ether at 0°. This material was used for the preparation of a natural angelate ester.[3] Note that esterification of an alcohol and angelic acid with DCC gives mixtures of angelate and tiglate esters.[4]

[1] S. M. Kupchan and A. Afonso, *J. Org.*, **25**, 2217 (1960).
[2] F. Bohlmann and B.-M. Tietze, *Ber.*, **103**, 561 (1970).
[3] P. J. Beeby, *Tetrahedron Letters*, 3379 (1977).
[4] W. M. Haskins and D. H. G. Crout, *J.C.S. Perkin I*, 538 (1977).

Oxalyl chloride–Aluminum chloride.

Aroyl chlorides. The reaction of oxalyl chloride with aluminum chloride in CH_2Cl_2 generates carbon monoxide and phosgene (*Caution !*) On addition of an alkyl- or a halobenzene at 0°, *p*-alkyl- or *p*-halobenzoyl chlorides are obtained in high yield and in excellent purity.[1]

$$(COCl)_2 + AlCl_3 \xrightarrow[-CO]{CH_2Cl_2} [COCl_2] \xrightarrow[-HCl]{} R-\underset{(65-85\%)}{\langle\;\rangle}-COCl$$

[1] M. E. Neubert and D. L. Fishel, *Org. Syn.*, submitted (1978).

Oxygen, 4, 362; 5, 482–486; 6, 426–430; 7, 258–260.

Oxygenation of diarylmethanes. Activated CH_2 groups of aromatic substrates are converted to C=O groups by oxygen in aqueous NaOH in the presence of phase-transfer catalysts.[1]

Examples:

Isoquinolones. When N-methylisoquinolinium iodide (1) in *t*-butyl alcohol is treated with solid KOH and then oxygenated, N-methylisoquinolone (2) is obtained in 82% yield. The oxidation is believed to involve the pseudo base **a**. The choice of solvent is important, but the substituent on nitrogen does not affect the yield. Also, the reaction appears to be general for heterocyclic iminium salts.[2]

Aryl α-hydroperoxy acids. These substances can be prepared easily as formulated in equation (I). It is essential to use *n*-butyllithium rather than LDA for the α-lithiation step since traces of amines are deleterious to the oxygenation step. Unfortunately only fair results obtain with nonaromatic acids. Of course O_2 can be replaced by other electrophiles [$(CH_3)_3SiCl$, $(C_6H_5)_2C=O$] in related syntheses.[3]

Dethioacetalization. Thioacetals are useful for protection of carbonyl groups, but deprotection is difficult because of their stability to acid and base. The most generally used method employs metal salts. A recent method is oxygenation under irradiation with a high-pressure mercury lamp and in some cases in the presence of benzophenone. The method is applicable to ethylene and dibenzyl dithioacetals. Yields of the resulting ketones are in the range 60–85%. This new photochemical reaction should be useful for compounds containing acid- or base-sensitive groups.[4]

[1] E. Alneri, G. Bottaccio, and V. Carletti, *Tetrahedron Letters*, 2117 (1977).
[2] S. Ruchirawat, S. Sunkul, Y. Thebtaranonth, and N. Thirasasna, *ibid.*, 2335 (1977).
[3] W. Adam and O. Cueto, *J. Org.*, **42**, 38 (1977).
[4] T. T. Takahashi, C. Y. Nakamura, and J. Y. Satoh, *J.C.S. Chem. Comm.*, 680 (1977).

Oxygen, singlet, 4, 362–363; **5**, 486–491; **6**, 431–436; **7**, 261–269.

Reaction with an arene oxide–oxepin (5, 489–490; **6**, 435–436). Holbert and Ganem[1] have synthesized the naturally occurring senepoxide as the DL-racemate **(5)** by photooxygenation of the arene oxide **(1)**, prepared as shown in equation (I). This chemical synthesis of **5** via an arene oxide may be similar to the biogenetic synthesis. Although a number of steps are involved, the yields are high except for

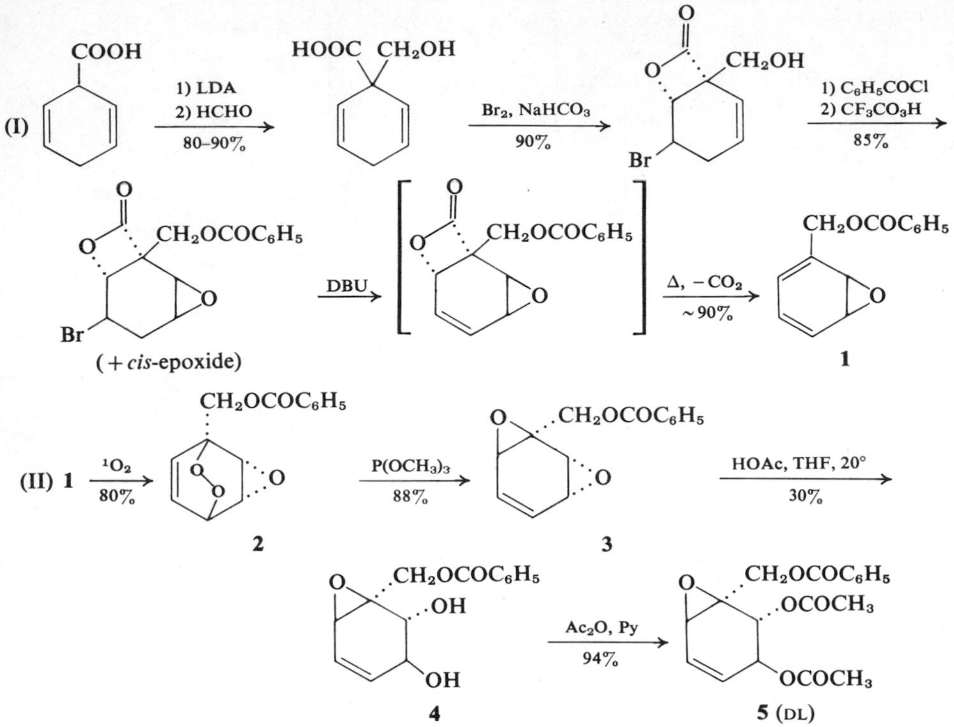

(I)

1) LDA
2) HCHO
80–90%

Br₂, NaHCO₃
90%

1) C₆H₅COCl
2) CF₃CO₃H
85%

(+ cis-epoxide)

DBU

Δ, – CO₂
~90%

1

(II) 1

¹O₂
80%

2

P(OCH₃)₃
88%

3

HOAc, THF, 20°
30%

4

Ac₂O, Py
94%

5 (DL)

the hydrolysis of the diepoxide **3**, which generates the desired diol epoxide **4** and two undesired 1,4-diols.

Photooxygenation of a β,β-dimethylstyrene. This reaction has been used in a synthesis of (±)-crotepoxide (**6**), a natural epoxide with activity against some

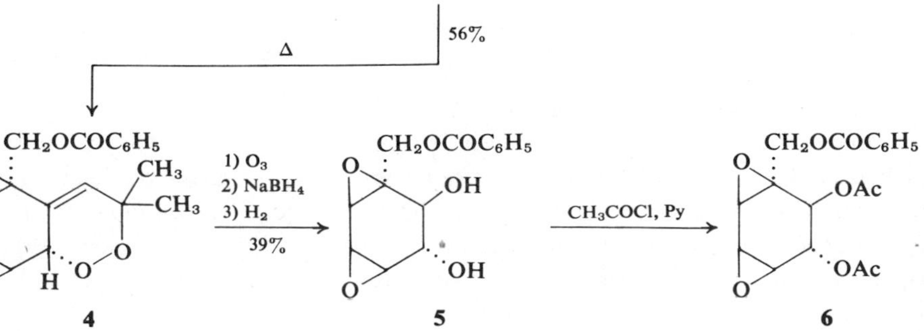

1

¹O₂
40%

2

+

33:67

3

56%

Δ

4

1) O₃
2) NaBH₄
3) H₂
39%

5

CH₃COCl, Py

6

forms of carcinoma.[2] The starting material (**1**) was chosen because the β,β-di-methyl groups improve the yield of the 1:2 mixture of 1,4-cycloadducts **2** and **3**. Remaining steps include thermal isomerization of **2** to the diepoxyperoxide **4**, cleavage of the double bond by ozonization followed by reduction with NaBH₄, and cleavage of the peroxy bond by hydrogenation to give **5**. The final step is acetylation to **6**.

   α-Hydroperoxy esters. Ketene methyl trimethylsilyl acetals such as **2**, prepared as shown, on singlet oxygenation give an α-silylperoxy ester (**3**) in high yield. Desilylation with methanol gives an α-hydroperoxy ester (**4**).

   Endoperoxides. The intermediacy of the endoperoxide **a** in the photooxy-genation of **1** has been established by diimide reduction and by reduction with thiourea (scheme I).[4]

Scheme (I)

Allylic hydroperoxides. The product ratios of the allylic hydroperoxides obtained on oxidation of alkenes with singlet oxygen differ significantly from those obtained by base-catalyzed isomerization of β-halo hydroperoxides, which involves perepoxide intermediates. A third mechanism must be operating in the reaction of triphenyl phosphite ozonide (**3**, 323–324; **4**, 559). This last reaction presumably proceeds by an ionic mechanism, since singlet oxygen is not formed at −70° from the ozonide.[5]

Example:

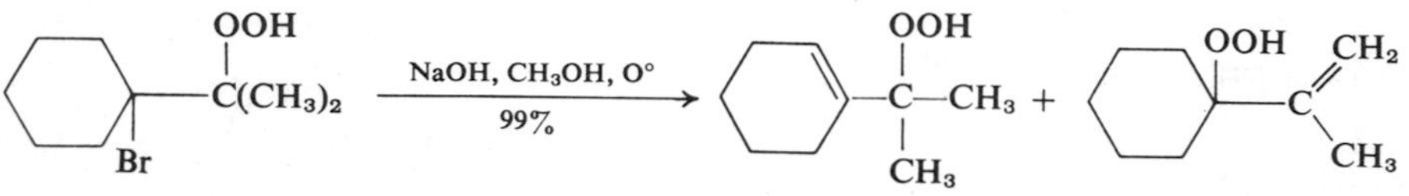

1 (94%)   2 (6%)

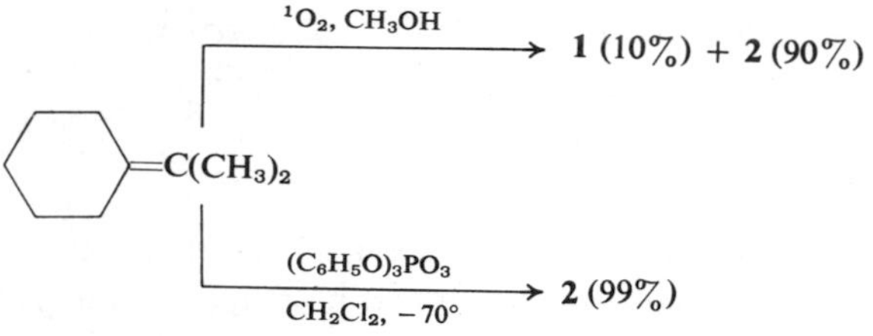

$^1O_2$, CH$_3$OH → 1 (10%) + 2 (90%)

(C$_6$H$_5$O)$_3$PO$_3$
CH$_2$Cl$_2$, −70° → 2 (99%)

Oxidative cleavage of N,N-dimethylhydrazones. This reaction can be conducted by photosensitized oxygenation followed by reduction (dimethyl sulfide) and hydrolysis. Yields of carbonyl compounds are in the range 50–85%. The mechanism is considered to involve an ene-type reaction (equation I).[6]

(I)

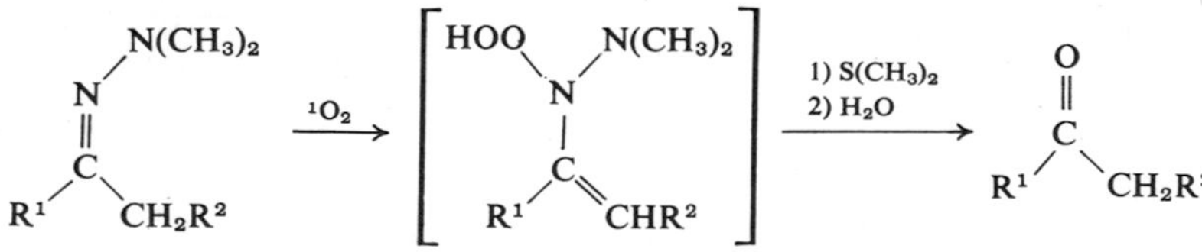

$\diagup$CHNO$_2$ → $\diagup$C=O. The transformation can be accomplished as shown

(I)

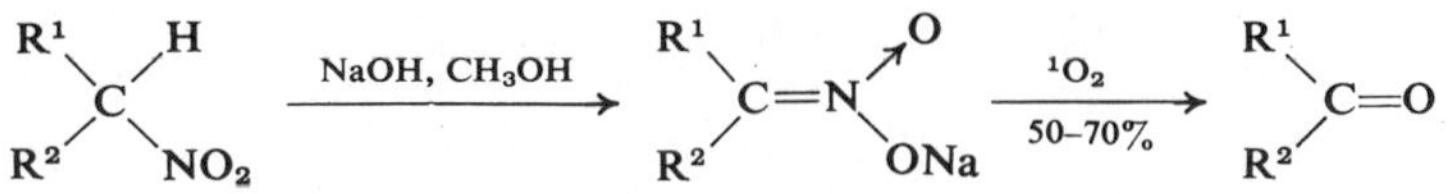

in equation (I) by reaction of singlet oxygen with a sodium nitronate. This method offers an alternative to use of ozone (5, 495) in the case of unsaturated substrates.[7]

Conversion of $\diagdown$CHCHO *into* $\diagdown$C=O. One commercial process for manufacture of progesterone (4) from stigmasterol involves degradation to the C_{22}-aldehyde, followed by oxidation under several conditions to the C_{20}-methyl ketone. This reaction has now been carried out on the protected C_{22}-aldehyde 1 by sensitized photooxidation in alkaline methanol, which gives 2 in high yield. The probable intermediate is a dioxetane formed from the enol of the aldehyde. Remaining steps from 2 to progesterone involve hydrolysis to pregnenolone (3) and Oppenauer oxidation to the Δ^4-3-ketone 4. Sensitized photooxidation of the Δ^4-3-ketone itself of the C_{22}-aldehyde also yields progesterone, but in only 60% yield because of further oxidation at C_6.[8]

α,β-*Unsaturated aldehydes and esters.* Conia *et al.*[9] have reported an interesting synthesis of α,β-unsaturated aldehydes and esters from ketones via the enol ether homologs, easily prepared by a Wittig reaction with methoxymethylenetriphenylphosphorane (1, 671; 6, 368). The complete sequence is shown in scheme (I). The ketone must bear an α-hydrogen that can be abstracted by singlet oxygen.

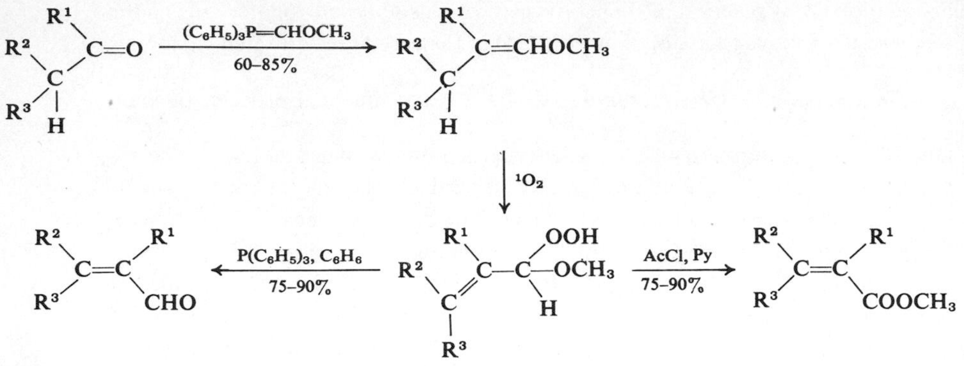

Scheme (I)

Examples:

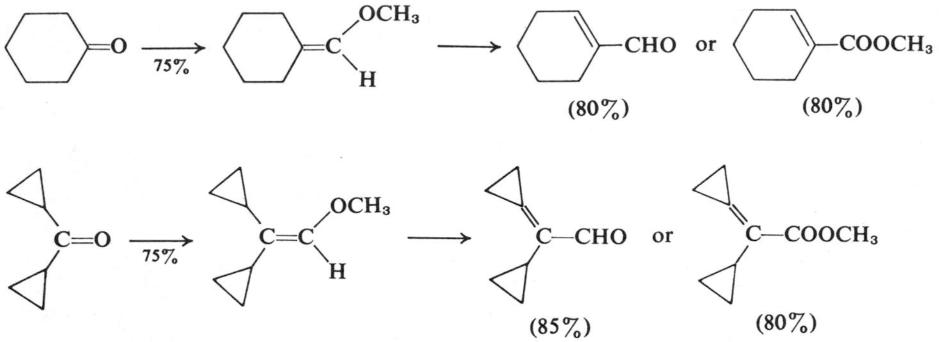

(80%)   (80%)

(85%)   (80%)

Keto thiolactones. Photosensitized oxygenation is the most satisfactory method for oxidative cleavage of bicyclic thioenol ethers (1) to keto thiolactones (2).[10]

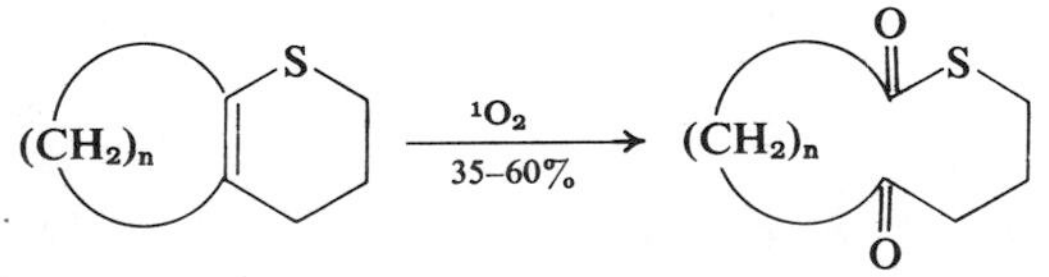

1 ($n = 4, 5, 6, 10$)   **2**

β-*Lactams* (**7**, 260; *cf. m*-Chloroperbenzoic acid, this volume). A new route from esters of azetidinecarboxylic acids is depicted in equation (I).[11]

(I)

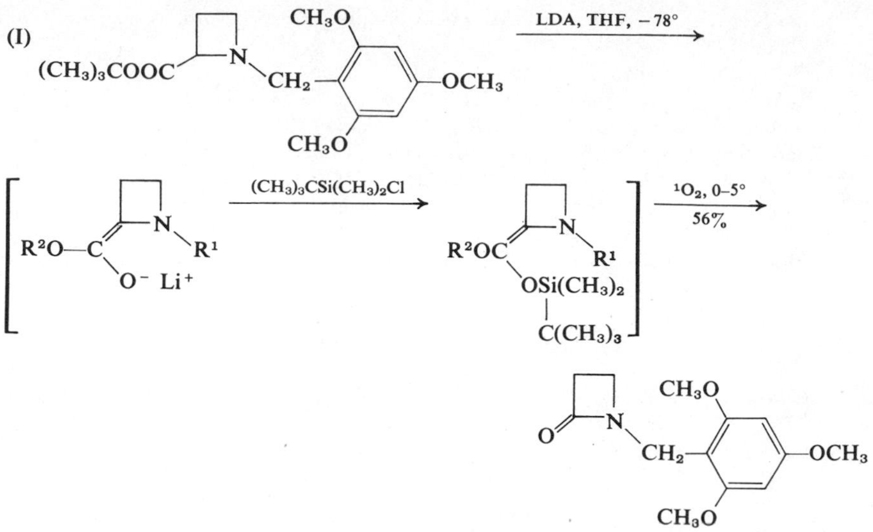

$(CH_3)_3COOC$ ... CH_3O ... CH_2 ... OCH_3 ... CH_3O

LDA, THF, $-78°$ →

R^2O-C ... N ... R^1, O^- Li^+

$(CH_3)_3CSi(CH_3)_2Cl$ →

R^2OC ... N ... R^1, $OSi(CH_3)_2$, $C(CH_3)_3$

$^1O_2$, 0–5°
56% →

O ... N ... CH_3O ... CH_2 ... OCH_3 ... CH_3O

Rearrangement of an endocyclic double bond to an exocyclic double bond. Büchi and co-workers[12] have reported a total synthesis of the norsesquiterpene khusimone (3), the last step of which required the isomerization of the more stable isomer, isokhusimone (1). This conversion was achieved in two steps by sensitized photooxidation of 1 followed by reduction with triethyl phosphite to give the allylic alcohols 2. Reduction of 2 with zinc and hydrogen chloride gave a separable mixture of 3 and epikhusimone 4.[13]

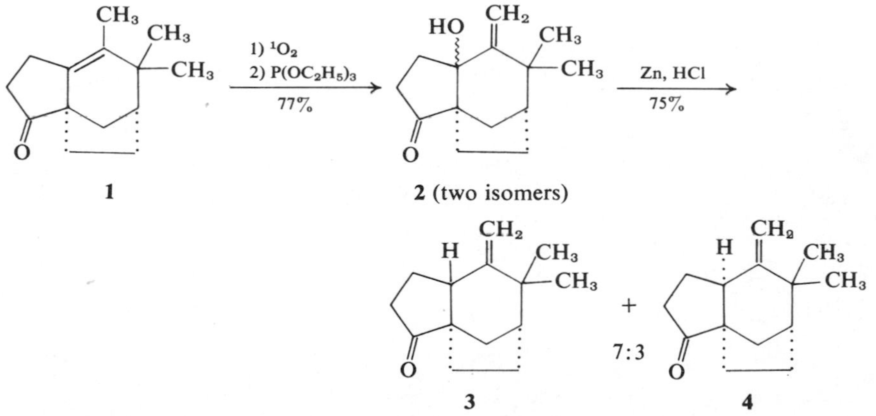

1

1) $^1O_2$
2) $P(OC_2H_5)_3$
77%

2 (two isomers)

Zn, HCl
75%

3   +   4

7:3

[1] G. W. Holbert and B. Ganem, *Am. Soc.*, **100**, 352 (1978); B. Ganem, G. W. Holbert, L. B. Weiss, and K. Ishizuma, *ibid.*, **100**, 6483 (1978).

[2] M. Matsumoto, S. Dobashi, and K. Kuroda, *Tetrahedron Letters*, 3361 (1977).

[3] W. Adam and J. del Fierro, *J. Org.*, **43**, 1159 (1978).

[4] W. Adam and I. Erden, *ibid.*, **43**, 2737 (1978).

[5] K. R. Kopecky, W. A. Scott, P. A. Lockwood, and C. Mumford, *Canad. J. Chem.*, **56**, 1114 (1978).

[6] E. Friedrich, W. Lutz, H. Eichenauer, and D. Enders, *Synthesis*, 893 (1977).

[7] J. R. Williams, L. R. Unger, and R. H. Moore, *J. Org.*, **43**, 1271 (1978).

[8] P. Sundararaman and C. Djerassi, *ibid.*, **42**, 3633 (1977).

[9] G. Rousseau, P. Le Perchec, and J. M. Conia, *Synthesis*, 67 (1978).

[10] H. C. Araújo and J. R. Mahajan, *ibid.*, 228 (1978).

[11] H. H. Wasserman, B. H. Lipschutz, and J. S. Wu, *Heterocycles*, **7**, 321 (1977).

[12] G. Büchi, A. Hauser, and J. Limacher, *J. Org.*, **42**, 3323 (1977).

[13] I. Elphimoff-Felkin, and P. Sarda, *Tetrahedron*, **33**, 511 (1977); *Org. Syn.*, **56**, 101 (1977).

Ozone, 1, 773–777; **4**, 363–364; **5**, 491–495; **6**, 436–441; **7**, 269–271.

Conversion of $\diagdown$CHNO$_2$ ***into*** $\diagdown$C=O (**5**, 495). An example of the ozonolysis

method for this conversion has been published in *Org. Syn.*[1]

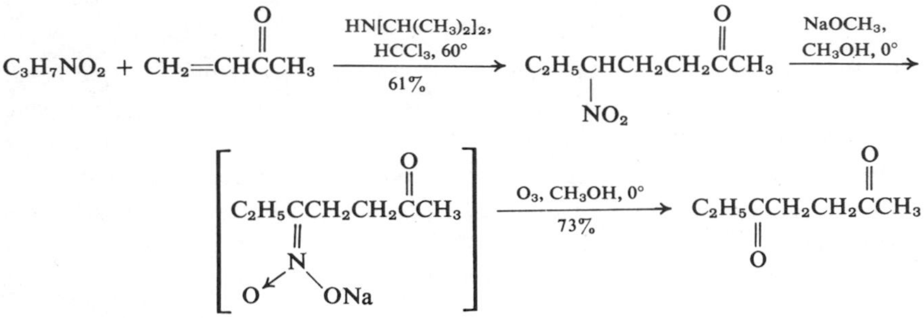

Oxidation of diene (**1**). After numerous experiments with OsO$_4$, singlet oxygen, and O$_3$, Dutch chemists found that the most satisfactory route to the

(I)

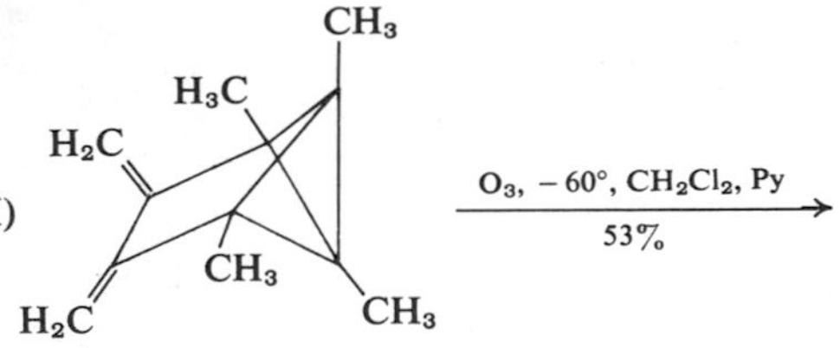

1

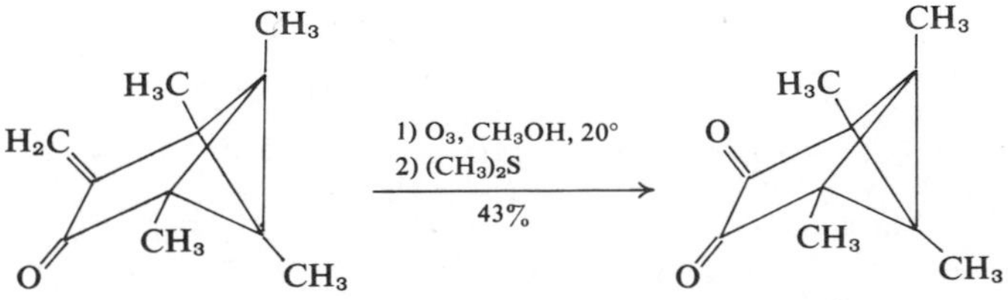

2   3

enone **2** and the dione **3** is that formulated in equation(I). Ozonization under usual conditions leads to by-products arising from Baeyer-Villiger products generated by the peracid formed during ozonization. Addition of pyridine limits this side reaction.[2]

Ozonolysis of vinylsilanes. Büchi and Wüest[3] ozonized the vinylsilane **1** expecting to obtain the norketone **2** as the major product. Instead the major product was an α-hydroperoxy aldehyde (**3**). They then investigated ozonolysis

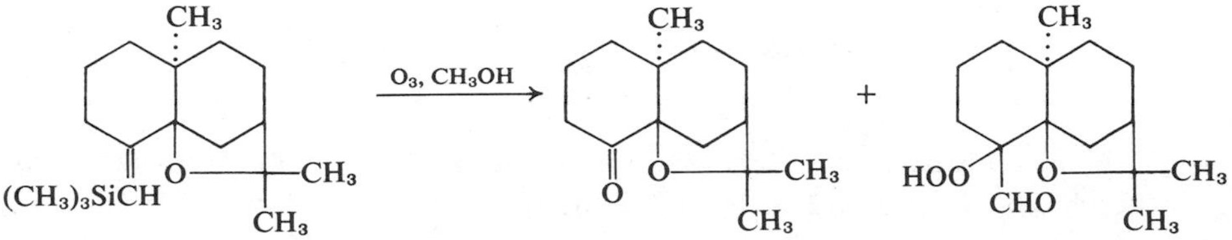

of simpler vinylsilanes and were able to deduce that the products are derived from intermediate dioxetanes and peroxides, usually obtained as products of photo-oxygenation of alkenes.

Aldehydes → esters. This conversion can be accomplished in one step by ozonization at $-78°$ of an aldehyde dissolved in an alcohol containing KOH. The KOH can be replaced with a lithium alkoxide in THF.[4]

Examples:

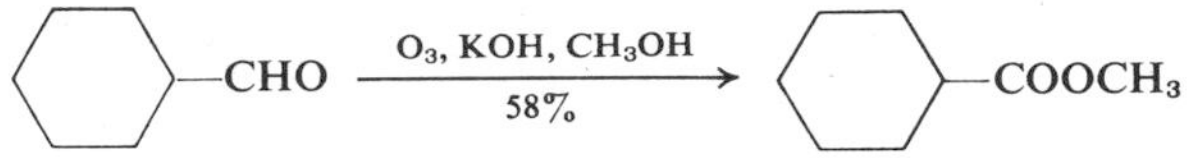

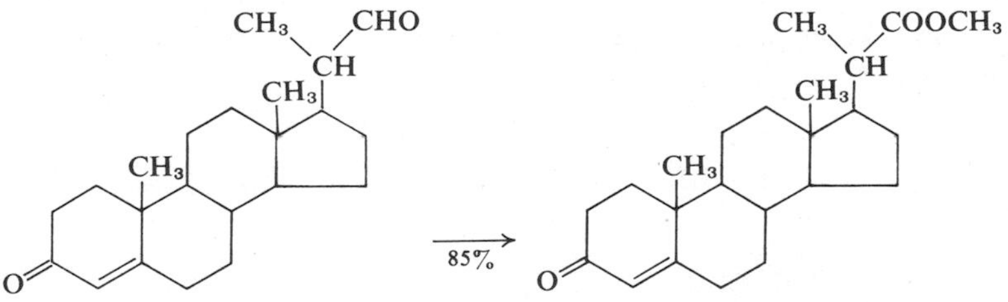

[1] J. E. McMurry and J. Melton, *Org. Syn.*, **56**, 36 (1977).
[2] R. F. Heldeweg, H. Hogeveen, and E. P. Schudde, *J. Org.*, **43**, 1912 (1978).
[3] G. Büchi and H. Wüest, *Am. Soc.*, **100**, 294 (1978).
[4] P. Sundararaman, E. C. Walker, and C. Djerassi, *Tetrahedron Letters*, 1627 (1978).

Ozone–Silica gel, 6, 440–441; **7**, 271–273.

$-NH_2 \rightarrow -NO_2$. Nitro compounds are obtained from primary amines on ozonization under usual conditions, but only in modest yields. However, this reaction is useful when conducted under dry conditions.[1]

Examples:

$$CH_3CH_2CH(CH_3)NH_2 \xrightarrow[70\%]{O_3, SiO_2} CH_3CH_2CH(CH_3)NO_2$$

$$C_6H_{11}NH_2 \xrightarrow[69\%]{} C_6H_{11}NO_2$$

$$C_6H_5CH_2NH_2 \xrightarrow[66\%]{} C_6H_5CH_2NO_2$$

Selective oxidations. Dry ozonization of cyclododecyl acetate (**1**) gives a mixture of three keto acetates, **2**, **3**, and **4**. Methylene groups remote from the binding site are attacked preferentially. This behavior is similar to oxidation of **1** by fungi.[2]

2 (15%)   **3 (27%)**   **4 (9%)**

Reaction with friedelane. Mazur ozonization of friedelane (**1**) gives the epoxide (**2**) as the major product; 19-oxofriedelane (**3**) is a minor product. The formation in addition of **4** and **5** is unexpected, since C_{15} and C_{16} are not tertiary or activated.[3]

1   **2 (48%)**

3 (2%)   4 (11%)   5 (11%)

Reaction with bicyclo[n.1.0]alkanes. The first step in the ozonization of bicyclo-[2.1.0]pentane (1) is considered to be cycloaddition across the central bond to give a 1,2,3-trioxane derivative (a), which is then converted to a mixture of cyclo-propylacetic acid (2) and succindialdehyde (3).[4]

1   a   2   2:5   3

[1] E. Keinan and Y. Mazur, *J. Org.*, **42**, 844 (1977).
[2] A. L. J. Beckwith and T. Duong, *J.C.S. Chem. Comm.*, 413 (1978).
[3] E. Akiyama, M. Tada, T. Tsuyuki, and T. Takahashi, *Chem. Letters*, 305 (1978).
[4] T. Preuss, E. Proksch, and A. de Meijere, *Tetrahedron Letters*, 833 (1978).

P

Palladium(II) acetate, 1, 778; **2,** 303; **4,** 365; **5,** 496–497; **6,** 442–443; **7,** 274–275.

α,β-*Enones.* Silyl enol ethers are converted into α,β-unsaturated carbonyl compounds by reaction with $Pd(OAc)_2$ in acetonitrile. $Pd(OAc)_2$ can be used in catalytic amounts if *p*-benzoquinone is also added for regeneration of the active Pd(II) species. The enones have the (E)-configuration regardless of the stereochemistry of the starting material. The last example indicates a useful extension of the reaction.[1]

Examples:

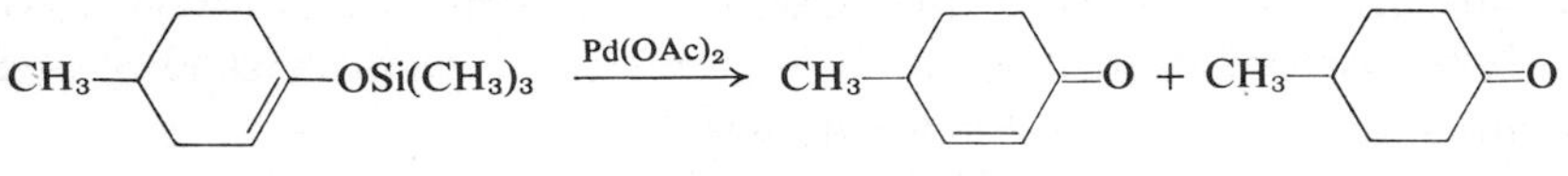

(91%) (8%)

(E and Z) (E, 97%)

(E and Z) (E, 92%)

91%

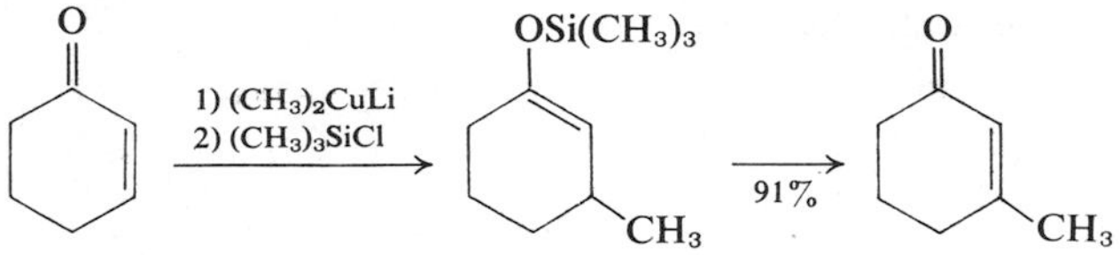

1,4-Dienes. Potassium alkenylpentafluorosilicates (**1**), prepared as shown, react with allylic halides under catalysis with Pd(II) salts to provide cross-coupled products (**2**).

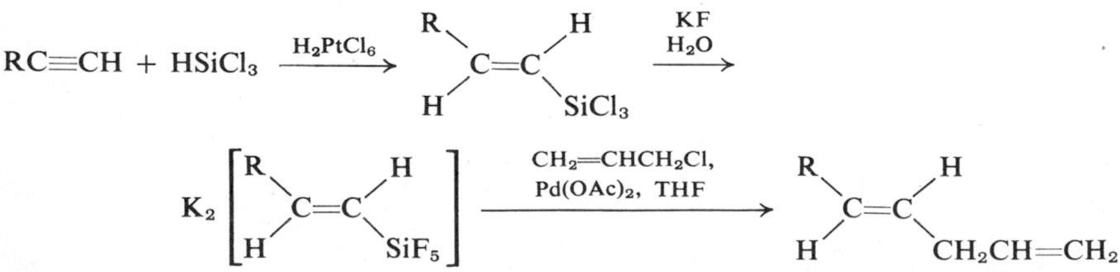

1 **2** (35–70%)

This synthesis was used to prepare (+)-methyl 11-hydroxy-*trans*-8-dodec-enoate (**1**), a precursor to the macrolide recifeiolide.[2]

$$HC{\equiv}C(CH_2)_6COOCH_3 \xrightarrow[62\%]{} K_2[F_5Si\diagdown\diagup\diagdown(CH_2)_6COOCH_3] \xrightarrow[55\%]{\substack{Pd(OAc)_2,\\ CH_2{=}CHCH_2Cl}}$$

1) Oxid. (67%)
2) NaBH₄ (93%)

1

(E)- and (Z)-Alkenes. Alkenylboranes (**1**), prepared by hydroboration of 1-alkynes, are converted into (E)-alkenes (**2**) by reaction with palladium(II) acetate (1 equiv.) and triethylamine (0.7–1.0 equiv.). The reaction is believed to involve *cis*-acetoxypalladation, migration with inversion of R′ from boron to carbon, and *cis* β-elimination (equation I).[3]

(I) $RC{\equiv}CH \xrightarrow{R'_2BH}$ [structure **1**] $\xrightarrow{Pd(OAc)_2,\ N(C_2H_5)_3,\ THF}$ [intermediates] $\xrightarrow{60-85\%}$ [structure **2**]

The reaction follows a different course when applied to alkenylboranes (**3**) prepared by hydroboration of internal alkynes (equation II). (Z)-Alkenes (**4**) are obtained and only catalytic amounts of $Pd(OAc)_2$ are necessary. An added base is not required because THF or acetone can serve as the proton source.[4]

(II)   [structure **3**] $\xrightarrow[70-95\%]{Pd(OAc)_2}$ [structure **4**]

Arylation of allylic alcohols (**7**, 274). This reaction has been applied in a synthesis of queen substance (**5**), in which the thiophene nucleus was used as a source of four methylene groups. Thus thienylation of α-methallyl alcohol with

1 in the presence of Pd(OAc)$_2$ gave **2** in 61% isolated yield. This was converted by known reactions into **5**.[5]

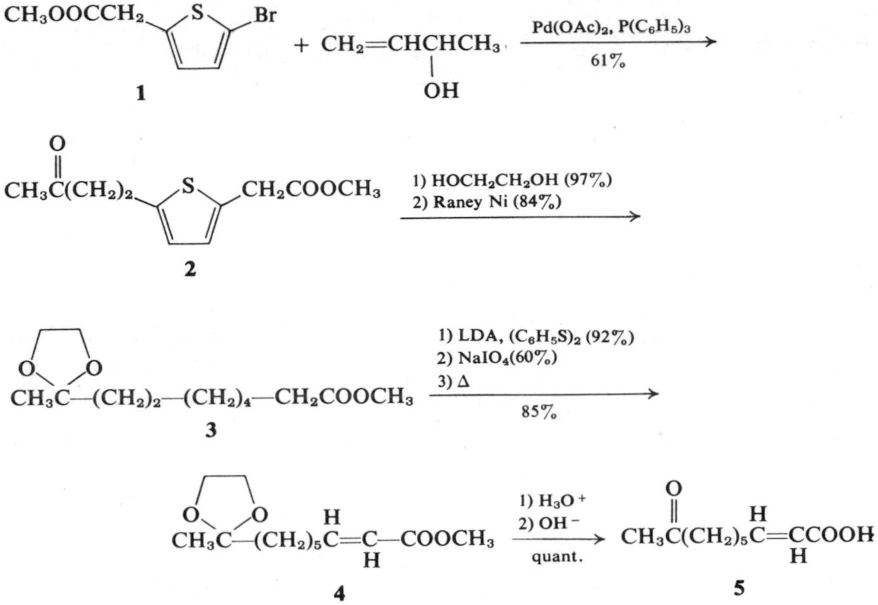

$$CH_3OOCCH_2 \overset{S}{\diagdown} Br + CH_2=CHCHCH_3 \overset{OH}{|} \xrightarrow[61\%]{Pd(OAc)_2, P(C_6H_5)_3}$$

1

$$\overset{O}{\underset{\|}{CH_3C}}(CH_2)_2 \overset{S}{\diagdown} CH_2COOCH_3 \xrightarrow{\substack{1) \ HOCH_2CH_2OH \ (97\%) \\ 2) \ Raney \ Ni \ (84\%)}}$$

2

$$CH_3C-(CH_2)_2-(CH_2)_4-CH_2COOCH_3 \xrightarrow[85\%]{\substack{1) \ LDA, \ (C_6H_5S)_2 \ (92\%) \\ 2) \ NaIO_4(60\%) \\ 3) \ \Delta}}$$

3

$$CH_3C-(CH_2)_5C=C-COOCH_3 \xrightarrow[quant.]{\substack{1) \ H_3O^+ \\ 2) \ OH^-}} \overset{O}{\underset{\|}{CH_3C}}(CH_2)_5C=CCOOH$$

4   **5**

Arylation of olefins (**6**, 156).[6] The vinylic substitution of aryl bromides with diacetatobis(triphenylphosphine)palladium(II) as catalyst is not satisfactory with bromides containing strongly electron-donating groups (OH, NH$_2$). Two solutions have been reported. One is to use an aryl iodide rather than the bromide and palladium acetate alone as catalyst.

Example:

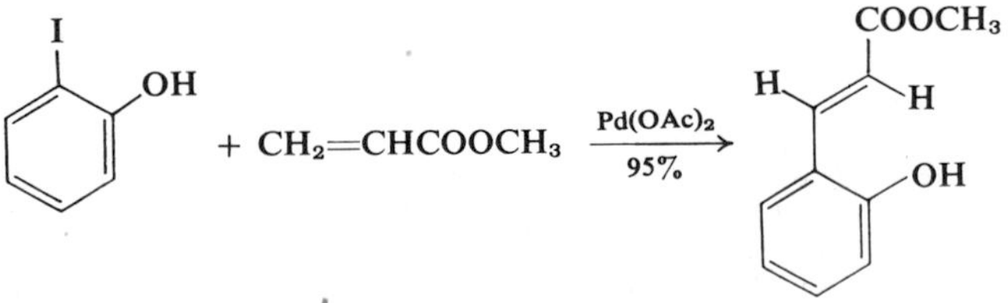

The other solution is to use tri-*o*-tolylphosphine in place of triphenylphosphine as the ligand. This variation was used for vinylic substitution reactions with various heterocyclic bromides. An example is the synthesis of nornicotine (**1**).

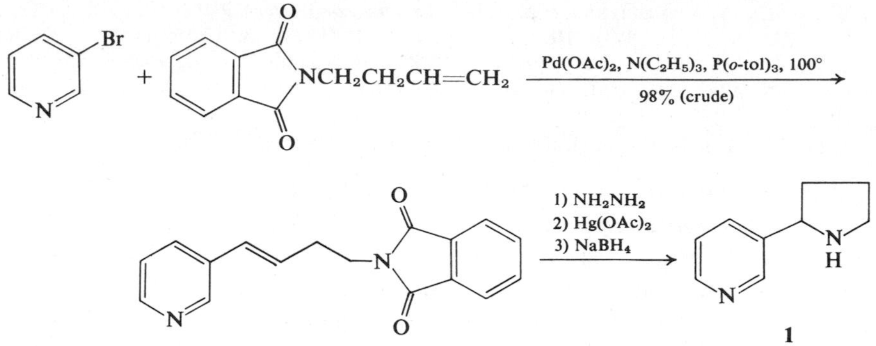

$$\xrightarrow[\text{98\% (crude)}]{\text{Pd(OAc)}_2,\ \text{N(C}_2\text{H}_5)_3,\ \text{P(o-tol)}_3,\ 100°}$$

1) NH$_2$NH$_2$
2) Hg(OAc)$_2$
3) NaBH$_4$

1

Vinylic arylation of 1,2-disubstituted olefins has been used as a route to 2-quinolones.

Example:

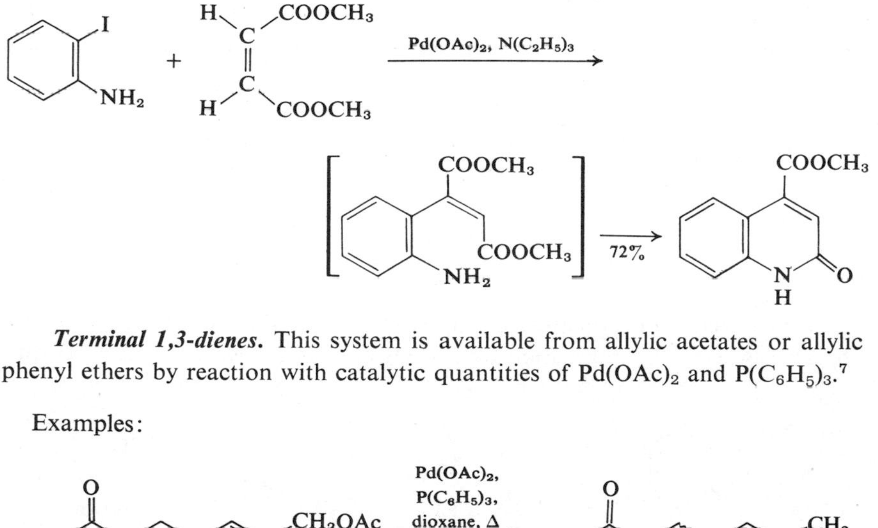

$$\xrightarrow{\text{Pd(OAc)}_2,\ \text{N(C}_2\text{H}_5)_3}$$

72%

Terminal 1,3-dienes. This system is available from allylic acetates or allylic phenyl ethers by reaction with catalytic quantities of Pd(OAc)$_2$ and P(C$_6$H$_5$)$_3$.[7]

Examples:

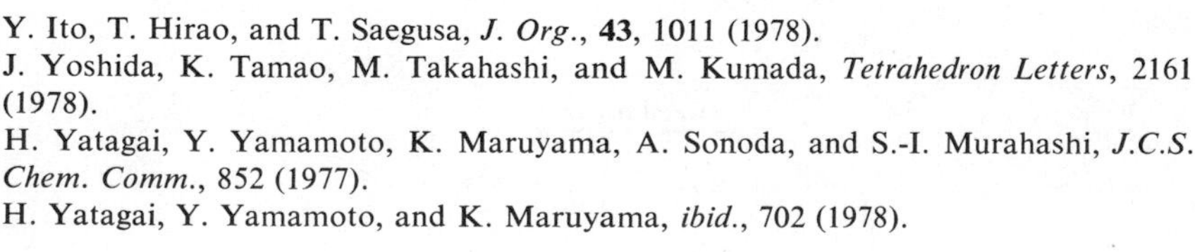

$$\xrightarrow[78\%]{\substack{\text{Pd(OAc)}_2, \\ \text{P(C}_6\text{H}_5)_3, \\ \text{dioxane, } \Delta}}$$

63%

[1] Y. Ito, T. Hirao, and T. Saegusa, *J. Org.*, **43**, 1011 (1978).
[2] J. Yoshida, K. Tamao, M. Takahashi, and M. Kumada, *Tetrahedron Letters*, 2161 (1978).
[3] H. Yatagai, Y. Yamamoto, K. Maruyama, A. Sonoda, and S.-I. Murahashi, *J.C.S. Chem. Comm.*, 852 (1977).
[4] H. Yatagai, Y. Yamamoto, and K. Maruyama, *ibid.*, 702 (1978).

[5] Y. Tamaru, Y. Yamada, and Z. Yoshida, *Tetrahedron Letters*, 919 (1978).

[6] C. B. Ziegler, Jr., and R. F. Heck, *J. Org.*, **43**, 2941 (1978); W. C. Frank, Y. C. Kim, and R. F. Heck, *ibid.*, **43**, 2947 (1978); C. B. Ziegler, Jr., and R. F. Heck, *ibid.*, **43**, 2949 (1978); N. A. Cortese, C. B. Ziegler, Jr., B. J. Hrnjez, and R. F. Heck, *ibid.*, **43**, 2952 (1978).

[7] J. Tsuji, T. Yamakawa, M. Kaito, and T. Mandai, *Tetrahedron Letters*, 2075 (1978).

Palladium catalysts, 1, 778–782; **2**, 203; **4**, 368–369; **5**, 499; **6**, 445–446; **7**, 275–277.

Dehydrogenation of polycyclic hydroarenes. This reaction catalyzed by Pt or Pd (mainly Pd on activated charcoal) has been reviewed (81 examples are cited).[1] The review notes that only one report[2] deals with use of soluble hydrogenation catalysts; in the limited cases examined, these catalysts were found to be more efficient than Pd/C.

Two cases have been reported by Crawford and co-workers[2] in which dehydrogenation is more successful with a combination of Pd/C and sulfur. An example is the conversion of **1** into **2**.

Pd/C–PtO$_2$. Hydrogenation of chrysene (**1**) over Pd/C affords 5,6-dihydrochrysene, whereas hydrogenation over PtO$_2$ affords 1,2,3,4-tetrahydrochrysene.

Hydrogenation over a mixed $Pd/C–PtO_2$ catalyst affords the hexahydrochrysene **2**. This product can be dehydrogenated by DDQ to 3,4,5,6-tetrahydrochrysene (**3**), which has been used to prepare the *anti*-diol epoxide **4**, of interest as a possible carcinogenic metabolite of chrysene.[4]

Transfer hydrogenation. Transfer hydrogenation from cyclohexene with palladium black or palladium charcoal is a convenient method for removal of protecting groups normally removed by catalytic hydrogenation in peptide synthesis (N-benzyloxycarbonyl, benzyl, and nitro groups). One advantage is that this method can be used with sulfur-containing peptides. *t*-Butoxycarbonyl groups are not affected under these conditions.[5]

Hydrogenation of a thiophene derivative. Thiophenes are known to be resistant to hydrogenation; even so, Hoffmann-LaRoche chemists[6] investigated this reduction in a recent synthesis of biotin (**3**), since this approach, if successful, would introduce the three *cis*-ring hydrogens of the vitamin in one operation. Surprisingly enough, hydrogenation (Pd/C) of the chosen intermediate (**1**) proceeded in 95% yield to give **2**. However, this hydrogenation is notably sub-

strate specific. The methyl ester of **1** is completely resistant. Presumably the COOH group can coordinate with the palladium and bring the ring into proximity to the catalyst. The ring closure step converting **2** to **3** is also novel and efficient.

Reductive cleavage of azo- and hydrazoarenes to anilines. Hydrogen transfer from refluxing cyclohexene with catalysis by Pd/C constitutes a useful method for this cleavage in 80–95% yield. Tetralin is a less satisfactory hydride donor.[7]

[1] P. P. Fu and R. G. Harvey, *Chem. Rev.*, **78**, 317 (1978).
[2] J. Blum and S. Biger, *Tetrahedron Letters*, 1825 (1970).
[3] H. S. Blair, M. Crawford, J. M. Spence, and C. V. Supanekar, *J. Chem. Soc.*, 3313 (1960); M. Crawford and C. V. Supanekar, *ibid.*, 2390 (1969).
[4] P. P. Fu and R. G. Harvey, *J.C.S. Chem. Comm.*, 585 (1978).
[5] G. M. Anantharamaiah and K. M. Sivanandaiah, *J.S.C. Perkin I*, 490 (1977).
[6] P. N. Confalone, G. Pizzolato, and M. R. Uskoković, *J. Org.*, **42**, 135 (1977).
[7] T.-L. Ho and G. A. Olah, *Synthesis*, 169 (1977).

Palladium(II) chloride, 1, 782; **3,** 303–305; **4,** 367–370; **5,** 500–503; **6,** 447–450; **7,** 277.

1,3-Dienes. Treatment of vinylmercurials, prepared from alkynes as shown, with $PdCl_2$ in benzene leads to unsymmetrical coupled 1,3-dienes. The yield is improved dramatically with added triethylamine. An example is formulated in equation (I).[1]

(I) 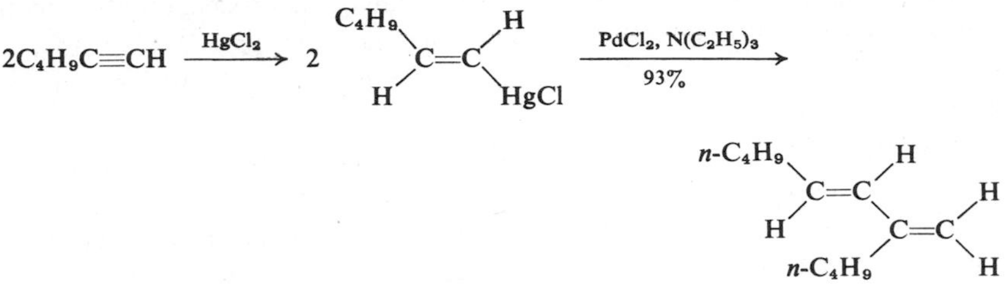 $2C_4H_9C{\equiv}CH$ $\xrightarrow{HgCl_2}$ 2 [structure: C_4H_9, H, H, HgCl on C=C] $\xrightarrow[93\%]{PdCl_2,\ N(C_2H_5)_3}$

[structure: n-C_4H_9, H, H, H, n-C_4H_9 diene]

Another example:

2 [structure: $(CH_3)_3C$, CH_3, H, HgCl on C=C] $\xrightarrow{59\%}$ [structure: $(CH_3)_3C$, CH_3, H, H, $(CH_3)_3C$, CH_3 diene]

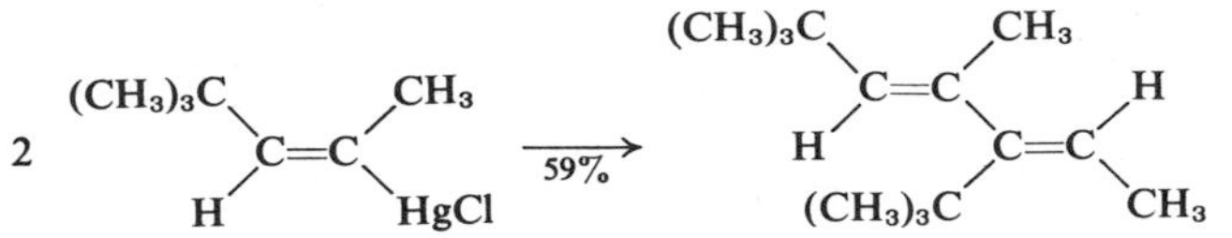

$PdCl_2$ can be used in catalytic amounts if 2 equiv. of $CuCl_2$ is also present.

α,β-Unsaturated ketosteroids. Dehydrogenation of 2- and 3-ketosteroids by $PdCl_2$ (1 equiv.) gives α,β-unsaturated ketones. The reaction involves a palladium complex with the enol tautomer, which can be isolated in some cases. No reaction is observed with 6-, 7-, 11-, 12-, 17-, or 20-ketosteroids. The position of dehydrogenation of 3-ketosteroids is governed by the A/B ring junction.[2]

Examples:

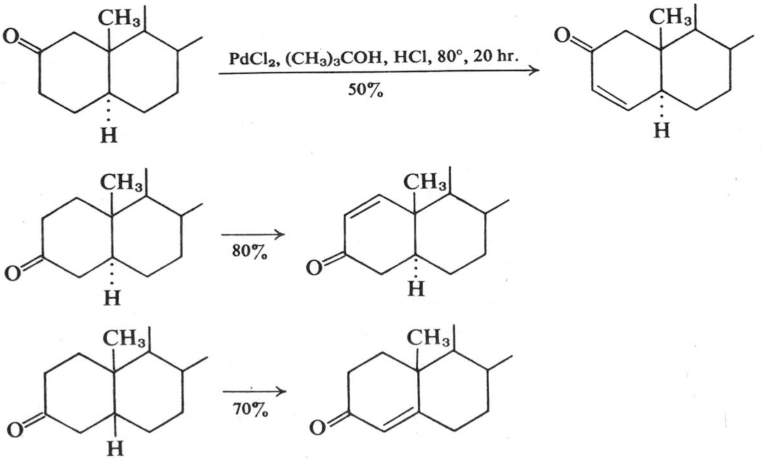

[structure: decalin ketosteroid with CH3 and H] $\xrightarrow[50\%]{PdCl_2,\ (CH_3)_3COH,\ HCl,\ 80°,\ 20\ hr.}$ [structure: unsaturated ketosteroid with CH3 and H]

[structure: decalin ketosteroid with CH3 and H] $\xrightarrow{80\%}$ [structure: unsaturated ketosteroid with CH3 and H]

[structure: decalin ketosteroid with CH3 and H] $\xrightarrow{70\%}$ [structure: unsaturated ketosteroid with CH3]

γ,δ-*Unsaturated sulfones.*[3] 1,3-Dienes such as **1** react with sodium alkane-sulfinates (**2**) in the presence of a stoichiometric amount of $PdCl_2$ to form π-allyl-palladium complexes (**3**). The complexes are converted selectively by dimethyl-glyoxime[4] in CH_3OH into γ,δ-unsaturated sulfones (**4**). Ordinarily this oxime reacts with π-allylpalladium complexes to give both possible olefins.

Example:

$$C_2H_5CH=CHCH=CH_2 + NaSO_2CH_2\overset{\overset{\displaystyle CH_3}{|}}{\underset{\underset{\displaystyle CH_3}{|}}{C}}C_6H_5 + PdCl_2 \xrightarrow[80\%]{HOAc}$$

1   2

$$C_2H_5CH\cdots CH\cdots CHCH_2SO_2CH_2\overset{\overset{\displaystyle CH_3}{|}}{\underset{\underset{\displaystyle CH_3}{|}}{C}}C_6H_5 \xrightarrow[89\%]{\overset{\overset{\displaystyle HON \; NOH}{|| \; ||}}{CH_3C-CCH_3}} C_2H_5CH=CHCH_2CH_2SO_2CH_2\overset{\overset{\displaystyle CH_3}{|}}{\underset{\underset{\displaystyle CH_3}{|}}{C}}C_6H_5$$

$ClPd/_2$

3   4

[1] R. C. Larock and B. Riefling, *J. Org.*, **43**, 1468 (1978).
[2] E. Mincione, G. Ortaggi, and A. Sirna, *Synthesis*, 773 (1977).
[3] Y. Tamaru, M. Kagotani, and Z. Yoshida, *J.C.S. Chem. Comm.*, 367 (1978).
[4] R. V. Lawrence and J. K. Ruff, *ibid.*, 9 (1976).

Palladium(II) chloride–Thiourea, 6, 450–451; **7,** 278–279.

α-*Methylene-δ-lactones.* The use of this catalyst for cyclocarbonylation of 3-butynols has been extended to the synthesis of an α-methylene-δ-lactone.[1] *exo-*Norbornene oxide (**1**) was converted into **2** by reaction with dimethylethynyl-

1   2   3

aluminum etherate.[2] This 4-pentynol on carbonylation in the presence of $PdCl_2$

and $H_2N\overset{\overset{\displaystyle S}{||}}{C}NH_2$ is converted into the α-methylene-δ-lactone **3** in reasonable yield.

[1] T. F. Murray, V. Varma, and J. R. Norton, *J. Org.*, **43**, 353 (1978).
[2] *Idem, J.C.S. Chem. Comm.*, 907 (1976).

Palladium hydroxide on charcoal, $Pd(OH)_2/C$. This catalyst was developed by W. M. Pearlman at Parke-Davis and has seen only limited use. Activated charcoal (68.6 g.) and $PdCl_2$(30 g). are mixed in 43 ml. of conc. HCl and 570 ml. of water

at 60°. Sodium hydroxide pellets (31 g.) are added at such a rate that the temperature does not rise above 80°. Solid NaHCO₃ (6.6 g.) is added and the mixture is
stirred for 12 hours. The catalyst is filtered and washed with H_2O (430 ml.)
and then with a mixture of H_2O (430 ml.) and HOAc (6 ml.). The catalyst is
filtered, dried *in vacuo* at 65°, and stored under nitrogen.[1]

Hydrogenation of a benzene ring. Hydrogenation of (R)-α-phenylglycine with
Pd/C in aqueous alkaline solution results in formation of phenylacetic acid, but
use of Pd(OH)₂/C leads to (R)-α-cyclohexylglycine (40–60% yield; 84% ee) and
cyclohexylacetic acid (35–55% yield).[2]

[1] R. G. Hiskey and R. C. Lorthrop, *Am. Soc.*, **83**, 4798 (1961).
[2] M. Tamura and K. Harada, *Syn. Comm.*, **8**, 345 (1978).

Peracetic acid, **1**, 785–791; **2**, 307–309; **3**, 219; **4**, 372; **5**, 505–506; **6**, 452–453.

Oxidation of 1,5-dihydroxynaphthalene. Juglone (**2**) has usually been prepared
by chromic acid oxidation of 1,5-dihydroxynaphthalene (**1**), but the yield by this
route is rather low. A new method involves oxidation of **1** with peracetic acid in
acetic acid to give a mixture of **2** and **3**, which is readily separable. The unstable
1,5-naphthoquinone **a** is considered to be an intermediate.[1]

Photosensitized oxygenation of **1** gives **2** in 70% yield, but this preparation of **2**
is not practicable on a large scale because of the high dilution required.

[1] C. Grundmann, *Synthesis*, 644 (1977).

Perbenzoic acid, **1**, 791–796; **3**, 219; **6**, 453; **7**, 279. Crystalline material can be
obtained by addition of benzoyl chloride (17.6 g.) to a solution of sodium
hydroxide (11 g.), magnesium sulfate heptahydrate (1 g.), and hydrogen peroxide
(15 ml. of 30% reagent) in water cooled to 5°. Dibenzoyl peroxide precipitates and
the temperature rises to 15°. The mixture is stirred at this temperature for another
hour. Dibenzoyl peroxide is removed by filtration and the filtrate is poured into
concentrated H_2SO_4 (15 g.) and ice. Perbenzoic acid (13 g.) precipitates; m.p.
40–42°, 99.5% pure. A totally aqueous medium is crucial for this method.[1]

The material reacts with glutaric and succinic anhydride in water to form
diacyl peroxides, $C_6H_5CO \cdot O \cdot O \cdot CO(CH_2)_nCOOH$, n = 2 or 3.

[1] P. R. H. Speakman, *Chem. Ind.*, 579 (1978).

$$\overset{\displaystyle OOH}{\underset{\displaystyle |}{}}$$

Peroxyacetimidic acid, $CH_3C{=}NH$, **1**, 469–470.

Epoxidation. Bach and Knight[1] recommend this reagent of Payne's for large-scale epoxidations, since it is inexpensive and safer to handle than peracids. An example cited is the epoxidation of *cis*-cyclooctene (4.4 moles) (equation I).

(I) $+ H_2O_2 + CH_3CN \xrightarrow[60.8\%]{}$

[1] R. D. Bach and J. W. Knight, *Org. Syn.*, submitted (1978).

Phase-transfer catalysts.

Methoxymethyl ethers of phenols. Phenols can be converted into methoxymethyl ethers ($ArOCH_2OCH_3$) in 80–95% yield by reaction with chloromethyl methyl ether under phase-transfer conditions.[1] Adogen 464 (**6**, 10) has been used as catalyst.

β,γ-Unsaturated ketones. Under phase-transfer conditions (Aliquat 336) cyanohydrins of aliphatic aldehydes react with allylic bromides to form β,γ-unsaturated ethers. On treatment with LDA (2 equiv.) these allylic ethers undergo a [2.3]sigmatropic rearrangement and elimination of lithium cyanide to give β,γ-unsaturated ketones.[2]

Methylenation of carbohydrates (*cf.* **6**, 10). Methylene acetals of carbohydrates can be prepared conveniently by reaction of *cis* and *trans vic*-diols with methylene bromide under phase-transfer conditions using tetra-*n*-butylammonium bromide.[3]
Examples:

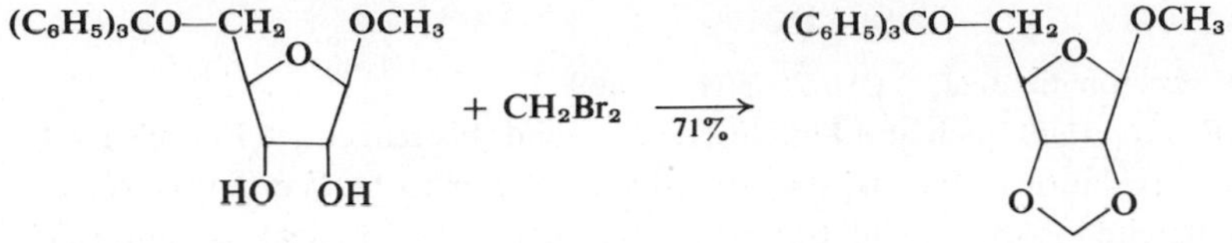

$(C_6H_5)_3CO-CH_2$ OCH_3

HO OH

+ CH_2Br_2 $\xrightarrow{71\%}$

$(C_6H_5)_3CO-CH_2$ OCH_3

Michael reactions of enol acetates. Certain Michael-type reactions of enol esters with 2-phenylpropionitrile and with chloroform can be carried out in a two-phase system with benzyltriethylammonium chloride.[4]
Examples:

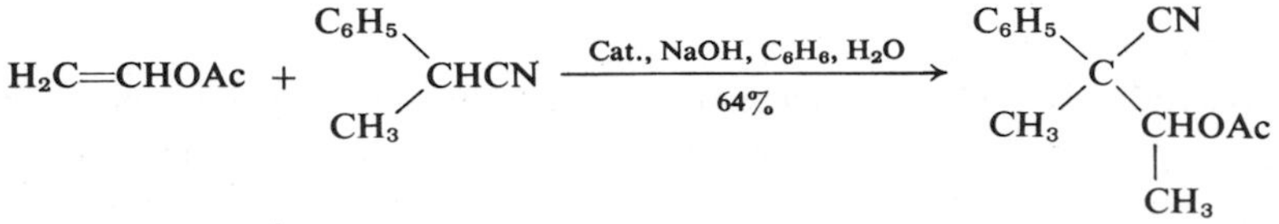

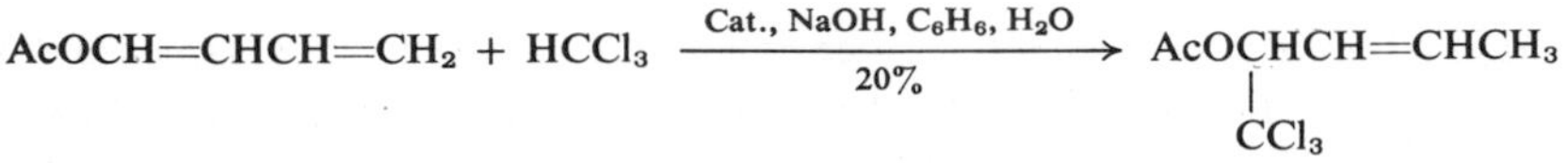

$H_2C=CHOAc$ + $\underset{CH_3}{\overset{C_6H_5}{>}}CHCN$ $\xrightarrow[64\%]{\text{Cat., NaOH, } C_6H_6, H_2O}$ $\underset{CH_3}{\overset{C_6H_5}{>}}\underset{CHOAc}{\overset{CN}{C}}$

CH_3

$AcOCH=CHCH=CH_2$ + $HCCl_3$ $\xrightarrow[20\%]{\text{Cat., NaOH, } C_6H_6, H_2O}$ $AcOCHCH=CHCH_3$

CCl_3

Dichloromethyl ketones. Dichlorocarbene generated by the Makosza technique reacts with acetals of type **1** to form ketals of dichloromethyl ketones (**2**). The same reaction is observed with dibromocarbene. The R group of **1** can be hydrogen, alkyl, or aryl.[5]

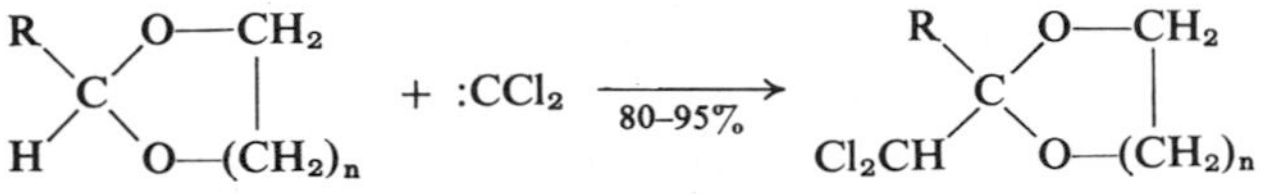

$\underset{H}{\overset{R}{>}}C\underset{O-(CH_2)_n}{\overset{O-CH_2}{<}}$ + $:CCl_2$ $\xrightarrow{80-95\%}$ $\underset{Cl_2CH}{\overset{R}{>}}C\underset{O-(CH_2)_n}{\overset{O-CH_2}{<}}$

1 (n = 1, 2)   **2**

N-Alkylation of β-lactams. This reaction can be carried out readily under phase-transfer conditions. Tetra-*n*-butylammonium bromide is somewhat more effective than tetraethylammonium bromide or benzyltriethylammonium chloride.[6]

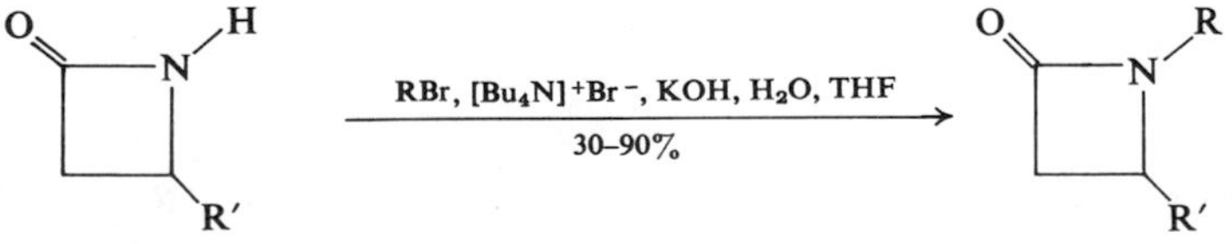

$\xrightarrow[30-90\%]{\text{RBr, } [Bu_4N]^+Br^-, KOH, H_2O, THF}$

(R' = CH=CH_2, C_6H_5)

Monoalkylation of primary amines. This reaction can be carried out in three steps. The primary amine (**1**) is phosphorylated under phase-transfer conditions to give a diethyl ester (**2**) of an N-alkylphosphoramidic acid (**6**, 42); this ester can be alkylated by a primary alkyl halide under phase-transfer catalysis to form **3**. The diethoxyphosphoryl group is then cleaved by hydrogen chloride in THF at 20° to give the secondary amine hydrochloride (**4**). Direct alkylation of primary amines usually affords mixtures of secondary and tertiary amines.[7]

$$R^1NH_2 + (C_2H_5O)_2P\overset{O}{\underset{H}{\big\backslash}} \xrightarrow[70-90\%]{\substack{C_6H_5CH_2N^+(C_2H_5)_3Cl^-, \\ CCl_4}} (C_2H_5O)_2P\overset{O}{\underset{NHR^1}{\big\backslash}} \xrightarrow[70-95\%]{\substack{R^2X, NaOH, H_2O, \\ (C_4H_9)_4N^+HSO_4^-}}$$

1   **2**

$$(C_2H_5O)_2P\overset{O}{\underset{\underset{R^2}{N}}{\big\backslash}}R^1 \xrightarrow[80-95\%]{HCl, THF} R^1NHR^2 \cdot HCl$$

3   **4**

Alkylation of amino acids. A general method for indirect alkylation of glycine involves preparation of the Schiff base **1**, which is an active methylene compound. This imine can be alkylated under the usual phase-transfer conditions to give

$$(C_6H_5)_2C=O + H_2NCH_2CN \xrightarrow[70\%]{BF_3 \cdot (C_2H_5)_2O, C_6H_5CH_3, \Delta}$$

$$(C_6H_5)_2C=NCH_2CN \xrightarrow[75-95\%]{RX, TEBA, NaOH, H_2O, C_6H_5CH_3}$$
1

$$(C_6H_5)_2C=N\overset{R}{\underset{}{C}}HCN \xrightarrow[\substack{2) 6N HCl, \Delta}]{1) 1N HCl, 25°} H_2N\overset{R}{\underset{}{C}}HCOOH$$
            2   **3**

2. These products are convertible by hydrolysis to alkylated derivatives (**3**) of glycine. Yields are higher if the hydrolysis is conducted in two stages.[8]

This indirect method gives higher yields than alkylation of $(C_6H_5)_2C=NCH_2$-$COOC_2H_5$, which must be conducted in 10% aqueous NaOH to avoid saponification of the ester group. As a consequence, a full equivalent of the phase-transfer reagent is necessary.[9]

Reissert reaction. The alkylation of Reissert compounds[10] with benzylic halides was originally conducted with NaH as base in DMF. Skiles and Cava[11] have examined three more modern systems: LDA and HMPT in THF, KOH and a crown ether, and phase-transfer conditions. Of these, the two last systems are clearly superior to the first; the phase-transfer system is somewhat superior to the crown ether system, both in yields and economic use of reagents. Cetyltrimethyl-ammonium bromide served as catalyst.

Example:

(R = H, OCH_3)

Metal carbonyl reactions. The first examples of phase-transfer catalysis in a metal carbonyl reaction were reported in 1977. Thus the reduction of nitroarenes by $Fe_3(CO)_{12}$ (**5**, 534) can be conducted under phase-transfer conditions in yields comparable to, or even superior to, those obtained under classical conditions (equation I). The actual reducing agent is believed to be sodium hydridoundeca-

(I) $ArNO_2 + Fe_3(CO)_{12} \xrightarrow[60-90\%]{[C_6H_5CH_2N(C_2H_5)_3]^+Cl^-,\ NaOH,\ H_2O,\ C_6H_6} ArNH_2$

carbonyltriferrate(I), $Na^+[HFe_3(CO)_{11}]^-$. The reduction can also be effected in a two-phase system with $Fe(CO)_5$ or $Fe_2(CO)_9$, but these reactions do not require a phase-transfer catalyst.[12]

Phase-transfer catalysis has also been found to be superior to NaK or Na/Hg for generation of the cyclopentadienyliron dicarbonyl anion from bis(dicarbonylcyclopentadienyliron), $[C_5H_5Fe(CO)_2]_2$.[13] Cetyltrimethylammonium bromide, $C_{16}H_{33}N(CH_3)_3^+Br^-$, and 18-crown-6 are about equally satisfactory as catalysts. The anion was used in a synthesis of fulvenes from thiobenzophenones (equation II).

(II)

Solid–liquid phase-transfer catalysis. Crown ethers have commonly been used as catalysts for reactions between a solid–liquid interface, and quaternary ammonium and phosphonium salts have been used only as catalysts for reactions in two-phase liquid–liquid reactions. However, several laboratories have reported that the latter catalysts are also satisfactory for two-phase solid–liquid reactions. Thus dichlorocarbene can be generated from chloroform and solid sodium hydroxide under catalysis from benzyltriethylammonium chloride in yields comparable to those of the classical Makosza method.[14] Another example of this type of catalysis is the oxidation of terminal and internal alkynes by solid potassium permanganate in CH_2Cl_2 with Adogen 464 as catalyst.[15] Aliquat 336 has been found to be as satisfactory as a crown ether for certain displacement reactions with NaOAc, KSCN, KNO_2, and KF in CH_3CN or CH_2Cl_2.[16]

[1] F. R. van Heerden, J. J. van Zyl, G. J. H. Rall, E. V. Brandt, and D. G. Roux, *Tetrahedron Letters*, 661 (1978).

[2] B. Cazes and S. Julia, *Bull. soc.*, 925, 931 (1977).

[3] K. S. Kim and W. A. Szarek, *Synthesis*, 48 (1978).

[4] M. Fedoryński, I. Gorzkowska, and M. Mąkosza, *ibid.*, 120 (1977).

[5] K. Steinbeck, *Tetrahedron Letters*, 1103 (1978).

[6] D. Reuschling, H. Pietsch, and A. Linkies, *ibid.*, 615 (1978).

[7] A. Zwierzak and J. Brylikowska-Piotrowicz, *Angew. Chem., Int. Ed.*, **16**, 107 (1977).

[8] M. J. O'Donnell and T. M. Eckrich, *Tetrahedron Letters*, 4625 (1978).

[9] M. J. O'Donnell, J. M. Boniece, and S. E. Earp, *ibid.*, 2641 (1978).

[10] M. Shamma, *The Isoquinoline Alkaloids*, Academic Press, New York, 1972.

[11] J. W. Skiles and M. P. Cava, *Heterocycles*, **9**, 653 (1978).

[12] H. des Abbayes and H. Alper, *Am. Soc.*, **99**, 98 (1977).

[13] H. Alper and H.-N. Paik, *ibid.*, **100**, 508 (1978).

[14] S. Julia and A. Ginebreda, *Synthesis*, 682 (1977).

[15] D. G. Lee and V. S. Chang, *ibid.*, 462 (1978).

[16] M. C. Vander Zwan and F. W. Hartner, *J. Org.*, **43**, 2655 (1978).

Phenyl benzenethiosulfonate, $C_6H_5\overset{\overset{O}{\|}}{\underset{\underset{O}{\|}}{S}}SC_6H_5$ (1). Mol. wt. 250.33, m.p. 44–45°.

The reagent is prepared by oxidation of diphenyl disulfide with H_2O_2 (72% yield)[1] or *m*-chloroperbenzoic acid (96% yield).[2]

1,2-Diketones.[2] The reagent is somewhat more reactive than diphenyl disulfide for sulfenylation of lithium enolates of ketones. The products can be converted into protected forms of 1,2-diketones or into the products of cleavage.

1,2-Carbonyl transposition.[3] The last steps in a total synthesis of *dl*-lycoramine (**5**) required a 1,2-transposition of the carbonyl group of **2**. This was accomplished by bissulfenylation of the lithium enolate of **2** with phenyl benzenethiosulfonate (**1**). The carbonyl group of the product was reduced and the

resulting alcohol was converted to the mesylate (**4**). Remaining steps are thioketal hydrolysis, reduction of the mesylate group, and finally reduction of the carbonyl group at C_3.

[1] H. J. Backer, *Rec. trav.*, **67**, 894 (1948).
[2] B. M. Trost and G. S. Massiot, *Am. Soc.*, **99**, 4405 (1977).
[3] A. G. Schultz, Y. K. Yu, and M. H. Berger, *ibid.*, **99**, 8065 (1977).

N-Phenylcampholylhydroxamic acid

(**1**). Mol. wt. 261.35. The chiral hydroxamic acid is prepared by reaction of 2 equiv. of phenylhydroxylamine with campholyl chloride in CH_3CN.

Asymmetric epoxidation of allylic alcohols. This reaction has been accomplished by use of the ligand **1** with VO(acac)$_2$ as catalyst and *t*-butyl hydroperoxide as the epoxidation reagent. Optical yields as high as 50% have been obtained; substantially lower inductions were obtained with cumene hydroperoxide. Molybdenum complexes with **1** give low asymmetric inductions.[1]

[1] R. C. Michaelson, R. E. Palermo, and K. B. Sharpless, *Am. Soc.*, **99**, 1990 (1977).

o-**Phenylenediamine, 1,** 834–837.

Bridged 1,5-benzodiazepines. *o*-Phenylenediamine reacts with **1**, 4,6-dimethyl-bicyclo[3.3.1]none-3,6-diene-2,8-dione, to give the diazepine **2** in 81% yield (pure).

This is a general reaction. Treatment of **2** with Meerwein's reagent results in N-alkylation and N,N′-dialkylation.[1]

[1] J. M. Mellor, J. H. A. Stibbard, and M. F. Rawlins, *J.C.S. Chem. Comm.*, 557 (1978).

Phenylene orthosulfite, **(1).** Mol. wt. 248.24, sensitive to H$_2$O, stable for a few days at −5°. This reagent is prepared by the reaction of SF$_4$ and the dilithium salt of catechol.

Dehydration. This sulfurane reacts with secondary alcohols to form alkenes (equations I and II). The results shown in the second equation favor a cyclic elimination mechanism.[1]

(either isomer)   1:1

[1] G. E. Wilson, Jr., and B. A. Belkind, *J. Org.*, **42**, 765 (1977).

(S)-(−)-1-Phenylethylamine (1), 1, 838; **2,** 272–273; **3,** 199–200.

Asymmetric amino acid synthesis. The reaction of aralkyl methyl ketones (**2**) with **1** as the chiral reagent in the presence of sodium cyanide in acetic acid affords optically pure α-methyl-α-aminonitriles (**3**) in 70–90% yield (equation I). When Ar = C_6H_5, the products (**3**) have the (R)-configuration; when the aryl group is

$$(I) \quad Ar{-}(CH_2)_n\overset{\overset{O}{\|}}{C}CH_3 + C_6H_5\underset{\underset{NH_2}{|}}{CH}CH_3 \xrightarrow[70-90\%]{NaCN, \ HOAc} Ar{-}(CH_2)_n{-}\underset{\underset{CH_3{-}\underset{C_6H_5}{\overset{|}{CH}}}{\overset{|}{NH}}}{\overset{\overset{CH_3}{|}}{\underset{}{C}}}{-}CN \xrightarrow{H_2SO_4}$$

$$2 \ (n = 1, 2) \qquad\qquad 1 \qquad\qquad\qquad\qquad\qquad\qquad\qquad 3$$

$$Ar{-}(CH_2)_n{-}\underset{\underset{NH_2}{|}}{\overset{\overset{CH_3}{|}}{\underset{}{\overset{*}{C}}}}{-}COOH$$

$$4$$

substituted by methoxy groups, the products have the (S)-configuration. The optically pure products (**3**) equilibrate in solution after 24 hours. Some can be hydrolyzed by concentrated sulfuric acid successfully to optically active α-methyl-α-amino acids.[1]

Another recent asymmetric amino acid synthesis involves the reaction of a methyl 2-isocyanoalkanoate (**5**) with **1** in the presence of n-butyllithium or potassium t-butoxide to form the metalated imidazolinones **6** in situ. Alkylation of **6** results in chiral 4,4-disubstituted imidazolinones (**7**) often in optical yields of 90–100%. The highest optical yields are obtained when R^1 has a higher priority than R^2. In this case **7** has the (R)-configuration. Reversal of the priority results in the (S)-configuration. Hydrolysis of **7** to the chiral amino acid **8** requires rather drastic conditions and is best conducted with base. This method was reported for the synthesis of twenty 2-imidazoline-5-ones of type **7**.[2]

$$\underset{5}{R^1\underset{\underset{NC}{|}}{CH}COOCH_3} + \underset{1}{H_2N^*\underset{\overset{|}{CH_3}}{CH}C_6H_5} \xrightarrow{n\text{-BuLi, THF}} \left[\underset{a}{R^1{-}\underset{\underset{Li}{|}}{\overset{\overset{NC}{|}}{C}}{-}CONH{-}\overset{*}{\underset{\overset{|}{CH_3}}{CH}}C_6H_5} \right] \longrightarrow$$

$$\left[\underset{6}{\underset{\underset{Li}{}}{R^1{-}}\overset{N}{\underset{O}{\diagdown}}{N{-}L^*}}\right] \xrightarrow[60-90\%]{R^2Br} \underset{7}{R^1{-}\underset{\underset{R^2}{}}{}\overset{N}{\underset{O}{\diagdown}}{N{-}L^*}} \xrightarrow[C_2H_5OH/H_2O]{KOH,} \underset{8}{R^2{\cdots}\underset{\overset{|}{R^1}}{\overset{\overset{NH_2}{|}}{C}}{-}COOH} + 1$$

$$\left(L^* = {}^*\overset{\overset{CH_3}{|}}{CH}C_6H_5 \right)$$

Resolution of allenic sulfones. The allenic sulfone buta-1,2-dienyl *p*-tolyl sulfone, $CH_3C_6H_4SO_2CH=C=CHCH_3$, can be resolved to the extent of 70% by treatment with 0.5 equiv. of (R)-α-phenylethylamine, α_D +40°, in ether at 25°. The recovered sulfone is optically active.[3]

[1] K. Weinges, K. Gries, B. Stemmle, and W. Schrank, *Ber.*, **110**, 2098 (1977).
[2] U. Schöllkopf, H. H. Hausberg, I. Hoppe, M. Segal, and U. Reiter, *Angew. Chem., Int. Ed.*, **17**, 117 (1978).
[3] M. Cinquini, S. Colonna, and F. Cozzi, *J.C.S. Perkin I*, 247 (1978).

α-Phenylglycine methyl ester, $H_2NCH(C_6H_5)COOCH_3$. Mol. wt. 165.19, m.p. 32°. Aldrich supplies the free acid.

β,γ-*Unsaturated ketones.* Steglich *et al.*[1] have developed a method for coupling a carboxylic acid and an allylic alcohol in which α-phenylglycine methyl ester serves as a building unit required for formation of an intermediate 3-oxazoline-5-one but is not part of the final product. The steps are reaction of a carboxylic acid with the amino acid to form an amide (**2**) and transesterification with an allylic alcohol to form **3**. On dehydration with phosgene or with triphenylphosphine and carbon tetrachloride, 3-oxazoline-5-ones (**4**) are formed, which are then converted on mild alkaline hydrolysis into β,γ-unsaturated ketones (**5**). The formation of **4** involves a Claisen-type rearrangement.

[1] N. Engel, B. Kübel, and W. Steglich, *Angew. Chem., Int. Ed.*, **16**, 394 (1977).

Phenyl isocyanate, 1, 842–843; **2**, 322–333; **4**, 378; **7**, 284–285.

N-Phenylcarbamates.[1] These useful derivatives of amines are prepared in low yield in the case of tertiary alcohols in the absence of a catalyst. Dibutyltin diacetate[2] exerts some catalytic effect, but use of the lithium alkoxide of the tertiary alcohol as catalyst raises yields of *t*-alkyl N-phenylcarbamates to about 80%.

[1] W. J. Bailey and J. R. Griffith, *J. Org.*, **43**, 2690 (1978).
[2] E. F. Cox and F. Hostettler, Abstracts, 135th Meeting of the *American Chemical Society*, April 1959.

Phenyllithium, C_6H_5Li, **1**, 845.

α,β-Disubstituted cycloenones. Conrad and Fuchs[1] have synthesized compounds of this type from a β-epoxy sulfone such as **1** by a four-step sequence that can be conducted without purification of intermediates. The first step involves reaction with 2 equiv. of an organolithium reagent (or a Grignard reagent) to

form the dianion *a*, which can be alkylated to form **2**. This alcohol is converted into the corresponding ketone by two-phase oxidation with chromic acid (**4**, 95–96). The final step involves elimination of benzenesulfinic acid by treatment with a base (**6**, 461) to give the enone **3**.

This sequence is applicable to a variety of substrates, alkyllithium reagents, and electrophilic species, and in general overall yields are in the range 45–90%.

[1] P. C. Conrad and P. L. Fuchs, *Am. Soc.*, **100**, 346 (1978).

Phenyl selenocyanate, C_6H_5SeCN. Mol. wt. 182.08, b.p. 125°/16 mm. Preparation.[1]

Selenol esters, $RCOSeC_6H_5$.[2] Carboxylic acids react with phenyl selenocyanate (1 equiv.) and tri-*n*-butylphosphine (2 equiv.) in methylene chloride or THF to form phenylselenol esters usually in high yield. Similarly, thiol esters,

$RCOSC_6H_5$, are formed from acids and phenyl thiocyanate.[3] In this reaction only 1.1 equiv. of the trialkylphosphine is necessary for satisfactory yields (80–95%).

[1] A. Behagel and H. Siebert, *Ber.*, **65**, 812 (1932).
[2] P. A. Grieco, Y. Yokoyama, and E. Williams, *J. Org.*, **43**, 1283 (1978).
[3] F. Challenger, C. Higginbottom, and A. Huntington, *J. Chem. Soc.*, 26 (1930).

4-Phenylseleno-1-pentene-3-one (1). Mol. wt. 239.17, b.p. 92–95°.

Preparation:

Divinyl ketone equivalent.[1] Two reactions using **1** as a divinyl ketone equivalent are shown in equations (I) and (II).

[1] S. Danishefsky and C. F. Yan, *Syn. Comm.*, **8**, 211 (1978).

2-Phenylselenopropionic acid, $CH_3CHCOOH$ (1). Mol. wt. 229.13, m.p. 49–50°.
         $|$
       SeC_6H_5

This reagent is prepared by the reaction of sodium selenophenoxide with sodium 2-bromopropanoate (79% yield).

α-Methylene-γ-lactones. The dianion of **1** has been used to synthesize both *trans-* and *cis-* α-methylene-γ-lactones (equations I and II).[1]

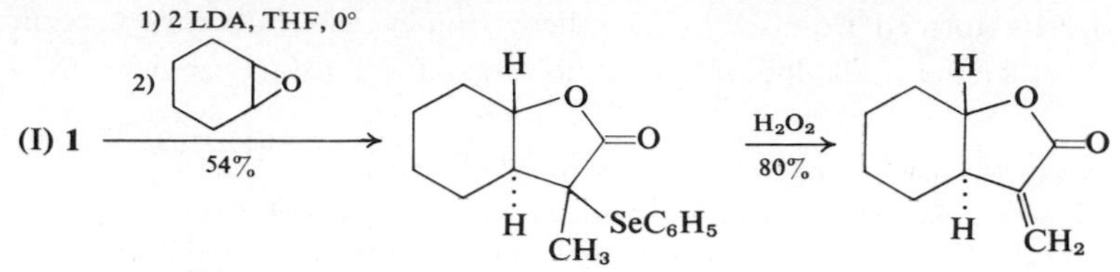

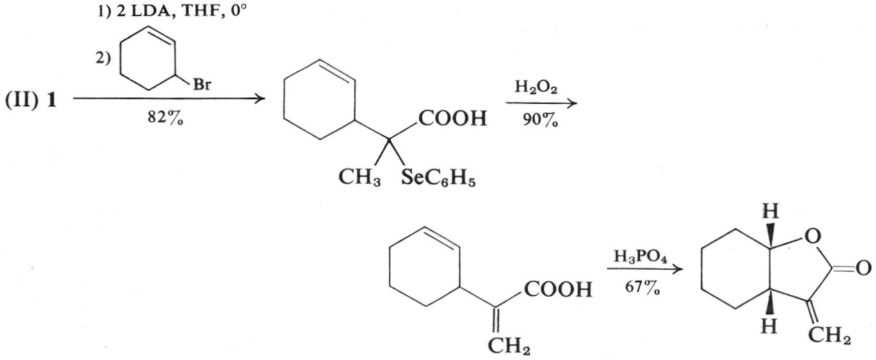

(I) 1   1) 2 LDA, THF, 0°

2) [epoxide structure]

54%

H_2O_2
80%

(II) 1   1) 2 LDA, THF, 0°

2) [cyclohexenyl bromide]

82%

H_2O_2
90%

H_3PO_4
67%

[1] N. Petragnani and H. M. C. Ferraz, *Synthesis*, 476 (1978).

1-Phenylseleno-2-trimethylsilyloxy-4-methoxy-1,3-butadiene (1). Mol. wt. 327.36.

Preparation:

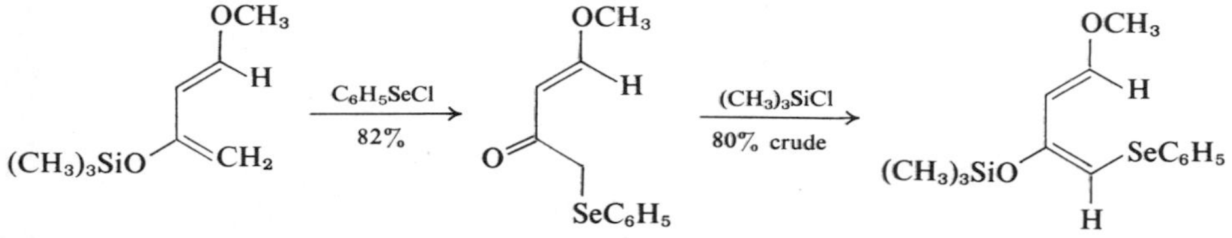

C_6H_5SeCl
82%

$(CH_3)_3SiCl$
80% crude

1

Cycloaddition.[1] The reagent has been used in Diels-Alder reactions to prepare 4-acyl-4-substituted cyclohexadienones after selenoxide elimination.

Examples:

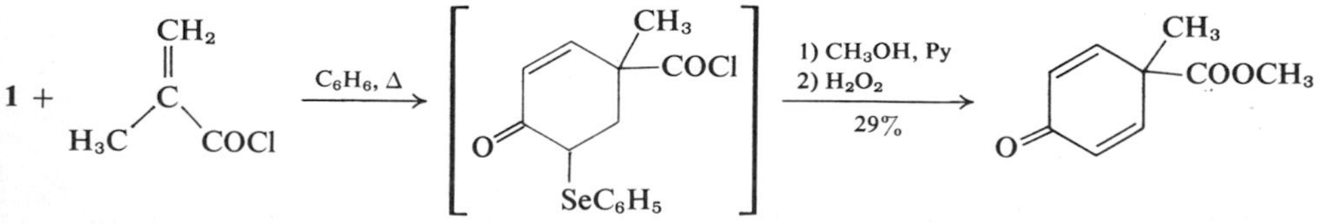

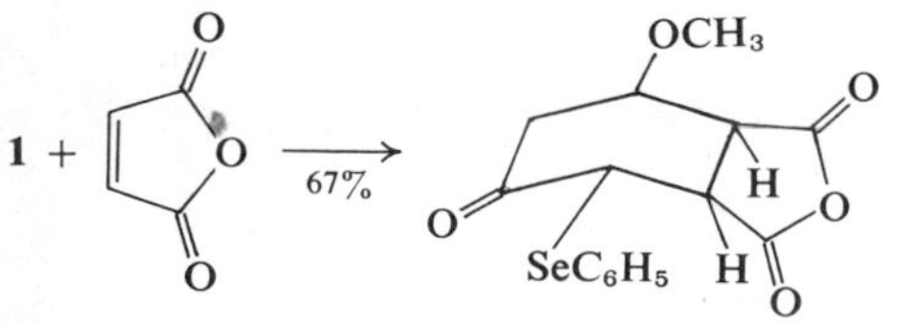

1 +

C_6H_6, Δ

1) CH_3OH, Py
2) H_2O_2
29%

1 +

67%

[1] S. Danishefsky, C. F. Yan, and P. M. McCurry, Jr., *J. Org.*, **42**, 1819 (1977).

N-Phenylthiosuccinimide, [structure] N—SC$_6$H$_5$ **(1).** Mol. wt. 207.25 m.p. 115–116°. The sulfenimide is prepared by reaction in DMF of succinimide and benzenesulfenyl chloride in the presence of triethylamine (75% yield).[1]

Thio ethers.[2] The reagent reacts with primary and secondary alcohols at 20° in the presence of tri-*n*-butylphosphine (triphenylphosphine is not so effective) to form phenyl thioethers, tri-*n*-butylphosphine oxide, and succinimide. The reaction proceeds with inversion. It is limited to N-(arylthio)succinimides, since N-(alkylthio)succinimides are desulfurized by phosphines.

$$1 + Bu_3P \longrightarrow \left[Bu_3P^+SC_6H_5{}^-N \right] \xrightarrow{ROH} RSC_6H_5 + Bu_3P{=}O + HN$$
$$(80-95\%)$$

Review of thioimides.[3]

[1] M. Behforouz and J. E. Kerwood, *J. Org.*, **34**, 51 (1969); *see also* W. Groebel, *Ber.*, **93**, 284 (1960).

[2] K. A. M. Walker, *Tetrahedron Letters*, 4475 (1977).

[3] P. T. S. Lau, *Eastman Org. Chem. Bulletin*, **46**, No. 2 (1975).

Phenyl vinyl sulfoxide, C$_6$H$_5\overset{\text{O}}{\underset{\|}{\text{S}}}CH=CH_2$ **(1).** Mol. wt. 152.20, stable for several days at 180°.

Preparation:

$$C_6H_5SOC_2H_5 + CH_2{=}CHMgBr \longrightarrow 1$$

Diels-Alder reactions. This vinyl sulfoxide undergoes Diels-Alder reactions at 110–130° usually with simultaneous extrusion of C$_6$H$_5$SOH. It thus behaves as an equivalent of acetylene, which is unreactive in [4 + 2]cycloadditions.[1]

Examples:

$$\xrightarrow[83\%]{1,\ 130°}$$

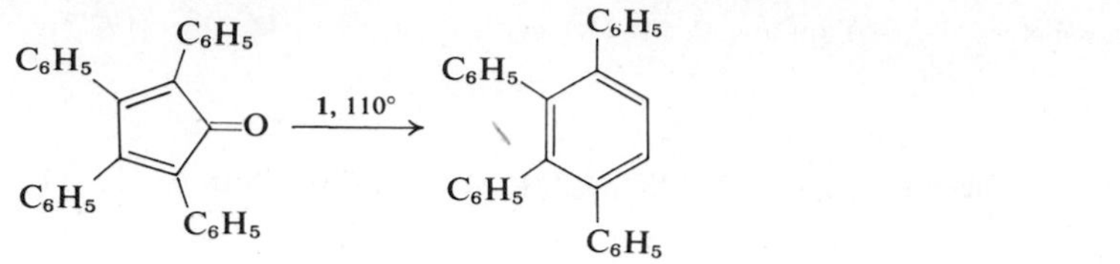

^{1} L. A. Paquette, R. E. Moerck, B. Harirchian, and P. D. Magnus, *Am. Soc.*, **100**, 1597 (1978).

Phosphorus(III) chloride, 1, 875–876.

$RCH_2NO_2 \rightarrow RC{\equiv}N.$ This conversion usually has been conducted via a Nef reaction to an aldehyde followed by formation and dehydration of an oxime. The transformation can be accomplished in one step by reaction of the primary nitro compound with PCl_3 and pyridine (40–75% yield). This method can also be used to convert allylic nitro compounds into α,β-unsaturated nitriles and to prepare aldehyde or ketone cyanohydrin acetates.[1]

Examples:

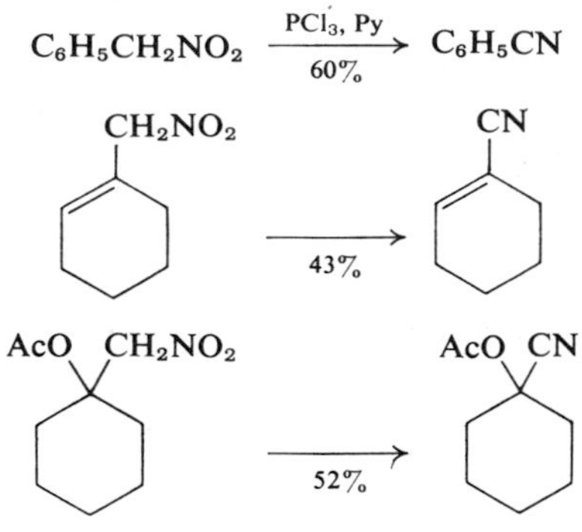

^{1} P. A. Wehrli and B. Schaer, *J. Org.*, **42**, 3956 (1977).

Phosphorus(V) oxide–Methanesulfonic acid, P_2O_5–CH_3SO_3H, **5,** 535; **7,** 292.

Dihydrophenanthrene synthesis. Evans *et al.*[1] have reported a new route to dihydrophenanthrene derivatives based on the condensation of *p*-quinone mono-ketals (**1**) or monosilyl cyanohydrin derivatives (**5,** 721–722) with the enolate of methyl 3-(3,4,5-trimethoxyphenyl)propionate (**2**) to give *p*-quinol ketals (**3**). These undergo acid-catalyzed cyclization to dihydrophenanthrene derivatives (**4**). A typical example is formulated in equation (I). The choice of the Lewis acid catalyst is sometimes critical for the cyclization. In the case of the cyclization shown above, use of $SnCl_4$, CF_3COOH and BF_3 etherate resulted in much lower

(I)

OCH₃ ... CH₃O— ... =O

+

OCH₃ / CH₃O— —OCH₃ / CH₂ / CH₂ / COOCH₃

$\xrightarrow{\text{LDA, THF}}$

1   **2**

OCH₃ / CH₃O— —OCH₃ / OCH₃ / CH₃O— / OH / COOCH₃

$\xrightarrow[\text{77\% overall}]{\text{P}_2\text{O}_5\text{–CH}_3\text{SO}_3\text{H}}$

OCH₃ / CH₃O— —OCH₃ / CH₃O— / COOCH₃

3   **4**

yields. At least one methoxy group in the phenyl group is also necessary for this cyclization to proceed.

[1] D. A. Evans, P. A. Cain, and R. Y. Wong, *Am. Soc.*, **99**, 7083 (1977).

Phosphorus(V) sulfide, 1, 870–871; **3**, 226–228; **4**, 389; **5**, 534–535.

Reduction of sulfoxides. Sulfoxides can be reduced to sulfides by P_2S_5 in methylene chloride at 25°. Yields are in the range 50–100%. The method is not applicable to sulfones.[1] Sulfilimines can also be reduced in this way to sulfides.[2]

$$R_2S{=}NTs(H) \xrightarrow[60-95\%]{P_2S_5} R_2S$$

[1] I. W. J. Still, S. K. Hasan, and K. Turnbull, *Synthesis*, 468 (1977).
[2] I. W. J. Still and K. Turnbull, *ibid.*, 540 (1978).

Phosphoryl chloride, 1, 876–882; **2**, 330–331; **3**, 228; **5**, 535–537; **7**, 292–293.

α,β-Unsaturated esters and lactones. Krief has reported a synthesis of alkenes by reaction of thionyl chloride–triethylamine with β-hydroxy selenides (**7**, 367). On extension of this reaction to preparation of α,β-unsaturated carbonyl compounds, difficulties were encountered in the synthesis of (Z)-isomers of α,β-unsaturated esters by *anti*-elimination from β-hydroxy selenides. Krief and Lucchetti[1] then found that phosphoryl chloride was a superior reagent in this and related cases.

Examples:

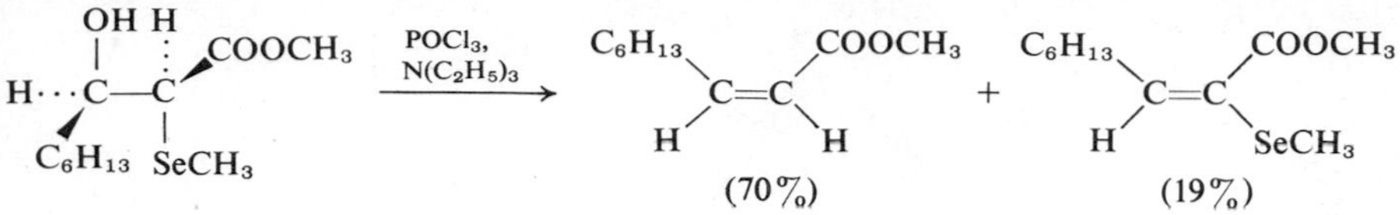

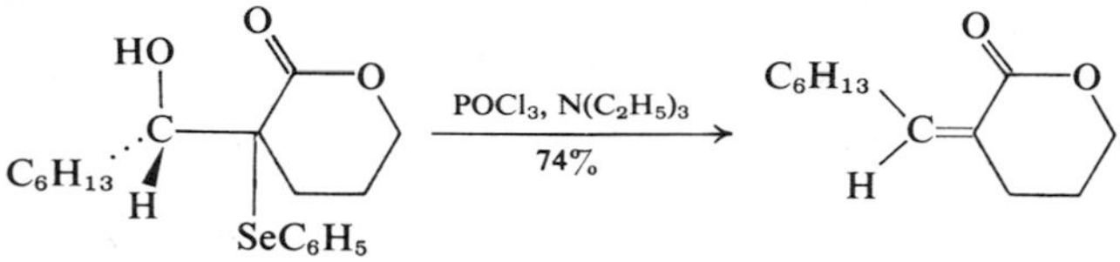

(70%)   (19%)

74%

[1] J. Lucchetti and A. Krief, *Tetrahedron Letters*, 2693 (1978).

Phthalide, (1), **5**, 537–538.

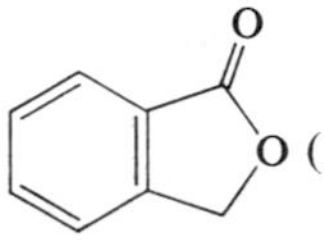

Naphthols. The anion of **1**, generated with LDA in THF at −40°, reacts with activated ethylenic compounds to give, after acid treatment to effect cyclization, tetralones. The products are dehydrated to naphthols by treatment with CF₃-COOH or BF₃ etherate.[1]

Examples:

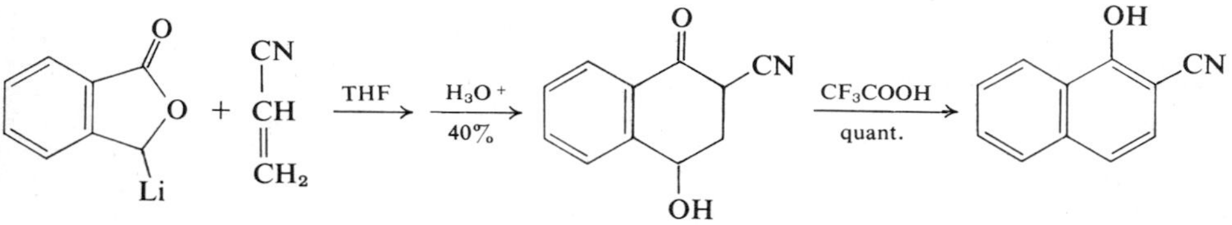

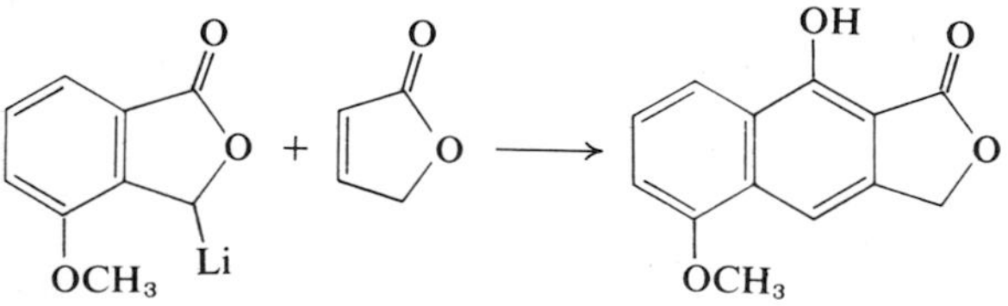

THF   H₃O⁺
        40%

CF₃COOH
quant.

[1] N. J. P. Broom and P. G. Sammes, *J.C.S. Chem. Comm.*, 162 (1978).

(S)-N-Phthaloylphenylalanyl chloride, (1 = RCOCl) (1).

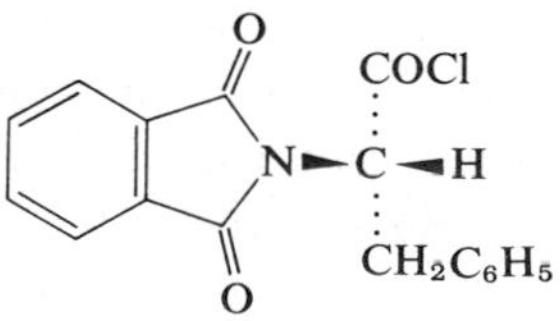

Mol. wt. 281.65, m.p. 83°, α_D −197°. Preparation.[1]

Resolution of **cis-2-cyclohexene-1,4-diol.**[2] The reagent has been used for the synthesis of optically pure **6** from the *meso*-diol **2**. Intermediates were purified

by crystallization. This synthesis is another example of a novel preparation of an optically active compound from an optically inactive substrate.[3]

[1] J. C. Sheehan, D. W. Chapman, and R. W. Roth, *Am. Soc.*, **74**, 3822 (1952).
[2] S. Terashima, M. Nara, and S. Yamada, *Tetrahedron Letters*, 1487 (1978).
[3] *Idem, ibid.*, 1001 (1977).

B-3α-Pinanyl-9-borabicyclo[3.3.1]nonane,

(**1**). Mol. wt. 258.26.

The reagent is prepared by hydroboration of (+)-α-pinene with 9-BBN in THF.

Asymmetric reduction.[1] This organoborane reduces aldehydes within minutes, but reduces ketones rather slowly. Of greater interest, the chiral organoborane from (+)-α-pinene reduces benzaldehyde-α-*d* quantitatively to (S)-(+)-benzyl-α-*d*-alcohol in 81.6% chemical yield (equation I). In principle the opposite enantiomer could be prepared when (−)-α-pinene is used.

(I)

$$1 + C_6H_5\overset{O}{\overset{\|}{C}}D \longrightarrow C_6H_5{-}\overset{OH}{\underset{H}{\overset{|}{C}}}{-}D$$

[1] M. M. Midland, A. Tramontano, and S. A. Zderic, *Am. Soc.*, **99**, 5211 (1977).

Pivaloyl chloride (Trimethylacetyl chloride), 1, 1229.

Selective esterification of sucrose. Under closely defined conditions, the reaction of sucrose (**1**) with pivaloyl chloride can result in two major derivatives,

2 and **3**, separable by crystallization from petroleum ether.[1] Few examples of selective esterification of sucrose are known.

Protection of phenolic OH.[2] This reagent has been used in the synthesis of 3-*t*-butyl-4-methoxyphenol (**4**), which has been found to decrease formation of carcinogenic metabolites of benzpyrene in an experimental organism.

[1] L. Hough, M. S. Chowdhary, and A. C. Richardson , *J.C.S. Chem. Comm.*, 664 (1978).
[2] L. K. T. Lam and K. Farhat, *Org. Prep. Proc. Int.*, **10**, 79 (1978).

Platinum catalysts, 1, 890–892; **3,** 332–333.

Cyclodehydrogenation. 4-Phenylphenanthrene (**1**) is converted into benz[*e*]-pyrene (**2**) when heated with platinum (10%) on activated carbon (E. Merck).[1]

Pt/C, 350°
68%

1   **2**

[1] P. Studt, *Ann.*, 530 (1978).

Potassium–Graphite, 4, 397; **7,** 296. Supplier: Alfa.

2-Alkenes. A recent synthesis of 2-alkenes from allyl phenyl sulfone (**1**) involves regioselective alkylation in the α-position to give a β,γ-unsaturated sulfone (**2**). These products are isomerized by potassium *t*-butoxide in THF at 0° to give (E)-α,β-unsaturated sulfones (**3**). These sulfones are not cleaved by aluminum amalgam, but are reduced by lithium–ethylamine. However, the preferred reagent is potassium–graphite. Both reagents unfortunately also induce partial (E)- to (Z)- isomerization, but C_8K is less prone to effect this undesirable side reaction.[1]

$$C_6H_5SO_2CH_2CH{=}CH_2 \xrightarrow[\substack{70-80\%}]{\substack{1)\ n\text{-BuLi} \\ 2)\ RX}} C_6H_5SO_2\overset{\overset{R}{|}}{C}HCH{=}CH_2 \xrightarrow[\text{quant.}]{KOC(CH_3)_3}$$

1   **2**

(E) **3**   **4** (Z/E ~ 35:65)

$$\xrightarrow[50-80\%]{C_8K,\ 20°} RCH{=}CHCH_3$$

Alkylation of nitriles and esters. C_8K metalates aliphatic nitriles and esters in THF at $-60°$. The resulting carbanions can be alkylated. The advantage of this new method is that dialkylated products are obtained only in traces.[2]

Examples:

$$CH_3CN \xrightarrow[42\%]{\substack{1)\ C_8K \\ 2)\ C_6H_5CH_2Cl}} C_6H_5(CH_2)_2CN$$

$$C_6H_5CH_2COOC_2H_5 \xrightarrow[70\%]{\substack{1)\ C_8K \\ 2)\ CH_3I}} \underset{\underset{CH_3}{|}}{C_6H_5CHCOOC_2H_5}$$

Review. Bergbreiter and Killough[3] have discussed the various uses of C_8K, particularly in comparison with the reactions of the soluble analog sodium naphthalenide. Perhaps the most useful role of C_8K is for the rapid formation of alkoxides from alcohols and of stabilized carbanions from carbonyl compounds. In these two reactions, it reacts more readily than potassium itself and can be separated from the products by filtration. The authors conclude that it has only limited value for reduction of alkyl and aryl halides and sulfonates, reactions accomplished more readily by other reagents. They also note that C_8K can be regarded as a suitable reagent for reactions in which sodium naphthalenide is useful.

[1] D. Savoia, C. Trombini, and A. Umani-Ronchi, *J.C.S., Perkin I,* 123 (1977).
[2] *Idem, Tetrahedron Letters,* 653 (1977).
[3] D. E. Bergbreiter and J. M. Killough, *Am. Soc.,* **100,** 2126 (1978).

Potassium 3-aminopropylamide, 6, 476; **7,** 296. This base can be prepared more conveniently by reaction of potassium amide (prepared from potassium and ammonia) with 1,3-diaminopropane with subsequent evaporation of excess ammonia *in vacuo.*[1] The Dutch chemists report further examples of use of the reagent for transforming alkynes and acetylenic alcohols into potassium acetylides with migration of the triple bond.

   ω-Alkynylorganoboranes. These compounds have been prepared by hydroboration and base-catalyzed isomerization as shown in equation (I). In this example, 9-BBN is used for hydroboration and potassium 3-aminopropylamide serves as base. The products undergo typical reactions of organoboranes, such as oxidation to 1-alkyne-ω-ols and alkyl transfer to α-bromo ketones.[2]

(I) $CH_2{=}CHCH_2C{\equiv}C(CH_2)_2CH_3 \xrightarrow[90-95\%]{R_2BH,\ 0°} R_2B(CH_2)_3C{\equiv}C(CH_2)_2CH_3 \xrightarrow{KAPA}$

$R_2B(CH_2)_6C{\equiv}CH \xrightarrow[95\%]{H_2O_2,\ OH^-} HO(CH_2)_6C{\equiv}CH$

$$\Big\downarrow \overset{\overset{O}{\|}}{(CH_3)_3CCCH_2Br}$$

$$\overset{\overset{O}{\|}}{(CH_3)_3CCCH_2(CH_2)_6C{\equiv}CH}$$

Elimination reactions of vinyl ethers and sulfides.[3] This base in 3-amino-propylamine at 25° converts these substrates rapidly into 1-alkynes and dienes.

Intermediate allenes and internal acetylenes are known to be isomerized by this base to 1-alkynes (**6**, 476). The reaction with vinyl sulfides is particularly useful for preparation of 1-alkynes.

Examples:

$$n\text{-}C_3H_7CH=\underset{\underset{SC_6H_5}{|}}{C}-C_4H_9\text{-}n \xrightarrow[83\%]{KAPA} CH_3(CH_2)_6C\equiv CH$$

[1] H. Hommes and L. Brandsma, *Rec. trav.*, **96**, 160 (1977).
[2] C. A. Brown and E. Negishi, *J.C.S. Chem. Comm.*, 318 (1977).
[3] C. A. Brown, *J. Org.*, **43**, 3083 (1978).

Potassium borohydride, 4, 398–399; **6,** 476–477.

$Δ^3$-***Chromenes.*** Benzopyrylium perchlorates such as **1** are reduced by this hydride in high yield to $Δ^3$-chromenes (**2**).[1] Reduction with formic acid and pyridine leads to a mixture of $Δ^2$- and $Δ^3$-chromenes.

1 (R = H, alkyl, aryl)   **2**

[1] P. Bouvier, J. Andrieux, H. Cunha, and D. Molho, *Bull. soc.*, 1187 (1977).

Potassium *t*-butoxide, 1, 911–927; **2,** 338–339; **3,** 233–234; **4,** 399–405; **5,** 544–552; **6,** 477–479; **7,** 296–298.

Dehydrobromination. Dehydrobromination of *meso*- and *dl*-2,3-dibromo-butane (**1** and **2**) and of (E)-2-bromobutene-2 (**4**) with solid potassium *t*-butoxide (heterogeneous) involves *anti*-elimination, as in homogeneous dehydrobromina-tions. (Z)-2-Bromobutene-2 (**3**) is stable under these conditions.[1]

                    1   **3**

2   **4**   **5**

Enynyl sulfides. These can be prepared by the one-pot procedure formulated in equation (I), which involves 1,4-elimination of RSH by base.[2]

$$(I)\ ClCH_2C{\equiv}CCH_2Cl \xrightarrow{\ 2RSNa,\ NH_3\ } [RSCH_2C{\equiv}CCH_2SR] \xrightarrow[85-90\%]{\ KOC(CH_3)_3,\ NH_3\ }$$

$$RSC{\equiv}CCH{=}CH_2 + KSR$$

Dehalogenation. Treatment of **1** with potassium *t*-butoxide in *t*-butyl alcohol at 80° (3 hours) yields **2** in 68% yield. This reaction is the first report of dechlorination with this base. The dibromide and diiodide corresponding to **1** are also

1   **2**

converted into **2** by the same method. The precursor (**1**) is available by addition of SCl_2 to cyclooctadiene-1,5.[3]

[1] M. J. Tremelling, S. P. Hopper, and P. C. Mendelowitz, *J. Org.*, **43**, 3076 (1978).
[2] R. H. Everhardus and L. Brandsma, *Synthesis*, 359 (1978).
[3] P. H. McCabe, C. M. Livingston, and A. Stewart, *J.C.S. Chem. Comm.*, 661 (1977).

Potassium carbonate, 5, 552–553.

Intramolecular Michael addition. Stork and co-workers[1] have examined a route to *cis*-hydrindanes by intramolecular Michael addition rather than vinylogous aldol condensation. Thus the 3,4-disubstituted cyclohexenone **1** undergoes Michael addition rather than aldol condensation when treated with K_2CO_3 at 20° (40 hours; the reaction is hindered by the β-methyl group). The method should be general.

1   **2**

[1] G. Stork, D. F. Taber, and M. Marx, *Tetrahedron Letters*, 2445 (1978).

Potassium cyanide, 5, 553; **7,** 299.

Homologated nitriles from aldehydes or ketones. This reaction can be carried out in two steps: conversion to the 2,4,6-triisopropylbenzenesulfonylhydrazone (**5,** 706) followed by reaction of the hydrazone with KCN (equation I).[1] The method is applicable to unsaturated ketones and to substrates containing a hydroxyl group.

$$(I) \quad \begin{array}{c} R^1 \\ \diagdown \\ R^2 \diagup \end{array} C{=}O + ArSO_2NHNH_2 \longrightarrow \begin{array}{c} R^1 \\ \diagdown \\ R^2 \diagup \end{array} C{=}NNHSO_2Ar \xrightarrow[\substack{50-75\% \\ \text{overall}}]{\substack{KCN, \\ CH_3OH}} \begin{array}{c} R^1 \\ \diagdown \\ R^2 \diagup \end{array} CHCN$$

[1] D. M. Orere and C. B. Reese, *J.C.S. Chem. Comm.*, 280 (1977).

Potassium cyanide–Acetone cyanohydrin, 1, 5.

Hydrocyanation. Hydrocyanation of α,β-unsaturated ketones can be effected by addition of the substrate in C_6H_6 or CH_3CN containing 18-crown-6 to dry KCN. Acetone cyanohydrin is then added and the two-phase system is stirred for 3–20 hours. No reaction is observed in the absence of the crown ether, acetone cyanohydrin, or potassium cyanide. Although the last reagent is only required in catalytic amounts, an equivalent is usually employed. The thermodynamic product predominates, particularly with C_6H_6 as solvent and at ambient temperatures.[1]

Example:

~1:4

[1] C. L. Liotta, A. M. Dabdoub, and L. H. Zalkow, *Tetrahedron Letters*, 1117 (1977).

Potassium cyanide–Chlorotrimethylsilane.

Cyanosilylation.[1] This reaction can be effected with potassium cyanide in combination with chlorotrimethylsilane in CH_3CN or DMF. Originally a crown ether was added, but it was later found to be detrimental in some cases. The rate can be enhanced by addition of zinc iodide (*cf.* **4,** 542–543).

$$\begin{array}{c} R^1 \\ \diagdown \\ R^2 \diagup \end{array} C{=}O + KCN + (CH_3)_3SiCl \xrightarrow{85-99\%} \begin{array}{c} OSi(CH_3)_3 \\ | \\ R^1{-}C{-}R^2 \\ | \\ CN \end{array}$$

[1] J. K. Rasmussen and S. M. Heilmann, *Synthesis*, 219 (1978).

Potassium dichromate, $K_2Cr_2O_7$. Mol. wt. 294.19.

Aldehydes. Primary alcohols (in CH_2Cl_2) can be oxidized to aldehydes with potassium dichromate in 9 M sulfuric acid in the presence of tetra-*n*-butyl-ammonium bisulfate as a phase-transfer catalyst. It is generally advantageous to use a slight excess of oxidant. The oxidation is complete in a few seconds at room temperature. Yields are generally high except for water-soluble aliphatic alcohols and α,β-unsaturated alcohols (the double bonds appear to be oxidized). Of course, secondary alcohols can also be oxidized to ketones in the same way.[1]

[1] D. Pletcher and S. J. D. Tait, *Tetrahedron Letters*, 1601 (1978).

Potassium ferricyanide, **1**, 929–933; **2**, 345; **4**, 406–407; **5**, 554–555; **6**, 480–482; **7**, 300–301.

Oxidation of acyloins. 1,2-Diketones are obtained in high yield by the reaction of acyloins in methanol with an aqueous solution of potassium ferricyanide and sodium hydroxide (four examples).[1]

[1] R. A. El-Zaru and A. A. Jarrar, *Chem. Ind.*, 741 (1977).

Potassium fluoride, **1**, 933–935; **2**, 346; **5**, 555–556; **6**, 481–482.

Phenacyl esters (**3**, 34; **5**, 154–155). KF is soluble in acetic acid and other liquid carboxylic acids owing to hydrogen bonding between F^- and the hydroxyl hydrogen.

This property can be used for a simple and efficient preparation of phenacyl esters, as formulated in equation (I). The reaction is usually complete within 10 minutes at 25° except in the case of sterically hindered acids.[1]

(I) $RCOOH + C_6H_5COCH_2Br + 2\,KF \xrightarrow[90-99\%]{DMF} RCOOCH_2COC_6H_5 + KBr + KHF_2$

Carbenes from silylvinyl triflates. Cunico and Han[2] have reported that the α-chlorovinylsilane (**1**) affords a vinylidenecarbene species (**a**) when treated with tetramethylammonium fluoride.

Silylvinyl triflates such as **3** are decomposed by fluoride ion even more readily.[3] Thus treatment of **3** at $-20°$ with KF, and a crown ether or with $R_4N^+F^-$ at $-20°$ also gives the carbene (**a**) in essentially quantitative yield.

The same paper also reports thermal decomposition of the tosylazoalkene $(CH_3)_2C{=}CHN{=}NTs$ at $25°$ to the same carbene, but in this case adducts with alkenes are obtained in only moderate yields (25–40%).

Cyclopentenones. The cyclization of 1,4-diketones to cyclopentenones is usually effected with a strong base such as potassium *t*-butoxide in *t*-butyl alcohol. The cyclization of **1** to **2** can also be effected with potassium fluoride and

dicyclohexyl-18-crown-6 in 30–40% yield in refluxing xylene.[4] In this particular case, this mild method offers no advantages and higher yields are obtained under classical conditions. However, the same paper reports successful cyclization of the very sensitive substrate **3** to **4** in 50% yield.

Fluorine substitution. 2′-Deoxy-2′-fluorocytidine (**2**) has not been available in reasonable amounts by a simple method. In a useful new route **1** is converted into **2** by KF activated by a crown ether (40% yield). The solvent (DMF) must be completely free from moisture.[5]

[1] J. H. Clark and J. M. Miller, *Tetrahedron Letters*, 599 (1977).
[2] R. F. Cunico and Y. K. Han, *J. Organometal. Chem.*, **105**, C29 (1976).

[3] P. J. Stang and D. P. Fox, *J. Org.*, **42**, 1667 (1977).
[4] W. G. Dauben and D. J. Hart, *ibid.*, **42**, 3787 (1977).
[5] R. Mengel and W. Guschlbauer, *Angew. Chem., Int. Ed.*, **17**, 525 (1978).

Potassium hexachloroosmiate(IV)–Zn. K_2OsCl_6 (mol. wt. 481.12) is available from Alfa.

Cyclic ketols. Rubezhov[1] has reported the reaction of dibenzylideneacetone (**1**) with this osmium salt and zinc to give **2** in almost quantitative yield. The reaction involves reductive coupling followed by intramolecular aldol condensation. Similar reactions are observed with other α,β-unsaturated ketones.

[1] A. Z. Rubezhov, *Tetrahedron Letters*, 2189 (1977).

Potassium hydride, 1, 935; **2**, 346; **4**, 409; **5**, 557; **6**, 482–483; **7**, 302–303.

Oxy-Cope rearrangement (**7**, 302). The Evans modification of the oxy-Cope rearrangement has been used in recent syntheses. It is an important step in a new synthesis by Still[1] of germacrane-type sesquiterpenes shown in equation (I).

Jung and Hudspeth[2] have extended this rearrangement to the nonbornenyl system. Thus treatment of **1** with KH in THF at 25° gives the rearranged ketone **2** in 72% isolated yield.

They then found that an aromatic ring can function as the olefinic component in this rearrangement and by this route prepared the tetracyclic diketo estrone derivative **4**.

Evans *et al.*[3] have observed dramatic rate enhancements in the thermal rearrangements of the potassium or sodium salts of 1,5-hexadiene-3-ols. DME is generally superior to THF as solvent.

Examples:

β-Elimination of a heteroatom bridge. A new route to condensed heterocycles is illustrated for synthesis of 4-carboethoxyoxindole (**2**) from the Diels-Alder adduct (**1**) of N-acetylpyrrole and 1,3-dicarboethoxyallene. Treatment of **1** with

potassium hydride in THF for 1 hour at 20° gives **2**, formed by a molecular rearrangement involving β-elimination of the heteroatom bridge.[4]

Other examples:

Alkylation of aldehydes. Formation of enolates of aldehydes by base presents a problem because of the competing aldol condensation. C-Alkylation of aldehydes can be effected by conversion to the potassium enolate with KH in THF and reaction with an activated primary bromide such as prenyl bromide or benzyl bromide (equation I). O-Alkylation becomes a competing reaction with

$$\text{(I)} \quad (CH_3)_2CHCHO \xrightarrow[\substack{88\%}]{\substack{1)\ KH,\ THF,\ 20° \\ 2)\ (CH_3)_2C=CHCH_2Br}} (CH_3)_2CCHO \atop \qquad\qquad\qquad\quad |\atop \qquad\qquad\qquad\quad CH_2CH=C(CH_3)_2$$

primary alkyl halides other than methyl iodide and with secondary alkyl halides.

This method is also useful for alkylation of β,γ-unsaturated aldehydes (equation II).[5]

$$\text{(II)} \quad \underset{\substack{|\quad\ |\\ H\ \ CH_3}}{C_2H_5C=C-CHCHO} \xrightarrow[\substack{92\%}]{\substack{1)\ KH \\ 2)\ (CH_3)_2C=CHCH_2Br}} \underset{\substack{|\quad\ |\\ H\ \ CH_3}}{C_2H_5C=C-CCHO} \overset{\substack{CH_2CH=C(CH_3)_2\\ |}}{}$$

Permethylation. Permethylation of ketones can be conducted in generally high yield by addition of the ketone to a THF suspension containing KH, followed by dropwise addition of methyl iodide.[6]

[1] W. C. Still, *Am. Soc.*, **99**, 4186 (1977).
[2] M. E. Jung and J. P. Hudspeth, *ibid.*, **100**, 4309 (1978).
[3] D. A. Evans, D. J. Baillargeon, and J. V. Nelson, *ibid.*, **100**, 2242 (1978).
[4] A. P. Kozikowski and M. P. Kuniak, *J. Org.*, **43**, 2083 (1978).
[5] P. Groenewegen, H. Kallenberg, and A. van der Gen, *Tetrahedron Letters*, 491 (1978).
[6] A. A. Millard and M. W. Rathke, *J. Org.*, **43**, 1834 (1978).

Potassium hydroxide, 5, 557–560; 6, 486; 7, 303–304.

Hydrolysis of amides[1] and esters.[2] Potassium *t*-butoxide in combination with water (2:1 equiv.) as a slurry in ether is an excellent reagent for hydrolysis of tertiary amides at room temperature (65–100% yield). One drawback is that primary and secondary amides are resistant to these conditions. The reagent is essentially finely divided anhydrous potassium hydroxide.[1] It is also useful for hydrolysis of esters at room temperature (70–100% yield). Simple esters are hydrolyzed in 1–2 hours, but longer reaction periods are required for hindered esters. Even methyl mesitoate can be saponified in 75% yield at room temperature in the course of 5–6 days.[2]

Sulfenylation of nitriles. Potassium hydroxide (solid) suspended in THF generates the α-carbanion of nitriles. Thus nitriles in THF can be sulfenylated generally in yields of 80–98% by treatment with KOH followed by addition of a disulfide (equation I).

(I) $C_6H_5CHR^1$ $\xrightarrow[50-98\%]{\text{KOH, THF, R}^2\text{S—SR}^2}$ $C_6H_5\overset{\overset{\displaystyle SR^2}{|}}{\underset{\underset{\displaystyle CN}{|}}{C}}R^1$

$\qquad\qquad\underset{\displaystyle CN}{|}$

$\qquad\quad (R^1 = H, CH_3, C_6H_5)\qquad\qquad\qquad (R^2 = CH_3, C_6H_5)$

Sulfides can also be prepared by the method shown in equation (II). Sulfenylation of a succinonitrile is difficult but is possible using dimethyl disulfide and methyl iodide (equation III).[3]

(II) $C_6H_5CH_2CN$ $\xrightarrow[50-98\%]{\text{KOH, THF, R}^2\text{S—SR}^2, \text{R}^3\text{X}}$ $C_6H_5\overset{\overset{\displaystyle SR^2}{|}}{\underset{\underset{\displaystyle CN}{|}}{C}}\text{—}R^3 + R^2SR^3$

(III) $C_6H_5\overset{\overset{\displaystyle CH_3}{|}}{\underset{\underset{\displaystyle CN}{|}}{C}}\text{—}CH_2CN$ $\xrightarrow[90\%]{\text{KOH, THF, CH}_3\text{SSCH}_3, \text{CH}_3\text{I}}$ $C_6H_5\overset{\overset{\displaystyle H_3C}{|}}{\underset{\underset{\displaystyle CN}{|}}{C}}\text{—}\overset{\overset{\displaystyle CH_3}{|}}{\underset{\underset{\displaystyle CN}{|}}{C}}\text{—}SCH_3 + S(CH_3)_2$

[1] P. G. Gassman, P. K. G. Hodgson, and R. J. Balchunis, *Am. Soc.*, **98**, 1275 (1976).
[2] P. G. Gassman and W. N. Schenk, *J. Org.*, **42**, 918 (1977).
[3] E. Marchand, G. Morel, and A. Foucaud, *Synthesis*, 360 (1978).

Potassium permanganate, 1, 942–952; **2**, 348; **4**, 412–413; **5**, 562–563.

Oxidation of **gem-*disulfides*.**[1] *gem*-Disulfides can be oxidized selectively to monosulfones with $KMnO_4$ in acetone at $0°$ (70–90% yield). Under the same conditions *vic*-disulfides are oxidized to disulfones.

Supported reagent. One limitation to potassium permanganate is the limited solubility in organic solvents. Solubilization with a crown ether is one solution (**4**, 143); another is impregnation on an inorganic solid support such as a molecular sieve, silica gel, and certain clays. Although the factors responsible for the enhanced reactivity of this form of $KMnO_4$ are not understood, such materials are useful for oxidation.[2]

Examples:

$(C_6H_5)_2CHOH \xrightarrow[100\%]{} (C_6H_5)_2C{=}O$

$CH_3(CH_2)_4CH_2OH \xrightarrow[29\%]{} CH_3(CH_2)_4CHO$

[1] M. Poje and K. Balenović, *Tetrahedron Letters*, 1231 (1978).
[2] S. L. Regen and C. Koteel, *Am. Soc.*, **99**, 3837 (1977).

Potassium persulfate, 1, 952–954.

Elbs oxidation (**1**, 952). The Elbs oxidation is a key step in a recent regio-specific synthesis of islandicin monomethyl ether (**3**), a possible precursor to anthracycline antibiotics (equation I).[1]

[1] R. D. Gleim, S. Trenbeath, F. Suzuki, and C. J. Sih, *J.C.S. Chem. Comm.*, 242 (1978).

Potassium selenocyanate, 6, 487–488. Additional supplier: Fluka.

Aryl selenocyanates.[1] These substances have usually been prepared by diazotization of an arylamine followed by reaction of the diazonium salt with KSeCN (*cf.* **6**, 420–421). Japanese chemists have reported an alternate method: reaction of an aryl iodide with KSeCN in HMPT at 100–120° with catalysis by CuI. Yields are only modest to good.

[1] H. Suzuki and M. Shinoda, *Synthesis*, 640 (1977).

Potassium superoxide, 6, 488–490; **7**, 304–307.

Oxidation of N-chloroamines to carbonyl compounds. Amines can be converted into carbonyl compounds by conversion to N-chloroamines (**1**, 92), which are not isolated but are oxidized directly with KO_2 in the presence of 18-crown-6. The intermediate imines are then hydrolyzed.[1]

Examples:

$$n\text{-}C_6H_{13}NH_2 \xrightarrow[49\%]{\substack{1)\ (CH_3)_3COCl \\ 2)\ KO_2}} n\text{-}C_5H_{11}CH{=}NH$$

$$[(CH_3)_2CHCH_2]_2NH \xrightarrow{(CH_3)_3COCl} [(CH_3)_2CHCH_2]_2NCl \xrightarrow[88\%]{KO_2}$$

$$(CH_3)_2CHCH{=}NCH_2CH(CH_3)_2$$

Singlet oxygen.[1] Oxygen is released during the reaction of potassium super-oxide with a diacyl peroxide (equation I). A significant fraction of the oxygen is in

(I) $$2KO_2 + RCOOCR \longrightarrow 2O_2 + 2RCOOK$$

the excited singlet state as shown by trapping experiments with 1,3-diphenyl-isobenzofurane and 1,2-dimethylcyclohexene.[2]

Prostaglandin endoperoxide. Upjohn chemists[3] have prepared the endoper-oxide PGH_2 methyl ester (6), from PGF_2 (1) by conversion to the ditosylate (with temporary protection of the 15-hydroxyl as the *t*-butyldimethylsilyl ether).

Replacement of the tosylate by bromine by an S_N2 reaction should proceed with inversion to give 3, but the major product is 4, in which all substituents on the cyclopentane ring are *trans*. Reaction of 3 with KO_2 in DMF containing dicyclo-hexyl-18-crown-6 (6, 488) gave the desired endoperoxide (6), but in low yield because of a competing intramolecular elimination process.

Reaction with hydrazo compounds. Potassium superoxide reacts with various hydrazo and related compounds in a variety of ways as shown in the examples. Oxidation of arylhydrazines also results in biaryls derived from the solvent and thus involves free aryl radicals.

The main interest in these reactions resides in the fact that hydrazo and related compounds are involved in biological processes.[4]

Examples:

$$C_6H_5NHNH_2 \xrightarrow[81\%]{KO_2,\ C_6H_5CH_3} C_6H_6$$

$$C_6H_5CH_2NHNH_2 \xrightarrow{KO_2,\ C_6H_6} \underset{(54\%)}{C_6H_5CH_3} + \underset{(15\%)}{C_6H_5COOH} + \underset{(6\%)}{C_6H_5CHO}$$

$$C_6H_5NHNHC_6H_5 \xrightarrow{98\%} C_6H_5N{=}NC_6H_5$$

Nucleophilic aromatic substitution. A few examples of this type of substitution have been observed using KO_2 solubilized with 18-crown-6 in DMSO, but not in benzene. The second reaction is unusual in that in this case the crown ether is not necessary.[5]

Examples:

Review.[6] The organic chemistry of superoxide has been reviewed (145 references).

[1] F. E. Scully, Jr., and R. C. Davis, *J. Org.*, **43**, 1467 (1978).
[2] W. C. Danen and R. L. Arudi, *Am. Soc.*, **100**, 3944 (1978).
[3] R. A. Johnson, E. G. Nidy, L. Baczynskyj, and R. R. Gorman, *ibid.*, **99**, 7738 (1977).
[4] C.-I. Chern and J. San Fillippo, Jr., *J. Org.*, **42**, 178 (1977).
[5] T. Yamaguchi and H. C. van der Plas, *Rec. trav.*, **96**, 89 (1977).
[6] E. Lee-Ruff, *Chem. Soc. Rev.*, **6**, 195 (1977).

Potassium (sodium) tetracarbonylhydridoferrate, 4, 268–269; **5,** 357–358; **6,** 483–486.

Methyl ketones (6, 484). Complete details for reductive deacylation of Knoevenagel products in ethanol are available.[1] When THF or acetone is used as solvent, the only reaction observed is reduction of the α,β-unsaturated double bond.

Examples:

$$CH_3CH_2CH_2CH=C(COCH_3)_2 \xrightarrow[69\%]{\substack{KHFe(CO)_4, \\ C_2H_5OH}} CH_3(CH_2)_4COCH_3$$

$$C_6H_5CH=C(COCH_3)_2 \xrightarrow[76\%]{C_2H_5OH} C_6H_5CH_2CH_2COCH_3$$

$$75\% \downarrow THF$$

$$C_6H_5CH_2CH(COCH_3)_2$$

$$C_6H_5CH=C\underset{COOC_2H_5}{\overset{COCH_3}{\big\langle}} \xrightarrow[90\%]{C_2H_5OH} C_6H_5CH_2CH\underset{COOC_2H_5}{\overset{COCH_3}{\big\langle}}$$

[1] M. Yamashita, Y. Watanabe, T. Mitsudo, and Y. Takegami, *Bull. Chem. Soc. Japan*, **51**, 835 (1978).

Potassium thiophenoxide, C_6H_5SK. Mol. wt. 148.27.

This reagent is prepared *in situ* from thiophenol and potassium *t*-butoxide.

Cleavage of cyclopropanes. Two laboratories[1,2] have reported a synthesis of prostaglandins based on the selective cleavage of a cyclopropane ring by potassium thiophenoxide. Thus reaction of **1** with this nucleophile gives the ring-opened product **2** with inversion of configuration. This sulfide is oxidized to the sulfoxide, which is subjected to the Evans rearrangement (6, 30–31) to give the allylic alcohol **3**. This product has been converted into 11-deoxyprostaglandin

E_2 and into PGA_2 by known methods. This synthetic scheme is noteworthy because it results in the desired configurations at C_8, C_{12}, and C_{15} and also in the (E)-configuration of the double bond.

The selective cleavage of a cyclopropane ring has also been used in a stereo-controlled total synthesis of the antitumor lactone vernolepin (10).[3] A key step involved reaction of 4 with sodium p-methoxythiophenoxide, which was used rather than sodium thiophenoxide because the corresponding sulfoxide is less prone to syn-elimination than a phenyl sulfoxide in the next step, 5 → 6. Remaining steps involved epoxidation and reduction to the diol (9). Conversion of 9 to 10 included hydrolysis, lactonization, and methylenation, accomplished in 22% yield. See structure on page 422.

[1] D. F. Taber, Am. Soc., 99, 3513 (1977).
[2] K. Kondo, T. Umemoto, Y. Takahatake, and D. Tunemoto, Tetrahedron Letters, 113 (1977); D. Tunemoto, N. Araki, and K. Kondo, ibid., 109 (1977).
[3] M. Isobe, H. Iio, T. Kawai, and T. Goto, Am. Soc., 100, 1940 (1978).

L-**Proline, 6**, 492–493; **7**, 307.

Asymmetric bromolactonization. A new synthesis of optically active α-hydroxy acids from α,β-unsaturated acids involves as the first step acylation of L-proline (1) with the acid chloride of the unsaturated acid, for example, tiglic acid (2).

The resulting unsaturated carboxylic acid (3) was then treated with NBS in DMF. This reaction gives the bromo lactone (4) and a diastereoisomer in the ratio 94.5:5.5 and is therefore highly stereoselective. The mixture can be separated to give optically pure 4 in 64% yield from 3. Remaining steps involve reduction to

Reaction scheme:

4 (CH=CH₂, CH[COOC(CH₃)₃]₂, C₂H₅OOC)

→ ArSNa, THF, 88% →

5 (CH=CH₂, CH[COOC(CH₃)₃]₂, SAr, C₂H₅OOC)

→ 1) ClC₆H₄CO₃H (88%), 2) P(OCH₃)₃ (84%) →

6 (CH=CH₂, OH, CH[COOC(CH₃)₃]₂, COOC₂H₅)

→ ClC₆H₄CO₃H, 81% →

7 (CH=CH₂, OH, CH[COOC(CH₃)₃]₂, COOC₂H₅)

→ NaBH₃CN, HMPT →

8 (CH=CH₂, OH, C[COOC(CH₃)₃]₂, OH, C₂H₅OOC)

→ BH₃, THF, −45°, 70% →

9 (CH=CH₂, OH, CH[COOC(CH₃)₃]₂, OH, C₂H₅OOC)

→ 1) OH⁻, 2) HCHO, HN(C₂H₅)₂, 22% →

10 (CH=CH₂, OH, CH₂, O, CH₂)

the lactone **5** and hydrolysis to the optically pure α-hydroxy carboxylic acid **6**. The overall optical yield of **6** from **3** is about 90%.[1]

This same asymmetric synthesis has been used to convert 3,4-dihydro-2-naphthoic acid (**7**) into R-(−)-**8** in about 95% optical purity[2]. This acid was prepared because it serves as a model for the A and B rings of anthracyclinones.

7   8 (α_D − 16.3°)

Asymmetric aldolization (6, 411). Hoffmann-La Roche chemists[3] have published a procedure for preparation of optically pure Wieland-Miescher ketone (**2**) in 100 g. quantities using the method of Eder and Hajos. Aldolization

1   2 (m.p. 50–51°,   3 (m.p. 50–51°,
                        α_D + 100°)   inactive.)

of **1** in the presence of L-proline results in **2** and **3** in 71% chemical yield with an enantiomeric excess of **2** of 70%. Such a mixture can be separated by one crystallization from ether at −21° seeded with a few crystals of pure **2**, but several crystallizations are used in practice; total yield of **2** is 50% and the yield of **3** is 18%.

Asymmetric hydrogenation. L-Proline has been converted into a series of (S)-2-(aminomethyl)pyrrolidines, such as (S)-2-(anilinomethyl)pyrrolidine (**1**), α_D + 19.7°, obtained as shown in equation (I). The product is an efficient chiral ligand for asymmetric reduction of various ketones by lithium aluminum hydride.

In ether solution at $-100°$ the complex reduces acetophenone to (S)-1-phenyl-ethanol in 92% optical purity. Additives such as TMEDA, DME, and $MgBr_2$ reduce the optical yields. All ketones of the type ArCOR are reduced in 50–85% optical yields, but optical yields are low in the reduction of purely aliphatic ketones.[4]

[1] S. Terashima and S. Jew, *Tetrahedron Letters*, 1005 (1977).
[2] S. Terashima, S. Jew, and K. Koga, *ibid.*, 4507 (1977).
[3] J. Gutzwiller, P. Buchschacher, and A. Fürst, *Synthesis*, 167 (1977).
[4] T. Mukaiyama, M. Asami, J. Hanna, and S. Kobayashi, *Chem. Letters*, 783 (1977); M. Asami, H. Ohno, S. Kobayashi, and T. Mukaiyama, *Bull. Chem. Soc. Japan*, **51**, 1869 (1978).

Pyridinium hydrochloride, 1, 964-966; **2,** 352-353; **3,** 239-240; **4,** 415-418; **5,** 566-567; **6,** 497-498; **7,** 308.

Fischer indole synthesis. This reaction can be carried out by reaction of a ketone (1 equiv.) and an arylhydrazine hydrochloride (1 equiv.) in refluxing pyridine (1–18 hours). Pyridinium chloride (1 equiv.) is formed and acts as catalyst for the cyclization to indoles. No prior formation of a hydrazone is required. In several instances, higher yields are obtained by this method than by use of other catalysts.[1]

Examples:

Aromatization.[2] Treatment of the cyclohexanone **1** with Py·HCl (30 minutes at 230°) gives the phenylacetic acid **2** in nearly quantitative yield. Surprisingly **2** is also obtained in the same way from **3**, but in lower yield. Lactones may be intermediates in this dehydration.

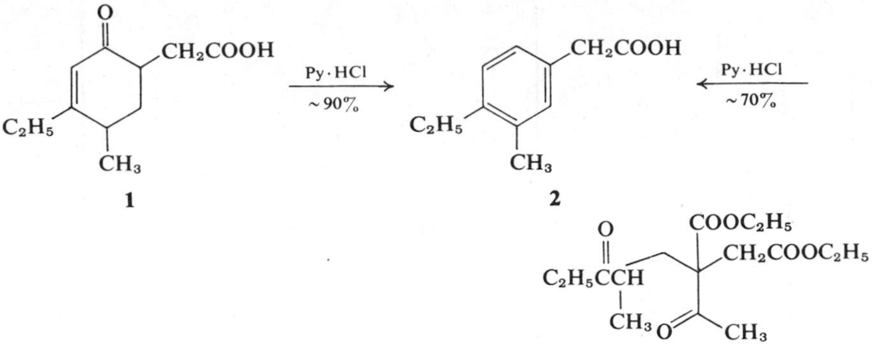

γ-*Chloro ketones*. α-Cyclopropyl ketones are cleaved by pyridinium chloride in CH_3CN to γ-chloro ketones in yields of 70–82%.[3]
Examples:

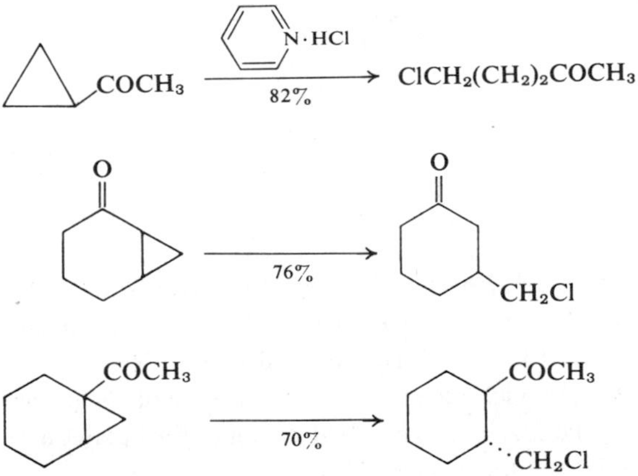

[1] W. M. Welch, *Synthesis*, 645 (1977).
[2] L. Baiocchi, M. Giannangeli, and M. Bonanomi, *Tetrahedron*, **34**, 951 (1978).
[3] N. Di Bello, L. Pellacani, and P. A. Tardella, *Synthesis*, 227 (1978).

Pyridinium chlorochromate (PCC), 6, 498–499; **7**, 308–309.

Oxidative cyclization. The earlier observation of conversion of (−)-citronellol to (−)-pulegone (**7**, 308) has been shown to be a general reaction for annelation of cyclic unsaturated alcohols or aldehydes. Several intermediates have been detected as shown for the oxidative cyclization of **1** to **2**.

One limitation is that this cyclization cannot be used for formation of cyclo-pentenones, even under forcing conditions. The reaction is also limited to prep-aration of β,β-disubstituted α,β-unsaturated cyclohexenones.[1]

Other examples:

Oxidation of **t-allylic alcohols.** Alcohols of this type, prepared by 1,2-addition of alkyllithium reagents to α,β-enones, are oxidized by this reagent to transposed 3-alkyl-α,β-unsaturated ketones.[2] The method is an alternative to that of Trost and Stanton (**6**, 31–32). Yields are excellent in the case of cyclic alcohols, but only moderate with acyclic alcohols. The third example formulates a modification for preparation of α,β-unsaturated aldehydes.

Oxidation of an allylic alcohol. This reagent has been used successfully to oxidize the benzyl ester of the β-lactam antibiotic clavulanic acid (1) to the labile allylic aldehyde 2. Attempted oxidation with the Pfitzner-Moffatt reagent resulted in formation of a 1,3-diene by a 1,4-elimination.[3]

Oxidation of enol ethers. The reagent oxidizes linear or cyclic enol ethers directly to esters or lactones in high yield. The reaction evidently involves attack of the double bond.[4]

Examples:

$$C_2H_5OCH{=}CH_2 \xrightarrow[75\%]{C_5H_5\overset{+}{N}HCrO_3Cl^-} CH_3COOC_2H_5$$

Oxidative deoximation. Aldehydes and ketones can be recovered from the oxime derivative by treatment with pyridinium chlorochromate (2 equiv.) in CH_2Cl_2 (12–24 hours, 25°). Addition of sodium acetate inhibits epimerization in the case of α-methyl ketones. Yields range from 30 to 85%.[5]

[1] E. J. Corey and D. L. Boger, *Tetrahedron Letters*, 2461 (1978).
[2] W. G. Dauben and D. M. Michno, *J. Org.*, **42**, 682 (1977).
[3] D. F. Corbett, T. T. Howarth, and I. Stirling, *J.C.S. Chem. Comm.*, 808 (1977).
[4] G. Piancatelli, A. Scettri, and M. D'Aurea, *Tetrahedron Letters*, 3483 (1977).
[5] J. R. Maloney, R. E. Lyle, J. E. Saavedra, and G. G. Lyle, *Synthesis*, 212 (1978).

Pyridinium *p*-toluenesulfonate, $CH_3C_6H_5SO_3^-$ (1). Mol. wt. 251.29, m.p.

120°. The salt is prepared in high yield by addition of *p*-toluenesulfonic acid to pyridine at 20°. It is soluble in CH_2Cl_2, $CHCl_3$, C_2H_5OH, and acetone.

Tetrahydropyranylation.[1] This weakly acidic salt is superior to *p*-TsOH as the catalyst for tetrahydropyranylation of alcohols, particularly acid-sensitive alcohols (allylic alcohols) and alcohols containing epoxide and ketal groups. Yields are practically quantitative. The reagent is equally satisfactory for deprotection of THP groups.

Protection of a hemiacetal group. A synthesis of (−)-canadensolide (2) from a product (3) derived from glucose, and therefore of known configuration,

2   3

required in the last steps protection of the hemiacetal group of 3 in order to effect carboethoxylation at the α-methylene position. Attempts to use the methyl or benzyl ether failed to be useful; reaction of 3 with dihydropyrane or ethyl vinyl ether with

3   4

5   6

p-TsOH as catalyst resulted in decomposition. The desired conversion to 4 was effected in high yield by catalysis with this salt. After ethoxycarbonylation to 5, the protective group was removed by treatment with hot ethanol containing the salt.[2]

[1] M. Miyashita, A. Yoshikoshi, and P. A. Grieco, *J. Org.*, **42**, 3772 (1977).
[2] R. C. Anderson and B. Fraser-Reid, *Tetrahedron Letters*, 3233 (1978).

Pyrrolidine, 1, 972–974; **2,** 354–355; **3,** 240; **7,** 309–310.

Enamines of $\Delta^{1,4}$-*3-ketosteroids.*[1] $\Delta^{1,4}$-3-Ketosteroids do not react with pyrrolidine under usual conditions, but when they are refluxed with this amine for 15 hours, dienediamines (**1**) are obtained by addition to the C_1–C_2 double bond followed by formation of the enamine of the 3-ketone. When these products are heated at 150° a trienamine (**2**) is formed, which is the true enamine of this system. Both **1** and **2** are stable to acid, but are hydrolyzed in about 90% yield

by NaOH at pH ~ 14; they are useful derivatives for protection of a $\Delta^{1,4}$-3-ketone. An 11-keto or 11β-hydroxyl group does not interfere with this conversion.

[1] R. Bucourt and J. Dube, *Bull. Soc.*, II-33 (1978).

Q

Quinine, 6, 501; **7,** 311.

Chiral epoxides. $(-)$-Benzylquininium chloride (**1**) can introduce a considerable degree of asymmetry in several methods for preparation of epoxides. Actually the earlier method (enone and H_2O_2, **7,** 311) is still the most effective; an enantiomeric excess of 55% has since been achieved by carrying out the epoxidation at 3° for 90 hours.[1]

Examples:

$$C_6H_5COCH\!=\!CHC_6H_5 \xrightarrow[66\%]{1,\,NaOCl} C_6H_5COCH\!-\!CHC_6H_5$$

$$(ee\ 25 \pm 5\%)$$

$$C_6H_5COCH_2Cl + p\text{-}ClC_6H_4CHO \xrightarrow[68\%]{1,\,NaOH} C_6H_5COCH\!-\!CHC_6H_4Cl\text{-}p$$

$$(ee\ 8.1\%)$$

$$C_6H_5COCH\!-\!CHC_6H_5 \xrightarrow[90\%]{1,\,NaOH} C_6H_5COCH\!-\!CHC_6H_5$$
$$\;\;\;\;\;\;OH\;\;\;Cl$$

$$(ee\ 6.1\%)$$

$$\begin{array}{c}CH_3\\|\\C_6H_5COCBr\\|\\CH_3\end{array} \xrightarrow[94\%]{1,\,NaCN} \begin{array}{c}CN\;\;\;CH_3\\C_6H_5C\!-\!C\\O\;\;\;\;CH_3\end{array}$$

$$(\alpha_D - 3°)$$

Asymmetric Michael addition. Michael addition of nitromethane to chalcone (equation I) in the presence of a chiral amine (quinine, N-methylephedrine) proceeds in methanol (but not in aprotic solvents) in 60–80% yield, but the optical yield at best is 1%. However, if these amines are converted into aminium fluorides, the addition takes place in aprotic solvents ($C_6H_5CH_3$, CH_3CN) in 50–100% yield; of more interest, asymmetric inductions of 20% can be obtained. $(-)$-Benzylquininium fluoride (**1**) is particularly effective.[2] The hydroxyl group in these salts is believed to play an important role in the stereochemical outcome of the reaction. Moreover, the fluoride ion is important as a strong base.

(I) C_6H_5C=$CCOC_6H_5$ + CH_3NO_2 $\xrightarrow{\quad 1 \quad}$ $C_6H_5CHCH_2COC_6H_5$

with H above and H below the first structure, and CH_2NO_2 below the product.

(ee ~ 10%)

Asymmetric addition of RSH to enones. Thiols in the presence of catalytic quantities of ($-$)-quinine add to cyclohexenone to give β-keto sulfides with relatively high asymmetric induction (6–50%). The absolute configurations are not known.[3]

(I) C_6H_5SH + [cyclohexenone structure] $\xrightarrow[94\%]{\text{Quinine, } C_6H_5CH_3, 22°}$ [cyclohexanone with SC_6H_5 structure]

(α_{578} +29.7°; ee 41%)

Asymmetric borohydride reduction. Colonna and Fornasier have examined the reduction of ketones with sodium borohydride under phase-transfer conditions in the presence of optically active ammonium salts containing at least one hydroxyl group. Of the seven catalysts tested ($-$)-benzylquininium chloride (1) (7, 311) was the most effective for asymmetric reduction of *t*-butyl phenyl ketone (pivalophenone) to the corresponding carbinol with optical yields as high as 32%. Two factors would appear to be important for this asymmetric reduction: the catalyst must be conformationally rigid and the hydroxyl group must be in the β-position to the onium function.[4]

[1] J. C. Hummelen and H. Wynberg, *Tetrahedron Letters*, 1089 (1978).
[2] S. Colonna, H. Hiemstra, and H. Wynberg, *J.C.S. Chem. Comm.*, 238 (1978).
[3] R. Helder, R. Arends, W. Bolt, H. Hiemstra, and H. Wynberg, *Tetrahedron Letters*, 2181 (1977).
[4] S. Colonna and R. Fornasier, *J.C.S. Perkin I*, 371 (1978).

Quinuclidine N-oxide, [structure] **(1).** Mol. wt. 127.19. This reagent can be prepared by oxidation of quinuclidine with *m*-chloroperbenzoic acid in $CHCl_3$ at 0–5°.[1]

Rubanes. The rubane skeleton, which includes quinine and related alkaloids, can be synthesized in one step from this N-oxide (1).[2] Treatment of 1 with *t*-butyllithium at $-78 \rightarrow 0°$ followed by reaction with 4-formylquinoline gives a separable mixture of the N-oxides of *erythro*- and *threo*-rubanol, 2 and 3, in 35% yield. Deoxygenation is effected in good yield with hexachlorodisilane (3, 148).

[1] J. C. Craig and K. K. Purushothaman, *J. Org.*, **35**, 1721 (1970).
[2] D. H. R. Barton, R. Beugelmans, and R. N. Young, *Nouv. J. Chim.*, **2**, 363 (1978).

R

Raney nickel, 1, 723–731; 2, 293–294; 5, 570–571; 6, 502; 7, 312.

β-*Lactams*. Desulfurization of mesoionic thiazole-4-ones (1) with Raney nickel in CH_3OH or THF results in ring contraction to β-lactams (2).[1]

1,3-*Cyclopentanedione*.[2] 4-Cyclopentene-1,3-dione (1) can be converted into 1,3-cyclopentanedione (2) by hydrogenation with partially deactivated neutral W-2 Raney nickel catalyst (1, 723).

Reduction of t-*azides to amines*. An improved method for conversion of tertiary alcohols (or their corresponding olefins) into amines is illustrated in equation (I) for the preparation of α,α-dimethylbenzylamine. This method of reduction can be used also for azides containing benzyloxy groups or double bonds.[3]

[1] T. Sheradsky and D. Zbaida, *Tetrahedron Letters*, 2037 (1978).
[2] J. Šraga and P. Hrnčiar, *Synthesis*, 282 (1977).
[3] D. Balderman and A. Kalir, *ibid.*, 24 (1978).

Rhodium catalysts, 1, 982–983; 4, 418–419; 6, 503.

Hydrogenation of cyclohexanones.[1] High stereoselectivity is observed in the hydrogenation of 4-t-butylcyclohexanone in ethanol or THF with rhodium metal as catalyst as shown in equation (I).

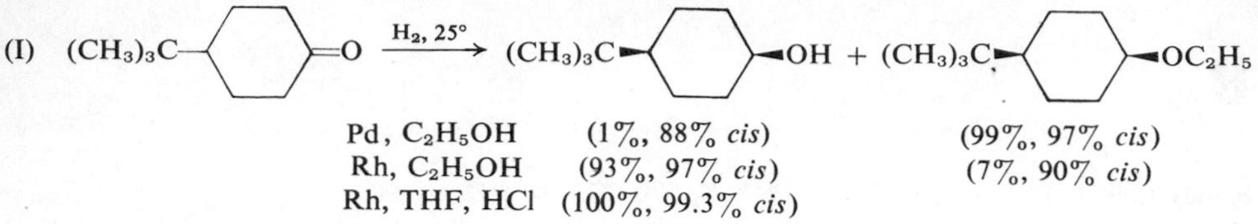
(I)
(CH3)3C
O
H2, 25°
(CH3)3C
OH
+ (CH3)3C
OC2H5
Pd, C2H5OH (1%, 88% cis) (99%, 97% cis)
Rh, C2H5OH (93%, 97% cis) (7%, 90% cis)
Rh, THF, HCl (100%, 99.3% cis)

This method is useful for synthesis of axial alcohols as shown in the steroid examples.

Examples:

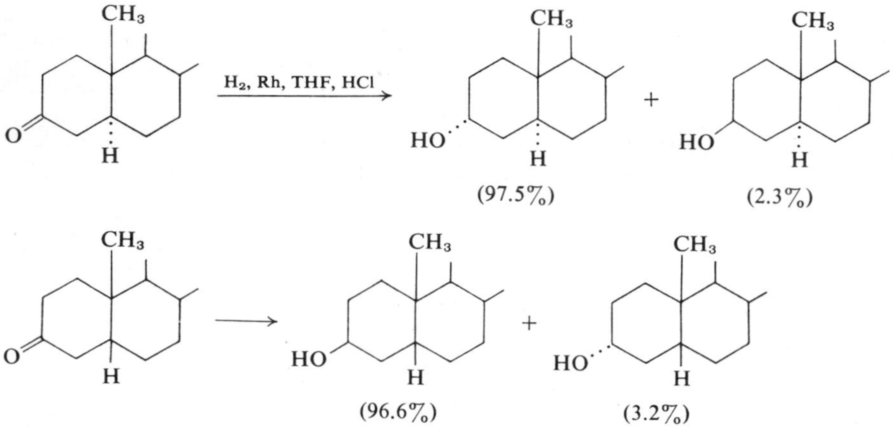
CH3
O
H
H2, Rh, THF, HCl
HO
H
(97.5%)
+
(2.3%)
(96.6%)
(3.2%)

Nishimura has reviewed stereoselective hydrogenation with group VIII metals.[2]

[1] S. Nishimura, M. Ishige, and M. Shiota, *Chem. Letters*, 535, 963 (1977).
[2] S. Nishimura, *Strem Chemiker*, **6**, 7 (1978).

Rhodium(II) carboxylates, 5, 571–572; **7**, 313.

Cyclopropenation. Cyclopropenes can be formed from alkynes by reaction with methyl diazoacetate using a rhodium(II) carboxylate as catalyst. The reaction is not particularly subject to steric hindrance, but polar groups (CH_2COOCH_3) inhibit cyclopropenation markedly. Insertion reactions compete with cyclopropenation in the case of acetylenic alcohols.[1]

Examples:

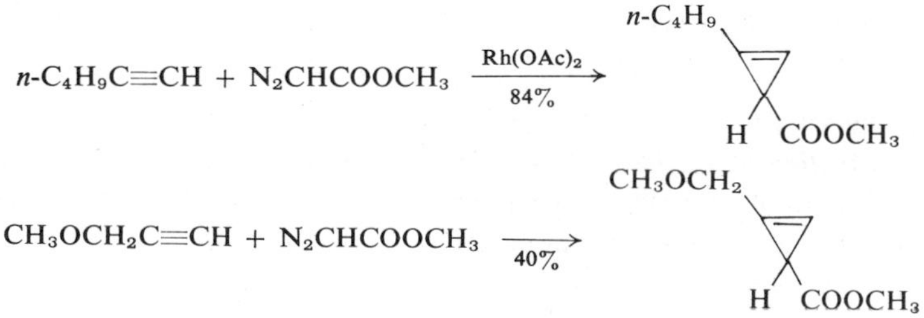
n-C4H9C≡CH + N2CHCOOCH3
Rh(OAc)2
84%
n-C4H9
H COOCH3
CH3OCH2C≡CH + N2CHCOOCH3
40%
CH3OCH2
H COOCH3

$$HC{\equiv}CCH_2OH \longrightarrow \quad + \quad HC{\equiv}CCH_2OCH_2COOCH_3$$

(12%)   (60%)

Thiophenium ylides. The reaction of dimethyl diazomalonate with thiophene with conventional copper catalysts is extremely slow, but does result in dimethyl thiophene-2-malonate (2) in about 35% yield. When rhodium(II) acetate is used as catalyst, the reaction at 20° is still slow (18 hours), but results in thiophenium bismethoxycarbonylmethylide (1) in 95% yield. This reaction is facilitated by halogen substituents; the ylides are not formed from thiophenes substituted with cyano, formyl, or acetyl groups.[2]

When 1 is heated at about 80° it rearranges to 2. This product can be obtained in one step by conducting the initial reaction with rhodium(II) acetate at the higher temperature. The rearrangement of 1 to 2 is intramolecular; a free carbene is not formed.[3]

[1] N. Petiniot, A. J. Anciaux, A. F. Noels, A. J. Hubert, and Ph. Teyssié, *Tetrahedron Letters*, 1239 (1978).

[2] R. J. Gillespie, J. Murray-Rust, P. Murray-Rust, and A. E. A. Porter, *J.C.S. Chem. Comm.*, 83 (1978).

[3] R. J. Gillespie, A. E. A. Porter, and W. E. Willmott, *ibid.*, 85 (1978).

Rhodium(III) chloride, 2, 357; **3,** 242–243; **6,** 504; **7,** 313–314.

Migration of double bonds. Rhodium trichloride trihydrate is useful for certain exocyclic–endocyclic migrations of a double bond that are usually difficult to achieve. Some examples are formulated.[1] The isomerization of ergosterol is interesting in that it yields a previously unknown coprostatrienol together with ergosterols B_1 and B_2.

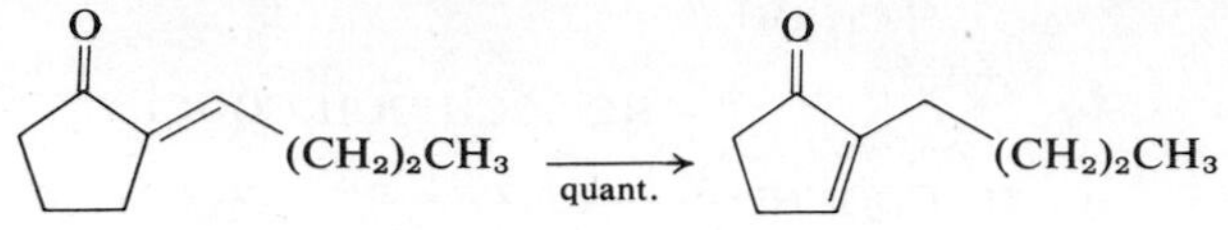

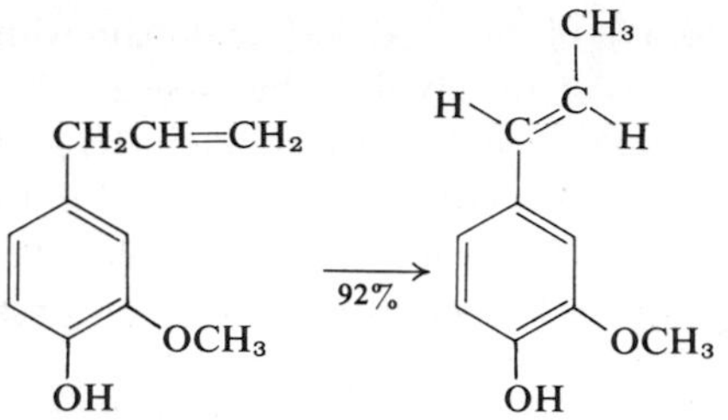

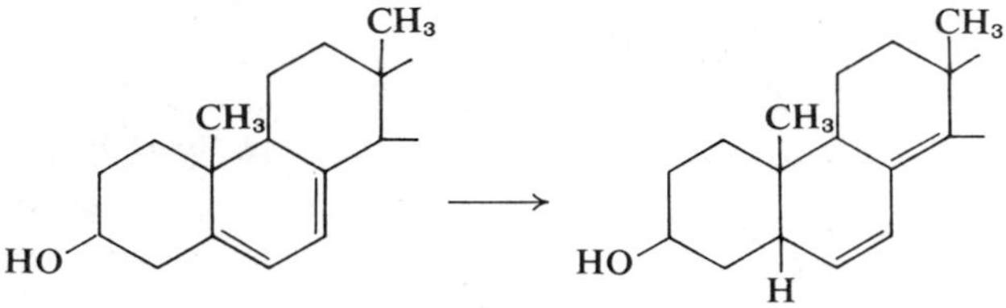

+ ergosterols B_2 and B_1

Aromatization. Cyclohexenones substituted with an unsaturated side chain are converted into phenols when heated in ethanol at 100° (sealed tube) with a catalytic amount of this transition metal salt. The double bond can be remote because $RhCl_3$ catalyzes isomerization of double bonds to give more stable systems (**3**, 242–243).

Examples:

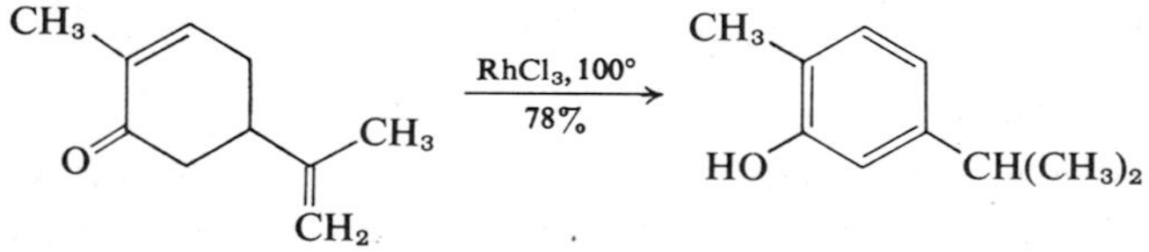

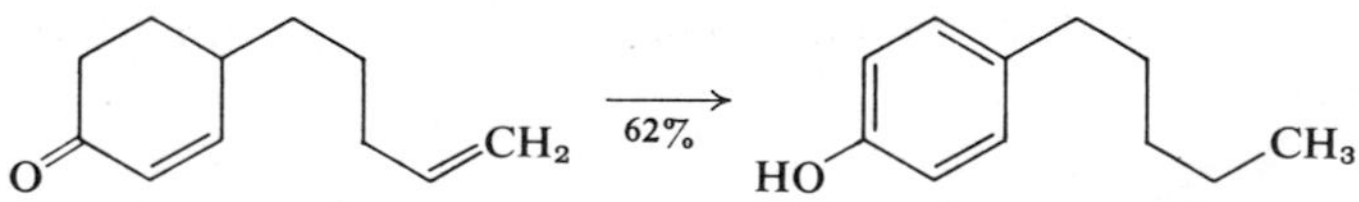

Related unsaturated imines are aromatized to aniline derivatives in the same way except that K_2CO_3 is added to prevent hydrolysis to the enone.[2]

Example:

Cleavage of allylamines. French chemists[3] recommend the allyl group for protection of amino groups. This group has been used extensively for protection of hydroxyl groups; and in this case deprotection is accomplished by isomerization to the enol ether followed by hydrolysis. Chlorotris(triphenylphosphine)-rhodium is the reagent of choice for isomerization of allyl ethers (**5**, 736). In the case of allylamines $RhCl_3 \cdot 3H_2O$ in ethanol–water is preferred because isomerization to the enamine is more rapid and hydrogenated products are not formed. The resulting enamines are rapidly hydrolyzed by dilute HCl in methanol. The allyl group offers advantages over the benzyl group, which requires more drastic conditions for cleavage (48% HBr at reflux) when hydrogenation is not feasible.

$$(C_6H_5)_2NCH_2CH{=}CH_2 \xrightarrow{RhCl_3 \cdot 3H_2O} [(C_6H_5)_2NCH{=}CHCH_3] \xrightarrow[\text{quant.}]{H^+} (C_6H_5)_2NH$$

[1] J. Andrieux, D. H. R. Barton, and H. Patin, *J.C.S. Perkin I*, 359 (1977).
[2] P. A. Grieco and N. Marinovic, *Tetrahedron Letters*, 2545 (1978).
[3] B. Moreau, S. Lavielle, and A. Marquet, *ibid.*, 2591 (1977).

Rhodium(III) oxide, Rh_2O_3. Mol. wt. 253.81. Suppliers: Alfa, Fluka.

Hydroformylation. Cobalt catalysts are commonly used in industry for this reaction, but rhodium catalysts are more active and more selective. Other less accessible rhodium catalysts have been found to be effective: $Rh_4(CO)_{12}$ and $HRh(CO)[P(C_6H_5)_3]_3$.[1]

Example:

[1] P. Pino and C. Botteghi, *Org. Syn.*, **57**, 11 (1977).

Ruthenium(III) chloride, 4, 421.

Oxygenation of amines and secondary alcohols. $RuCl_3$ (hydrated) catalyzes the oxygenation of primary and secondary amines and secondary alcohols in toluene (100°). Addition of water inhibits formation of alkenes in the case of alcohols.

Examples:

$$C_6H_5CH_2NH_2 + O_2 \xrightarrow{RuCl_3} C_6H_5C{\equiv}N + C_6H_5CONH_2$$
$$(53\%) \qquad\qquad (30\%)$$

$$CH_3(CH_2)_3\underset{\underset{NH_2}{|}}{C}HCH_3 + O_2 \longrightarrow CH_3(CH_2)_3\underset{\underset{O}{\|}}{C}CH_3 + CH_3(CH_2)_3\underset{\underset{NH}{\|}}{C}CH_3$$
$$(39\%) \qquad\qquad (31\%)$$

$$(76\%) \qquad (6\%)$$

$$CH_3(CH_2)_5\underset{\underset{OH}{|}}{C}HCH_3 \xrightarrow{70\%} CH_3(CH_2)_5\underset{\underset{O}{\|}}{C}CH_3$$

[1] R. Tang, S. E. Diamond, N. Neary, and F. Mares, *J.C.S. Chem. Comm.*, 562 (1978).

Ruthenium tetroxide, 1, 986–989; **2**, 357–359; **3**, 243–244; **4**, 420–421; **6**, 504–506; **7**, 315.

Sulfoximines. S,S-Dimethyl-N-sulfonyl- and N-acysulfilimines are oxidized to sulfoximines by RuO_4, generated *in situ* from RuO_2 and $NaIO_4$.[1]

$$(CH_3)_2S{=}NR \xrightarrow[85-95\%]{RuO_4, CH_2Cl_2} (CH_3)_2\overset{\overset{O}{\|}}{S}{=}NR$$

$$(R = SO_2Ar, COAr, COOR)$$

[1] H. S. Veale, J. Levin, and D. Swern, *Tetrahedron Letters*, 503 (1978).

S

Samarium(II) iodide, SmI_2. Mol. wt. 404.17. The salt can be prepared in quantitative yield by reaction of the metal with solutions of 1,2-diiodoethane in THF:

$$Sm + ICH_2CH_2I \longrightarrow SmI_2 + CH_2{=}CH_2$$

The same reaction can be used to prepare ytterbium diiodide, YbI_2.

Reactions.[1] SmI_2 (or YbI_2) in THF–CH_3OH reduces α,β-unsaturated acids and esters to saturated acids and esters in high yield. Of more interest, aldehydes are much more easily reduced than ketones so that selective reduction is possible. In the presence of an alkyl halide ketones undergo a Grignard-like reaction:

$$n\text{-}C_6H_{13}COCH_3 + CH_3I \xrightarrow[95\%]{SmI_2} n\text{-}C_6H_{13}\overset{\displaystyle CH_3}{\underset{\displaystyle OH}{\overset{|}{\underset{|}{C}}}}\text{---}CH_3$$

[1] J. L. Namy, P. Girard, and H. B. Kagan, *Nouveau J. Chim.*, **1**, 5 (1977).

Selenium, 1, 990–992; **4,** 222; **5,** 575; **6,** 507–509.

Review. Clive[1] has reviewed new uses for this metal for synthesis of seleno-ketones, ureas, carbonates, formanilides, and carbamates (16 references).

[1] D. L. J. Clive, *Tetrahedron*, **34**, 1121 (1978).

Selenium dioxide, 1, 992–1000; **2,** 360–362; **3,** 245–247; **4,** 422–424; **5,** 575–576; **6,** 509–510; **7,** 319.

Removal of colloidal selenium (**1,** 992–993). The red colloidal form of Se, often obtained in oxidations with SeO_2, when heated in DMF precipitates as black granules, which are removable by filtration.[1]

SeO_2–pyridine allylic oxidation.[2] Allylic oxidation of substrates containing ethylene ketal derivatives with SeO_2 by the usual procedure is attended by extensive loss of the protecting group. The problem can be overcome by addition of pyridine to the reaction. The second example illustrates allylic oxidation of a substrate containing a tetrahydropyranyl ether group.

Examples:

$$\overset{CH_3}{\underset{CH_3}{\diagdown}}\!\!\!\diagup\!\!\!\diagdown\!\!\!\overset{CH_3}{\diagup}\!\!\!\diagdown\!\!\!\diagup\!\!OTHP \xrightarrow[35\%]{} HOCH_2\!\!\!\diagup\!\!\!\overset{CH_3}{\diagdown}\!\!\!\diagup\!\!\!\overset{CH_3}{\diagdown}\!\!\!\diagup\!\!OTHP$$

[1] S. R. Milstein and E. A. Coats, *Aldrichim. Acta*, **11**, 10 (1978).
[2] F. Camps, J. Coll, and A. Parente, *Synthesis*, 215 (1978).

Selenium dioxide–Hydrogen peroxide.

Allylic oxidation.[1] Selenium dioxide is the classical reagent for allylic hydroxylation (**1**, 994–996), but it has the drawback that selenium compounds are about as toxic as arsenic compounds. However, SeO_2 need be used in only catalytic amounts if used in conjunction with H_2O_2 (50%). Thus β-pinene (**1**) can be converted into *trans*-pinocarveol (**2**) in 49–55% yield by oxidation with SeO_2–H_2O_2 in *t*-butyl alcohol. This product has been prepared in somewhat higher yield

by isomerization of the oxide of α-pinene with lithium diethylamide (**4**, 298).[2] The two methods are complementary in that the overall transformation differs (equation I).

[1] J. M. Coxon, E. Dansted, and M. P. Hartshorn, *Org. Syn.* **56**, 25 (1977).
[2] J. K. Crandall and L. C. Crawley, *ibid.*, **53**, 17 (1973).

Silver acetate, 1, 1002–1004; 2, 362–363; 6, 511.

Ring contraction of cyclobutyl ketones. The cyclobutyl ketone **1** has been transformed into the cyclopropyl ketone **3** by bromination and treatment of the epimeric bromo ketones **2** with silver acetate in acetic acid. The desired product (**3**) is accompanied by the acetoxy ketone **4**, a product of ring cleavage.[1]

This ring contraction was used in a biogenetic-type synthesis of illudadiene (**6**) from **5**.

        5   **6**

[1] Y. Ohfune, S. Misumi, A. Furusaki, H. Shirahama, and T. Matsumoto, *Tetrahedron Letters*, 279 (1977).

Silver carbonate, 1, 1005; **2**, 363; **4**, 425; **6**, 511.

    Di-t-alkyl ethers. These ethers can be prepared in about 30–75% yield by reaction of *t*-alkyl chlorides with silver carbonate (1.1 equiv.) in pentane at 20°. Silver(I) oxide is somewhat less satisfactory than Ag_2CO_3; HgO, ZnO, $ZnCO_3$, PbO_2, and Tl_2O_3 are not useful in this reaction. The yields of ethers are very dependent on the degree of bulkiness of the R groups.[1]

[1] H. Masada and T. Sakajira, *Bull. Chem. Soc. Japan*, **51**, 866 (1978).

Silver carbonate–Celite, 2, 363; **3**, 247–249; **4**, 425–428; **5**, 577–580; **6**, 511–514; **7**, 319–320.

    Oxidation of a 1,4-diol to a lactone (**6**, 512). The epoxy diol **1** is oxidized by Ag_2CO_3–Celite (15 equiv.) in refluxing benzene to the epoxy lactone **2**. The reaction was used in a total synthesis of *dl*-cerulenin (**4**), an antibiotic and also an inhibitor of lipid synthesis. The aminolysis of **2** gives the amide alcohol **3** in almost quantitative yield. The last step in the synthesis is oxidation with pyridinium chlorochromate.[1]

[1] R. K. Boeckman, Jr., and E. W. Thomas, *Am. Soc.*, **99**, 2805 (1977).

Silver chromate–Iodine, Ag_2CrO_4–I_2. Silver chromate is prepared[1] by reaction of silver nitrate with potassium chromate in water. Silver chromate precipitates and is washed with water and dried *in vacuo* at 90°.

α-*Iodo ketones*. α-Iodo ketones are usually prepared by reaction of enol acetates with NIS. A new method involves reaction of an alkene with silver chromate and iodine in the presence of pyridine. The actual reagent is presumably a mixed anhydride of hypoiodous acid and chromic acid. Dichloromethane is the most useful solvent.[2]

Examples:

$$C_6H_{13}CH{=}CH_2 \xrightarrow[74\%]{} C_6H_{13}\overset{\overset{\displaystyle O}{\|}}{C}CH_2I$$

$$C_6H_5CH{=}CH_2 \xrightarrow[86\%]{} C_6H_5\overset{\overset{\displaystyle O}{\|}}{C}CH_2I$$

[1] R. B. Turner, V. R. Mattox, L. L. Engel, R. F. McKenzie, and E. C. Kendall, *J. Biol. Chem.*, **166**, 345 (1946).
[2] G. Cardillo and M. Shimizu, *J. Org.*, **42**, 4268 (1977).

Silver cyanide, 1, 1006; **5,** 581; **7,** 320.

N-Acyl isocyanides.[1] These unstable compounds can be prepared readily by treatment of carboxylic acid iodides with silver cyanide in CH_2Cl_2 at 20°. N-Benzoyl isocyanide was formed in this way in 73% yield.

$$\overset{\overset{\displaystyle O}{\|}}{R}CI + AgCN \longrightarrow \overset{\overset{\displaystyle O}{\|}}{R}CN{=}C$$

[1] G. Höfle and B. Lange, *Angew. Chem., Int. Ed.*, **16**, 262 (1977).

Silver(I) oxide, 1, 1011; **2,** 368; **3,** 252–254; **4,** 430–431; **5,** 583–585; **6,** 515–518; **7,** 321–322.

o-Quinone methides. *o*-Quinone methides have been postulated as intermediates in phenol oxidations; two crystalline substances of this type have now been characterized. One (**2**) is obtained by oxidation of the phenol **1** with silver oxide; the other (**4**) is obtained by similar oxidation of **3**. Silver oxide can be replaced by DDQ, but then purification of the methides is complicated by the presence of DDQH.[1]

Diels-Alder reactions of o-benzoquinones. 3-Substituted catechols (**1**) in the presence of an oxidizing reagent (Ag_2O, MnO_2, NiO_2) undergo Diels-Alder reactions with 2,3-dimethylbutadiene to give bis adducts (**2**). The product (**2**, R = COOH) has been converted into the phenanthrenequinone **3**.[2]

[1] L. Jurd, *Tetrahedron*, **33**, 163 (1977).
[2] R. Al-Hamdany and B. Ali, *J.C.S. Chem. Comm.*, 397 (1978).

Silver tetrafluoroborate, 1, 1015–1018; **2**, 365–366; **3**, 250–251; **4**, 428–429; **5**, 587–588; **6**, 519–520.

  Rearrangement of Diels-Alder adducts from 2-chloroacrylonitrile. 2-Chloro-acrylonitrile has been used as a ketene equivalent in Diels-Alder reactions (**3**,

66–67; **4**, 76–77). The adducts **2** and **3** from the Diels-Alder reaction of 2-chloro-acrylonitrile with 5-methoxy-4,7-dihydroindane are rearranged uniquely by silver ion into novel tricyclic bicyclo[3.2.1]octane derivatives **4** and **5**.[1]

[1] Y. Yamada, M. Kimura, H. Nagaoka, and K. Ohnishi, *Tetrahedron Letters*, 2379 (1977).

Silver(I) trifluoroacetate, $AgOCOCF_3$, **1**, 1018–1019; **7**, 323–324.

Synthesis of esters and lactones (**6**, 581–582). Masamune *et al.*[1] have reported further studies on the synthesis of esters by the reaction of a 2-methylpropane-2-thioester with an alcohol by activation of **1** with a metal salt (equation I). Originally mercury(II) trifluoroacetate and mercury(II) methanesulfonate were

(I)

$$R^1C(\!=\!O)\!-\!SC(CH_3)_3 + R^2OH \xrightarrow{Hg(II)} R^1C(\!=\!O)\!-\!OR^2$$

used to effect activation of the thioester for this $S \to O$ conversion. This method is fairly general for synthesis of macrolides, but cannot be used when R^1 or R^2 contains activated double bonds (which undergo oxymercuration) or other electron-rich centers. Fortunately other less thiophilic metal cations can also be used and are often as effective as Hg(II) salts. Examples are salts of Cu(I), Cu(II), and Ag(I). Actually silver tetrafluoroborate had already been used for this purpose (**6**, 519–520).

From this study of possible variations, Masamune was able to achieve the first partial synthesis of the macrolides cytochalasin A and B, **2** and **3**, by use of silver(I) trifluoroacetate for activation of a benzenethiolate ester, $RCOSC_6H_5$.

Masamune has reviewed macrolide synthesis.[2]

[1] S. Masamune, Y. Hayase, W. Schilling, W. K. Chan, and G. S. Bates, *Am. Soc.*, **99**, 6756 (1977).
[2] S. Masamune, *Aldrichim. Acta*, **11**, 23 (1978).

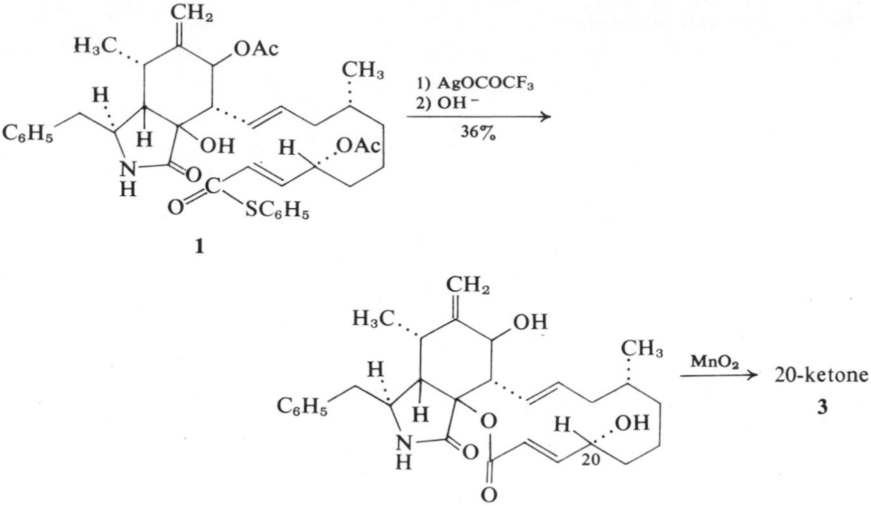

1

2

Simmons-Smith reagent, 1, 1019–1022; **2,** 371–372; **3,** 255–256; **4,** 436–437; **5,** 588–589; **6,** 521–523.

Reaction with allylic alcohols. French chemists[1] have observed notable stereo-specificity in the reaction of the Simmons-Smith reagent with acyclic (Z)-allylic alcohols (equation I). In the reaction of the (E)-isomers, both *erythro-* and *threo-*cyclopropyl alcohols are formed, with a slight bias for the former products.

(I)

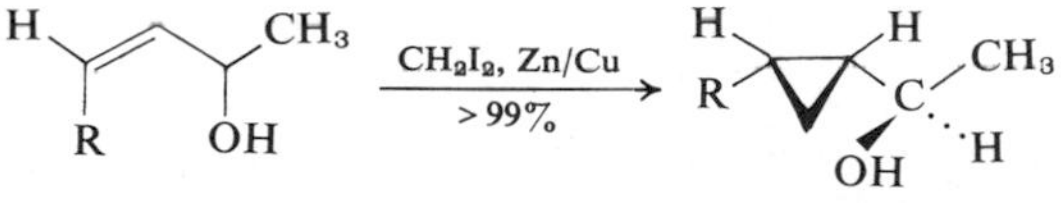

(*erythro*)

[1] M. Ratier, M. Castaing, J.-Y. Godet, and M. Pereyre, *J. Chem. Res. (M),* 2309 (1978).

Sodium, dispersed on charcoal or graphite. Sodium has less tendency to be inter-calated in graphite than potassium (**4,** 397), but it can be absorbed on charcoal or graphite to produce a very reactive, pyrophoric form. This material (and also C_8K) can be used to generate enolates of ketones that can differ somewhat in behavior from enolates generated in solution. In general, such enolates are mono-alkylated to a greater extent than usual. The main competing side reaction is reduction, but this reaction is less important than it is with C_8K.[1]

[1] H. Hart, B. Chen, and C. Peng, *Tetrahedron Letters,* 3121 (1977).

Sodium–Chlorotrimethylsilane.

Trimethylsilyl enol ethers of acyltrimethylsilanes. These useful silanes (3) can be prepared from benzenethiol esters (1) by conversion into the silyl enol ethers

(2) in the usual way followed by treatment with sodium dispersion and chlorotrimethylsilane in refluxing benzene. This reductive silylation is not applicable to alkanethiol esters.[1]

[1] I. Kuwajima, M. Kato, and T. Sato, *J.C.S. Chem. Comm.*, 478 (1978).

Sodium aluminum chloride, 1, 1027–1029; 2, 372; 4, 438; 6, 524–525.

Cyclodehydrogenation. 1-Phenylpyrene (1) is converted into indeno[1.2.3-*cd*]-pyrene (2) when heated at 270° for 4 minutes with equal weights of NaCl and AlCl$_3$ and 10% of activated carbon.[1]

[1] P. Studt, *Ann.*, 528 (1978).

Sodium amide, 1, 1034–1041; 2, 373–374; 4, 439; 5, 591–593; 6, 525–526.

1-Ethoxy-1-butyne (5, 592). The preparation of 1-ethoxy-1-butyne from chloroacetaldehyde diethyl acetal has been published.[1]

Haller-Bauer cleavage.[2] 2-Benzoyl-2-phenylbicyclo[1.1.1]pentane (1) is cleaved to 2-phenylbicyclo[1.1.1]pentane (2) by sodium amide in refluxing benzene.[3] The product can be converted into the carboxylic acid 3.

[1] M. S. Newman and W. M. Stalick, *Org. Syn.*, **57**, 65 (1977).
[2] K. E. Hamlin and A. W. Weston, *Org. React.*, **9**, 1 (1957).
[3] E. C. Alexander and T. Tom, *Tetrahedron Letters*, 1741 (1978).

$$1 \quad \xrightarrow[\text{quant.}]{\text{NaNH}_2} \quad \text{H}_2\text{NCOC}_6\text{H}_5 \; + \quad 2 \quad \xrightarrow[> 85\%]{\text{RuO}_4} \quad 3$$

Sodium benzeneselenolate, 5, 273; **6**, 548–549.

Uncomplexed reagent. Liotta et al.[1] have generated this reagent by the two processes shown in equation (I); this material is insoluble in ethanol–THF, unlike the reagent generated from diphenyl diselenide and sodium borohydride. Evidently the latter reagent is complexed with borane: $\text{Na}^+[\text{Se}(\text{C}_6\text{H}_5)\text{BH}_3]^-$. In fact material prepared according to equation (I) can be converted into the complexed reagent by reaction with diborane and ethanol in THF.

(I) $\text{C}_6\text{H}_5\text{SeSeC}_6\text{H}_5 \xrightarrow{\text{Na, THF}} \text{Na}^{+-}\text{SeC}_6\text{H}_5 \xleftarrow{\text{NaH, THF}} \text{C}_6\text{H}_5\text{SeH}$

Ester cleavage.[1] The uncomplexed reagent solubilized in THF with HMPT cleaves esters to the corresponding carboxylic acids, in essentially quantitative yield, in reactions conducted at reflux for 7–24 hours.

Cleavage of lactones.[2,3] Two groups have reported cleavage of lactones; one group used uncomplexed reagent solubilized with either HMPT or 18-crown-6 at 67°. The other used complexed reagent in DMF at 110–120°. Essentially similar results are obtained by both approaches. Five- and six-membered lactones are cleaved readily in high yield; larger lactones require longer reaction times. Lactones substituted at the carbinolic carbon (last example) resist this cleavage. The products can be converted by usual methods into ω-vinylcarboxylic esters (first example).

Examples:

no reaction

Fragmentation of chiral selenoxides. Selenoxide fragmentation of 7β-phenyl-selenocholesteryl benzoate (**1**) gives about equal amounts of **2** and **3**; fragmentation of the epimeric phenylselenide **4** gives only one product (**5**). The differing behavior is explained as a result of chirality at selenium of the intermediate selenoxides.[4]

[1] D. Liotta, W. Markiewicz, and H. Santiesteban, *Tetrahedron Letters*, 4365 (1977).
[2] D. Liotta and H. Santiesteban, *ibid.*, 4369 (1977).
[3] R. M. Scarborough, Jr., and A. B. Smith III, *ibid.*, 4361 (1977).
[4] W. G. Salmond, M. A. Barta, A. M. Cain, and M. C. Sobala, *ibid.*, 1683 (1977).

Sodium bis-(2-methoxyethoxy)aluminum hydride (SMEAH), 3, 260–261; **4**, 441–442; **5**, 596; **6**, 528–529; **7**, 327–329.

α-*Hydroxy aldehydes.* A new synthesis of these compounds from the next lower homologous aldehyde or ketone involves conversion to the cyanohydrin. The hydroxyl group is then protected as the acetal; partial reduction to the imide and hydrolysis then affords the α-hydroxy aldehyde. Sodium bis-(2-methoxy-ethoxy)aluminum hydride is superior to lithium aluminum hydride for the reduction step. An example of the method is shown in equation (I).[1]

[1] M. Schlosser and Z. Brich, *Helv.*, **61**, 1903 (1978).

(I)

$$\underset{\substack{(CH_3)_2CH \quad\quad CN}}{\overset{\substack{H \quad\quad OH}}{C}} \xrightarrow[\substack{97\%}]{\substack{C_2H_5OCH=CH_2,\ HCl,\ 25°}}$$

$$\underset{\substack{(CH_3)_2CH \quad\quad CN}}{\overset{\substack{OC_2H_5\\ |\\ CHCH_3\\ H \quad\quad O}}{C}} \xrightarrow[\substack{70\%}]{\substack{SMEAH,\ ether,\ H^+}} \underset{\substack{(CH_3)_2CH \quad\quad CHO}}{\overset{\substack{OC_2H_5\\ |\\ CHCH_3\\ H \quad\quad O}}{C}} \xrightarrow[\substack{90\%}]{\substack{p\text{-}TsOH,\ CH_3OH}}$$

$$\underset{\substack{(CH_3)_2CH \quad\quad CHO}}{\overset{\substack{H \quad\quad OH}}{C}}$$

Sodium borohydride, 1, 1049–1055; **2,** 377–378; **3,** 262–264; **4,** 443–444; **5,** 597–601; **6,** 530–534; **7,** 329–331.

Cleavage of epoxides.[1] *vic,cis*-Epoxy alcohols are inert to sodium borohydride, but methoxyhydrins can be obtained by reaction of some *vic,trans*-epoxy alcohols with sodium borohydride in methanol. This cleavage requires a quasi diaxial orientation of the OH group with the epoxide.

Examples:

$$\xrightarrow[\substack{quant.}]{\substack{NaBH_4,\ CH_3OH}}$$

$$\xrightarrow{\ \ \ \ }$$

$$\xrightarrow[\substack{(slow)}]{\ \ \ \ }$$

Selective reduction of aldehydes.[2] An aldehyde group can be reduced selectively in the presence of a ketone group by NaBH₄ in the presence of an added thiol (ethyl mercaptan, *t*-butyl mercaptan; *see* **6,** 532).

Reduction of a mesylate. The last step in a total synthesis of the lignan ($\pm$)-schizandrin (**2**) involved reduction of the secondary hydroxyl group of **1**. This was accomplished by reduction of the monomesylate of **1** with $NaBH_4$ in DMF at 80°. Reduction of the mesylate with lithium triethylborohydride or with

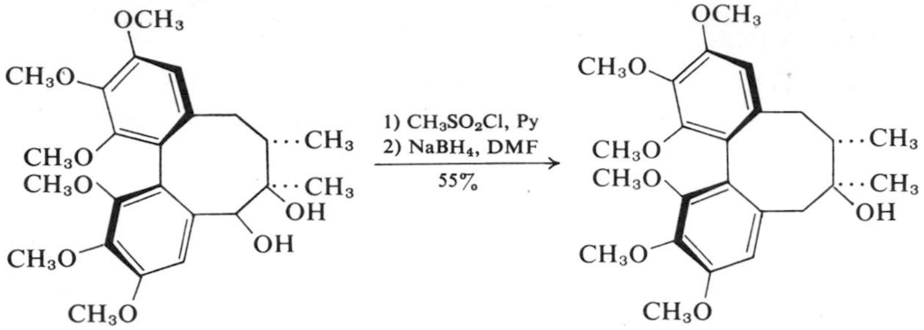

1) CH_3SO_2Cl, Py
2) $NaBH_4$, DMF
55%

1   **2**

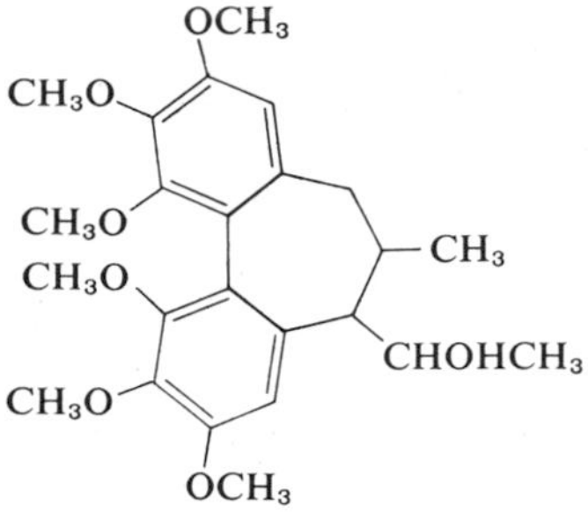

3

lithium aluminum hydride gave exclusively the Wagner-Meerwein rearranged alcohol **3**.[3]

Reductive denitration of α-nitroepoxides. Carbohydrates containing this grouping are reduced by $NaBH_4$ to α-hydroxy-β-deoxy derivatives in high yields. The presence of a nitro group controls the regioselectivity of attack by the reducing agent so that the β-carbon atom is attacked; thus the rule of diaxial opening of epoxide rings is not necessarily followed. Differing behavior of substrates indicates that the reactivity depends on the facility of approach by the nucleophile.[4]

Examples:

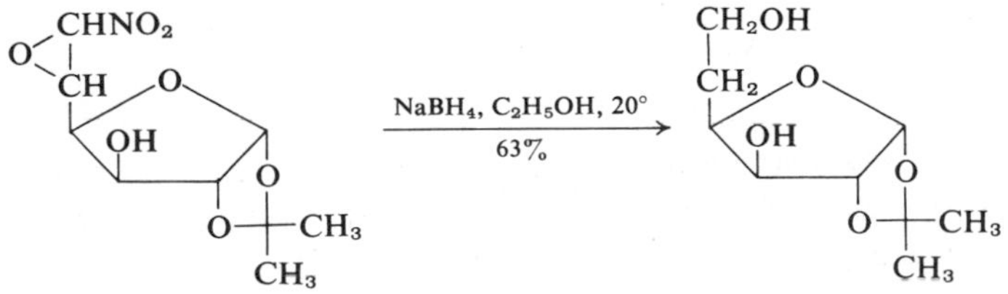

$NaBH_4$, C_2H_5OH, 20°
63%

Aryl sulfamates.[5] Aryl esters of sulfamic acid (3) can be prepared conveniently by reduction of aryloxysulfonyl azides[6] (2) with sodium borohydride in THF at 0–5°.

$$Ar—OSO_2Cl \xrightarrow[90–98\%]{NaN_3} Ar—OSO_2N_3 \xrightarrow[50–75\%]{NaBH_4,\ THF} Ar—OSO_2NH_2$$

$$\mathbf{1} \qquad\qquad\qquad \mathbf{2} \qquad\qquad\qquad \mathbf{3}$$

RCOCl → RCHO. Acid chlorides are reduced at 0° to aldehydes in 60–75% yield by sodium borohydride in the presence of cadmium chloride and DMF. Yields are low in the absence of either the salt or DMF. Some of the corresponding alcohol is usually formed.[7]

Reduction of di- and triarylmethanols. These substrates can be reduced to hydrocarbons by $NaBH_4$ in an acidic medium such as TFA generally in high yield.[8]

[1] M. Weissenberg, P. Krinsky, and E. Glotter, *J.C.S. Perkin I*, 565 (1978).
[2] Y. Maki, K. Kikuchi, H. Sugiyama, and S. Seto, *Tetrahedron Letters*, 263 (1977).
[3] E. Ghera and Y. Ben-David, *J.C.S. Chem. Comm.*, 480 (1978).
[4] H. H. Baer and C. B. Madumelu, *Canad. J. Chem.*, **56**, 1177 (1978).
[5] M. Hedayatullah and A. Guy, *Synthesis*, 357 (1978).
[6] *Idem, Tetrahedron Letters*, 2455 (1975).
[7] R. A. W. Johnstone and R. P. Telford, *J.C.S. Chem. Comm.*, 354 (1978).
[8] G. W. Gribble, R. M. Leese, and B. E. Evans, *Synthesis*, 172 (1977).

Sodium borohydride–Cerium(III) chloride.

1,2-Reduction of enones. Sodium borohydride in combination with catalytic amounts of cerium chloride reduces α,β-unsaturated ketones selectively to allylic alcohols. Reactions proceed within 5 minutes even at 20°.[1,2]
Examples:

Similar reduction of prostanoids is regioselective but not stereoselective. Reduction of PGA$_2$ methyl ester (1) affords an equimolar mixture of the diols 2a and 2b.[2]

2a (R^1 = H; R^2 = OH)
2b (R^1 = OH; R^2 = H)

[1] J.-L. Luche, *Am. Soc.*, **100**, 2226 (1978).
[2] J.-L. Luche, L. Rodriquez-Hahn, and P. Crabbé, *J.C.S. Chem. Comm.*, 601 (1978).

Sodium borohydride–Tin(IV) chloride etherate. The tin complex is prepared by addition of SnCl$_4$ dropwise to ether; the complex precipitates and is isolated by filtration. It is hygroscopic and should be used immediately.

Reduction of amides. The Borch method for reduction of *sec-* and *t*-amides to amines (2, 430–431) is improved by use of NaBH$_4$ and SnCl$_4$·2(C$_2$H$_5$)$_2$O in the ratio 4:1. Under these conditions even primary amides are reduced to amines.[1] Examples:

$$C_6H_5\overset{\overset{\displaystyle O}{\|}}{C}NH_2 \xrightarrow{(C_2H_5)_3O^+BF_4^-} C_6H_5\overset{\overset{\displaystyle OC_2H_5}{|}}{C}{=}NH \xrightarrow[> 50\%]{NaBH_4,\ SnCl_4 \cdot 2(C_2H_5)_2O} C_6H_5CH_2NH_2$$

$$\sim 79\%$$

[1] Y. Tsuda, T. Sano, and H. Watanabe, *Synthesis*, 652 (1977).

Sodium cyanide, 4, 446–447; **5,** 606–607; **6,** 535–536; **7,** 333.

γ-*Keto nitriles* (**4,** 447; **6,** 535). Stetter[1] has reviewed a wide range of reactions that involve addition of aldehydes to activated double bonds and that result in γ-diketones and γ-keto esters. A route to γ-keto nitriles involves addition of aromatic aldehydes to α,β-unsaturated nitriles with catalysis with sodium cyanide.

Selective ring opening of cyclopropanes. Ring opening by nucleophiles of suitably activated cyclopropanes has gained importance during recent years. Sodium or potassium cyanide can be used for this purpose. An example is the conversion of **1** or **2** into **3**, *anti*-7-cyano-*endo*-5-bromobicyclo[2.2.1]heptane-2-one, by reaction with potassium cyanide in the presence of catalytic amounts of sodium methoxide (equation I). The intermediate **2** can be isolated in high yield by reaction of **1** with 1 equiv. of sodium hexamethyldisilazide (**4,** 407) in ether.[2]

(I)

A related selective ring cleavage has been used in a novel synthesis of methyl jasmonate (**9**) from methyl 2-oxobicyclo[3.1.0]hexane-1-carboxylate (**4**). Thus reaction of **4** with sodium cyanide in DMSO at 10–20° results in selective cleavage of the cyclopropane ring. Use of acetone cyanohydrin catalyzed by triethylamine results in cleavage of both rings.[3]

(II)

4 → NaCN, DMSO 69% → **5** → BrCH₂C≡CC₂H₅, K₂CO₃ 85% →

6 → LiI, DMSO 85% → **7** → H₂, Lindlar cat. 90% →

8 → CH₃OH, HCl 84% → **9**

[1] H. Stetter, *Angew. Chem., Int. Ed.*, **15**, 639 (1976).
[2] J. C. Gilbert, T. Luo, and R. E. Davis, *Tetrahedron Letters*, 2545 (1975).
[3] K. Kondo, Y. Takahatake, K. Sugimoto, and D. Tunemoto, *ibid.*, 907 (1978).

Sodium cyanoborohydride, 4, 448–451; **5,** 607–609; **6,** 537–538; **7,** 334–335.

Reduction of dimethyl acetals and ketals. These substances are reduced by NaCNBH₃ and HCl (gas) in methanol at 0° to methyl ethers.[1]

Examples:

$$CH_3(CH_2)_8CH(OCH_3)_2 \xrightarrow[83\%]{NaCNBH_3,\ HCl,\ CH_3OH,\ 0°} CH_3(CH_2)_8CH_2OCH_3$$

$$CH_2=CHCH=CHCH(OCH_3)_2 \xrightarrow[46\%]{} CH_2=CHCH=CHCH_2OCH_3$$

[1] D. A. Horne and A. Jordan, *Tetrahedron Letters*, 1357 (1978).

Sodium dicarbonylcyclopentadienyl ferrate (1), 5, 610; **6,** 538–539.

β-*Lactams*. This reagent can be used to prepare cationic alkene complexes (2)[1] which are susceptible to nucleophilic attack by heteroatom containing molecules.

Migration of the resulting alkyl group to CO occurs with retention of configuration and can be induced by oxidation with Ag_2O or PbO_2. Further oxidation can lead to β-lactams, as is formulated in the example below. This sequence can also be used for synthesis of bicyclic β-lactams.[2]

[1] A. Cutler, D. Erntholt, W. P. Giering, P. Lennon, S. Raghu, A. Rosan, M. Rosenblum, J. Tanerede, and D. Wells, *Am. Soc.*, **98**, 3495 (1976).

[2] P. K. Wong, M. Madhavarao, D. F. Marten, and M. Rosenblum, *ibid*, **99**, 2823 (1977).

Sodium O,O-diethyl phosphorotelluroate, $(C_2H_5O)_2P(O)TeNa$ (**1**). Mol. wt. 287.70, unstable to air. The salt is prepared by addition of Te metal to $(C_2H_5O)_2PONa$ in C_2H_5OH.

Deoxygenation of epoxides. The reaction of a terminal epoxide and **1** in ethanol results in an exothermic reaction with formation of an olefin:

$$\underset{\text{R}}{\triangle\!\!\!\text{O}} \; + \; 1 \longrightarrow CH_2{=}CHR + Te + (C_2H_5O)_2PO_2Na$$
$$(70\text{–}90\%)$$

Internal epoxides react only slowly and, therefore selective deoxygenation of diepoxides is possible. The reaction can be carried out as a catalytic process by addition of $(C_2H_5O)_2PONa$ (1–2 equiv.) to a solution of the epoxide and Te (0.1 equiv.) in C_2H_5OH. The selenium salt corresponding to **1** is much less reactive in this deoxygenation and is less suitable for catalytic deoxygenations.[1]

[1] D. L. J. Clive and S. M. Menchen, *J.C.S. Chem. Comm.*, 658 (1977).

Sodium 4,6-diphenyl-1-oxido-2-pyridone, (**1**). Mol. wt.

285.29. This stable, crystalline salt is prepared by reaction of 1-hydroxy-4,6-diphenyl-2-pyridone[1] with $NaOCH_3$ in CH_3OH (1 hour reflux).

$RCH_2Br(Cl) \rightarrow RCHO$. Primary halides react with **1** in ethanol at 0° or at reflux to form the intermediates **2**, which when heated at 180–200° yield the aldehyde **3** and 4,6-diphenyl-2-pyridone (**4**). Yields are usually high in both steps (equation I).[2]

(I) $1 + RCH_2X \xrightarrow[75-99\%]{C_2H_5OH}$

                    2   **3**   **4**

[1] I. E.-S. El-Khaly, F. K. Rafla, and M. M. Mishrikey, *J. Chem. Soc.* (*C*), 1578 (1970).
[2] M. J. Cooke, A. R. Katritzky, and G. H. Millet, *Heterocycles*, **7**, 227 (1977).

Sodium dithionite, 1, 1081–1083; **5**, 615–617; **7**, 336.

Intramolecular Marschalk cyclization.[1] This reaction is a key step in a total synthesis of daunomycinone (**3**), the aglycone of an anthracycline antibiotic.[2] Thus treatment of **1** with $Na_2S_2O_4$ and NaOH in dioxane at 25°–90° results in

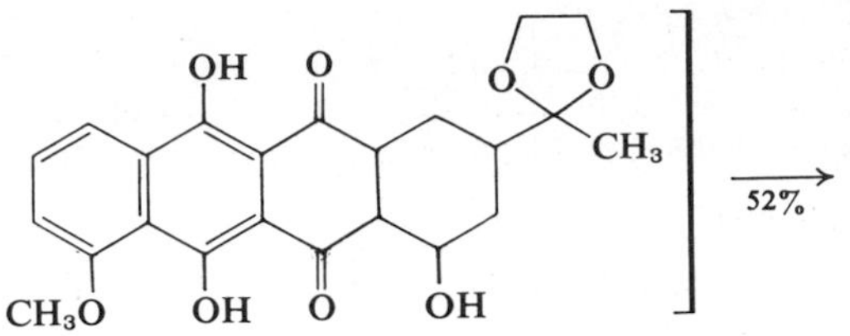

b

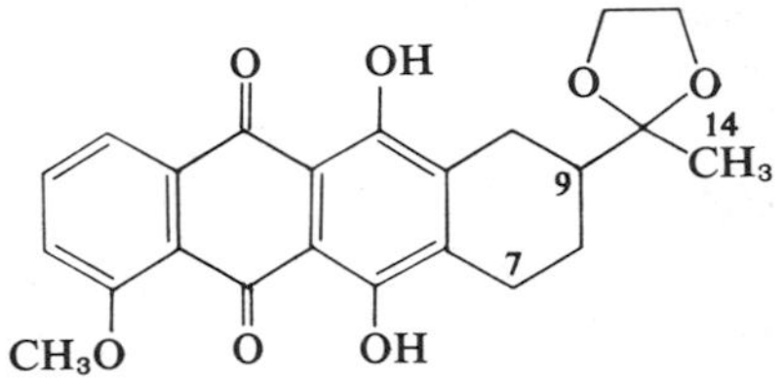

2

the desired product (**2**) in 52% yield; deketalization gives (±)-7,9-dideoxy-daunomycinone, which has been hydroxylated at C_7 and C_9 to give **3**.[3]

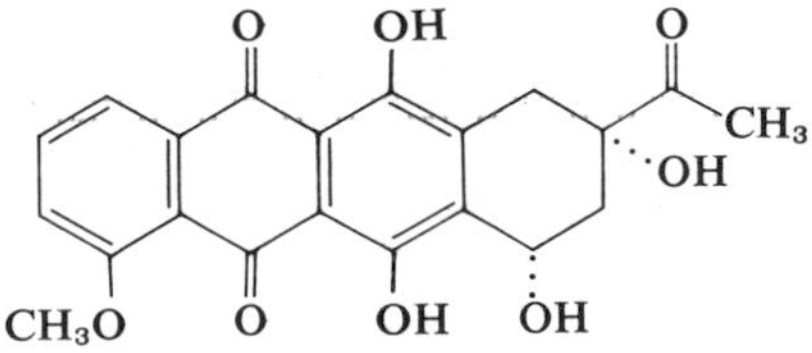

3

Reduction of aldehydes and ketones.[4] $Na_2S_2O_4$ is a useful alternative to the more expensive metal hydrides commonly used for this reduction. If the substrate is soluble in water, this medium is used; mixtures of H_2O with dioxane or DMF are also used. Sodium hydrogen carbonate is added to keep the reaction mixture basic.

Examples:

$$n\text{-}C_5H_{11}CHO \xrightarrow[67\%]{Na_2S_2O_4} n\text{-}C_5H_{11}CH_2OH$$

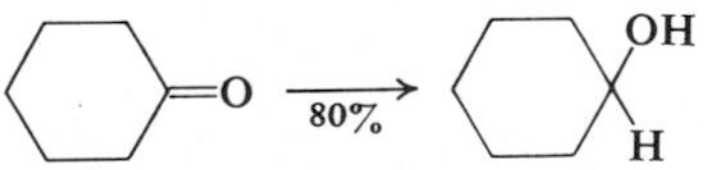

$$C_6H_5COCH_3 \xrightarrow[94\%]{} C_6H_5CHOHCH_3$$

$$n\text{-}C_6H_{13}COCH_3 \xrightarrow[89\%]{} n\text{-}C_6H_{13}CHOHCH_3$$

[1] C. Marschalk, *Bull. soc.*, **6**, 655 (1939).
[2] F. Suzuki, S. Trenbeath, R. D. Gleim, and C. J. Sih, *Am. Soc.*, **100**, 2272 (1978).
[3] A. S. Kende, Y. G. Tsay, and J. E. Mills, *ibid.*, **98**, 1967 (1976).
[4] J. G. de Vries, T. J. van Bergen, and R. M. Kellogg, *Synthesis*, 246 (1977).

Sodium formate, HCOONa.

Reduction of aryl bromides.[1] In a new method for this reaction, sodium formate serves as the hydride donor and tetrakis(triphenylphosphine)palladium(0) as catalyst. Sodium formate is generally superior to sodium methoxide because of ease of handling and compatibility with functional groups.

Examples:

[1] P. Helquist, *Tetrahedron Letters*, 1913 (1978).

Sodium hydride, **1**, 1075–1081; **2**, 380–383; **4**, 452–455; **5**, 610–614; **6**, 591–592.

Cyclization of N-allylamides to ε-lactams. Conia *et al.*[1] have reported cyclization of N-allylamides of type **1** to ε-lactams **2** and **3**. Certain conditions are mandatory for this cyclization. R^3 can be H or an alkyl group; R^1 and R^2 cannot be hydrogen. The nitrogen atom cannot be trisubstituted. A slight excess of base is required and no cyclization is observed when NaH is replaced by LiH, KH, or $NaNH_2$.

N-Allylamides of type **4** are cyclized thermally to γ-lactams (**5**).[2]

4　　　　　　**5**

[1] M. Bortolussi, R. Block, and J. M. Conia, *Tetrahedron Letters*, 2289 (1977).
[2] *Idem, Bull. soc.*, 2731 (1975).

Sodium hydrogen selenide, 6, 542–543.

Diselenides. Aromatic aldehydes are converted into diselenides in about 80% yield when treated with hydrogen selenide in the presence of morpholine (1 equiv.) followed by reduction with NaBH$_4$ (0.25 equiv.). However, H$_2$Se is extremely toxic. Fortunately the same reaction can be effected in comparable yield with NaHSe and piperidine hydrochloride (equation I).[1]

(I)　$ArCHO + NaHSe + HCl \cdot HN$ $\xrightarrow[\sim 80\%]{NaBH_4}$ $ArCH_2SeSeCH_2Ar$

Protection of carboxyl groups. 2-Bromoethyl and 2-chloroethyl esters of carboxylic acids are cleaved in 90–99% yield by NaHSe in C$_2$H$_5$OH (20°, 1 hour).[2]

[1] J. W. Lewicki, W. H. H. Günther, and J. Y. C. Chu, *J. Org.*, **43**, 2672 (1978).
[2] T.-L. Ho, *Syn. Comm.*, **8**, 301 (1978).

Sodium hydrogen telluride, NaHTe. The reagent is prepared *in situ* from Te powder and NaBH$_4$.[1]

Debromination.[2] *vic*-Dibromides are converted into alkenes when refluxed with the reagent in ethanol (2 hours). Yields are in the range 76–89%.
Examples:

Reduction of $\overset{\diagdown}{\underset{\diagup}{C}}=\overset{\diagup}{\underset{\diagdown}{C}}.$ On attempted debromination of **1** with NaTeH (1 equiv.) the product was **2**; yield 45%. Reduction with 2 equiv. of NaTeH increased the yield to 74%. The inference that **2** was formed by reduction of **a** was confirmed by a similar reduction of related alkenes.[3]

Examples:

$$C_6H_5CH=CHCOOCH_3 \xrightarrow[89\%]{} C_6H_5CH_2CH_2COOCH_3$$

[1] D. H. R. Barton and S. W. McCombie, *J.C.S. Perkin I*, 1574 (1975).
[2] K. Ramasamy, S. K. Kalyanasundaram, and P. Shanmugam, *Synthesis*, 311 (1978).
[3] *Idem, ibid.*, 545 (1978).

Sodium hydroxide, 1, 1083; **5**, 616–617; **7**, 336.

Bamford-Stevens reaction. Tosylhydrazones undergo the Bamford-Stevens reaction in a two-phase system (aqueous NaOH–dioxane) to give diazo compounds. Depending on the structure, the diazo compounds can be isolated or converted into alkenes or products of cyclization or cycloaddition.[1]

Examples:

$$C_6H_5CH=CHC=NNHTs \xrightarrow[95\%]{} $$

$$(C_6H_5)_2C=NNHTs \longrightarrow$$

$$C_6H_5CH=NNHTs \xrightarrow[59\%]{NaOH,\ DME,\ (C_4H_9)_4N^+Cl^-} C_6H_5CHN_2$$

$\diagdown CHNO_2 \rightarrow \diagdown C{=}O$. The Nef reaction[2] is the classical method for carrying out this transformation (equation I). Israeli chemists have devised a useful and

$$(I) \quad 2 \; \begin{array}{c} R^1 \\ \diagdown \\ R^2 \diagup \end{array}\!\!CHNO_2 \;\; \xrightarrow{2NaOH} \;\; \left[2 \; \begin{array}{c} R^1 \\ \diagdown \\ R^2 \diagup \end{array}\!\!C{=}\overset{+}{N} \begin{array}{c} O^- \\ \diagup \\ \diagdown \\ O^- \end{array} \right] \;\; \xrightarrow{2H_2SO_4}$$

$$2 \; \begin{array}{c} R^1 \\ \diagdown \\ R^2 \diagup \end{array}\!\!C{=}O + H_2O + 2NaHSO_4 + N_2O$$

mild version of this reaction that utilizes a basic silica gel. This material is prepared by mixing silica gel with a methanolic solution of NaOH, evaporation, and activation by heat (400°). The nitro compound is mixed with the gel for some hours at 25° or for a shorter period at 80°. The aldehyde or ketone is then eluted with ether. Yields are generally $>80\%$.[3]

Examples:

$$CH_3(CH_2)_5CH_2NO_2 \;\; \xrightarrow{87\%} \;\; CH_3(CH_2)_5CHO$$

$$CH_3COCH_2CH_2CH(NO_2)CH_3 \;\; \xrightarrow{81\%} \;\; CH_3COCH_2CH_2COCH_3$$

$$C_6H_5CH_2CH_2NO_2 \;\; \xrightarrow{60\%} \;\; C_6H_5CH_2CHO$$

This version of the Nef reaction was used to convert 1-nitromethylcyclohexene to 1-cyclohexenecarboxaldehyde in 82% yield (equation II).[4]

(II)

[1] A. Jończyk, J. Włostowska, and M. Mąkosza, *Bull. Soc. Chim. Belg.*, **86**, 739 (1977).
[2] N. E. Noland, *Chem. Rev.*, **55**, 137 (1955).
[3] E. Keinan and Y. Mazur, *Am. Soc.*, **99**, 3861 (1977).
[4] J. L. Hogg, T. E. Goodwin, and D. W. Nave, *Org. Prep. Proc. Int.*, **10**, 9 (1978).

Sodium hypochlorite, 1, 1084–1087; **2**, 67; **3**, 45, 243; **4**, 456; **5**, 617; **6**, 543.

Arene oxides. Many arenes (and azaarenes) can be converted directly into arene oxides by reaction in an organic solvent with aqueous sodium hypochlorite in the presence of a phase-transfer catalyst such as tetra-*n*-butylammonium hydrogen sulfate. The products generally result from reaction at the most reactive (K) region. Some of the oxides (and the yields) that have been prepared by this simple method are formulated. Preliminary results indicate that this method is

(90%)   (19%)   (10%)

applicable also to simple alkenes. A free-radical is believed to be involved. Arene oxides have been implicated in carcinogenesis and this reaction could be similar to enzymic oxidation of arenes.[1]

Epoxidation. Marmar[2] epoxidized two α,β-unsaturated ketones, benzalaceto-phenone and *trans*-dibenzoylethylene, in yields of >90% by using ordinary commercial hypochlorite bleach (Clorox) in pyridine solution. 1,4-Naphthoquin-one is also epoxidized in 71.5% yield with sodium hypochlorite in dioxane. In this case addition of pyridine results in further oxidation.

This method has attracted little attention but was a key reaction used by Tishler and co-workers[3] in a total synthesis of the antibiotic cerulenin (4), which contains a sensitive 4-keto-2,3-epoxy amide system. Thus the butenolide 1 under-goes selective epoxidation with sodium hypochlorite in pyridine to give the hy-droxy acid **a**, which lactonizes on warming to the epoxy lactone 2. Ammonolysis followed by oxidation with Collins reagent gives (±)-cerulenin (4).

Reaction with arylacetic acids. Alkaline sodium hypochlorite solution oxidizes arylacetic acids to an aldehyde and/or a carboxylic acid containing one carbon atom less than the starting material.[4]

Examples:

$$C_6H_5CH_2COOH \xrightarrow[40\%]{NaOCl, OH^-, 100°} C_6H_5CHO$$

$$p\text{-}ClC_6H_4CH_2COOH \xrightarrow[89\%]{} p\text{-}ClC_6H_4COOH$$

[1] S. Krishnan, D. G. Kuhn, and G. A. Hamilton, *Am. Soc.*, **99**, 8121 (1977).
[2] S. Marmar, *J. Org.*, **28**, 250 (1968).
[3] A. A. Jakubowski, F. S. Guziec, Jr., and M. Tishler, *Tetrahedron Letters*, 2399 (1977).
[4] F. Kaberia and B. Vickery, *J.C.S. Chem. Comm.*, 459 (1978).

Sodium methoxide, 1, 1091–1094; **2**, 385–386; **3**, 259–260; **4**, 457–459; **5**, 617–620.

1,2-Diones. Alkyl aryl ketones (**1**) can be converted into aryl-α',α'-dichloroalkyl ketones (**2**) by conversion to a ketimine (**a**) followed by chlorination with NCS in CCl$_4$; α,α-dichloro ketones (**2**) are obtained on hydrolysis.[1] When treated with sodium methoxide in methanol **2** is converted into a mixture of the isomers **3** and **4** (probably through an epoxide intermediate). Both isomers are converted into α-aryl-1,2-diones (**5**) on acid hydrolysis. Overall yields vary from 72 to 84% when all intermediates are isolated, but are somewhat higher when the reaction is carried out as a one-pot operation.[2]

[1] N. De Kimpe, R. Verhé, L. DeBuyck, and N. Schamp, *Syn. Comm.*, **8**, 75 (1978).
[2] *Idem, J. Org.*, **43**, 2933 (1978).

Sodium naphthalenide, 1, 711–712; **2**, 289; **4**, 349–350; **5**, 468–470; **7**, 340.

Activated metals. Magnesium and zinc have been prepared in a very active and dispersed form by reduction of $MgCl_2$ or $ZnCl_2$ with this radical anion in THF. The reaction is almost instantaneous at 20° as evidenced by the discharge of the green color of the reagent.[1]

Reduction–decyanation. *vic*-Cyanohydrins can be converted to alkenes by conversion to a methylthiomethyl ether followed by oxidation to the sulfone. Reduction of this derivative with sodium naphthalenide results in an alkene

(equation I). This elimination reaction follows a *syn* pathway, as shown by reduction of the *trans*-isomer (equation II).[2]

Reduction of isocyanides. Sodium naphthalenide reduces isocyanides to the corresponding hydrocarbons with fewer rearranged products than noted with lithium or sodium in NH_3 or THF. The only drawback is that the reduction can result in racemization.[3]

[1] R. T. Arnold and S. T. Kulenovic, *Syn. Comm.*, **7**, 223 (1977).
[2] J. A. Marshall and L. J. Karas, *Am. Soc.*, **100**, 3615 (1978).
[3] G. E. Niznik and H. M. Walborsky, *J. Org.*, **43**, 2396 (1978).

Sulfur dioxide, 1, 1122; **2**, 292; **4**, 469; **5**, 633; **6**, 558; **7**, 346–347.

Nitrile oxides → isocyanates. This rearrangement has been accomplished thermally, but is accompanied by dimerization of the oxide.[1] It can be carried out more satisfactorily in refluxing benzene in the presence of catalytic amounts of

SO_2, which undergoes a 1,3-cycloaddition with nitrile oxides to form an intermediate 1,3-dioxa-2,4-thiazole 2-oxide (2), which loses SO_2 with formation of an isocyanate.[2]

$$RC{\equiv}N \rightarrow O + SO_2 \longrightarrow R\!\!-\!\!\underset{\textbf{2}}{\overset{N-O}{\underset{O}{\diagup}}}\!\!S{=}O \xrightarrow{C_6H_6,\ \Delta} O{=}C{=}N{-}R + SO_2$$

1   **2**   **3 (70–95%)**

Sulfines. Sulfines are generally prepared by oxidation of thiocarbonyl compounds with peracids. Two other methods have been reported recently. One is the reaction of SO_2 with sufficiently reactive phosphorous ylides. One example is shown in equation (I).[3] The second, more general method is based on the Peterson

(I) $(C_6H_5)_2C{=}P(C_6H_5)_3 + SO_2 \xrightarrow[50\%]{C_6H_6,\ 0°} (C_6H_5)_2C{=}S{\overset{O}{\diagup}}$

alkylidenation reaction (**5**, 724) of α-silyl carbanions. An example is shown in equation (II).[4]

(II)

(80%)

Isomerization of alkenes. Exocyclic alkenes are rearranged in high yield to the more stable endocyclic isomers (β-pinene $\rightarrow$ α-pinene; methylenecyclopentane $\rightarrow$ 1-methylcyclopentene) in SO_2. A reversible ene reaction followed by a retroene reaction has been suggested as a possible mechanism.[5]

The isomerization is completely suppressed in the presence of water or D_2O. However, another reaction becomes possible: exchange of the allylic hydrogens with D_2O. Two examples are shown in equations (I) and (II).[6]

(I)

(II)

[1] C. Grundmann, P. Kochs, and J. R. Boal, *Ann.*, **761**, 162 (1972).
[2] G. Trickes and H. Meier, *Angew. Chem., Int. Ed.*, **16**, 555 (1977).
[3] B. Zwanenburg, C. G. Venier, P. A. T. W. Porskamp, and M. van der Leij, *Tetrahedron Letters*, 807 (1978).

[4] M. van der Leij, P. A. T. W. Porskamp, B. H. M. Lammerink, and B. Zwanenburg, *ibid.*, 811 (1978).
[5] M. M. Rogić and D. Masilamani, *Am. Soc.*, **99**, 5219 (1977).
[6] D. Masilamani and M. M. Rogić, *ibid.*, **100**, 4634 (1978).

Sulfur trioxide–Trimethylamine, 1, 1128.

Protection of an enediol.[1] As expected, the primary hydroxyl group of ascorbic acid can be selectively protected as an ether or ester; however, selective

oxidation presents a problem because the enediol group is also susceptible to oxidation, particularly in an alkaline medium. The difficulty can be circumvented by conversion of **1** to potassium ascorbate 2-sulfate (**2**)[2] by reaction of the sodium salt of **1** with the sulfur trioxide–trimethylamine complex followed by conversion to the potassium salt. This derivative is oxidized to the carboxylic acid **3** by the method of Heyns (**1**, 432). The protective group is hydrolyzed by trifluoroacetic acid to give **4**.[3] This product is postulated to be a metabolite of ascorbic acid.

[1] Contributed by Theodora M. Greene.
[2] B. M. Tolbert, M. Downing, R. W. Carlson, M. K. Knight, and E. M. Baker, *Ann. N.Y. Acad. Sci.*, **258**, 48 (1975).
[3] H. A. Stuber and B. M. Tolbert, *Carbohydrate Res.*, **60**, 251 (1978).

T

Tetra-*n*-butylammonium fluoride (TBAF), 4, 477–478; **5,** 645; **7,** 353–354.

Aldol condensation.[1] In the presence of catalytic amounts of this salt, enol silyl ethers of ketones react with aldehydes to give the aldol products in high yield. Ketones do not undergo this reaction. KF in combination with dicyclohexyl-18-crown-6 is also an effective catalyst.

Examples:

Acylation of hydroxyl groups. Hydroxyl groups of nucleosides are often not acylated in high yield under standard conditions (acid chloride or anhydride in pyridine). However, the reaction occurs in high yield with an anhydride in THF at 22° when catalyzed by fluoride ion. The same conditions also allow replacement

467

of a silyl protecting group by an acyl group. The reaction is successful even with highly hindered hydroxyl groups.[2]

Alkylation of purines and pyrimidines. This salt causes rapid alkylation of purine and pyrimidine bases with alkyl halides or trialkyl phosphates (THF, 20°).[3]

Examples:

(80%)   (20%)

[1] R. Noyori, K. Yokoyama, J. Sakata, I. Kuwajima, E. Nakamura, and M. Shimizu, *Am. Soc.*, **99**, 1265 (1977).

[2] S. L. Beaucage and K. K. Ogilvie, *Tetrahedron Letters*, 1691 (1977).

[3] K. K. Ogilvie, S. L. Beaucage, and M. F. Gillen, *ibid.*, 1663 (1978).

Tetra-*n*-butylammonium permanganate, $(C_4H_9)_4N^+MnO_4^-$. Mol. wt. 361.42, m.p. 120–121°. The reagent is prepared by the reaction of tetrabutylammonium bromide with potassium permanganate in water. The crystalline precipitate is dried *in vacuo*. It is stable for prolonged periods and slightly soluble in benzene, soluble in acetone, CH_2Cl_2, and pyridine.

Oxidation.[1] The reagent is convenient for oxidation of aldehydes to carboxylic acids in pyridine solution in high yield (85–99%). Other reported oxidations are: benzyl alcohol → benzoic acid (98%); benzhydrol → benzophenone (97%); *cis*-stilbene → benzoic acid (98%). The advantage of this reagent over $KMnO_4$ is that water, which decomposes $KMnO_4$ slowly, is avoided.

[1] T. Sala and M. V. Sargent, *J.C.S. Chem. Comm.*, 253 (1978).

Tetracarbonylcyclopentadienylvanadium, $(\eta^5\text{-}C_5H_5)V(CO)_4$. Mol. wt. 228.08, m.p. 138°. Supplier: Alfa.

Reduction of halides. This complex can be converted into the anionic hydride **1** as shown in equation (I). The salt (**1**) precipitates on addition of petroleum ether and is obtained in about 70% yield. This species reduces a variety of organic halides. It reduces alkyl halides at room temperature in the order RI > RBr >

(I) $(C_5H_5)V(CO)_4$ $\xrightarrow{\text{Na/Hg, THF}}$ $Na_2^{2+}[(C_5H_5)V(CO)_3]^{2-}$ $\xrightarrow[-\text{NaOH}]{\text{H}_2\text{O}}$

$Na^+[(C_5H_5)V(CO)_3H]^-$ $\xrightarrow[-\text{NaCl}]{[(C_6H_5)_3P]_2N^+Cl^-}$ $[(C_6H_5)_3P]_2N^+[(C_5H_5)V(CO)_3H]^-$

1

RCl, with similar rates for primary, secondary, and tertiary halides. *gem*-Dibromocyclopropanes are reduced rapidly to monobromocyclopropanes, which are then inert to further reduction at 25°. Acyl and benzyl halides are reduced very rapidly. The reactivity of **1** is similar to that of trialkyltin hydrides, and reductions with **1** probably also proceed by a radical chain mechanism.[1]

Examples:

$n\text{-}C_8H_{17}Br$ $\xrightarrow[75\%]{\text{1, THF, 25°}}$ $n\text{-}C_8H_{18}$

$C_6H_5CH_2COCl$ $\xrightarrow{100\%}$ $C_6H_5CH_2CHO$

$(CH_3)_3CBr$ $\xrightarrow{100\%}$ $(CH_3)_3CH$

$C_6H_5CH_2Br$ $\xrightarrow{90\%}$ $C_6H_5CH_3$

$C_4H_9CHBrCH_2Br$ $\xrightarrow{100\%}$ $C_4H_9CH{=}CH_2$

C_6H_5Br $\xrightarrow{43\%}$ C_6H_6

[1] R. J. Kinney, W. D. Jones, and R. G. Bergman, *Am. Soc.*, **100**, 635 (1978).

Tetracarbonyldi-μ-chlorodirhodium [Bis(chlorodicarbonylrhodium)], $[Rh(CO)_2Cl]_2$. Mol. wt. 388.76, m.p. 121°. Supplier: Alfa, PCR.

Reaction with arene oxides.[1] The simplest arene oxide, benzene oxide, reacts with chloroform and with methanol in the presence of 20 mole % of this rhodium

complex as shown in equation (I). The oxide derived from phenanthrene is converted mainly into a phenolic product (equation II). Oxides that cannot be converted directly into phenolic products are deoxygenated quantitatively (equation III).

(II)

$$\xrightarrow{\text{Cat., CHCl}_3}$$

(82%) + (18%)

(III)

$$\xrightarrow[100\%]{\text{Cat., HCCl}_3}$$

1,3-Dienes; biaryls. Vinylmercuric chlorides can be dimerized to 1,3-dienes by stoichiometric amounts of palladium(II) chloride and lithium chloride (Li$_2$-PdCl$_4$)[2]. More recently, this reaction has been improved by use of this rhodium(I) complex in combination with lithium chloride. This system can function in catalytic amounts. RhCl$_3$ and LiCl are somewhat less effective. An example is the synthesis of *trans,trans*-5,7-dodecadiene (equation I).[3]

(I)   2   CH$_3$(CH$_2$)$_3$\ /H
           C=C
        H/ \HgCl
$$\xrightarrow[95\%]{\text{[Rh(CO)}_2\text{Cl]}_2\text{, LiCl}}$$

CH$_3$(CH$_2$)$_3$\ /H
           C=C
        H/ \C=C
              /H
           H/ \(CH$_2$)$_3$CH$_3$
                          + HgCl$_2$ + Hg

Arylmercuric chlorides are also dimerized to biaryls by this catalytic system; in this case HMPT is the preferred solvent. Yields of biaryls range from 40 to 90%.

[1] R. W. Ashworth and G. A. Berchtold, *Tetrahedron Letters*, 343 (1977).
[2] R. C. Larock, *J. Org.*, **41**, 2241 (1976).
[3] R. C. Larock and J. C. Bernhardt, *ibid.*, **42**, 1680 (1977).

Tetraethylammonium fluoride, (C$_2$H$_5$)$_4$NF. Mol. wt. 149.25. Eastman supplies the dihydrate. Anhydrous material is obtained by drying over P$_2$O$_5$.

Protection of amino groups. The β-(trimethylsilyl)ethoxycarbonyl group is deblocked by tetraethylammonium fluoride (equation I). KF is less effective in

(I)   RNHCO$_2$CH$_2$CH$_2$Si(CH$_3$)$_3$ $\xrightarrow{\text{F}^-,\ \text{CH}_3\text{CN}}$

RNH$_2$ + CO$_2$ + CH$_2$=CH$_2$ + (CH$_3$)$_3$SiF

this reaction. The protecting group is sensitive to acids, but is stable to hydrogenation and common basic reagents.[1] The group is introduced by the reaction of the amine with β-trimethylsilylethyl chloroformate or the corresponding azidoformate.[1]

[1] L. A. Carpino, J.-H. Tsao, H. Ringsdorf, E. Fell, and G. Hettrick, *J.C.S. Chem. Comm.*, 358 (1978).

Tetraisobutyldialuminoxane, $(i\text{-}C_4H_9)_2Al\text{---}O\text{---}Al(i\text{-}C_4H_9)_2$ **(1).** Mol. wt. 298.43. This reagent is prepared by addition of 0.5 equiv. of water to diisobutylaluminum hydride in THF ($-78°$) followed by concentration *in vacuo*.

Cyclization of terpene derivatives.[1] Japanese chemists have devised a method for biomimetic conversion of acyclic terpene allylic phosphates into cyclic compounds based on organoaluminum compounds. Of several reagents tested of this type, tetraisobutyldialuminoxane (1) proved to be most selective. Two of the cyclizations realized with 1 are formulated in equations (I) and (II).

(I)

Neryl diethyl phosphate   Limonene   Terpinolene

(II)

α-Chamigrene

Cyclization of an acyclic terpene derivative to a seven-membered ring has also been achieved.[2] Thus the silyl enol ether **2** related to nerol is converted into karahanaenone (3) in 95% yield by methylaluminum bis(trifluoroacetate) in CH_2Cl_2 (0 $\rightarrow$ 25°).

2   **3**

[1] Y. Kitagawa, S. Hashimoto, S. Iemura, H. Yamamoto, and H. Nozaki, *Am. Soc.*, **98**, 5030 (1976); *see also* H. Yamamoto and H. Nozaki, *Angew. Chem., Int. Ed.*, **17**, 169 (1978).

[2] S. Hashimoto, A. Itoh, Y. Kitagawa, H. Yamamoto, and H. Nozaki, *Am. Soc.*, **99**, 4192 (1977).

Tetrakis(triphenylphosphine)palladium(0), 6, 571–573; **7,** 357–358.

Alkenes from vinyl iodides and Grignard reagents. Cross coupling of vinyl iodides with Grignard reagents to form alkenes is possible with catalysis by this palladium complex. Yields are 80–87% (six examples). Of more interest, the reaction occurs almost exclusively ($\geq 97\%$) with retention of configuration (equation I).

$$\text{(I)} \qquad \begin{array}{c} R^1 \\ \diagdown \\ R^2 \end{array} C{=}C \begin{array}{c} H \\ \diagup \diagdown \\ I \end{array} + RMgX \xrightarrow[80-87\%]{Pd(0)} \begin{array}{c} R^1 \\ \diagdown \\ R^2 \end{array} C{=}C \begin{array}{c} H \\ \diagup \diagdown \\ R \end{array}$$

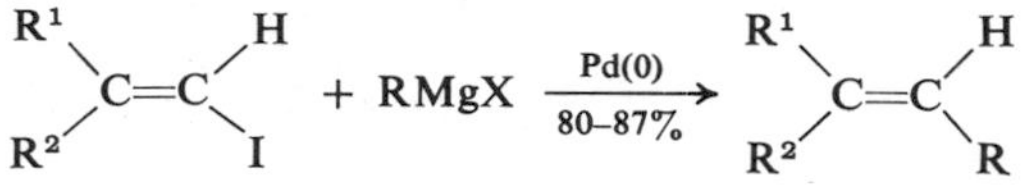

This Pd-catalyzed reaction can be extended to a synthesis of 1,3-dienes from vinylic iodides and vinylic Grignard reagents. This coupling also occurs with retention of configuration (equation II).[1]

$$\text{(II)} \qquad \begin{array}{c} R^1 \\ \diagdown \\ R^2 \end{array} C{=}C \begin{array}{c} H \\ \diagup \diagdown \\ I \end{array} + \begin{array}{c} XMg \\ \diagdown \\ R^3 \end{array} C{=}C \begin{array}{c} R^5 \\ \diagup \diagdown \\ R^4 \end{array} \xrightarrow{75-87\%} \begin{array}{c} R^1 \\ \diagdown \\ R^2 \end{array} C{=}C \begin{array}{c} H \\ \diagup \diagdown \\ C{=}C \end{array} \begin{array}{c} R^5 \\ \diagup \diagdown \\ R^4 \end{array}$$

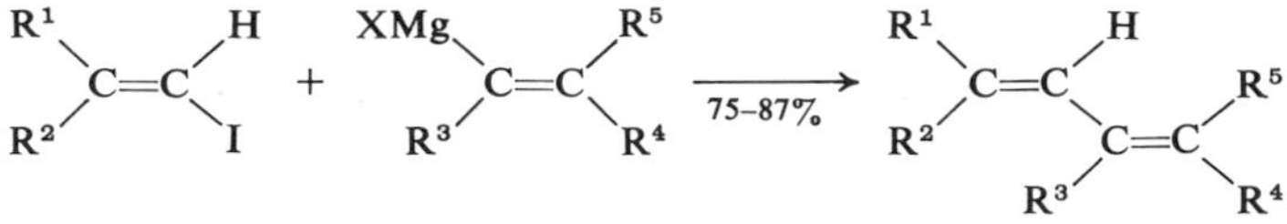

Conjugated enynes. Negishi *et al.*[2] have developed an efficient synthesis of both terminal and internal conjugated enynes by the reaction of an alkenyl iodide or bromide with an alkynylzinc chloride catalyzed by this Pd(0) complex.

$$\text{(I)} \qquad \begin{array}{c} R^1 \\ \diagdown \\ R^2 \end{array} C{=}C \begin{array}{c} R^3 \\ \diagup \diagdown \\ I(Br) \end{array} + ClZnC{\equiv}CR^4 \xrightarrow[70-85\%]{Pd(0),\ THF} \begin{array}{c} R^1 \\ \diagdown \\ R^2 \end{array} C{=}C \begin{array}{c} R^3 \\ \diagup \diagdown \\ C{\equiv}CR^4 \end{array}$$

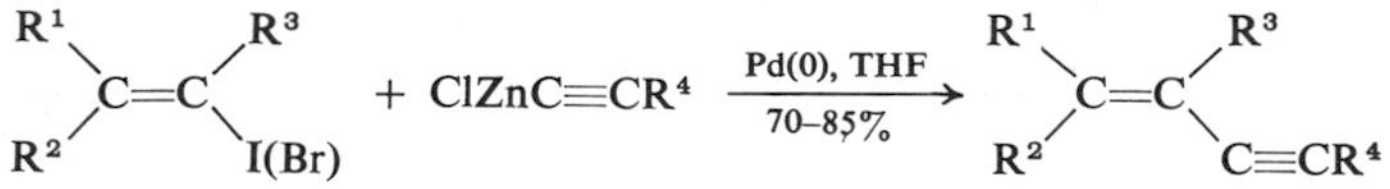

Alkynylzinc chlorides are readily available by treatment of lithium acetylide (**6,** 324) or an ethynyllithium with anhydrous zinc chloride in THF. Tetrakis-(triphenylphosphine)nickel(0) (**6,** 570) is a much less efficient catalyst than the Pd(0) system. Dienes and diynes are formed in only traces ($< 5\%$).

Arylalkynes.[3] The above reaction has been extended to a synthesis of aryl alkynes by cross-coupling of a 1-alkynylzinc chloride with an aryl iodide or bromide (equation I) with tetrakis(triphenylphosphine)palladium(0) as catalyst. This method was used for the synthesis of 1-(2′-thienyl)-1-propyne (**1**), which on

$$\text{(I)} \qquad\qquad RC{\equiv}CZnCl + ArI(Br) \xrightarrow[55-80\%]{Pd[P(C_6H_5)_3]_4} RC{\equiv}CAr$$

formylation (**7**, 301) is converted into junipal (**2**), an odiferous constituent of *Doedalea juniperina*.

$$CH_3C{\equiv}CZnCl \;+\; \underset{S}{\bigcirc}{-}I \;\xrightarrow[92\%]{Pd(0)}$$

$$\underset{S}{\bigcirc}{-}C{\equiv}CCH_3 \;\xrightarrow[66\%]{\substack{1)\ n\text{-BuLi, }0^\circ \\ 2)\ DMF,\ -78^\circ}}\; OHC{-}\underset{S}{\bigcirc}{-}C{\equiv}CCH_3$$

1   **2**

Unsymmetrical biaryls and diarylmethanes. Negishi *et al.*[4] have used a nickel-(0) complex or a palladium(0) complex as catalyst for the cross-coupling of aryl- or benzylzinc halides with aryl halides at room temperature to form biaryls or diarylmethanes. The zinc derivatives are prepared by reaction of aryl- or benzyl-lithium with zinc chloride or bromide.

Examples:

$$C_6H_5ZnCl \;+\; I{-}\langle\ \rangle{-}OCH_3 \;\xrightarrow[87\%]{Pd(0)}\; \langle\ \rangle{-}\langle\ \rangle{-}OCH_3$$

$$C_6H_5CH_2ZnBr \;+\; I{-}\langle\ \rangle{-}NO_2 \;\xrightarrow[88\%]{}\; C_6H_5CH_2{-}\langle\ \rangle{-}NO_2$$

Vinyl nitriles. Vinyl bromides are converted into vinyl nitriles by reaction with KCN in benzene in the presence of this Pd(0) catalyst and 18-crown-6. The reaction is highly stereospecific.[5]

$$\underset{R^2}{\overset{R^1}{\diagdown}}C{=}C\underset{Br(Cl)}{\overset{R^3}{\diagup}} \;+\; KCN \;\xrightarrow[85-95\%]{Pd[P(C_6H_5)_3]_4,\ crown\ ether}\; \underset{R^2}{\overset{R^1}{\diagdown}}C{=}C\underset{CN}{\overset{R^3}{\diagup}}$$

Cyclization via π-allylpalladium complexes. Trost and Verhoeven[6] have reported several cyclizations involving α-alkylation of sulfones by means of π-allylpalladium complexes (equations I and II).

(I)

$$\xrightarrow[67-75\%]{NaH,\ THF,\ [(C_6H_5)_3P]_4Pd}$$

(II)

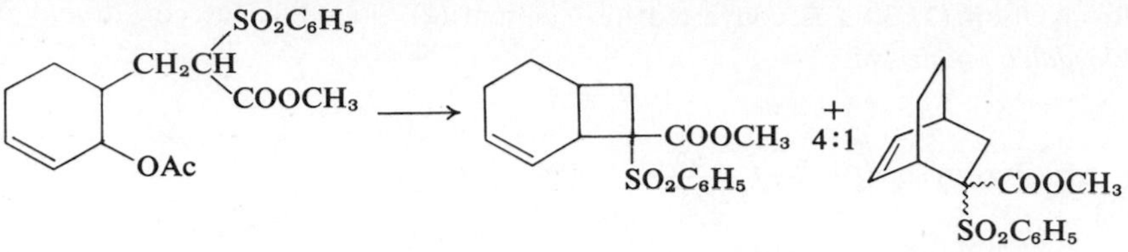

This cyclization was applied successfully to a synthesis of the 16-membered macrolide exaltolide (**3**) and of the macrolide reifeiolide (**6**).[7]

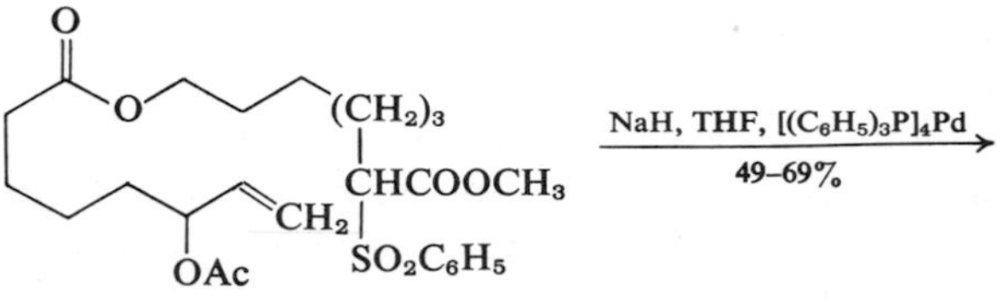

1

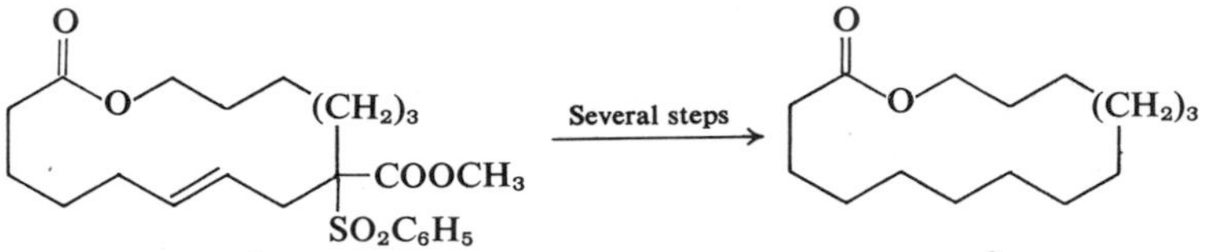

2   **3**

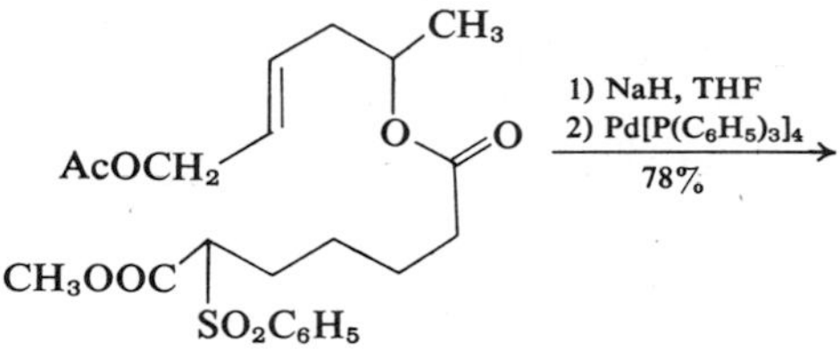

4

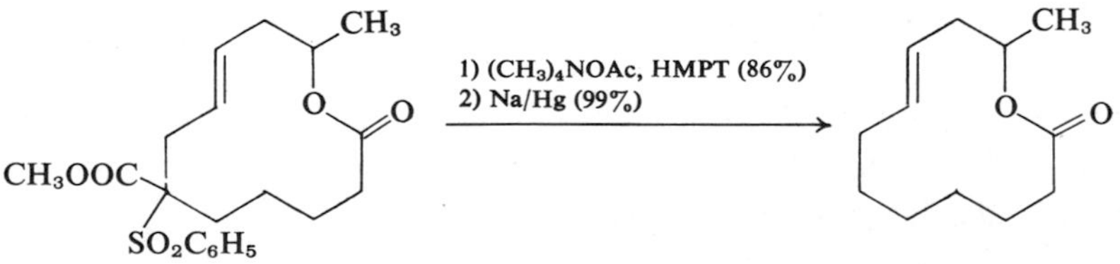

5   **6**

Japanese chemists[8] have reported a stereoselective synthesis of humulene (**9**), an 11-membered cyclic sesquiterpene, by intramolecular α-alkylation of a keto group by a π-allylpalladium complex. The starting material (**7**) was derived from geranyl acetate. In this case, the efficiency of the cyclization was markedly improved by addition of a diphosphine ligand such as 1,3-bis(diphenylphosphino)-

$$7 \xrightarrow[\text{45\%}]{\substack{\text{1) NaH, THF} \\ \text{2) Pd(0)}}}$$

$$8 \xrightarrow{\text{Several steps}} 9$$

propane, $(C_6H_5)_2PCH_2CH_2CH_2P(C_6H_5)_2$. THF–HMPT proved most satisfactory as the solvent.

Steroid side chain. π-Allylpalladium complexes have been used to attach various side chains to C_{17}-keto steroids. The stoichiometric reaction using Na_2PdCl_4 results in unnatural steroids at C_{20}; in the catalytic reaction with Pd(0) the natural configuration at C_{20} results. An example is shown in equation (I). In this reaction

(I)

$$\xrightarrow[\text{83\%}]{\text{NaCH(COOCH}_3)_2,\ \text{Pd(0)},\ \text{P(C}_6\text{H}_5)_3}$$

the catalyst is $Pd[P(C_6H_5)_3]_4$ with added triphenylphosphine. This stereocontrolled alkylation was used to synthesize 5α-cholestanone from testosterone.[9]

[1] H. P. Dang and G. Linstrumelle, *Tetrahedron Letters*, 191 (1978).
[2] A. O. King, N. Okukado, and E. Negishi, *J.C.S. Chem. Comm.*, 683 (1977).
[3] A. O. King, E. Negishi, F. J. Villiani, Jr., and A. Silveira, *J. Org.*, **43**, 358 (1978).
[4] E. Negishi, A. O. King, and N. Okukado, *ibid.*, **42**, 1821 (1977).
[5] K. Yamamura and S.-I. Murahashi, *Tetrahedron Letters*, 4429 (1977).

[6] B. M. Trost and T. R. Verhoeven, *Am. Soc.*, **99**, 3867 (1977).

[7] *Idem, Tetrahedron Letters*, 2275 (1978).

[8] Y. Kitagawa, A. Itoh, S. Hashimoto, H. Yamamoto, and H. Nozaki, *Am. Soc.*, **99**, 3864 (1977).

[9] B. M. Trost and T. R. Verhoeven, *ibid.*, **100**, 3435 (1978).

Tetramethylammonium tetracarbonylhydridoferrate, $(CH_3)_4N \cdot HFe(CO)_4$ **(1).** Mol. wt. 243.04, somewhat sensitive to air. The salt is obtained by reaction of $Fe(CO)_5$ with KOH and $(CH_3)_4NBr$ in H_2O.

Reduction of acid chlorides. Acid chlorides are reduced to aldehydes by this salt in CH_2Cl_2 generally in yields of 80–100%. The reagent is related to potassium

$$\text{(I)} \quad 2RCOCl + 3(CH_3)_4N \cdot HFe(CO)_4 \xrightarrow[\text{75–100\%}]{CH_2Cl_2}$$

$$2RCHO + 2(CH_3)_4NCl + CO + (CH_3)_4N \cdot HFe_3(CO)_{11}$$

(sodium) tetracarbonylhydridoferrate (**4**, 268–269; **5**, 357–358; **6**, 483–486), but is less expensive and more easily prepared. Yields of aldehydes are generally high but this method gives low yields of α,β-unsaturated aldehydes. The reducing agent apparently also attacks aromatic nitro groups.[1]

[1] T. E. Cole and R. Pettit, *Tetrahedron Letters*, 781 (1977).

Thallium(I) cyanide. TlCN. Mol. wt. 230.39. The reagent can be prepared in quantitative yield by reaction of dry HCN with thallium(I) phenoxide in ether.

The reagent has been used for preparation of α-ketonitriles, cyanoformates, and cyanotrimethylsilane.[1]

Examples:

$$C_6H_5COCl + TlCN \xrightarrow[74\%]{\text{Ether}} C_6H_5COCN$$

$$ClCOOCH_3 + TlCN \xrightarrow[66\%]{CH_3COOC_2H_5} NCCOOCH_3$$

$$ClSi(CH_3)_3 + TlCN \xrightarrow[84\%]{} (CH_3)_3SiCN$$

[1] E. C. Taylor, J. G. Andrade, K. C. John, and A. McKillop, *J. Org.*, **43**, 2280 (1978).

Thallium(III) nitrate (TTN), 4, 492–497; **5**, 656–657; **6**, 578–579; **7**, 362–365.

Ring expansion. Taylor *et al.*[1] have extended the TTN-oxidation of styrenes to ring expansion of cyclic aralkyl ketones. A Wittig reaction followed by oxidative rearrangement with TTN results in **3** or **4**, in which a methine group has been inserted between the aryl ring and the C=O of **1**.

Oxidation of chalcones. Chalcones (**1**) are oxidized by TTN in acidic aqueous glyme to benzils (**2**) (**4**, 493) with migration of the Ar group. Oxidation in acidic methanol traps the intermediate aldehyde as the acetal (**3**).[2] If the oxidation is carried out on the preformed acetal (**4**) only **5** is obtained, formed with migration of the Ar′ group, provided that the migratory aptitude of the Ar′ is moderate to high. When the migratory aptitude of the Ar group is greater than that of the Ar′ group, complex mixtures are formed and this reaction is not useful. When the oxidation of **1** is carried out in the presence of trimethyl orthoformate about equal amounts of **3** and **5** are obtained.[3]

[1] E. C. Taylor, C.-S. Chiang, and A. McKillop, *Tetrahedron Letters*, 1827 (1977).
[2] A. McKillop, B. P. Swann, M. E. Ford, and E. C. Taylor, *Am. Soc.*, **95**, 3641 (1973).
[3] E. C. Taylor, R. A. Conley, D. K. Johnson, and A. McKillop, *J. Org.*, **42**, 4167 (1977).

Thallium(III) perchlorate, Tl(ClO$_4$)$_3$, **5,** 657–6$\underline{5}$8; **7,** 365.

Alkyl-p-benzoquinones. Oxidation of *p*-alkylphenols (1) with this salt in the presence of perchloric acid results in 2-alkyl-1,4-benzoquinones (2) in 65–70% yield (three examples).[1]

| 1 | a | b | 2 |

Remote oxidation of estrone. Reaction of estrone (1) with this thallium(III) salt gives as the major products 2 and 3.[1] The *p*-quinol 3 had been obtained earlier by oxidation of 1 with thallium(III) trifluoroacetate (4, 500–501). The ketone 2 is not formed via 3, which is stable to the oxidant, but it can be formed by oxidation of 9,11-dehydroestrone in the same way.[2]

2 (40%)   +   3 (20%)

[1] Y. Yamada and K. Hosaka, *Synthesis,* 53 (1977).
[2] Y. Yamada, K. Hosaka, T. Sawahata, Y. Watanabe, and K. Iguchi, *Tetrahedron Letters,* 2675 (1977).

Thallium(III) trifluoroacetate, 3, 286–289; **4,** 496–501; **5,** 658–659; **6,** 579–581; **7,** 365.

Biaryls. Symmetrical biaryls can be prepared in 60–99% yield by treatment of various arenes substituted with electron-donating groups with thallium(III) trifluoroacetate in TFA or in CCl$_4$ or CH$_3$CN containing BF$_3$ etherate. This synthesis supplements the Ullmann reaction, which is inefficient when the aromatic halides are substituted with electron-donating groups. The method is also applicable to intramolecular coupling.[1]

Examples:

(R = H, CH₃, X)

This aryl coupling reaction has been shown to be useful for nonphenolic oxidative coupling for synthesis of alkaloids; for example, the aporphine alkaloid (±)-ocoteine (**2**) can be obtained in 46% yield by reaction of **1** with TTFA.[2]

Oxidative phenol-benzyl coupling. The key step in a new total synthesis of
($\pm$)-picropodophyllone (**3**) involves oxidation of the phenol **1** with thallium(III)
trifluoroacetate in the presence of BF_3 etherate in CH_2Cl_2 at 20°. After work-up,
which included bisulfite reduction and methylation, the diester (**2**) was obtained
in 55% yield. Some evidence suggests that the cyclization proceeds through a *p*-
quinone methide (**a**).

Oxidation of the methyl ether of **1** with VOF_3 results in a dibenzocycloocta-
diene.[3]

Aryl fluorides. These halides are available from arenes by conversion to an
arylthallium(III) bis(trifluoroacetate) followed by reaction with KF, and finally
reaction of the arylthallium(III) difluoride with BF_3 in ether or cyclohexane.
Yields are comparable to those obtained by the Schiemann reaction.[4]

$$(I) \qquad ArH \xrightarrow{\text{TTFA}} ArTl(OCOCF_3)_2 \xrightarrow{\text{KF}} ArTlF_2 \xrightarrow[\substack{45-70\% \\ \text{overall}}]{BF_3,\ C_6H_{12}} ArF$$

Transannular cyclization. Treatment of 1,5-cyclooctadiene with thallium(III)
trifluoroacetate in CH_2Cl_2 at room temperature leads to the *cis*-bicyclo[3.3.0]-
octenol derivatives **2** and **3**.[5]

[1] A. McKillop, A. G. Turrell, and E. C. Taylor, *J. Org.*, **42**, 764 (1977).
[2] E. C. Taylor, J. G. Andrade, and A. McKillop, *J.C.S. Chem. Comm.*, 538 (1977).

[3] A. S. Kende, L. S. Liebeskind, J. E. Mills, P. S. Rutledge, and D. P. Curran, *Am. Soc.*, **99**, 7082 (1977); A. S. Kende and P. S. Rutledge, *Syn. Comm.*, **8**, 245 (1978).
[4] E. C. Taylor, E. C. Bigham, D. K. Johnson, and A. McKillop, *J. Org.*, **42**, 362 (1977).
[5] Y. Yamada, A. Shibaba, K. Iguchi, and H. Sanjoh, *Tetrahedron Letters*, 2407 (1977).

Thionyl chloride, 1, 1158–1163; **2**, 412; **3**, 290; **4**, 503–505; **5**, 663–667; **6**, 585; **7**, 366–367.

$CH_2 \rightarrow$ $C{=}O$. α-Chlorosulfenyl chlorides are formed in 85–95% yield by reaction of activated methylene compounds with thionyl chloride in the presence of pyridine. The reaction is reversed on simple hydrolysis, but if the sulfenyl chlorides are first converted into morpholides and then hydrolyzed with dilute HCl, ketones are obtained in satisfactory yield. Actually the three-step conversion can be carried out without isolation of intermediates. The method is particularly useful for preparation of α-diketones.[1]

Example:

[1] K. Oka and S. Hara, *Tetrahedron Letters*, 695 (1977).

Titanium(0), 7, 368–369.

Reductive coupling of ketones (**7**, 368–369). Coupling of 1-methyl-2-adamantanone with Ti(0), prepared from TiCl$_3$ and potassium, results in formation of the highly hindered *trans*-1-methyl-2-adamantylidene-1-methyladamantane in 52% yield. The *cis*-isomer is not formed because it has much higher energy. Probably olefins more highly strained than **1** cannot be obtained by this reductive coupling. For example, di-*t*-butyl ketone cannot be coupled with lower valent titanium salts.[1]

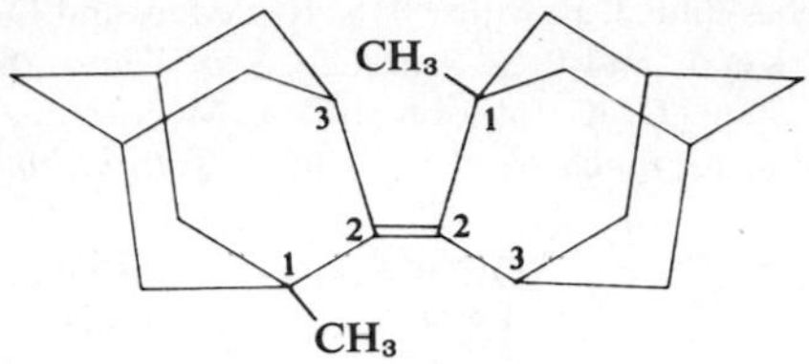

1

Conversion of ketones into olefins (3, 98). This reaction can be effected by conversion to the thermodynamically controlled enol phosphate with diethyl phosphorochloridate in TMEDA followed by reduction. Originally lithium in NH_3 or $C_2H_5NH_2$ was used for this step. Titanium metal is more effective, except for ketones conjugated to aromatic rings.[2]

Examples:

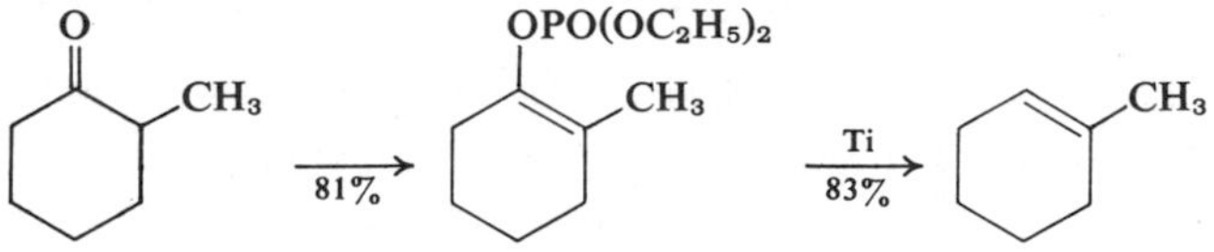

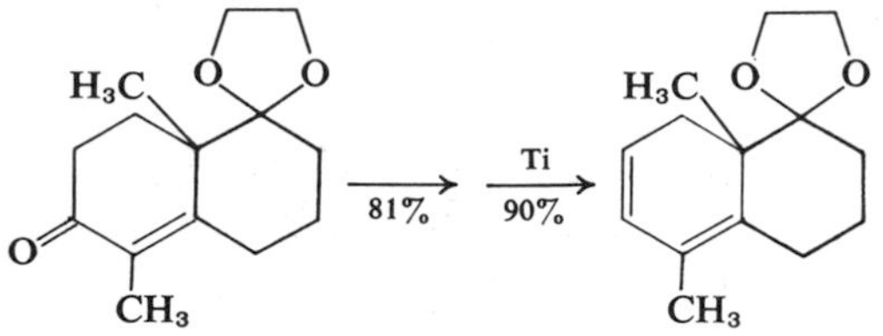

[1] D. Lenoir and R. Frank, *Tetrahedron Letters*, 53 (1978).
[2] S. C. Welch and M. E. Walters, *J. Org.*, **43**, 2715 (1978).

Titanium(III) chloride, 2, 415; **4**, 506–508; **5**, 669–671; **6**, 587–588; **7**, 369.

Cleavage of oximes and semicarbazones.[1] Steroidal oximes and semicarbazones are cleaved in high yield by $TiCl_3$ in buffered dioxane solution (pH ~ 6). Oximes are cleaved at room temperature; steam-bath temperatures are preferred for semicarbazones.

Reduction of azides. Heterocyclic azides are reduced to amines by $TiCl_3$ in aqueous ethanol or aqueous ethanol–DMF (reflux, 20–50 minutes). The actual yield is quantitative, but isolated yields are only moderate.[2]

Examples:

$$CH_3-\underset{\underset{O}{\|}}{\overset{\overset{O}{\|}}{C_6H_4}}-S-N_3 \xrightarrow{78\%} CH_3-\underset{\underset{O}{\|}}{\overset{\overset{O}{\|}}{C_6H_4}}-S-NH_2$$

[1] V. V. Vakatkar, J. G. Tatake, and S. V. Lunthankar, *Chem. Ind.*, 742 (1977).
[2] B. Stanovnik, M. Tišler, S. Polanc, and M. Gračner, *Synthesis*, 65 (1978).

Titanium(III) chloride–Zinc/copper couple.

Cycloalkenes. McMurry and Kees[1] have described an intramolecular coupling of dicarbonyl compounds to cycloalkenes. In the intermolecular version of this reaction titanium(0) powder obtained by reduction with potassium or lithium is used (7, 368). In the intramolecular reaction a superior coupling reagent is prepared by reduction of $TiCl_3$ with a zinc/copper couple[2] in DME.

Examples:

$$\underset{CH_2COC_6H_5}{\overset{CH_2COC_6H_5}{|}} \xrightarrow[87\%]{Ti\ cat.,\ DME} $$

$$C_4H_9CO(CH_2)_6COC_4H_9 \xrightarrow{67\%}$$

$$C_4H_9CO(CH_2)_8COC_4H_9 \xrightarrow{75\%}$$

$$HCO(CH_2)_{12}CHO \xrightarrow{71\%}$$

The value of this cyclization is that yields are satisfactory for all ring sizes, unlike the Thorpe-Ziegler cyclization or the acyloin cyclization, in which yields are low in the formation of 7–11 membered rings.

[1] J. E. McMurry and K. L. Kees, *J. Org.*, **42**, 2655 (1977).
[2] Zinc dust in deoxygenated water is treated with $CuSO_4$; the black slurry is washed with water, acetone, and ether. The dried slurry is stable indefinitely when stored under N_2.

Titanium(IV) chloride, 1, 1169–1171; 2, 414–415; 3, 291; 4, 507–508; 5, 671–672; 6, 590–596; 7, 370–372.

Dimerization of ketene silyl acetals. Ketene silyl acetals when treated with $TiCl_4$ in CH_2Cl_2 dimerize to succinates. No reaction occurs with $TiCl_3$, and yields of dimers are low with other metal salts.[1]

Examples:

$$(CH_3)_2C=C\begin{smallmatrix}OCH_3\\OSi(CH_3)_3\end{smallmatrix} \xrightarrow[80\%]{TiCl_4} \begin{smallmatrix}(CH_3)_2C-COOCH_3\\|\\(CH_3)_2C-COOCH_3\end{smallmatrix}$$

$$\begin{smallmatrix}H\\C_6H_5\end{smallmatrix}C=C\begin{smallmatrix}OCH_3\\OSi(CH_3)_3\end{smallmatrix} \xrightarrow{73\%} \begin{smallmatrix}C_6H_5\\|\\H-C-COOCH_3\\|\\H-C-COOCH_3\\|\\C_6H_5\end{smallmatrix}$$

Knoevenagel condensation; α-methylene-γ-lactones. A new synthesis[2] of these lactones is based on the Knoevenagel condensation by the procedure of Lehnert (**4,** 507–508). Condensation of **1** with Meldrum's acid (**2**) gives **3**, which is hydrolyzed by acid mainly to an α-carboxy-γ-lactone (**5**). Lactones of this type have been converted into α-methylene-γ-lactones (**6**).[3] The lactones **5** and **6** have the *cis* ring junction.

1 (*n* = 0, 1, 2)   **2**   **3**

4   **5** (44–100%)   **6**

β-Lactams. The reaction of ketene silyl acetals and Schiff bases in the presence of TiCl$_4$ results in β-amino esters, which can be cyclized by base to β-lactams. In some cases β-lactams are obtained directly as in the second example.[4]
Examples:

$$(CH_3)_2C=C\begin{smallmatrix}OCH_3\\OSi(CH_3)_3\end{smallmatrix} + C_6H_5CH=NC_6H_5 \xrightarrow[85\%]{\substack{1)\ TiCl_4,\ CH_2Cl_2\\2)\ H_2O}}$$

$$\begin{smallmatrix}C_6H_5CH-NHC_6H_5\\|\\(CH_3)_2C-COOCH_3\end{smallmatrix} \xrightarrow[95\%]{LDA,\ THF} \begin{smallmatrix}C_6H_5CH-NC_6H_5\\|\qquad\ |\\(CH_3)_2C-C{=}O\end{smallmatrix}$$

$$(CH_3)_2C=C\underset{OSi(CH_3)_3}{\overset{OCH_3}{}} + C_6H_5CH=NCH_3 \xrightarrow[72\%]{\substack{1)\ TiCl_4 \\ 2)\ H_2O}} \underset{(CH_3)_2C—C\diagdown_O}{\overset{C_6H_5CH—NHCH_3}{}}$$

γ-Keto aldehydes. In the presence of this Lewis acid (1.1 equiv.) α-siloxy-allylsilanes (1) react with acid chlorides to give, after hydrolysis, γ-keto aldehydes (2).[5]

Example:

$$CH_2=CHCH_2OSi(CH_3)_3 \xrightarrow[72\%]{\substack{1)\ sec\text{-}BuLi,\ THF,\ HMPT \\ 2)\ ClSi(C_2H_5)_3}}$$

$$\underset{\overset{|}{OSi(CH_3)_3}}{CH_2=CHCHSi(C_2H_5)_3} \xrightarrow[43\%]{\substack{1)\ (CH_3)_2CHCOCl,\ TiCl_4 \\ 2)\ H_2O}} (CH_3)_2CH\overset{\overset{\displaystyle O}{\|}}{C}CH_2CH_2CHO$$

        1   **2**

Alkylation of trimethylsilyl enol ethers. Trimethylsilyl enol ethers of ketones can be alkylated under Friedel-Crafts conditions by *t*-butyl chloride in CH_2Cl_2 at $-23°$ in the presence of 1 equiv. of a Lewis acid. The order of effectiveness of catalysts is $TiCl_4 > ZnCl_2 > SnCl_4 > AlCl_3$.[6]

Examples:

$$+ (CH_3)_3CCl \xrightarrow[86\%]{TiCl_4,\ CH_2Cl_2,\ -40°}$$

$$+ (CH_3)_3CCl \xrightarrow[67\%]{}$$

$$\underset{\overset{|}{C_6H_5C=CH_2}}{\overset{OSi(CH_3)_3}{}} \xrightarrow[43\%]{} C_6H_5\overset{\overset{\displaystyle O}{\|}}{C}CH_2C(CH_3)_3$$

(E)-α,β-Unsaturated aldehydes. The reaction of dichloromethyl methyl ether with vinylsilanes catalyzed by $TiCl_4$ leads to α,β-unsaturated aldehydes (equation I).[7]

(I) $RCH=CHSi(CH_3)_3 + Cl_2CHOCH_3 \xrightarrow{TiCl_4,\ CH_2Cl_2}$

$$\left[\underset{\overset{|}{Cl}}{RCH=CHCHOCH_3} \right] \xrightarrow[70-85\%]{H_2O} RCH=CHCHO$$

This reaction has since been extended to 1,2-disubstituted vinylsilanes (equation II).[8] The resulting aldehydes all have the (E)-configuration. This reaction was used in a synthesis of nuciferal (2).

(II) $R^1CH{=}CR^2Si(CH_3)_3$ + Cl_2CHOCH_3 $\xrightarrow{70-90\%}$

$$\underset{H}{R^1}\diagup C{=}C\diagup\underset{CHO}{R^2}$$

$$\xrightarrow[48\%]{Cl_2CHOCH_3,\ TiCl_4}$$

1   2

Enamines. Use of TiCl₄ for preparation of enamines of acyclic ketones was first reported by White and Weingarten (**2**, 414). A recent paper has determined optimal conditions: a sixfold excess of morpholine and 6–7 mole% of TiCl₄. The reaction is conducted in refluxing hexane. Under these conditions yields of enamines from alkyl methyl ketones are 94–100% after 4–5 hours. The same paper compares this method with use of molecular sieves as water scavenger (**4**, 345). The latter method is slower unless the molecular sieve is activated by an acid and it fails with hindered ketones such as *t*-butyl methyl ketones.[9]

[1] S. Inaba and I. Ojima, *Tetrahedron Letters*, 2009 (1977).
[2] E. Campaigne and J. C. Beckman, *Synthesis*, 385 (1978).
[3] P. A. Grieco and K. Hiroi, *J.C.S. Chem. Comm.*, 500 (1973).
[4] I. Ojima, S. Inaba, and K. Yoshida, *Tetrahedron Letters*, 3643 (1977).
[5] A. Hosomi, H. Hashimoto, and H. Sakurai, *J. Org.*, **43**, 2551 (1978).
[6] T. H. Chan, I. Paterson, and J. Pinsonnault, *Tetrahedron Letters*, 4183 (1977); M. T. Reetz and W. F. Maier, *Angew Chem., Int. Ed.*, **17**, 48 (1978).
[7] K. Yamamoto, O. Nunokawa, and J. Tsuji, *Synthesis*, 721 (1977).
[8] K. Yamamoto, J. Yoshitake, N. T. Qui, and J. Tsuji, *Chem. Letters*, 859 (1978).
[9] R. Carlson, R. Phan-Tan-Luu, D. Mathieu, F. S. Ahouande, A. Babadjamian, and J. Metzger, *Acta Chem. Scand.*, **B32**, 335 (1978).

Titanium(IV) chloride-Lithium aluminum hydride, 6, 596; 7, 372-373.

Desulfurization of thioketals.[1] A recent synthesis of 25-hydroxycholesterol (**5**) from the *i*-steroid **1** involved transformation of the side chain as shown to give **3**. The final steps then required desulfurization of the dithioketal and conversion of the *i*-ether group into the Δ⁵-3β-ol. Raney nickel did reduce the dithioacetal group, but also reduced the *i*-ether group. Titanium(IV) chloride in combination with LiAlH₄ led to the desired selective reduction to **4**. The final step involved

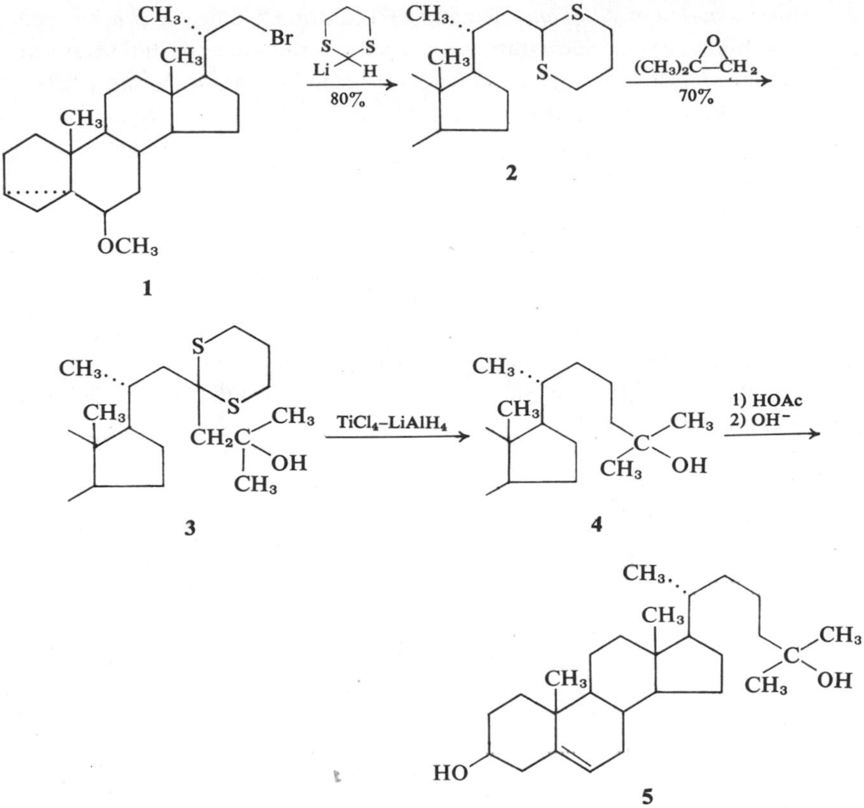

2

3   TiCl$_4$–LiAlH$_4$ 4   1) HOAc 2) OH$^-$

5

treatment with acetic acid and hydrolysis of the resulting 25-hydroxycholesteryl acetate.

[1] W. G. Salmond and K. D. Maisto, *Tetrahedron Letters*, 987 (1977).

Titanium(IV) chloride–Zinc.

McMurry olefin synthesis. Lenoir[1] has published a variation of the McMurry reaction for reductive coupling of ketones (**6**, 589). The actual reagent presumably is also TiCl$_2$, but is produced by reduction of TiCl$_4$ in THF or dioxane with zinc. This reagent in the presence of pyridine converts ketones into tetrasubstituted alkenes. The reaction is most satisfactory with symmetrical ketones; mixtures of (E)- and (Z)-isomers obtain from unsymmetrical ketones, with the latter predominating. Strongly hindered ketones are reduced by the reagent to the secondary alcohol.

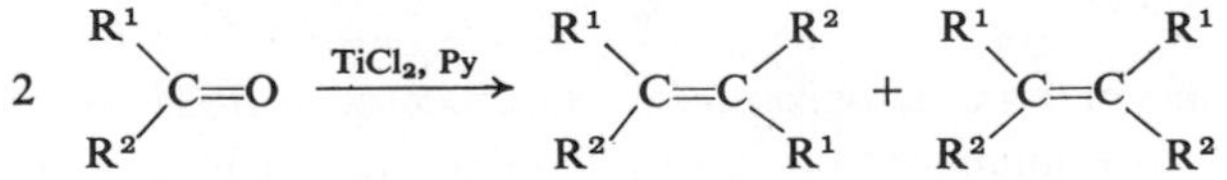

Deoxygenation of sulfoxides. Titanium(II) chloride, formed *in situ* by reduction of titanium(IV) chloride with zinc dust, reduces sulfoxides to sulfides at room temperature almost instantly. Ether and methylene chloride are the preferred solvents. Yields are in the range 85–95%.[2]

[1] D. Lenoir, *Synthesis*, 553 (1977).
[2] J. Drabowicz and M. Mikołajczyk, *ibid.*, 138 (1978).

p-Toluenesulfinic acid (TsH), 7, 374.

Olefins from vinylsilanes. This reaction is generally conducted with HCl, HBr, HI, or CF_3SO_2OH. The cleavage with HI proceeds with retention of configuration (**6**, 281). Büchi and Wüest have found that *p*-toluenesulfinic acid is usually very effective for this reaction, which is conducted in moist acetonitrile. Catalytic amounts of the acid suffice if the reaction is conducted at reflux. This cleavage is not stereospecific.[1]

Examples:

[1] G. Büchi and H. Wüest, *Tetrahedron Letters*, 4305 (1977).

p-Toluenesulfonic acid, 1, 1172–1178; 4, 508–510; 5, 673–675; 6, 597; 7, 374–375.

Translactonization. Internal translactonization provides a useful route to thermodynamically more stable lactones from less stable lactones. Thus treatment of the 9-membered lactone **1** with a trace of *p*-toluenesulfonic acid in methylene chloride at 23° for 2 hours results in formation of the 12-membered lactone **2** in 93% yield. 1,5-Diazabicyclo[4.3.0]nonene-5 (DBN) is a considerably less efficient catalyst in this and related reactions.

From a study of translactonization of lactones of various ring size, Corey *et al.*[1] were able to summarize the most favorable ring expansions as indicated.

$$8 \longrightarrow 11 \longrightarrow 13 \text{ or } 14$$
$$9 \longrightarrow 12 \longrightarrow 14 \text{ or } 15$$
$$10 \longrightarrow 13$$

[1] E. J. Corey, D. J. Brunelle, and K. C. Nicolaou, *Am. Soc.*, **99**, 7359 (1977).

p-**Toluenesulfonyl chloride, 1**, 1179–1184; **3**, 292; **4**, 510–511; **5**, 676–677; **6**, 598.

Oxiranes from **vic-***diols.* Epoxidation with a peracid of a double bond in the presence of a triple bond is often not successful. However, 1-alkynyloxiranes can be obtained in yields of about 80% by treatment of *vic*-diols of type **1** with 1 equiv. of tosyl chloride in monoglyme followed by reaction with powdered sodium

$$C_6H_5CH{-}CHC{\equiv}CH \xrightarrow[80\%]{TsCl,\ NaOH} C_6H_5CH{-}CHC{\equiv}CH$$
$$\underset{HO\ \ \ \ OH}{|\ \ \ \ \ |} \qquad\qquad \underset{O}{\diagdown\diagup}$$
$$\textbf{1} \qquad\qquad\qquad\qquad\qquad \textbf{2}$$

hydroxide. Under these conditions cyclization is faster than tosylation of the second hydroxyl group. The method appears to be satisfactory as a general route to oxiranes.[1]

Examples:

$$n\text{-}C_4H_9C{\equiv}CCH{-}CHC{\equiv}CC_4H_9\text{-}n \xrightarrow[85\%]{} n\text{-}C_4H_9C{\equiv}CCH{-}CHC{\equiv}CC_4H_9\text{-}n$$
$$\underset{HO\ \ \ \ OH}{|\ \ \ \ \ |} \qquad\qquad\qquad \underset{O}{\diagdown\diagup}$$

$$CH_2{=}CHCH{-}CHCH{=}CH_2 \xrightarrow[66\%]{} CH_2{=}CHCH{-}CHCH{=}CH_2$$
$$\underset{HO\ \ \ \ OH}{|\ \ \ \ \ |} \qquad\qquad\qquad \underset{O}{\diagdown\diagup}$$

$$C_6H_5CH{-}CHC_6H_5 \xrightarrow[72\%]{} C_6H_5CH{-}CHC_6H_5$$
$$\underset{HO\ \ \ \ OH}{|\ \ \ \ \ |} \qquad\qquad\qquad \underset{O}{\diagdown\diagup}$$

[1] S. Holand and R. Epsztein, *Synthesis*, 706 (1977).

p-**Toluenesulfonylhydrazine, 1**, 1185–1187; **2**, 417–418; **3**, 293; **4**, 511–512; **5**, 678–681; **6**, 598–600; **7**, 375–376.

Shapiro reaction. (**2**, 418–419; **6**, 598–600). Several laboratories have used LDA instead of an alkyllithium for decomposition of tosylhydrazones of ketones to olefins. Trisubstituted alkenes can be prepared by this modification in moderate yields from tosylhydrazones that contain only tertiary α-hydrogens. This modification also favors formation of the (Z)-disubstituted olefin.[1]

Examples:

$$\overset{\displaystyle NNHTs}{\overset{\|}{C_6H_5CCH(CH_3)_2}} \xrightarrow[57\%]{LDA,\ TMEDA} C_6H_5CH{=}C(CH_3)_2$$

$$\underset{\text{NNHTs}}{n\text{-}C_3H_7\overset{\|}{C}C_3H_7\text{-}n} \xrightarrow[55\%]{} n\text{-}C_3H_7CH\!=\!CHCH_2CH_3$$
$$(Z/E = 92\!:\!8)$$

Grieco and Nishizawa[2] used LDA as base for a Shapiro reaction in the presence of a lactone group to convert **1** into **2**. The carbonyl group of the lactone was protected as the enolate.

1) TsNHNH$_2$
2) LDA
65%

1   **2**

Piers *et al.*[3] used sodium bis(trimethylsilyl)amide as base for the decomposition of a hydrazone (**3**) containing a nonenolizable keto group.

[(CH$_3$)$_3$Si]$_2$NNa, THF, 125°
~40%

3   **4**

Homoallylic alcohols. Dianions of tosylhydrazones are trapped with aldehydes and ketones to form β-hydroxytosylhydrazones in good to excellent yields. These products on treatment with 3 equiv. of *n*-butyllithium undergo elimination to homoallylic alcohols. An example is shown in equation (I). Attempts to

$$(I)\quad \underset{\text{NNHTs}}{C_2H_5\overset{\|}{C}CH_3} \xrightarrow[61\%]{\substack{1)\ 2n\text{-BuLi} \\ 2)\ CH_3COCH_3}} \underset{\substack{\text{NNHTs} \\ \\ \text{OH}}}{C_2H_5\overset{\|}{C}CH_2\underset{|}{C}(CH_3)_2} \xrightarrow[75\%]{3n\text{-BuLi}}$$

$$\underset{\substack{| \\ \text{OH}}}{CH_3CH\!=\!CHCH_2\overset{}{C}(CH_3)_2}$$
(E and Z)

generate β-hydroxy ketones from the hydrazones under acidic conditions fail because of a retro-aldol reaction; no reaction occurs under basic conditions.[4]

β,γ-*Unsaturated esters.* Bunnell and Fuchs[5] have extended the Shapiro reaction to a synthesis of β,γ-unsaturated esters from the tosylhydrazones of

β-keto esters. This reaction requires 3 equiv. of base (LDA). When the reaction is quenched with an alkyl halide, α-alkylated-β,γ-unsaturated esters are obtained.

Examples:

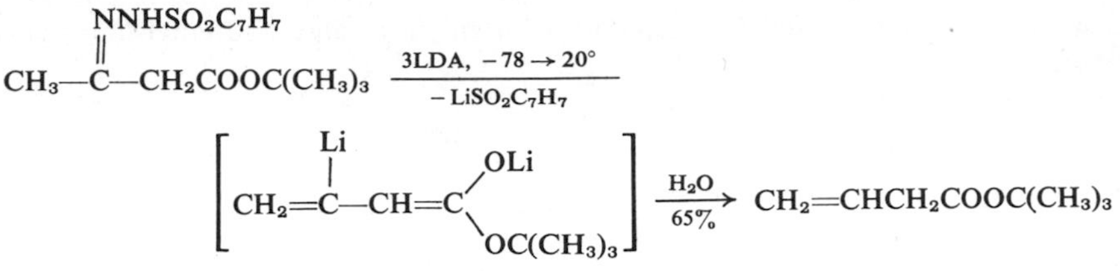

a

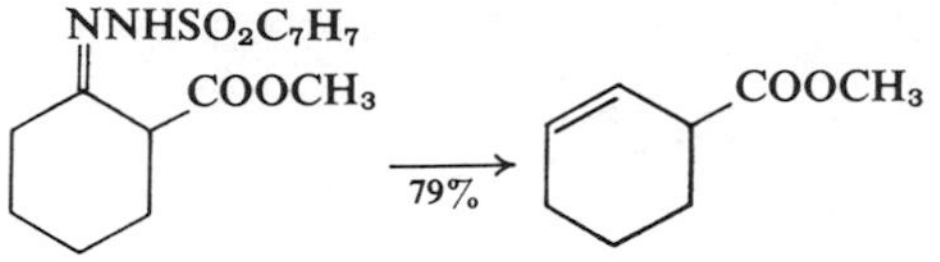

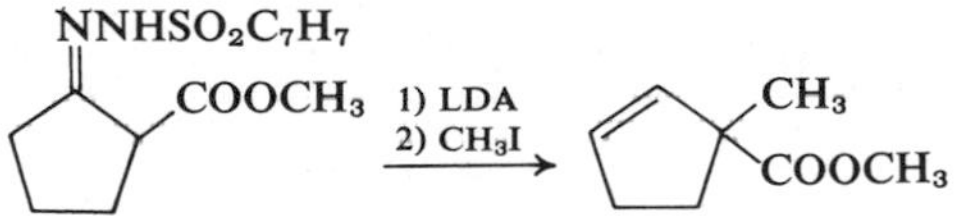

79%

1) LDA
2) CH$_3$I

1,2-*Transposition* of C=O groups. A method for 1,2-transposition of a carbonyl group is shown in equation (I). A method for 1,2-transposition of a carbonyl group with alkylation is illustrated in equation (II).[6]

(I)

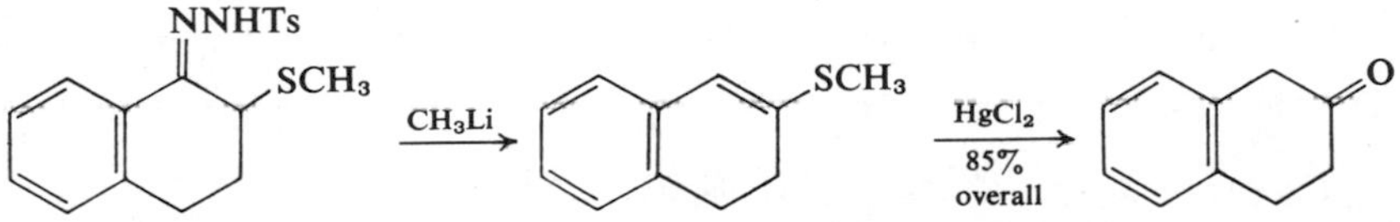

CH$_3$Li

HgCl$_2$
85%
overall

1   2   3

(II)

+ (C$_6$H$_5$)$_3$P=CH$_2$ 87%

n-BuLi, THF

HgCl$_2$
85%

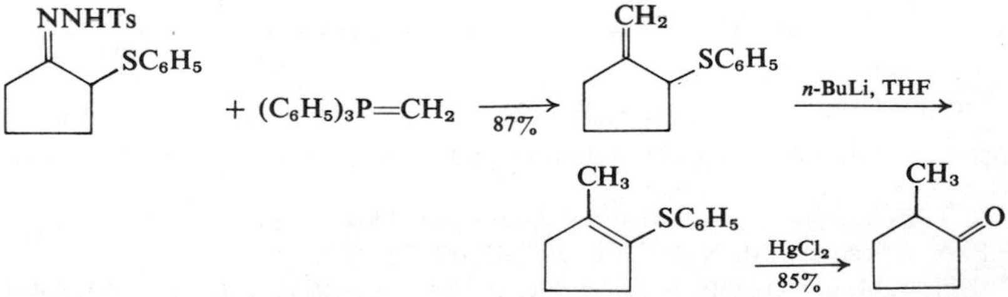

Vinylsilanes.[7] Tosylhydrazones of ketones are converted into vinylsilanes by treatment with *n*-butyllithium (4 equiv.) in TMEDA followed by reaction of the resulting carbanion with (CH$_3$)$_3$SiCl:

$$\underset{RCCH_2R^1}{\overset{O}{\parallel}} \longrightarrow \underset{RCCH_2R^1}{\overset{NNHTS}{\parallel}} \xrightarrow{n\text{-}C_4H_9Li,\ TMEDA} \underset{RC=CHR^1}{\overset{Li}{\vert}} \xrightarrow{(CH_3)_3SiCl} \underset{RC=CHR^1}{\overset{Si(CH_3)_3}{\vert}}$$

The $Si(CH_3)_3$ group is bonded to the carbon of the original carbonyl group. The same method can be used for preparation of vinylgermanes and vinylstannanes.

Examples:

$$RCHO \rightarrow RCH_2R^1.$$ This transformation can be accomplished by reductive alkylation of tosylhydrazones of aldehydes with alkyllithium reagents (equation

$$\text{(I)} \quad RCH=NNHTs \xrightarrow{R^1Li} \left[\underset{RCH=NNTs}{\overset{Li}{\vert}} \xrightarrow{R^1Li} \right.$$

$$\left. \underset{RR^1CHN-NTs}{\overset{Li\ Li}{\vert\ \vert}} \xrightarrow{-LiTs} RR^1CHN=NLi \right] \xrightarrow[40-60\%]{-N_2,\ H_2O} RR^1CH_2$$

I). The 2:1 adduct, RR^1CHCH_2R, is also formed as a minor product. Yields of the products RR^1CH_2 are only moderate and variable, but the procedure is simple.[8]

[1] K. J. Kolonko and R. H. Shapiro, *J. Org.*, **43**, 1404 (1978).
[2] P. A. Grieco and M. Nishizawa, *ibid.*, **42**, 1717 (1977).
[3] E. Piers, M. B. Geraghty, R. D. Smillie, and M. Soucy, *Canad. J. Chem.*, **53**, 2849 (1975).
[4] M. F. Lipton and R. H. Shapiro, *J. Org.*, **43**, 1409 (1978).
[5] C. A. Bunnell and P. L. Fuchs, *Am. Soc.*, **99**, 5184 (1977).
[6] S. Kano, T. Yokomatsu, T. Ono, S. Hibino, and S. Shibuya, *J.C.S. Chem. Comm.*, 414 (1978).

[7] R. T. Taylor, C. R. Degenhardt, W. P. Melega, and L. A. Paquette, *Tetrahedron Letters*, 159 (1977).
[8] E. Vedejs and W. T. Stolle, *ibid.*, 135 (1977).

Tosylmethyl isocyanide, 4, 514–516; **5,** 684–685; **6,** 600; **7,** 377. The preparation[1] and use of this reagent for synthesis of nitriles[2] have been published.

Ketone synthesis. $TsCH_2N{=}C$ can be mono- or dialkylated either under classical conditions (NaH, DMSO, RX) or under phase transfer conditions (CH_2Cl_2, aqueous NaOH, n-Bu_4NI, RX). The products on acid hydrolysis are converted into ketones. Both symmetrical and unsymmetrical ketones can be prepared in this way. In this synthesis, the overall result is dialkylation of form-aldehyde (from which TsMIC is prepared).[3]

$$(I) \quad TsCH_2N{=}C \xrightarrow{R^1X} \underset{R^1}{TsCHN{=}C} \xrightarrow{R^2X} \underset{R^2}{\overset{R^1}{TsCN{=}C}} \xrightarrow{H_3O^+}$$

$$\left[\underset{R^2}{\overset{R^1}{\diagdown}} C{=}NCHO \right] \xrightarrow[40-70\%]{} \underset{R^2}{\overset{R^1}{\diagdown}} C{=}O$$

Symmetrical and unsymmetrical α-diketones can be prepared by acylation of a monoalkylated derivative of $TsCH_2N{=}C$.[4]

$$(II) \quad \underset{R^1}{TsCHN{=}C} \xrightarrow{n\text{-BuLi, }R^2COCl} \underset{Ts}{\overset{N{=}C}{R^1C{-}COR^2}} \xrightarrow[50-70\%]{H_3O^+} \overset{O\ \ O}{R^1C{-}CR^2}$$

Imidazoles. These heterocycles can be obtained by base-induced cycloaddition of $TsCH_2N{=}C$ to aldimines in either a one- or two-step reaction (equation I).[5]

$$(I) \quad R^1CH{=}NR^2 + TsCH_2N{=}C \xrightarrow[40-95\%]{K_2CO_3,\ CH_3OH,\ DME} \underset{R^2}{\overset{N}{R^1 \diagup N}} + TsH$$

$$\xrightarrow[]{NaH,\ DME} \underset{R^2}{\overset{Ts}{R^1 \cdots N}} \xrightarrow[\substack{20-65\% \\ overall}]{K_2CO_3,\ CH_3OH}$$

1,3-Thiazoles. This heterocyclic system can be prepared by reaction of tosylmethyl isocyanide with carbon disulfide under phase-transfer conditions; the tetrabutylammonium salt of 5-mercapto-4-tosyl-1,3-thiazole that is formed can be alkylated or acylated in high yields.[6]

$$\text{TsCH}_2\text{NC} + \text{CS}_2 \xrightarrow[\text{95\%}]{\substack{(n\text{-}C_4H_9)_4N^+Br^-,\\ \text{CHCl}_3,\ 10\%\ \text{NaOH}}}$$

[1] B. E. Hoogenboom, O. H. Oldenziel, and A. M. van Leusen, *Org. Syn.*, **57**, 102 (1977).
[2] O. H. Oldenziel, J. Wildeman, and A. M. van Leusen, *ibid.*, **57**, 8 (1977).
[3] O. Possel and A. M. van Leusen, *Tetrahedron Letters*, 4229 (1977).
[4] D. van Leusen and A. M. van Leusen, *ibid.*, 4233 (1977).
[5] A. M. van Leusen, J. Wildeman, and O. H. Oldenziel, *J. Org.*, **42**, 1153 (1977).
[6] A. M. van Leusen and J. Wildeman, *Synthesis*, 501 (1977).

2,4,6-Tri-t-butylphenoxy-N,N-dimethylcarbamate

1

Mol. wt. 333.50. Tri-*t*-butylphenol is converted into the chloro carbonate by reaction of the sodium salt (NaH) with phosgene in the presence of a trace of DMF. This product is converted into **1** by reaction with dimethylamine.

   Dimethylaminomethylation.[1] The reagent is converted into the lithium deriva-tive **2** by reaction with *sec*-butyllithium in the presence of TMEDA in THF (20°). The products (**3**) obtained by reaction of **2** with carbonyl compounds are cleaved

2

3   **4**

to tertiary amines (**4**) by lithium aluminum hydride with quantitative recovery of the tri-*t*-butylphenol. The reagent **2** also reacts with alkyl halides to form the expected carbamates in 30–85% yield.

[1] D. Seebach and T. Hassel, *Angew. Chem., Int. Ed.*, **17**, 274 (1978).

Tri-*n*-butylstannyllithium, $(n\text{-}C_4H_9)_3SnLi$ **(1)**. Mol. wt. 296.97. The reagent is prepared most conveniently by deprotonation of tri-*n*-butyltin hydride with LDA in THF at 0° (about quantitative yield).

RCH(OR')Li reagents.[1] The reagent **(1)** reacts with an aldehyde at −78°

(I) $RCHO \xrightarrow{1, -78°} RCH\overset{OH}{\underset{SnBu_3}{|}}$ $\xrightarrow[>90\%]{CH_3\overset{Cl}{\overset{|}{C}}HOC_2H_5,\ C_6H_5N(CH_3)_2}$

2

$$RCH\overset{\overset{CH_3}{|}\ OCHOC_2H_5}{\underset{SnBu_3}{|}} \xrightarrow{n\text{-BuLi, THF, } -78°} RCH\overset{\overset{CH_3}{|}\ OCHOC_2H_5}{\underset{Li}{|}}$$

3   **4**

to form tri-*n*-butylstannylcarbinols **(2)**, which are rather unstable on prolonged standing. They are therefore converted into the ethoxyethyl ethers **(3)** by reaction with α-chloroethyl ethyl ether. This reaction occurs at 0° in CH_2Cl_2 in the presence of a base. The ethers are stable and can be stored for months. On treatment with *n*-butyllithium they are converted into the α-alkoxy organolithium reagent **4**.

Use of reagents of type **4** in synthesis is shown in scheme (I) for a simple synthesis of dendrolasin **(7)** and (±)-9-hydroxydendrolasin **(6)** from furane-3-carboxaldehyde and geranyl chloride.

Scheme (I)

Conjugate addition. Trialkylstannyllithium reagents prepared in ether solution undergo predominantly 1,2-addition to cyclohexenones, but solutions prepared in THF undergo conjugate addition to almost all enones, even hindered ones. The intermediate lithium enolates can be alkylated with reactive alkyl halides. These reactions are useful because secondary alkylstannanes are converted into the corresponding carbonyl compound by oxidation with $CrO_3 \cdot 2Py$. Tertiary alkylstannanes are also oxidized by $CrO_3 \cdot 2Py$, but mixtures of alcohols and products of dehydration are formed.

Examples:

This general method was used for a synthesis of dihydrojasmone from 2-cyclopentenone (equation **I**).[2]

Cleavage of 2-bromoethyl esters. These esters are cleaved by consecutive treatment with trimethylstannyllithium and tetra-*n*-butylammonium fluoride (equation **I**).[3]

[1] W. C. Still, *Am. Soc.*, **100**, 1481 (1978).
[2] *Idem, ibid.*, **99**, 4836 (1977).
[3] T.-L. Ho, *Syn. Comm.*, **8**, 359 (1978).

Tri-*n*-butyltin hydride, 1, 1192–1193; **2,** 424; **3,** 294; **4,** 518–520; **5,** 685–686; **6,** 604; **7,** 379–380.

Reduction of bridgehead bromides.[1] Bridgehead bromides are reduced to the hydrocarbons in 90–100% yields by this hydride with irradiation with ultraviolet light. Yields are lower in the absence of irradiation. This photochemical process is not as effective with bridgehead chlorides.

Alkenes from **vic-*diols.*** Cyclic and acyclic *vic*-diols can be converted into olefins by conversion to bisdithiocarbonates (dixanthates) followed by reaction with tri-*n*-butyltin hydride (equation I).[2] Five examples of the reaction in the carbohydrate field were cited.

(I)

Alkenes from β-*hydroxy sulfides.*[3] This elimination was first observed on attempted reduction of the xanthate ester group of **1** with tri-*n*-butyltin hydride; the product, however, was the alkene **2**. This is a general reaction of β-hydroxy and β-chloro sulfides. It has wide application since β-hydroxy sulfides are readily

available by sulfenylation of ketones followed by reduction. For example, 3-cholestanone was converted in this way into Δ²-cholestene (82% yield).

β-Hydroxy sulfones also undergo this elimination. An example is shown in equation (I). Note that only the (E)-isomer is formed.

Reduction of 4-chloroazetidinones. These substances (1) are readily reduced to azetidinones (2) by n-Bu$_3$SnH and AIBN in toluene (65°). The products have been prepared previously in rather low yield by Raney nickel desulfurization of penicillins. The reduction is not useful for 3-phthalimido derivatives, probably because of steric effects.[4]

Example:

[1] T.-Y. Luh and L. M. Stock, *J. Org.*, **42**, 2790 (1977).
[2] A. G. M. Barrett, D. H. R. Barton, R. Bielski, and S. W. McCombie, *J.C.S. Chem. Comm.*, 866 (1977).
[3] B. Lythgoe and I. Waterhouse, *Tetrahedron Letters*, 4223 (1977).
[4] C. A. Whitesitt and D. K. Herron, *Tetrahedron Letters*, 1737 (1978).

Tri-μ-carbonylhexacarbonyldiiron, Fe$_2$(CO)$_9$, **1**, 259–260; **2**, 139–140; **3**, 101; **4**, 157–158; **5**, 221–224; **6**, 195–198; **7**, 110–111.

Reaction with α,α′-dibromo ketones. Mechanistic aspects of this reaction have been discussed in detail.[1] Use of this reaction to form 4-cycloheptenones[2] and troponoids (**5**, 222–223; **6**, 195–196),[3] tropane alkaloids (**6**, 223–224),[4] 3-aryl-cyclopentanones,[5] and cyclopentenones (**4**, 158)[6] has been described in detail.

Coupling of acid chlorides; symmetrical ketones. The reaction of α-mono-substituted acetyl chlorides with 1 equiv. of Fe$_2$(CO)$_9$ results in symmetrical ketones in fair yield.[7] This coupling is not observed with other iron carbonyls; Ni(CO)$_2$[P(C$_6$H$_5$)$_3$]$_2$ does effect this coupling, but in lower yield.

$$RCH_2COCl + Fe_2(CO)_9 \xrightarrow[25-70\%]{Ether, \Delta} RCH_2\overset{\displaystyle O}{\overset{\displaystyle \|}{C}}CH_2R$$

[1] R. Noyori, Y. Hayakawa, H. Takaya, S. Murai, R. Kobayashi, and N. Sonoda, *Am. Soc.*, **100**, 1759 (1978).
[2] H. Takaya, S. Makino, Y. Hayakawa, and R. Noyori, *ibid.*, **100**, 1765 (1978).

[3] *Idem, ibid.*, **100**, 1778 (1978).
[4] Y. Hayakawa, Y. Baba, S. Makino, and R. Noyori, *ibid.*, **100**, 1786 (1978).
[5] Y. Hayakawa, K. Yokoyama, and R. Noyori, *ibid.*, **100**, 1791 (1978).
[6] *Idem, ibid.*, **100**, 1799 (1978).
[7] T. C. Flood and A. Sarhangi, *Tetrahedron Letters*, 3861 (1977).

2,2,2-Trichloro-*t*-butyloxycarbonyl chloride (1). Mol. wt. 239.92, b.p. 77–81°/ 12 mm.

Preparation:

$$Cl_3C-\overset{\overset{\displaystyle CH_3}{|}}{\underset{\underset{\displaystyle CH_3}{|}}{C}}-OH + COCl_2 \xrightarrow[89\%]{CH_2Cl_2,\ -20°} Cl_3C-\overset{\overset{\displaystyle CH_3}{|}}{\underset{\underset{\displaystyle CH_3}{|}}{C}}-OCOCl$$

1

Protection of amino groups.[1] Amino acids are converted to TCBOC-protected amino acids by reaction with **1** under the usual Schotten-Baumann conditions (80–95% yield). The products are nicely crystalline and are readily soluble in organic solvents. The TCBOC group is stable to $1N$ NaOH (40°, 2 hours) and to TFA (20°, 2 hours). The protecting group can be cleaved, as expected, with zinc in acetic acid or by reaction with lithium cobalt(I)–phthalocyanine.

[1] H. Eckert, M. Listl, and I. Ugi, *Angew. Chem., Int. Ed.*, **17**, 361 (1978).

Trichlorosilane–*t*-Amines, 3, 298–299; **4**, 525–526; **5**, 688.

ArCOOH → ArCH₃ (**4**, 526). The procedure for reduction of 2-biphenyl-carboxylic acid to 2-methylbiphenyl has been published. The overall yield is 74–80%.[1]

Cleavage of carbamates to carbinols.[2] Carbamates are cleaved to alcohols and isocyanates by treatment with trichlorosilane and triethylamine in a dry

$$(I) \qquad R O \overset{\overset{\displaystyle O}{||}}{C} NHR^1 \xrightarrow[70-95\%]{HSiCl_3,\ N(C_2H_5)_3} ROH + R^1N=C=O$$

solvent (usually benzene) under reflux for 4 hours. The reaction is independent of the structure of the alcohol, and no racemization or rearrangement is noted, both of which have been observed when base or acid was used.

[1] G. S. Li, D. F. Ehler, and R. A. Benkeser, *Org. Syn.*, **56**, 83 (1977).
[2] W. H. Pirkle and J. R. Hauske, *J. Org.*, **42**, 2781 (1977).

1,1,1-Trichloro-3,3,3-trifluoroacetone, 6, 606–607. The reagent can be prepared readily by reaction of chloropentafluoroacetone (PCR) and aluminum chloride (55–69% yield). It trifluoroacetylates amino groups of amino acids under neutral and mild conditions.[1]

[1] C. A. Panetta and T. G. Casanova, *J. Org.*, **35**, 4275 (1970); C. A. Panetta, *Org. Syn.*, **56**, 122 (1977).

Triethanolamine borate, (1). Mol. wt. 156.98, m.p. 236.5–237.5°.

Supplier: Aldrich. Preparation.[1]

Monoalkylation of ketones. Polyalkylated products are usually obtained as by-products of attempted monoalkylation of lithium ketone enolates. Polyalkylation can be suppressed by addition of triethylboron, but this substance is spontaneously flammable. The safer boron derivative 1 is also effective, but since it is sparingly soluble in THF, DMSO is also added to the reaction. The position of alkylation can be controlled also by use of the kinetically generated enolate or the more stable equilibrium enolate.[2]

1) LDA, 1 hr.
2) 1
3) CH₃I

4 (63%) + 3 (9.2%)

[1] H. C. Brown and E. A. Fletcher, *Am. Soc.*, **73**, 2808 (1951).
[2] M. W. Rathke and A. Lindert, *Syn. Comm.*, **8**, 9 (1978).

Triethyloxonium tetrafluoroborate, 1, 1210–1212; **2,** 430–431; **3,** 303; **4,** 527–529; **5,** 691–693; **7,** 386–387.

Esterification (**5,** 693). The esterification of 4-acetoxybenzoic acid with Meerwein's reagent (85–95% yield) has been published. The report includes results for the esterification of eight other acids; a note states that phenols can be converted into ethers in this way.[1]

Homologation of ketones with diazoacetic esters. Under catalysis with this Lewis acid or antimony pentachloride, diazoacetic esters react with ketones to form the homologated β-keto esters (equation I).[2] Reactions of this type have

$$(I) \quad R_2CO + N_2CHCOOR^1 \xrightarrow[\text{50–90\%}]{(C_2H_5)_3O^+BF_4^-,\ CH_2Cl_2,\ 0°} RCOCHR + N_2$$

with a COOR¹ group on the CHR.

generally been carried out with diazomethane,[3] but usually in only moderate yields. The mechanism of this newer method has been studied with the hope of controlling the stereoselectivity. At this time, it appears that insertion into the least substituted C(O)—C bond is favored, but in all cases only one residue is introduced.[4]

$$CH_3CH_2COCH_2CH_3 \; + \; N_2CHCOOC_2H_5 \xrightarrow[86\%]{} CH_3CH_2COCH\overset{\displaystyle COOC_2H_5}{\underset{\displaystyle |}{C}}HCH_2CH_3$$

$+ \; N_2CHCOOC_2H_5 \xrightarrow[90\%]{}$

$(C_2H_5)_3O^+BF_4{}^-, CH_2Cl_2$

[1] D. J. Raber, P. Gariano, Jr., A. O. Brod, A. L. Gariano, and W. C. Guida, *Org. Syn.*, **56**, 59 (1977).
[2] W. L. Mock and M. E. Hartman, *J. Org.*, **42**, 459 (1977).
[3] C. D. Gutsche, *Org. React.*, **8**, 364 (1954).
[4] W. L. Mock and M. E. Hartman, *J. Org.*, **42**, 466 (1977).

Triethyl phosphite, 1, 1212–1216; **2,** 432–433; **3,** 304; **4,** 529–530; **5,** 693; **6,** 612; **7,** 387.

Reduction of α-keto acids. α-Keto acids can be reduced selectively to α-hydroxy acids by reaction with triethyl phosphite in acetonitrile overnight followed by alkaline hydrolysis. Four examples were reported, and yields ranged from 70 to 95%. A possible mechanism is shown in the formulation.[1]

$$RC\underset{\displaystyle COOH}{\overset{\displaystyle =O}{\underset{|}{}}} + P(OC_2H_5)_3 \longrightarrow \left[R\bar{C}\overset{+}{OP}(OC_2H_5)_3 \; \longrightarrow \; RCH\overset{+}{OP}(OC_2H_5)_3 \right] \xrightarrow{NaOH}$$

with COOH and COO⁻ substituents

$$RCHOH + 3C_2H_5OH + (HO)_3P{=}O$$

COOH

(70–95%)

[1] T. Saegusa, S. Kobayashi, Y. Kimura, and T. Yokoyama, *J. Org.*, **42**, 2797 (1977).

Triethylsilane–Boron trifluoride, 7, 387–388.

Deoxygenation of carbonyl compounds.[1] This system, which has been shown to reduce alcohols to hydrocarbons, also reduces aliphatic and aromatic ketones and aromatic aldehydes to hydrocarbons. Yields are higher when the BF₃ is scrubbed by HF before use. The reaction proceeds stepwise, since intermediate alcohols can be isolated. A large excess of the silane is required.

[1] J. L. Fry, M. Orfanopoulos, M. G. Adlington, W. R. Dittman, Jr., and S. B. Silverman, *J. Org.*, **43**, 374 (1978).

Triethylsilane–Trifluoroacetic acid, 5, 695.

Reduction of an aralkyl keto group. A recent example of this transformation is the conversion of **1** into **2** in 95% yield (pure) with 2.5 equiv. of triethylsilane and trifluoroacetic acid.[1]

$$(C_2H_5)_3SiH, CF_3COOH \xrightarrow{\quad 95\% \quad}$$

1   **2**

[1] J. S. Swenton and P. W. Raynolds, *Am. Soc.*, **100**, 6188 (1978).

2-Triethylsilyl-1,3-butadiene (1). Mol. wt. 168.36, b.p. 30–33°/0.7 mm.
Preparation:

Diels-Alder reactions. This diene is considerably more reactive than 1-tri-alkylsilylbutadienes. The silyl group exerts only a moderate effect on the regio-selectivity (last example), but this effect can be increased by the addition of BF_3 etherate.[1]

Examples:

1 + ... $\xrightarrow{75\%}$ ...

1 + ... $\xrightarrow{87\%}$ ...

1 + ... $\xrightarrow{77\%}$ ...

$$+ \quad 3.3:1$$

[1] D. G. Batt and B. Ganem, *Tetrahedron Letters*, 3323 (1978).

Trifluoroacetic acid, 1, 1219–1221; **2,** 433–434; **3,** 305–308; **4,** 530–532; **5,** 695–700; **6,** 613–615; **7,** 388–389.

Rearrangement of cyclobutanones. α-Phenylcyclobutanones can be rearranged to β-tetralones by acids. This cyclization was used to rearrange **1**, obtained by cycloaddition of 1-naphthylketene to 1,3-cyclopentadiene, to the steroid **2**.[1]

Polyene cyclization (**3,** 305–307; **4,** 531–532; **5,** 696–697; **6,** 613–614; **7,** 389). Johnson *et al.*[2] have noted that a terminal trimethylsilylacetylenic group influences the structure of the final steroid in biomimetic polyene cyclizations. Thus **1** is cyclized to a D-homosteroid (**2**) in contrast to the cyclization of **3** to **4**.

[1] L. H. Dao, A. C. Hopkinson, and E. Lee-Ruff, *Tetrahedron Letters*, 1413 (1978).
[2] W. S. Johnson, T. M. Yarnell, R. F. Myers, and D. R. Morton, *ibid.*, 2549 (1978).

Trifluoroacetic anhydride–Pyridine.

Nitriles. Primary amides and aldoximes are dehydrated to nitriles in excellent yields (usually $\sim 90\%$) by this system at a temperature below 5° (45 minutes).[1]

[1] F. Campagna, A. Carotti, and G. Casini, *Tetrahedron Letters*, 1813 (1977).

Trifluoroacetic anhydride–Sodium iodide.

Reduction of sulfoxides to sulfides. This reduction can be conducted at 0° by addition of TFAA to an acetone solution of a sulfoxide and sodium iodide. Yields are 90–98%. The reduction involves acylation of the sulfoxide to form an acyloxysulfonium salt, which is converted into a sulfide by iodide anion (equation I).[1]

$$\text{(I)} \quad R^1 \overset{\overset{\displaystyle O}{\|}}{S} R^2 \xrightarrow{\text{(CF}_3\text{CO)}_2\text{O, 0°}} \left[\begin{array}{c} R^1 \overset{+}{S} R^2 CF_3 COO^- \\ | \\ OCOCF_3 \end{array} \right] \xrightarrow{2\text{NaI}} R^1 SR^2 + I_2 + 2CF_3 COONa \quad (90–98\%)$$

Deoxygenation of epoxides. Epoxides react with trifluoroacetic anhydride (1 equiv.) and sodium iodide (1 equiv.) in CH_3CN–THF (1:1) to give a β-iodotrifluoroacetate; when this product is treated with sodium iodide (3 equiv.) in the same solvent system, an olefin, sodium trifluoroacetate, and iodine are formed. The olefin has the same geometry as the starting epoxide; yields of the olefin are 80–95%. Presumably, trifluoroacetyl iodide is generated *in situ* from the anhydride and NaI.[2]

[1] J. Drabowicz and S. Oaė, *Synthesis*, 404 (1977).
[2] P. E. Sonnet, *J. Org.*, **43**, 1841 (1978).

Trifluoromethanesulfonic acid (Triflic acid), 4, 533; 5, 701–702; 6, 617–618.

Review. The chemical uses of this acid and its derivatives have been reviewed (302 references).[1]

[1] R. D. Howells and J. D. McCown, *Chem. Rev.*, **77**, 69 (1977).

9-Trifluoromethylsulfonyl-9-borabicyclo[3.3.1]nonane (9-BBN triflate). Mol. wt. 254.09, b.p. 38°/0.03 mm. The triflate is prepared by reaction of 9-BBN and triflic acid in hexane at 20° (80% yield).

Directed cross-aldol reactions. The reaction of methyl ketones with other carbonyl compounds occurs regiospecifically when this triflate and a tertiary base are used.[1]

Example:

$$C_6H_5CH_2CH_2COCH_3 + C_6H_5CHO \xrightarrow{\text{9-BBN triflate, N(C}_2\text{H}_5)_3}$$

$$C_6H_5CH_2\overset{\displaystyle |}{\underset{\displaystyle C_6H_5CHOH}{C}}HCOCH_3 + C_6H_5CH_2CH_2\overset{\overset{\displaystyle O}{\|}}{C}CH_2\overset{\overset{\displaystyle C_6H_5}{|}}{C}HOH$$

$$\quad\quad (88\%) \quad\quad\quad\quad\quad\quad (2\%)$$

[1] T. Inoue, T. Uchimaru, and T. Mukaiyama, *Chem. Letters*, 153 (1977).

Trifluoroperacetic acid, 1, 821–827.

γ-Hydroxy-α-nitroalkenes. Oximes can be oxidized to nitroalkanes by trifluoroperacetic acid in refluxing acetonitrile containing a buffer such as sodium hydrogen phosphate or sodium bicarbonate. The reaction is particularly useful for preparation of secondary nitroparaffins (**1**, 822).[1] This reaction has been extended to oxidation of α,β-epoxy ketoximes (**6**, 214) to give γ-hydroxy-α-nitroalkenes.[2]

Examples:

Trifluoroacetylation. Trifluoroacetylation of a hydroxyl group was first observed as a side reaction in the epoxidation of an unsaturated hindered alcohol with trifluoroperacetic acid at room temperature. This reaction is apparently general; thus trifluoroacetylation was observed in the case of cyclohexylmethanol (55%) and 3β-cholestanol (60%). The esters are easily hydrolyzed by chromatography on silica gel.[3]

[1] W. D. Emmons and A. S. Pagano, *Am. Soc.*, **77**, 4557 (1955).
[2] T. Takamoto, Y. Ikeda, Y. Tachimori, A. Seta, and R. Sudoh, *J.C.S. Chem. Comm.*, 350 (1978).
[3] G. W. Holbert and B. Ganem, *J.C.S. Chem. Comm.*, 248 (1978).

Trifluorovinyllithium, 6, 622.

2-Fluoro-2-alkenals. These α,β-unsaturated aldehydes can be prepared as shown in equation (I).[1]

[1] R. Sauvêtre, D. Masure, C. Chuit, and J. F. Normant, *Synthesis*, 128 (1978).

Trimethylaluminum, 5, 707–708; **6**, 622–623.

N-Alkylation of amides. This reaction can be carried out in two steps: condensation with formaldehyde[1] followed by reaction of the resulting methylol with a trialkylaluminum.[2]

(I) $R^1CNH_2 + HCHO \longrightarrow R^1CNHCH_2OH \xrightarrow[20-90\%]{AlR_3, C_6H_6, \Delta} R^1CNHCH_2R$

Trimethylaluminum complexes with metallocene dichlorides. Zirconocene dichloride and trimethylaluminum form an organometallic reagent of uncertain structure that reacts with 1-octyne at 20–25° to form 2-methyl-1-octene and 2-nonene in the ratio 95:5 (equation I).

(I) $C_6H_{13}C{\equiv}CH + Cl_2ZrCp_2 + (CH_3)_3Al \xrightarrow{100\%}$

$$C_6H_{13}C{=}CH_2 + C_6H_{13}CH{=}CHCH_3$$
$$\underset{CH_3}{|} \qquad 95:5$$

The stereochemistry of this new carbometalation was established by the reaction of phenylacetylene and of 2-deuteriophenylacetylene (equations II and III).

(II) $C_6H_5C{\equiv}CH \xrightarrow{(CH_3)_3Al, Cl_2ZrCp_2} \xrightarrow{D_2O}$

$$\begin{array}{c} C_6H_5 \diagdown {=}C \diagup H \\ CH_3 \diagup \diagdown D \end{array}$$
$$(\sim 96\% \ E)$$

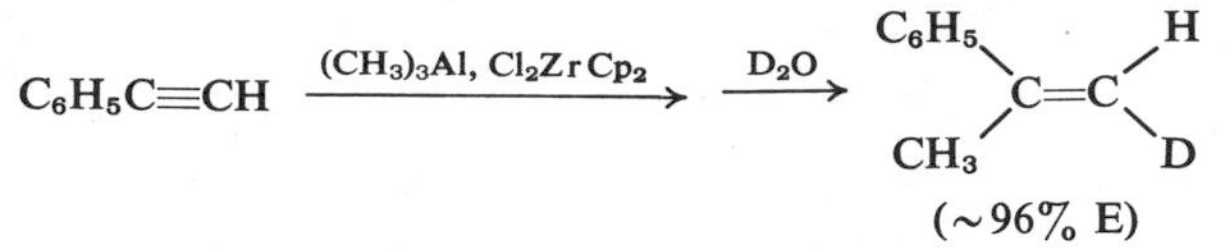

(III) $C_6H_5C{\equiv}CD \longrightarrow \xrightarrow{H_2O}$

$$\begin{array}{c} C_6H_5 \diagdown {=}C \diagup D \\ CH_3 \diagup \diagdown H \end{array}$$
$$(> 98\% \ Z)$$

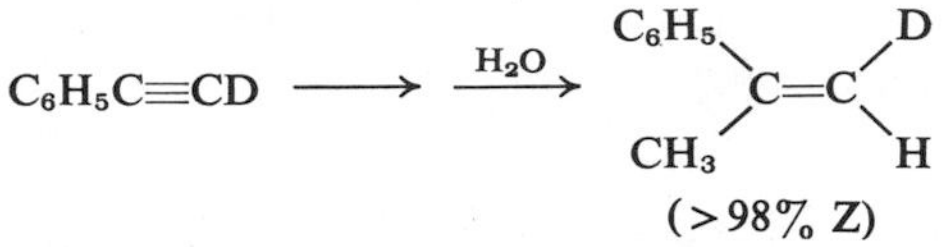

The reaction is equally applicable and stereospecific for internal acetylenes; other trialkylalanes can also be used.[3]

The intermediate alkenylzirconium compounds can be used in a cross-coupling reaction with alkenyl, aryl, or alkynyl halides in the presence of Pd or Ni catalysts and $ZnCl_2$.[4]

Examples:

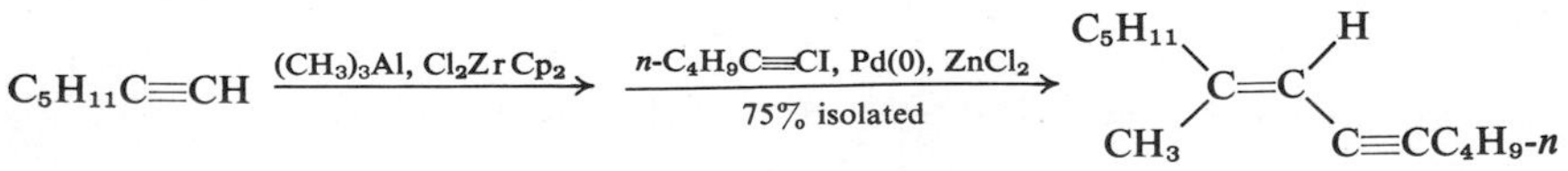

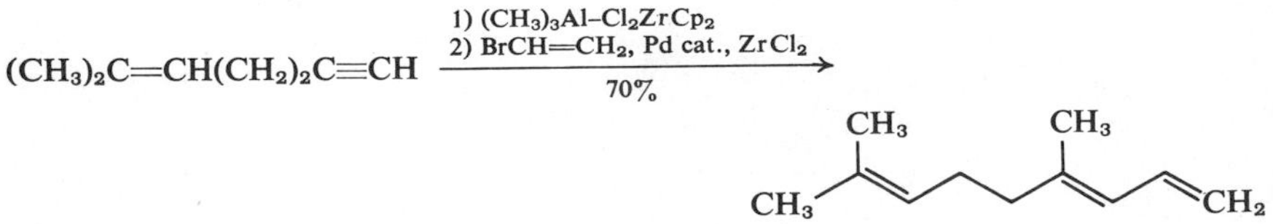

$C_5H_{11}C{\equiv}CH \xrightarrow{(CH_3)_3Al, Cl_2ZrCp_2} \xrightarrow[75\% \ isolated]{n\text{-}C_4H_9C{\equiv}CI, Pd(0), ZnCl_2}$

$$\begin{array}{c} C_5H_{11} \diagdown {=}C \diagup H \\ CH_3 \diagup \diagdown C{\equiv}CC_4H_9\text{-}n \end{array}$$

$(CH_3)_2C{=}CH(CH_2)_2C{\equiv}CH \xrightarrow[70\%]{\substack{1) (CH_3)_3Al-Cl_2ZrCp_2 \\ 2) BrCH{=}CH_2, Pd \ cat., ZrCl_2}}$

Negishi *et al.*[5] have extended the reaction of alkynes with trimethylaluminum and zirconocene dichloride to a corresponding addition reaction with trimethyl-aluminum–titanocene dichloride. This latter reagent is somewhat more reactive. It reacts with diphenylacetylene to form (Z)-α-methylstilbene (2) in high yield. The yield of 2 is < 30% when the corresponding zirconocene reagent is used.

(I) $C_6H_5C\equiv CC_6H_5$ $\xrightarrow[84\%]{Al(CH_3)_3-TiCp_2Cl_2, \ CH_2Cl_2, \ 20°}$ $\xrightarrow{H_2O}$

$$\begin{array}{c} C_6H_5 \quad\quad C_6H_5 \\ \diagdown C=C \diagup \\ CH_3 \quad\quad\quad H \end{array}$$

2

Unexpectedly the reaction of this complex with 5-decyne under the same conditions leads to the allene 3, formed by dehydrometalation of the initial addition product (equation II).

(II) $CH_3(CH_2)_3C\equiv C(CH_2)_3CH_3$ $\xrightarrow[92\%]{H_2O}$

$$\begin{array}{c} CH_3(CH_2)_3 \quad\quad\quad (CH_2)_2CH_3 \\ \diagdown C=C=C \diagup \\ CH_3 \quad\quad\quad\quad\quad H \end{array}$$

3

Yields are low in reactions with 1-alkynes unless $ZnCl_2$ is added to form an intermediate alkynylzinc chloride. Under these conditions 1-alkenes can be obtained in 75–85% yield (equation III).

(III) $CH_3(CH_2)_3C\equiv CZnCl$ $\xrightarrow[75\%]{H_2O}$

$$\begin{array}{c} CH_3(CH_2)_3 \\ \diagdown C=CH_2 \\ CH_3 \diagup \end{array}$$

[1] H. E. Zaugg and W. B. Martin, *Org. React.*, **14**, 52 (1965).
[2] A. Basha and S. M. Weinreb, *Tetrahedron Letters*, 1465 (1977).
[3] D. E. Van Horn and E. Negishi, *Am. Soc.*, **100**, 2252 (1978).
[4] E. Negishi, N. Okukado, A. O. King, D. E. Van Horn, and B. I. Spiegel, *ibid.*, **100**, 2254 (1978).
[5] D. E. Van Horn, L. F. Valente, M. J. Idacavage, and E. Negishi, *J. Organometal. Chem.*, **156**, C20 (1978).

Trimethylamine N-oxide, 1, 1230–1231; **2**, 434; **3**, 309–310; **6**, 624–625; **7**, 392.

Oxidation of organoboranes (**6**, 624). Oxidation of organoboranes with this oxide is a useful alternative to alkaline hydrogen peroxide.[1] Hydrocarbon or ethereal solvents can be used, but because of limited solubility of the reagent, vigorous stirring is necessary. The rates of reaction vary with the alkyl groups in the organoborane: tertiary > cyclic secondary > acyclic secondary > *n*-primary > branched primary > vinyl.

[1] G. W. Kabalka and S. W. Slayden, *J. Organometal. Chem.*, **125**, 273 (1977).

Trimethyl orthoformate, 3, 313; **4**, 540; **5**, 714.

Acetals.[1] Trimethyl orthoformate when absorbed on montmorillonite clay

K-10[2] is a very effective reagent for rapid conversion of carbonyl compounds to dimethyl acetals in high yields (90–100%). This reagent is superior with respect to yields to the Amberlyst-15–triethyl orthoformate reagent (**5**, 356).

[1] E. C. Taylor and C.-S. Chiang, *Synthesis*, 467 (1977).
[2] Girdler catalyst K-10, Girdler Chemicals, Inc., Louisville, Ky.

(−)-**4,6,6-Trimethyl-1,3-oxathiane**, (**1**). Mol. wt. 145.24,

α_D − 30.4°. This oxathiane was prepared in several steps from (−)-3-benzylthio-butyric acid (in 44% enantiomeric excess).

Diastereoselective reaction of a Grignard reagent. Eliel and co-workers[1] have reported a virtually completely asymmetric synthesis of (S)-(+)-atrolactic acid methyl ester (**6**) from this chiral starting material (44% ee). Metalation and reaction with C_6H_5CHO gives only **2**, in which the entering group has the equatorial configuration. The (S)-configuration at C_4 therefore induces the (R)-configuration at C_2 in **2**. This is then oxidized to the ketone **3**, which has the same ee as **1**. Treatment of **3** with methylmagnesium iodide gives a single product (**4**), which is methylated and cleaved to the aldehyde **5**. This is converted into **6** in two steps. Again the ee of **6** is the same as **1**. Therefore the Grignard reaction must proceed in about quantitative optical yield and affords a striking example of the Prelog-Cram rule.[2]

[1] E. L. Eliel, J. K. Koshimies, and B. Lohri, *Am. Soc.*, **100**, 1614 (1978).
[2] D. J. Cram and F. A. Abd Elhafez, *ibid.*, **74**, 5828 (1952); V. Prelog, *Helv.*, **36**, 308 (1953).

Trimethyloxosulfonium hydroxide, $(CH_3)_2\overset{\overset{\displaystyle O}{\|}}{S}\text{—}CH_3^-OH$. The reagent was first prepared by Kuhn and Trischmann.[1] In a modified method, trimethyloxosulfonium iodide is treated with (excess) silver oxide in aqueous methanol.[2]

Methylation.[2] The reagent converts pyrimidines and pyrimidine nucleosides into the corresponding N-methylated derivatives in excellent yields.

[1] R. Kuhn and H. Trischmann, *Ann.*, **611**, 117 (1958).
[2] K. Yamauchi, K. Nakamura, and M. Kinoshita, *J. Org.*, **43**, 1593 (1978).

Trimethylsilylacetyl chloride, $(CH_3)_3SiCH_2COCl$ (**1**). Mol. wt. 148.67. This acid chloride is prepared by reaction of oxalyl chloride with trimethylsilylacetic acid (**6**, 631–632).

α,β-Unsaturated thiol esters. α-Silyl thiol esters (**2**) are available from the reaction of **1** with mercaptans. These esters, in the presence of LDA or trityllithium, react with carbonyl compounds in THF to form α,β-unsaturated thiol esters (**3**). One noteworthy feature is the stereoselectivity: **3** has the (E)-configuration (>95% of product).[1]

[1] D. H. Lucast and J. Wemple, *Tetrahedron Letters*, 1103 (1977).

3-Trimethylsilyl-1-cyclopentene (**1**). Mol. wt. 104.67.
Preparation:

Cyclopentene derivatives. These compounds are formed by reaction of **1** in the presence of $TiCl_4$ with aldehydes, ketones, α-keto esters, or α,β-unsaturated ketones. The allylsilane reacts with acyl chlorides in the presence of $AlCl_3$ to form cyclopentene-2-yl ketones, which are rearranged on silica gel to the α,β-unsaturated ketones.[1]

Examples:

$$1 + n\text{-}C_3H_7CHO \xrightarrow[]{TiCl_4,\ CH_2Cl_2} \xrightarrow[78\%]{H_2O}$$

$$1 + C_6H_5COCOOC_2H_5 \xrightarrow[90\%]{}$$

$$1 + C_6H_5COCl \xrightarrow[72\%]{AlCl_3} \left[\right] \xrightarrow[72\%]{H_2SiO_3}$$

[1] I. Ojima, M. Kumagai, and Y. Miyazawa, *Tetrahedron Letters*, 1385 (1977).

2-Trimethylsilylethanol, $(CH_3)_3SiCH_2CH_2OH$. Mol. wt. 118.25, b.p. 50–52°/10 mm. Supplier: Fluka.

Protection of carboxyl groups. Trimethylsilylethyl esters, —COOCH₂-$CH_2Si(CH_3)_3$, can be prepared from N-benzyloxycarbonylamino acids and this reagent with DCC in the presence of pyridine (65–95% yield). The esters are stable under usual conditions of peptide synthesis, but are readily cleaved by fluoride ion, preferably by tetra-*n*-butylammonium fluoride in DMF.[1]

This protecting group has been used in a synthesis of the mold metabolite curvularin (**6**), shown briefly in scheme (I). The intermediate (**3**) contains two different ester groups, but only the desired group is cleaved by fluoride ion to give **4**. Remaining steps are an intramolecular acylation and hydrogenolysis of the benzyl groups.[2]

Scheme (I)

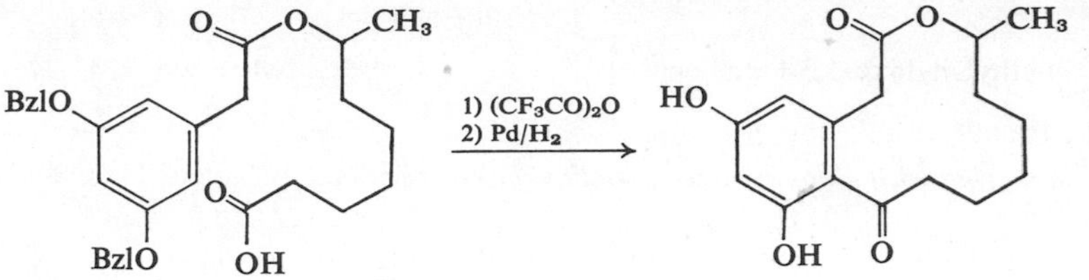

4 → **6**

Scheme (I)—*continued*

[1] P. Sieber, *Helv.*, **60**, 2711 (1977).
[2] H. Gerlach, *ibid.*, **60**, 3039 (1977).

α-Trimethylsilyl-2-methylbenzothiazole, —$CH_2Si(CH_3)_3$ **(1)**.   Mol.
wt. 221.40. This reagent (denoted as $BTCH_2TMS$) may be prepared *in situ* by treatment of α-lithio-2-methylbenzothiazole with chlorotrimethylsilane at $-78°$ (45 minutes) and then at $-23°$ (45 minutes) in ether.[1]

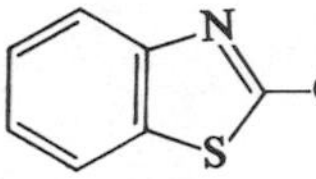

   Spiroannelation. Cyclic ketones readily condense with the α-lithio derivative of **1** giving α,β-unsaturated benzothiazoles (β,β-disubstituted enal equivalents). These in turn may be converted into spiro derivatives as shown in equation (I).[2]

(I)

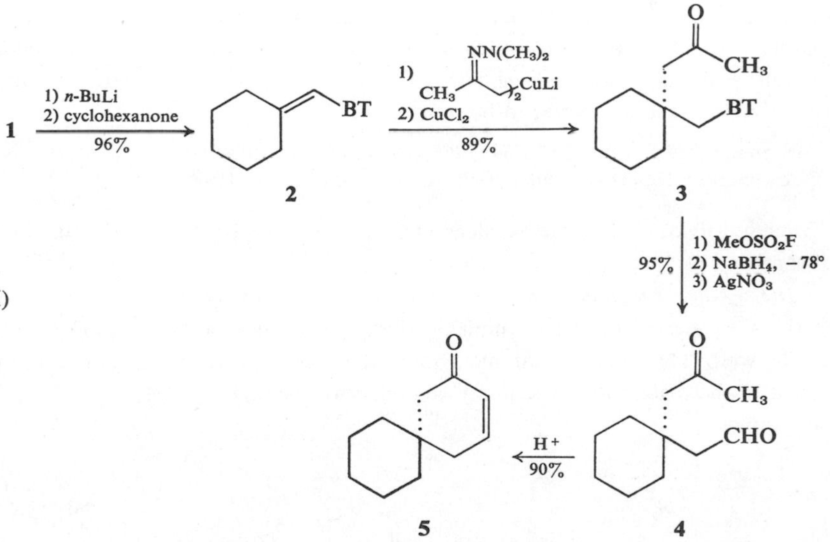

[1] E. J. Corey and D. L. Boger, *Tetrahedron Letters*, 5 (1978).
[2] *Idem, ibid.*, 13 (1978).

1-Trimethylsilyloxy-1,3-butadiene, (1). Mol. wt. 142.27, b.p. 131°. Preparation.[1]

Reaction with dimethyl acetylenedicarboxylate.[2] The butadiene (1) reacts with

dimethyl acetylenedicarboxylate to give **2** as the initial adduct. When heated **2** loses trimethylsilanol to give dimethyl phthalate (**3**). A more interesting reaction is the oxidation of **2** to **4** with either Collins reagent or DDQ.

Other trimethylsiloxy-substituted butadienes also undergo this Diels-Alder reaction and also eliminate trimethylsilanol readily. However, the oxidative aromatization reaction seems to be more limited.

[1] J. Dunoguès, Belgium Pat. 670769 (1966); *C.A.*, **65**, 5487d (1966).
[2] K. Yamamoto, S. Suzuki, and J. Tsuji, *Chem. Letters*, 649 (1978).

2-Trimethylsilyloxy-1,3-cyclohexadiene (1). Mol. wt. 166.30, b.p. 33–37°/0.1 mm. Preparation, **6**, 521.

Diels-Alder reaction. This 1,3-diene is useful for [4 + 2] cycloaddition reactions as shown in the examples. Alkyl substituted derivatives of **1** serve equally well. The second example shows the importance of the temperature, at least in the reaction with dimethyl acetylenedicarboxylate.[1]

Examples:

$$1 + \underset{\text{COOCH}_3}{\overset{\text{COOCH}_3}{\underset{\text{C}}{\overset{\text{C}}{\parallel}}}} \xrightarrow[91\%]{\text{C}_6\text{H}_6, \Delta} \quad (\text{CH}_3)_3\text{SiO} \cdots \text{COOCH}_3 \xrightarrow[84\%]{\text{H}_3\text{O}^+}$$

74% | C₆H₅CH₃, Δ

[1] G. M. Rubottom and D. S. Krueger, *Tetrahedron Letters*, 611 (1977).

4-Trimethylsilyloxyvaleronitrile (1). Mol. wt. 171.3, b.p. 65–67°/3 mm.
    Preparation:

$$\text{LiCH}_2\text{CN} \xrightarrow[78\%]{\begin{array}{c} 1)\ \overset{\text{CH}_3}{\triangle}, \text{THF} \\ 2)\ (\text{CH}_3)_3\text{SiCl} \end{array}} \quad \underset{(\text{CH}_3)_3\text{SiO}}{\overset{\text{CH}_3}{\diagdown}}\text{CHCH}_2\text{CH}_2\text{CN}$$

1

1,4-Diketones. The reagent (1) reacts with Grignard reagents to form γ-hydroxy ketones, which can be oxidized directly with Jones reagent to give 1,4-diketones.[1]

$$1 + \text{RCH}_2\text{MgBr} \xrightarrow{\text{Ether}} \xrightarrow[60-95\%]{\text{H}_3\text{O}^+} \underset{\text{OH}}{\text{CH}_3\text{CHCH}_2\text{CH}_2\overset{\text{O}}{\overset{\parallel}{\text{C}}}\text{CH}_2\text{R}} \xrightarrow[45-85\%]{\text{Jones oxid.}} \text{CH}_3\overset{\text{O}}{\overset{\parallel}{\text{C}}}\text{CH}_2\text{CH}_2\overset{\text{O}}{\overset{\parallel}{\text{C}}}\text{CH}_2\text{R}$$

[1] S. Murata and I. Matsuda, *Synthesis*, 221 (1978).

Trimethylsilylpotassium, 7, 402–403.

    (E) → (Z) *Isomerization* (7, 402–403). This method for isomerization of (E)- to (Z)-olefins was used in a synthesis of (Z)-20(22)-didehydrocholesterol (4) from pregnenolone (1). Wittig reaction of **1** led to the undesired (E)-20(22)-olefin (2). This was epoxidized and then treated with trimethylsilylpotassium; the overall yield was 52%.[1]

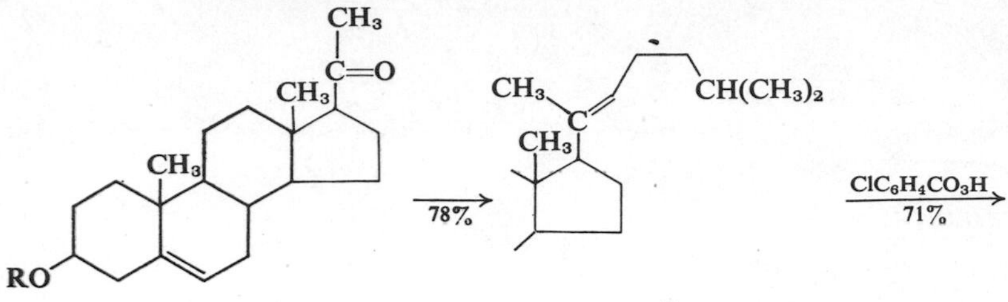

1

2

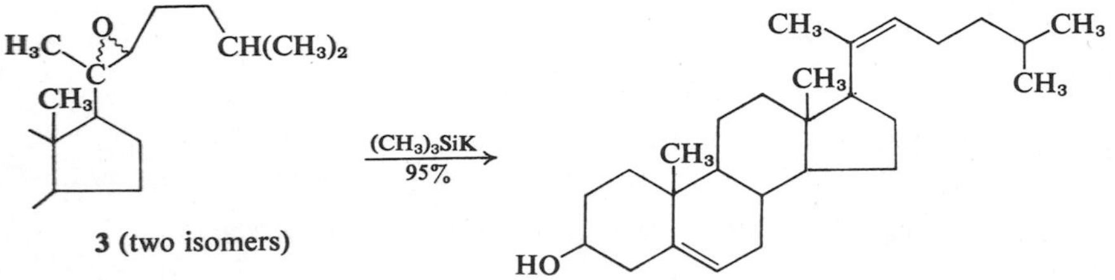

3 (two isomers)

4

Trimethylsilylarenes.[2] Aryl chlorides and bromides on reaction with trimethylsilylpotassium, trimethylsilylsodium, or trimethylsilyllithium (7, 400) are
converted into trimethylsilylarenes in about 90% yield together with some of the
product of reduction (ArH, 5–8%). In the case of aryl iodides, arenes are obtained
in about 25% yield. This trimethylsilyldehalogenation is probably a free-radical
process.

$$ArCl(Br) \xrightarrow[\text{KOCH}_3, \text{ HMPT}]{(CH_3)_3SiSi(CH_3)_3,} ArSi(CH_3)_3 + ArH$$
$$(85-90\%) \quad (5-8\%)$$

[1] M. Koreeda, N. Koizumi, and B. A. Teicher, *J.C.S. Chem. Comm.*, 1035 (1976).
[2] M. A. Shippey and P. B. Dervan, *J. Org.*, **42**, 2654 (1977).

Trimethylsilyl trifluoromethanesulfonate, $CF_3\overset{\overset{O}{\|}}{\underset{\underset{O}{\|}}{S}}$—$OSi(CH_3)_3$ **(1)**. Preparation.[1]

Silylation of nitriles.[2] The combination of this reagent (**1**) and triethylamine
has been used for silylation of nitriles. The actual reagent is probably the salt
$(C_2H_5)_3\overset{+}{N}Si(CH_3)_3CF_3SO_3^-$, which can be isolated.
    Examples:

$$CH_3CH_2C\equiv N + 1 + N(C_2H_5)_3 \xrightarrow[33\%]{\text{Ether, } 5°} CH_3\underset{\underset{Si(CH_3)_3}{|}}{\overset{\overset{Si(CH_3)_3}{|}}{C}}C\equiv N$$

$$N \equiv CCH_2C \equiv N + 1 \xrightarrow[80\%]{} \underset{\underset{Si(CH_3)_3}{|}}{N \equiv CCHC \equiv N}$$

$$C_6H_5CH_2C \equiv N + 1 \xrightarrow[48\%]{} \underset{\underset{Si(CH_3)_3}{|}}{C_6H_5CHC \equiv N}$$

α-*Trimethylsilyl esters.* These esters can be prepared in 40–65% yield by reaction of esters with this reagent and triethylamine in ether at 0–5°.[3]

$$R^1CH_2COOR^2 + CF_3SO_2OSi(CH_3)_3 \xrightarrow[40-65\%]{\substack{N(C_2H_5)_3, \\ ether, 0-5°}} \underset{\underset{Si(CH_3)_3}{|}}{R^1CHCOOR^2}$$

[1] M. Schmeisser, P. Sartori, and B. Lippsmeier, *Ber.*, **103**, 868 (1970).
[2] H. Emde and G. Simchen, *Synthesis*, 636 (1977).
[3] *Idem, ibid.*, 867 (1977).

α-Trimethylsilylvinylmagnesium bromide, $CH_2 = C \underset{MgBr}{\overset{Si(CH_3)_3}{\diagdown}}$ **(1).**

α-*Methylenebutyrolactones.* In the presence of CuI as catalyst, **1** reacts with epoxides to give homoallylic alcohols (**2**). These can be converted into α-methylene-γ-lactones (**4**) by replacement of —Si(CH₃)₃ by —Br (**3**) followed by reaction with nickel carbonyl and potassium acetate as base.[1]

[1] I. Matsuda, *Chem. Letters*, 773 (1978).

Trimethylzinclithium, $(CH_3)_3ZnLi$. Trialkylzinclithium reagents are obtained by reaction of zinc chloride (or bromide) and an alkyllithium in a 1:3 ratio:

$$3RLi + ZnCl_2 \xrightarrow{THF} R_3ZnLi + 2LiCl$$

These reagents undergo conjugate addition exclusively to α,β-unsaturated ketones. The reactivity, but not the mode of addition, depends on the size of the

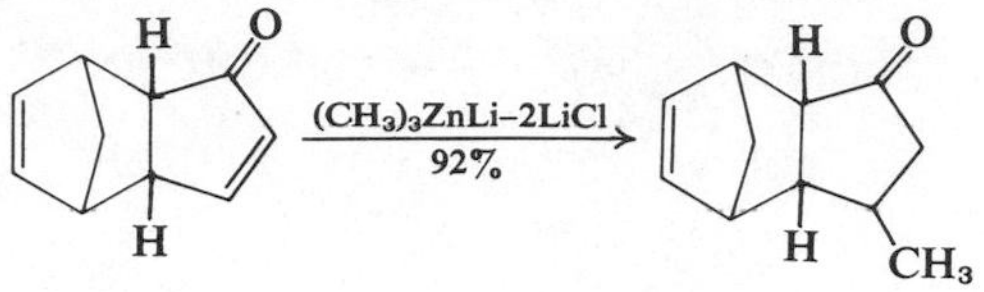

R group; when R is *t*-butyl in the reaction cited, the yield of adduct obtained in the same time is only 58%.

[1] M. Isobe, S. Kondo, N. Nagasawa, and T. Goto, *Chem. Letters*, 679 (1977).

Triphenylphosphine–Carbon tetrachloride, 1, 1274; **2**, 445; **3**, 320; **4**, 551–552; **5**, 727; **6**, 644–645; **7**, 404.

Arene imines. Blum *et al.*[1] have converted arene oxides into N-alkyl arene imines by reaction with an amine followed by cyclodehydration (method of Appel and Kleinstück, **5**, 727). The reaction is illustrated for preparation of phenanthrene-9,10-imines (equation I). For a different route to arene imines *see* **6**, 541.

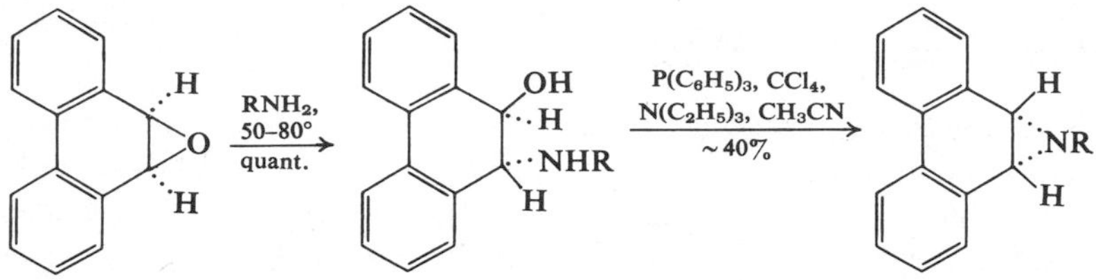

Allylic chlorides (**4**, 552). This system[2] is useful for regioselective conversion of allylic alcohols into allylic chlorides. The method, however, is not so useful for preparation of allylic chlorides with boiling points near those of carbon tetrachloride (77°) and chloroform (62°). The isolation of the product can be simplified by use of hexachloroacetone (b.p. 202°) rather than carbon tetrachloride. Yields and regioselectivity are still high in this variation.[3]

Example:

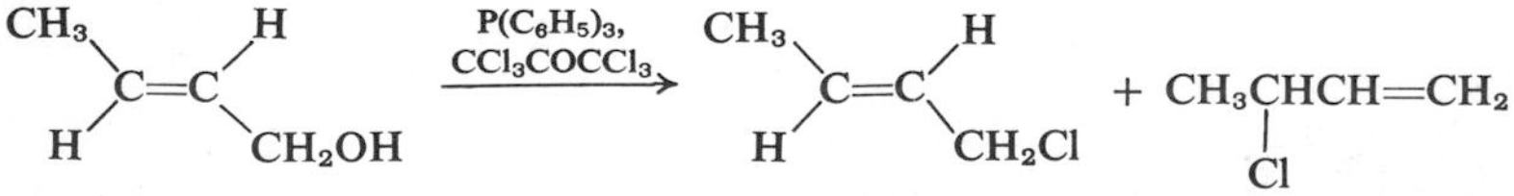

[1] Y. Ittah, I. Shahak, and J. Blum, *J. Org.*, **43**, 397 (1978).
[2] R. Appel, *Angew. Chem., Int. Ed.*, **14**, 801 (1975).
[3] R. M. Magid, O. S. Fruchey, and W. L. Johnson, *Tetrahedron Letters*, 2999 (1977).

Triphenylphosphine–Diethyl azodicarboxylate, 4, 553–555; **5,** 727–728; **6,** 645; **7,** 404–406.

Alkenes. β-Hydroxy carboxylic acids are converted into alkenes in fair to good yield by treatment with triphenylphosphine and diethyl azodicarboxylate (equation I).[1] The reaction involves loss of CO_2 and H_2O (decarboxylative dehydration).

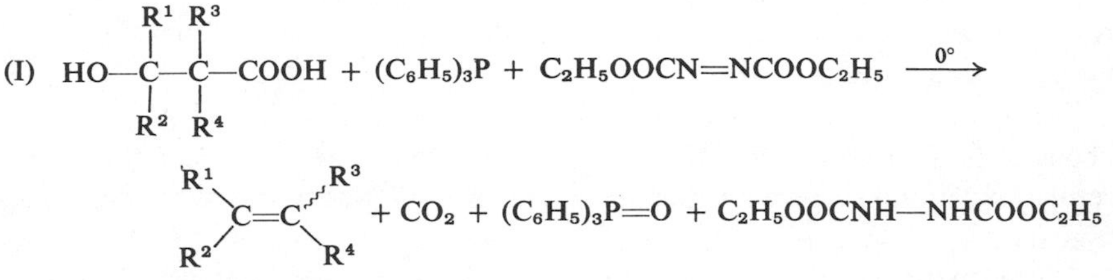

(I) $HO\!-\!\overset{R^1}{\underset{R^2}{C}}\!-\!\overset{R^3}{\underset{R^4}{C}}\!-\!COOH + (C_6H_5)_3P + C_2H_5OOCN\!=\!NCOOC_2H_5 \xrightarrow{\;0°\;}$

$\overset{R^1}{\underset{R^2}{}}C\!=\!C\overset{R^3}{\underset{R^4}{}} + CO_2 + (C_6H_5)_3P\!=\!O + C_2H_5OOCNH\!-\!NHCOOC_2H_5$

(55–75%)

Macrocyclic lactones. Seebach et al.[2] have synthesized the diketo dilactone (−)-pyrenophorin (3) by cyclodimerization of the dithiane derivative 1 with triphenylphosphine and diethyl azodicarboxylate. The product 2 on hydrolysis of the dithiane groups (HgO–BF$_3$ etherate) yields the corresponding diketone (3).

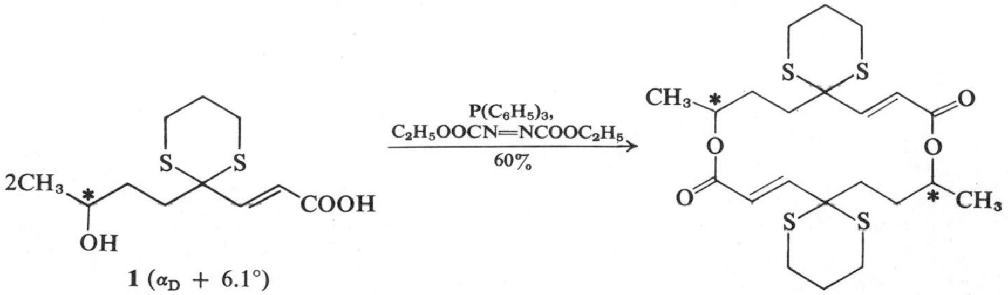

1 (α_D + 6.1°)

2 (α_D −109°)

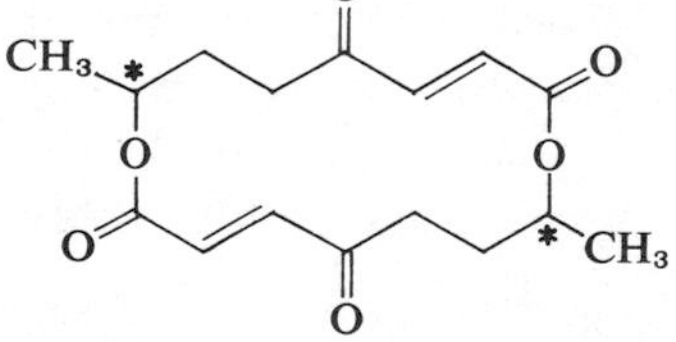

3 (α_D −54.5°)

[1] J. Mulzer and G. Brüntrup, *Angew. Chem., Int. Ed.*, **16**, 255 (1977).
[2] D. Seebach, B. Seuring, H.-O. Kalinowski, W. Lubosch, and B. Renger, *ibid.*, **16**, 264 (1977).

Triphenylphosphine–Iodine, 7, 407.

Deoxygenation of sulfoxides. This reaction can be carried out with triphenyl-phosphine activated by iodine (1 equiv.) in refluxing acetonitrile in 10–60 minutes. Sodium iodide is added to increase the rate. Sulfides are obtained in 70–95% yield.[1]

[1] G. A. Olah, B. G. B. Gupta, and S. C. Narang, *Synthesis*, 137 (1978).

Triphenylphosphine–Thiocyanogen. The reagent is prepared by addition of Br_2 to lead thiocyanate suspended in CH_2Cl_2 (0°); the solution of thiocyanogen is decanted and stirred with triphenylphosphine at $-40°$.

The reagent converts primary alcohols into thiocyanates, RSCN, in yields of about 60–95%. Tertiary alcohols are converted into isothiocyanates, RNCS, and secondary alcohols are converted into mixtures of thiocyanates and isothio-cyanates.

Indoles and pyrroles are cyanated by the reagent in fairly good yields[1]:

1,1-Disubstituted thioureas. Secondary amines (1) are converted into these thioureas (2) by freshly prepared $(C_6H_5)_3P(SCN)_2$ in CH_2Cl_2.[2]

Acyl isothiocyanates. The reaction of the reagent with carboxylic acids gives, as expected, acyl isothiocyanates, RCONCS, in 40–55% yield. A more interesting example is the reaction with salicylic acid in CH_2Cl_2 at $-40°$ to give 2-thio-3,4-dihydro-2H-1,3-benzoxazine-4-one (1) in 70% yield.[3]

[1] Y. Tamura, T. Kawasaki, M. Adachi, M. Tanio, and Y. Kita, *Tetrahedron Letters*, 4417 (1977).

[2] Y. Tamura, M. Adachi, T. Kawasaki, and Y. Kita, *ibid.*, 1753 (1978).

[3] Y. Tamura, T. Kawasaki, M. Tanio, and Y. Kita, *Chem. Ind.*, 806 (1978).

Triphenyl phosphite ozonide, 3, 323–324; **4,** 559; **7,** 408.

α-*Peroxylactones.* These very sensitive substances, which are intermediates in chemiluminescent reactions, have recently been prepared by cyclodehydration of α-hydroperoxy acids with dicyclohexylcarbodiimide under carefully controlled conditions (equation I).[1]

(I)

$$R_2C=C=O + (C_6H_5O)_3PO_3 \xrightarrow{CS_2, -20°}$$

These compounds have also been prepared more simply by reaction of triphenyl phosphite ozonide with the more readily available ketenes (equation II). The ozonide is known to generate singlet oxygen, which reacts with the ketene to form an intermediate that collapses to the α-peroxylactone.[2]

(II) $R_2C=C=O + (C_6H_5O)_3PO_3 \xrightarrow{CS_2, -20°}$

[1] W. Adam, A. Alzérreca, J.-C. Liu, and F. Yany, *Am. Soc.*, **99,** 5768 (1977).

[2] N. J. Turro, Y. Ito, M.-F. Chow, W. Adam, O. Rodriguez, and F. Yany, *ibid.*, **99,** 5836 (1977).

2,4,6-Triphenylpyrylium iodide (1). Mol. wt. 436.28.

Preparation:

Iodides from amines. Alkyl-, benzyl-, and pyridylamines react with **1** to form pyridinium iodides (**2**), which are converted into iodides (**3**) and 2,4,6-triphenylpyridine (**4**) when heated under vacuum.[1]

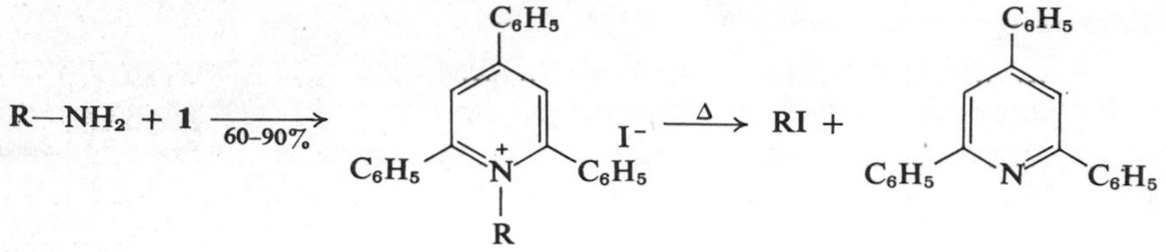

$$R-NH_2 + 1 \xrightarrow[60-90\%]{} \quad \xrightarrow{\Delta} RI +$$

2   3 (50–85%)   4

[1] N. F. Eweiss, A. R. Katritzky, P.-L. Nie, and C. A. Ramsden, *Synthesis*, 634 (1977).

2,4,6-Triphenylpyrylium tetrafluoroborate (1). Mol. wt. 396.18, m.p. 253–255°, yellow.

Preparation[1]:

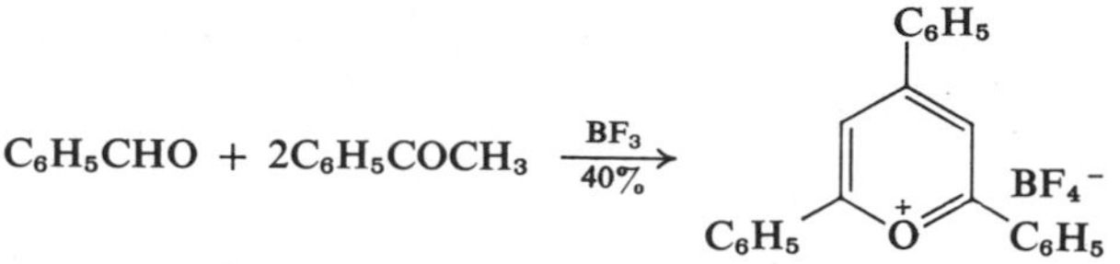

$$C_6H_5CHO + 2C_6H_5COCH_3 \xrightarrow[40\%]{BF_3}$$

1

Conversion of amines into esters.[2] Primary amines have been converted into acetates and benzoates (**2**) by a two-step process shown in equation (I), in which $R^1 = CH_3$ or C_6H_5.

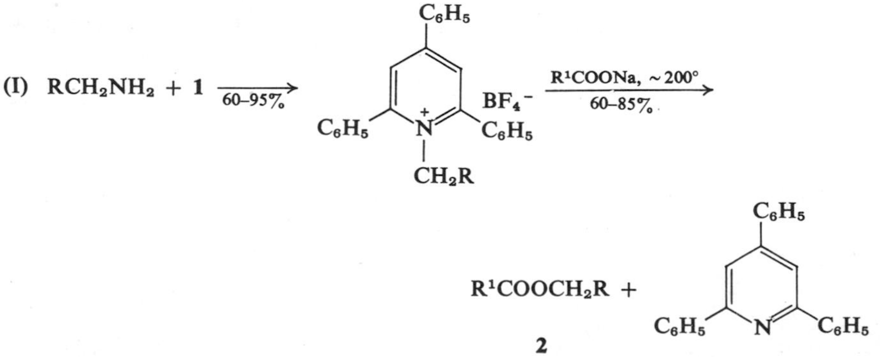

$$(I) \quad RCH_2NH_2 + 1 \xrightarrow[60-95\%]{} \quad \xrightarrow[60-85\%]{R^1COONa, \sim 200°}$$

$$R^1COOCH_2R +$$

2

A similar method has been used to convert hydrazides to isocyanates (equation II).[3]

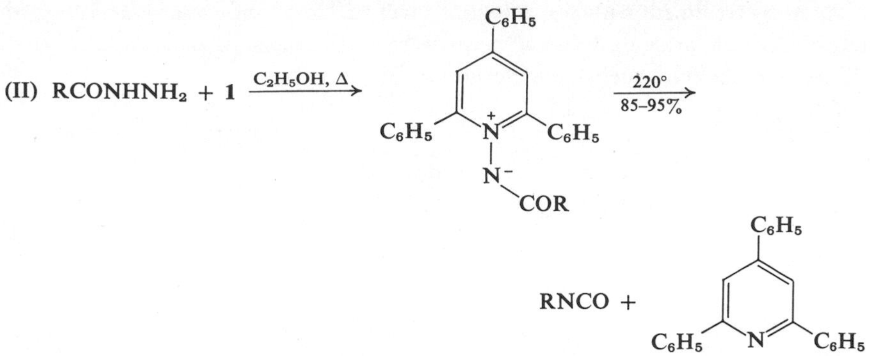

(II) RCONHNH$_2$ + 1 $\xrightarrow{C_2H_5OH, \Delta}$

$\xrightarrow[85-95\%]{220°}$

RNCO +

[1] R. Lombard and J.-P. Stephen, *Bull. soc.*, 1458 (1958).
[2] U. Gruntz, A. R. Katritzky, D. H. Kenny, M. C. Rezende, and H. Sheikh, *J.C.S. Chem. Comm.*, 701 (1977).
[3] J. B. Bapat, R. J. Blade, A. J. Boulton, J. Epsztajn, A. R. Katritzky, J. Lewis, P. Molina-Buendia, P. L. Nie, and C. A. Ramsden, *Tetrahedron Letters*, 2691 (1976).

2,4,6-Triphenylpyrylium thiocyanate,

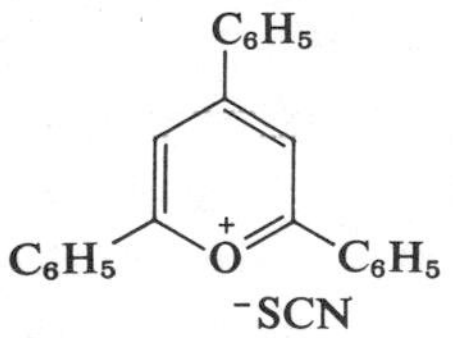

(1). Mol. wt. 399.53,

m.p. 192°, orange-red. The salt is obtained by treatment of the perchlorate salt with sodium acetate and then with ammonium thiocyanate.[1]

Thiocyanates. Primary aliphatic amines are converted into thiocyanates by the reaction formulated in equation (I).[2]

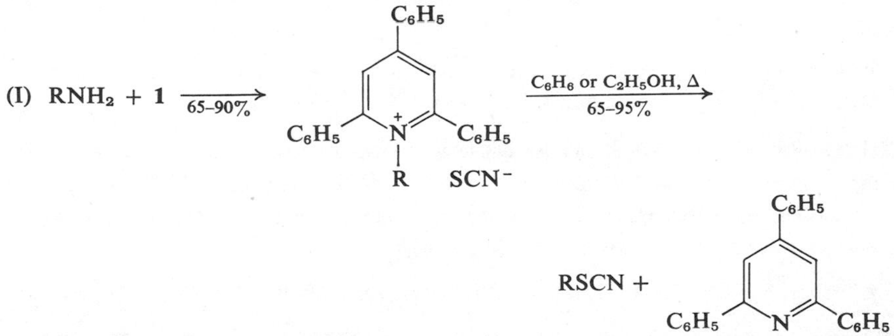

(I) RNH$_2$ + 1 $\xrightarrow[65-90\%]{}$

$\xrightarrow[65-95\%]{C_6H_6 \text{ or } C_2H_5OH, \Delta}$

RSCN +

[1] A. T. Balaban and M. Paraschiv, *Rev. Roumaine Chim.*, **19**, 1731 (1974).
[2] A. Katritzky, U. Gruntz, N. Mongelli, and M. C. Rezende, *J.C.S. Chem. Comm.*, 133 (1978).

Triphenyltin hydride, 1, 1250–1251; **3,** 448; **4,** 559; **5,** 734; **6,** 649.

Reduction of phenyl selenides and selenoketals. The C$_6$H$_5$Se group can be reduced by treatment with 2–4 equiv. of this hydride in refluxing toluene. Use of tri-*n*-butyltin hydride in refluxing benzene requires longer reaction times. Under the same conditions phenyl sulfides are not reduced.

Examples:

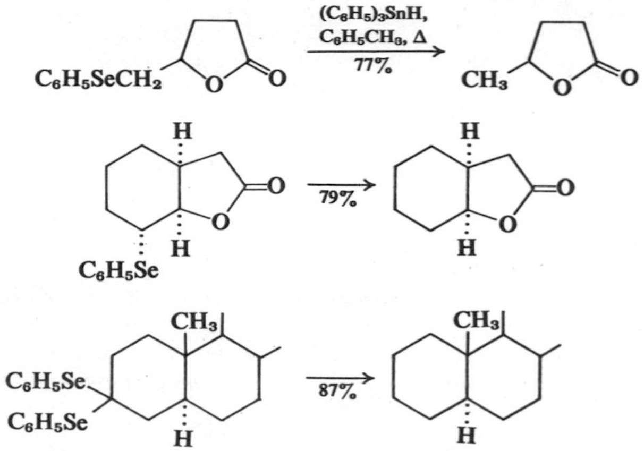

Since selenides can be prepared directly from primary alcohols (C_6H_5SeCN, **7**, 253) and selenoketals are readily available from ketones (**6**, 361–362), this process represents a new method for reduction of ketones and alcohols to hydrocarbons.[1]

[1] D. L. J. Clive, G. Chittattu, and C. K. Wong, *J.C.S. Chem. Comm.*, 41 (1978).

Tripotassium nonachloroditungstate(III), $K_3W_2Cl_9$. Mol. wt. 744.11. Preparation.[1]

Reduction of sulfoxides.[2] This complex ion of tungsten, as well as related complexes such as K_3MoCl_6 (available from Climax Molybdenum) and Cs_3Mo_2-Cl_8H, selectively reduces sulfoxides to sulfides under mild conditions and in high yield (90–100%).

[1] R. Sailliant, J. L. Hayden, and R. A. D. Wentworth, *Inorg. Chem.*, **6**, 1497 (1967).
[2] R. G. Nuzzo, H. J. Simon, and J. San Filippo, Jr., *J. Org.*, **42**, 568 (1977).

Tris(ethylthio)borane, $B(SC_2H_5)_3$. Mol. wt. 194.19, b.p. 78–81.5°/1.3 mm. This reagent is prepared by reaction of bis(ethylthio)lead with BCl_3 (80% yield).

Thioesters. The reagent converts carboxylic acids, RCOOH, into ethyl thioesters, $RCOSC_2H_5$ (about 75–80% yield).[1]

[1] A. Pelter, T. E. Levitt, K. Smith, and A. Jones, *J.C.S. Perkin I*, 1672 (1977).

Tris(phenylseleno)borane, $(C_6H_5Se)_3B$. Mol. wt. 478.99, m.p. 152–153°, stable. This reagent is prepared by reaction of selenophenol with boron tribromide in CS_2 (2 days, 20°).[1]

Selenoacetals. This stable reagent is more convenient than C_6H_5SeH for conversion of carbonyl compounds into selenoacetals. Addition of some TFA is usually advantageous. Yields are 50–90%.[2]

[1] M. Schmidt and H. D. Block, *J. Organometal. Chem.*, **25**, 17 (1970).
[2] D. L. J. Clive and S. M. Menchen, *J.C.S. Chem. Comm.*, 356 (1978).

Tris(triphenylphosphine)nickel(0), 6, 653–654.

Ullmann reaction. The Semmelhack-Kende variation of the Ullmann reaction (**4**, 33; **6**, 654) can be made catalytic with respect to the Ni(0) catalyst if the reaction is carried out in the presence of zinc to regenerate Ni(0) from the Ni(II) species formed during the reaction. The optimum molar ratio of $ArBr/Zn/P(C_6H_5)_3/NiCl_2[P(C_6H_5)_3]_2$ is $1:1:0.4:0.5$. DMF is the preferred solvent. Yields are also higher in the catalytic reaction; the yield of biphenyl from bromobenzene is 89%. The same paper also reports that when potassium iodide is added the reaction proceeds at room temperature to give biphenyl in 81% after 24 hours.[1]

[1] M. Zembayashi, K. Tamao, J. Yoshida, and M. Kumada, *Tetrahedron Letters*, 4089 (1977).

Tris(trimethylsilyloxy)ethylene (1). Mol. wt. 292.58.

Preparation[1]:

$$HOCH_2COOH \xrightarrow[\text{ClSi(CH}_3)_3\text{, Py}]{\text{HN[Si(CH}_3)_3]_2,} (CH_3)_3SiOCH_2COOSi(CH_3)_3 \xrightarrow[\text{ClSi(CH}_3)_3]{\text{LiN[Si(CH}_3)_3]_2,}}$$

$$\underset{(CH_3)_3SiO}{\overset{H}{\diagdown}} C{=}C[OSi(CH_3)_3]_2$$

1

Hydroxymethyl ketones.[1] Preparation of ketones of this type by reaction of acid chlorides and diazomethane followed by hydrolysis of the diazo ketone[2] is not practical for large scale work. The desired transformation can be carried out with the new reagent **1**, a derivative of ketene. A typical example is formulated in equation (I). Aromatic acid chlorides react more slowly with **1** than aliphatic acid

$$(I)\quad CH_3(CH_2)_6COCl + 2\ 1 \xrightarrow{SnCl_4} \underset{R}{\overset{(CH_3)_3SiO}{\diagdown}} C{=}C \underset{OSi(CH_3)_3}{\overset{COOSi(CH_3)_3}{\diagup}} \xrightarrow{H_3O^+}$$

$$\left[\underset{\underset{OH}{|}}{\overset{\overset{O}{\|}}{RC{-}CHCOOH}} \right] \xrightarrow[\substack{84\% \\ \text{pure}}]{-CO_2} CH_3(CH_2)_6\overset{\overset{O}{\|}}{C}CH_2OH$$

chlorides; a hindered acid chloride such as 1-adamantanecarbonyl chloride does not react at all after 4.5 hours at 95°.

[1] A. Wissner, *Tetrahedron Letters*, 2749 (1978).
[2] M. Steiger and T. Reichstein, *Helv.*, **20**, 1164 (1937).

Tritylpotassium (Potassium triphenylmethide), $(C_6H_5)_3CK$. This strong base can be generated by addition of triphenylmethane dissolved in DME to potassium hydride which has been treated with a small quantity of DMSO.[1] It has also been generated from triphenylmethane and potassium hydride in the presence of a crown ether.[2]

It is a useful base for effecting formation of potassium enolates. It was used for methylation of isobutyrophenone to give pivalophenone in 76% yield.[1]

[1] J. W. Huffman and P. G. Harris, *Syn. Comm.*, **7**, 137 (1977).
[2] E. Buncel and B. Menon, *J.C.S. Chem. Comm.*, 648 (1976).

Trityl tetrafluoroborate, 1, 1256–1258; **2**, 454; **4**, 565–567; **6**, 657; **7**, 414–415.

Alkenes from organolithium and organomagnesium compounds. This reagent abstracts the β-hydrogen of organolithium and organomagnesium compounds at low temperatures. Trialkylboranes also have this property, but the reaction is somewhat slower. The yields of alkenes increase within the series primary < secondary < tertiary. Formation of the less stable Hofmann alkene is favored.[1]

Examples:

$$CH_3(CH_2)_8CH_2MgCl \xrightarrow[28\%]{(C_6H_5)_3C^+BF_4^-} CH_3(CH_2)_7CH{=}CH_2 + (C_6H_5)_3CH$$

$$\underset{\overset{|}{C_6H_5}}{\overset{\overset{C_6H_5}{|}}{CH_3(CH_2)_4C{-}Li}} \xrightarrow[65\%]{(C_6H_5)_3C^+BF_4^-} CH_3(CH_2)_3CH{=}C(C_6H_5)_2$$

Trisubstituted alkenes can be obtained in a one-pot reaction by addition of an alkyllithium to activated disubstituted 1-alkenes followed by abstraction of the hydrogen atom β to lithium by reaction with triphenylborane (equation I).[2]

$$(I)\quad (C_6H_5)_2C{=}CH_2 + n\text{-}C_4H_9Li \longrightarrow$$

$$\left[\underset{\overset{|}{Li}}{(C_6H_5)_2C{-}CH_2(CH_2)_3CH_3} \right] \xrightarrow{B(C_6H_5)_3} (C_6H_5)_2C{=}CH(CH_2)_3CH_3 + LiB(C_6H_5)_3H$$

α,β-*Unsaturated ketones.* Trialkylsilyl enol ethers of ketones are oxidized by this reagent or by trityl methyl ether and BF_3 etherate in CH_2Cl_2 at 25° to give, after hydrolytic work-up, α,β-unsaturated ketones. Triphenylsilyl enol ethers and ethyl enol ethers are not oxidized under these conditions. The same oxidation can be used with trimethylsilyl enol ethers of esters. Yields of enones are only moderate because the reaction does not proceed to completion even under drastic conditions. The oxidation involves abstraction of an allylic hydrogen from the enol.[3]

Examples:

Oxidation of ols and diols (7, 414–415). Even unprotected *sec*-alcohols can be oxidized to ketones by trityl tetrafluoroborate (2 equiv.) in CH_2Cl_2 at 25° (10–13 hours). Typical yields are in the range 65–85%. Primary, secondary diols can be oxidized selectively to hydroxy ketones by the reagent (2.1–2.2 equiv.) under the same conditions (yields 55–80%).

The main limitation to this simplified procedure is that sterically hindered alcohols are not oxidized.[4]

[1] M. T. Reetz and W. Stephan, *Angew. Chem., Int. Ed.*, **16**, 44 (1977).
[2] M. T. Reetz and D. Schinzer, *ibid.*, **16**, 44 (1977).
[3] M. E. Jung, Y.-G. Pan, M. W. Rathke, D. F. Sullivan, and R. P. Woodbury, *J. Org.*, **42**, 3961 (1977).
[4] M. E. Jung and R. W. Brown, *Tetrahedron Letters*, 2771 (1978).

Tungsten carbonyl, $W(CO)_6$, **6**, 658.

Dehalogenative coupling. Benzyl halides and *gem*-dihalides undergo dehalogenative coupling in the presence of 2 equiv. of $W(CO)_6$, $W(CO)_5P(C_6H_5)_3$, $W(CO)_6$–$AlCl_3$, or WCl_6–$LiAlH_4$. In one case a carbene intermediate has been trapped. Some alcohols can be coupled with these tungsten reagents.[1]

Examples:

$$2C_6H_5CHCl_2 \longrightarrow$$

(~60%)   (10%)

$$(C_6H_5)_2CCl_2 \xrightarrow{80\%} (C_6H_5)_2C{=}C(C_6H_5)_2$$

$$C_6H_5CH_2OH \xrightarrow{75\%} C_6H_5CH_2CH_2C_6H_5$$

[1] Y. Fujiwara, R. Ishikawa, and S. Teranishi, *Bull. Chem. Soc. Japan*, **51**, 589 (1978).

V

Vanadyl trichloride, 3, 331–332; **5**, 744.

Oxidative coupling. Schwartz et al.[1] have examined various reagents for intramolecular coupling of diphenolic, monophenolic, and nonphenolic 1,3-diarylpropanes. Vanadyl trichloride was found to be effective in all three cases. Thallium (III) trifluoroacetate (**6**, 580) and silver(II) trifluoroacetate were found useful for monophenolic coupling.

Examples:

[1] M. A. Schwartz, B. F. Rose, R. A. Holton, S. W. Scott, and B. Vishnuvajjala, *Am. Soc.*, **99**, 2571 (1977).

Vanadyl trifluoride, 5, 745–746; **6**, 660; **7**, 418–421.

Oxidation of a 1,4-diarylbutane. The diarylbutane **1**, of known stereochemistry, is oxidized by VOF_3 to the *cis*-dimethyldibenzooctadiene (**2**) as shown by

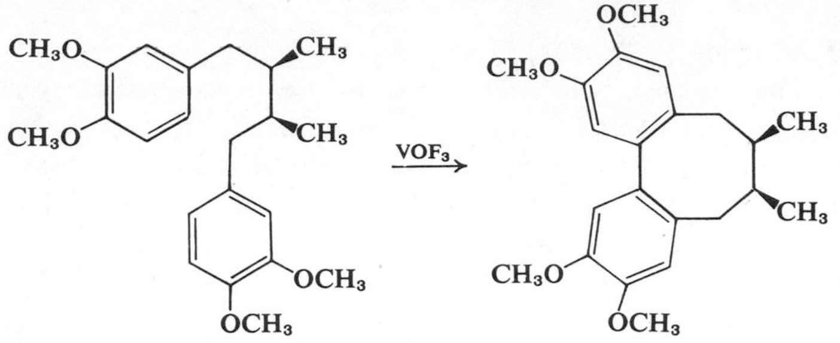

1   **2**

NMR spectra. The reaction is also used for a synthesis of a related lignan, (+)-deoxyschizandrin (**3**), shown schematically in equation (I).[1]

(I)

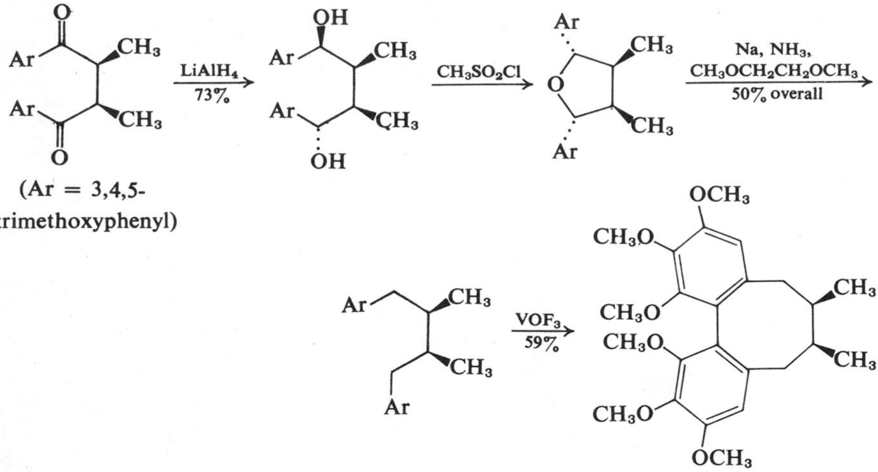

(Ar = 3,4,5-trimethoxyphenyl)

3

Phenanthrene synthesis. Stilbene derivatives are oxidatively cyclized to phenanthrenes by vanadyl trifluoride in $TFA–CH_2Cl_2$ at $0°$.[2] An example is the conversion of **1** into **2**. This reaction had been conducted by photocyclization,

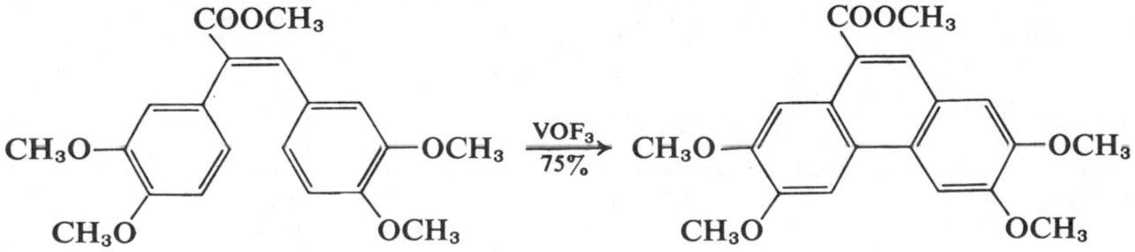

1   **2**

but in lower yield (31%).[3] The oxidative cyclization is also applicable to synthesis of benzophenanthrenes. It was used in a synthesis of the ring system of phenanthroindolizidine alkaloids, **3 → 4**; yield 59%. (±)-Tylophorine (**5**) was obtained by reduction of **4**.

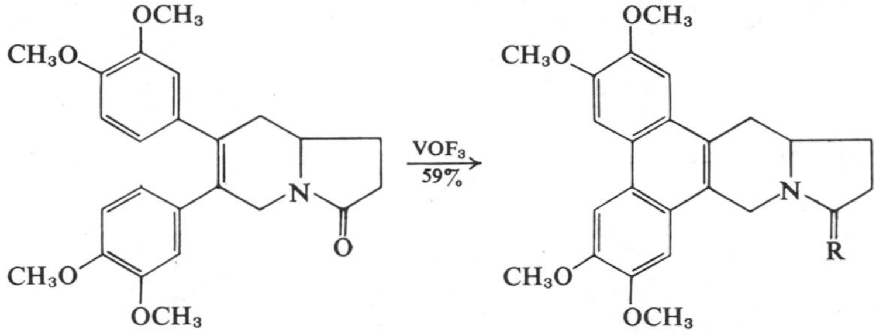

3

4 (R = O)
5 (R = H₂)

[1] T. Biftu, B. G. Hazra, and R. Stevenson, *J.C.S. Chem. Comm.*, 491 (1978).
[2] A. J. Liepa and R. E. Suammons, *ibid.*, 826 (1977).
[3] R. B. Herbert and C. J. Moody, *ibid.*, 121 (1970).

Vilsmeier reagent, 1, 284–289; **2**, 154; **3**, 116; **4**, 186; **5**, 251; **6**, 220; **7**, 422–424.

Formylation of trienes. Formylation of alkenes and dienes with the Vilsmeier reagent has been of limited value, but Dutch chemists find that formylation of trienes can be useful. Formylation occurs mainly at C_1 (equation I); the reaction has been used for synthesis of dehydrocitral from 2,6-dimethyl-2,5-heptadienone (equation II).[1]

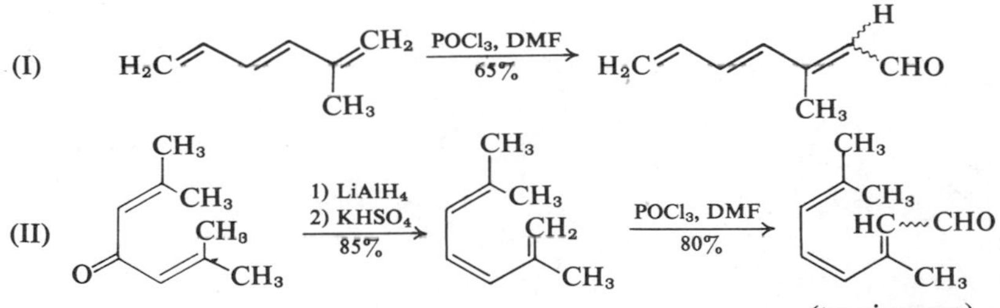

(two isomers)

Cycloheptatriene (**1**) when heated with POCl₃ and DMF is converted into 2-formylheptatriene (**2**) and bitropyl (**3**).[2]

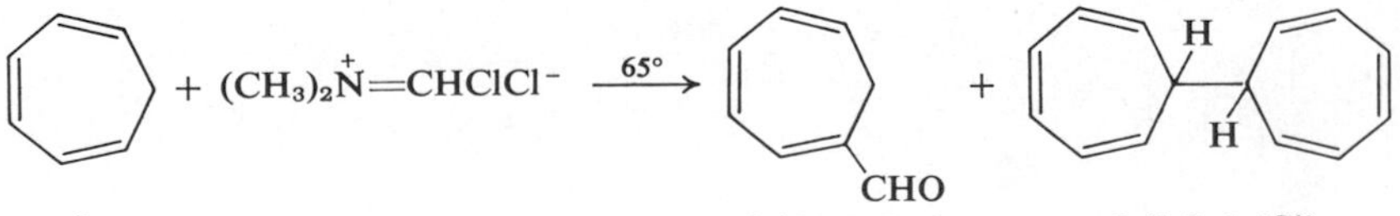

1

2 (18–30%)

3 (1.2–1.5%)

[1] P. C. Traas, H. J. Takken, and H. Boelens, *Tetrahedron Letters*, 2129 (1977).
[2] T. Asao, S. Kuroda, and K. Kato, *Chem. Letters*, 41 (1978).

Vinyl chloroformate (VOCCI), $\overset{O}{\underset{\|}{Cl}}COCH=CH_2$ (1). Mol. wt. 106.81, b.p. 67–69°.
Preparation[1]:

$$COCl_2 + HOCH_2CH_2OH \longrightarrow (CH_2OCOCl)_2 \xrightarrow[44\%]{\Delta} 1$$

Protection of amino groups (VOC-amino acids).[2] VOC-protected amino

acids ($H_2C=CHO\overset{O}{\underset{\|}{C}}-NHR$) are useful for peptide synthesis. The protective
group can be removed with HCl–ethanol or HBr–HOAc; by bromination–
alcoholysis; by Hg(OAc)$_2$ in aqueous HOAc.

N-Dealkylation of t-amines.[3] An example of this reaction is shown in equation (I).

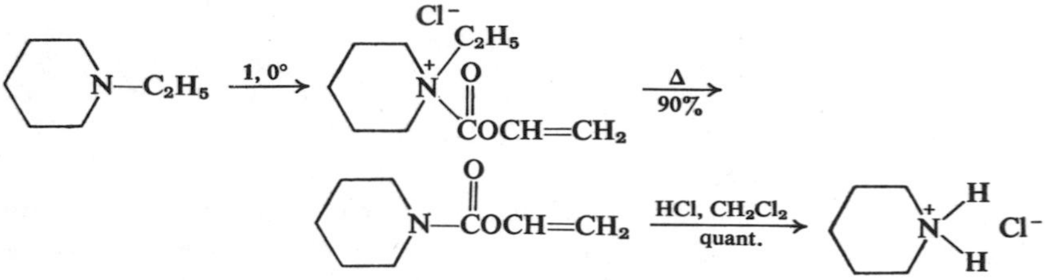

The reagent has also been used in an improved synthesis of naloxone, a narcotic
antagonist used for heroin overdose (equation II).

3,14-Diacetyl-
oxymorphone

Naloxone

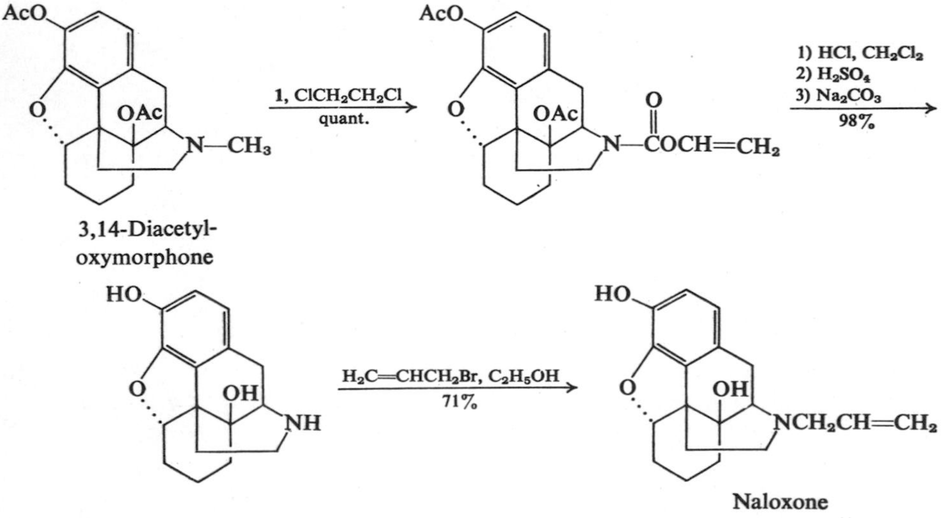

Protection of —OH.[4] VOC esters of phenols and alcohols can be prepared by standard methods, and they are readily hydrolyzed by Na_2CO_3 in aqueous dioxane, conditions that do not affect N-VOC groups. Unlike N-VOC groups, VOC esters are stable to HCl in dioxane at 25° and fairly stable to 50% aqueous HBF_4. The hydrolytic selectivity is illustrated by the preparation of nalorphine, a heroin antagonist, from morphine (equation I) in 77% overall yield.

(I)

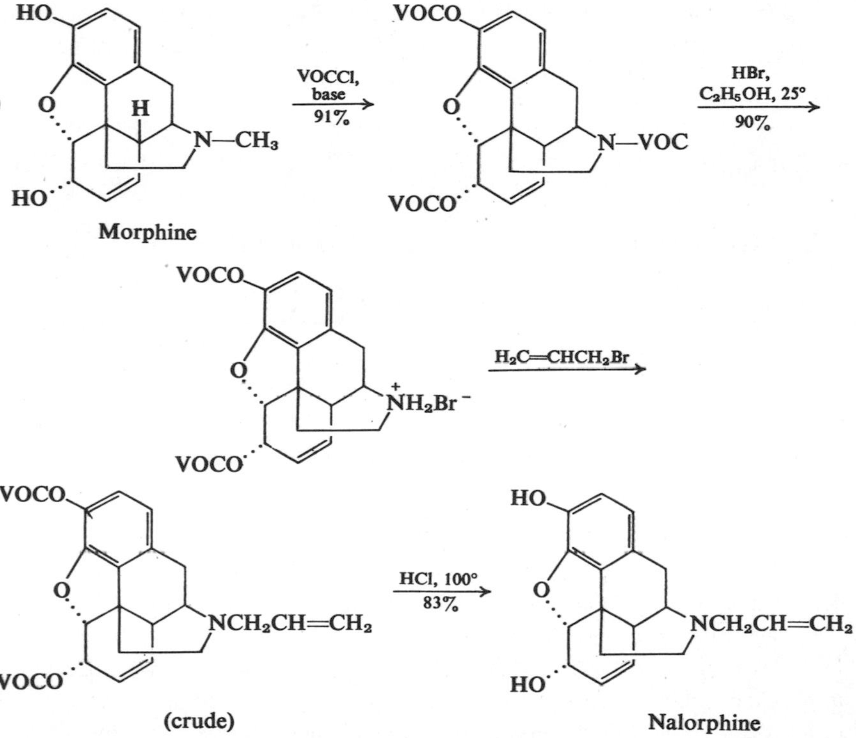
HO
O
H
N—CH3
HO
Morphine
VOCCl,
base
91%
VOCO
O
N—VOC
VOCO
HBr,
C2H5OH, 25°
90%
VOCO
O
NH2Br⁻
VOCO
H2C=CHCH2Br
VOCO
O
NCH2CH=CH2
VOCO
(crude)
HCl, 100°
83%
HO
O
NCH2CH=CH2
HO
Nalorphine

[1] L.-H. Lee, *J. Org.*, **30**, 3943 (1965).
[2] R. A. Olofson, Y. S. Yamamoto, and D. J. Wancowicz, *Tetrahedron Letters*, 1563 (1977).
[3] R. A. Olofson, R. C. Schnur, L. Bunes, and J. P. Pepe, *ibid.*, 1567 (1977).
[4] R. A. Olofson and R. C. Schnur, *ibid.*, 1571 (1977).

Z

Zinc, 1, 1276–1284; 2, 459–462; 3, 334–337; 4, 574–577; 5, 753–756; 6, 672–675; 7, 426–428.

Three-component condensation. β-Hydroxy nitriles such as 1 can be obtained in excellent yield by warming a mixture of an alkyl halide, acrylonitrile, and a carbonyl compound in the presence of zinc powder.[1]

$$(CH_3)_2CHI + CH_2{=}CHCN + (CH_3)_2CO \xrightarrow[98\%]{Zn, \Delta} (CH_3)_2CHCH_2\overset{CN}{\underset{}{C}}H{-}\overset{OH}{\underset{}{C}}(CH_3)_2$$

1

Two variations have also been reported (equations I and II).

(I) $(CH_3)_2CHI + CH_2{=}CHCN + (CH_3CO)_2O \xrightarrow[81\%]{Zn, CH_3CN}$

$$(CH_3)_2CHCH_2\overset{NC}{\underset{}{C}}{=}\overset{OCOCH_3}{\underset{}{C}}CH_3$$

(II) $CH_2{=}CHCN + CH_3\overset{O}{\overset{\|}{C}}CH_2CH_2CH_2I \xrightarrow[42\%]{Zn}$

Vitamin K. Vitamin K can be obtained in 60–65% yield in one step by reaction of 2-methyl-1,4-naphthoquinone with phytyl chloride in THF in the presence of zinc dust. Other metals are less satisfactory.[2]

Reduction of anthraquinones to anthracenes.[3] This reaction can be effected by dropwise addition of 80% HOAc to a mixture of the quinone and zinc dust in pyridine. Yields vary, but can be as high as 95%. The method is an adaptation of a procedure used by Kuhn and Winterstein[4] for reduction of bixins.

[1] T. Shono, I. Nishiguchi, and M. Sasaki, *Am. Soc.*, **100**, 4314 (1978).
[2] Y. Tachibana, *Chem. Letters*, 901 (1977).
[3] J. T. Traxler, *Syn. Comm.*, **7**, 161 (1977).
[4] R. Kuhn and A. Winterstein, *Ber.*, **65**, 646 (1932).

Zinc–Chlorotrimethylsilane, 7, 429–430.

Desulfurization. Japanese chemists[1] have found that α-phenylthio ketones are desulfurized by zinc and chlorotrimethylsilane in ether as shown in equation (I).

(I)

$\xrightarrow[84\%]{Zn, (CH_3)_3SiCl}$

The mechanism is not known. In any case this method was used in a synthesis of 11-deoxyprostaglandin E and the 8,12-diepimer (equation II).

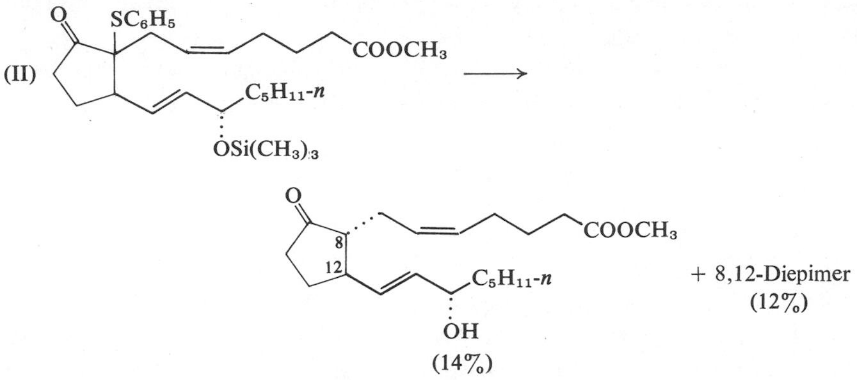

(II)

+ 8,12-Diepimer
(12%)

(14%)

[1] S. Kurozumi, T. Toru, M. Kobayashi, and S. Ishimoto, *Syn. Comm.*, **7**, 427 (1977).

Zinc–Copper couple, 1, 1292–1293; **5,** 758–760; **7,** 428–429.

Reaction of an α,α′-dihaloketone with isoprene.[1] Dehalogenation of the α,α′-dihaloketone (**1**) with a zinc–copper couple on alumina (N_2)[2] in the presence of isoprene results in formation of **2, 3,** and **4** in the approximate ratio 2:2:1. The product (**2**) is a natural terpene, karahaenone. The zinc oxyallyl cation (**a**) is

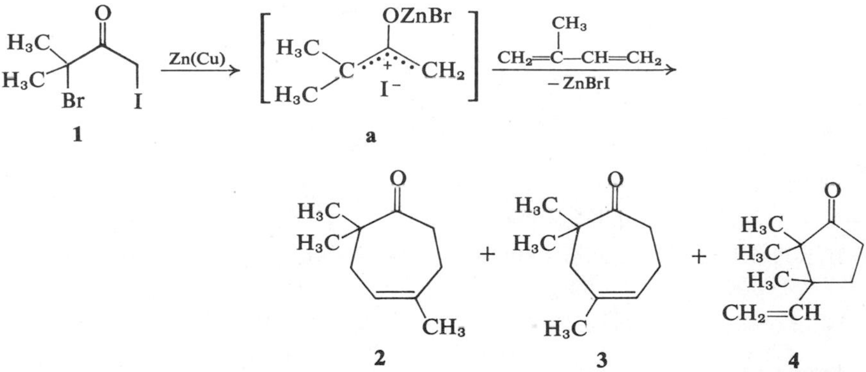

considered a key intermediate. This synthesis is related to the cycloheptenone synthesis of Noyori (**4,** 157–158; **5,** 221–224), in which $Fe_2(CO)_9$ is used for dehalogenation of α,α′-dibromoketones, and indeed use of this iron carbonyl in the above reaction also results in formation of the same three products with a somewhat greater proportion of the five-membered ketone (**4**).

Reformatsky reaction. The Reformatsky reaction of ethyl bromoacetate with several aldehydes and ketones proceeds in improved yields when the zinc is replaced by Le Goff's zinc–copper couple (**1,** 1020). THF is usually superior to ether as solvent.[3]

Reduction of halides. Stephenson and co-workers[4] have prepared a Zn–Cu couple by reaction of zinc dust, suspended in H_2O, with $CuCl_2$ dissolved in dilute hydrochloric acid (O_2-free system). The resulting black solid is washed with H_2O, acetone, and finally ether.

The couple in the presence of water reduces halides in ethereal solvents in moderate to high yields at room temperatures. It has been used primarily to provide deuterated compounds.

[1] R. Chidgey and H. M. R. Hoffmann, *Tetrahedron Letters*, 2633 (1977).
[2] J. G. Vinter and H. M. R. Hoffmann, *Am. Soc.*, **96**, 5466 (1974).
[3] E. Santaniello and A. Manzocchi, *Synthesis*, 698 (1977).
[4] L. M. Stephenson, R. V. Gemmer, and S. P. Current, *J. Org.*, **42**, 212 (1977).

Zinc–Potassium cyanide, 6, 674.

Reduction of $C{\equiv}C$ (6, 674). Reduction of **1** with lithium aluminum hydride–sodium methoxide in THF gives the (E,E)-trienol (**2**) exclusively. Reduction with

zinc in the presence of KCN gives a 1:2 mixture of **2** and **4**. However, if **1** is converted into the silyl ether **3** and then reduced and desilylated, the (Z,E)-trienol **4** is the major product.[1]

[1] W. Oppolzer, C. Fehr, and J. Warneke, *Helv.*, **60**, 48 (1977).

Zinc amalgam, 1, 1287–1289; 2, 462–463.

Reductive cleavage of allylic alcohols, ethers, or acetates (4, 574–575). An example of this reaction has been published in detail in *Org. Syn.* and a number of other examples are tabulated. Formation of the less stable olefin is a result of protonation of an intermediate allylic zinc chloride at the more substituted end of the system.

[1] I. Elphimoff-Felkin and P. Sarda, *Org. Syn.*, **56**, 101 (1977).

Zinc bromide, 2, 463–464.

Cyclization of l-isopulegol (1). This substance is cyclized to *l*-menthol (2) stereoselectively with $ZnBr_2$ in benzene at 5–10°. Previous methods resulted in mixtures that required extensive purification.[1]

1   2 (70%)   3 (trace)

[1] Y. Nakatani and K. Kawashima, *Synthesis*, 147 (1978).

Zinc carbonate, $ZnCO_3$. Mol. wt. 125.38. Preparation.[1]

2,4-Dienones.[2] A new synthesis of 2,4-dienones (4) involves reaction of an allylic alcohol (1) with an allenyl phenyl sulfoxide (2)[3] in benzene (ice bath). The resulting enol ether (3) is treated with this base, which induces Claisen rearrangement and elimination of benzenesulfenic acid, with formation of the dienone 4 (equation I).

[1] C. F. Hüttig, A. Zörner, and O. Hnevkovsky, *Monatsh.*, **72**, 31 (1939).
[2] R. C. Cookson and R. Gopalan, *J.C.S. Chem. Comm.*, 608 (1978).
[3] Preparation: L. Horner and V. Binder, *Ann.*, **757**, 33 (1972).

Zinc chloride, 1, 1289–1292; **2,** 464; **3,** 338; **5,** 763–764; **6,** 676; **7,** 430.

Carbodiimide peptide synthesis. Racemization often presents a problem in the carbodiimide method (**1,** 233). Several additives have been recommended, such as N-hydroxysuccinimide and 1-hydroxybenzotriazole (**3,** 156; **5,** 342). Some Lewis acids have also been found to suppress racemization; $SbCl_3$ is very effective, but yields of peptides are low. The most useful additive appears to be $ZnCl_2$, and this salt also increases the yield.[1]

Oxetanes; β-hydroxy esters. Ketene acetals (**1**) react with aldehydes and ketones in the presence of zinc chloride to form 2,2-dialkoxyoxetanes (**2**). This reaction fails when $R^1 = R^2 = H$.[2] The products are hydrolyzed under weakly acidic conditions to β-hydroxy esters (**3**).[3] The sequence can be carried out in one pot. Yields of **3** are 40–80% when $R^4 = H$. Yields are lower in the case of ketones because the cycloaddition reaction is reversible and faster than hydrolysis with ketones. The method offers an alternative to the Reformatstky reaction, which often gives low yields with simple aldehydes.

Diselenoacetals and diselenoketals.[4] Anhydrous zinc chloride is a mild but very effective reagent for the preparation of diselenoacetals and diselenoketals by the reaction of aldehydes and ketones with alkane- or areneselenols (equation I). Concentrated sulfuric acid is a satisfactory catalyst when RSeH is benzene-selenol (70–75% yields).

(I)

The reaction of an aldehyde and benzeneselenol with gaseous hydrogen bromide in benzene in the presence of anhydrous calcium chloride results in an α-bromoalkyl phenyl selenide (**1**) in high yield. The corresponding anion reacts with aldehydes or ketones to form a β-hydroxy selenide (**2**). which can serve as a precursor to epoxides (**6,** 29) and olefins (**6,** 85–86).

(II) $R^1CHO + C_6H_5SeH + HBr \longrightarrow$ $\underset{\underset{1}{\overset{\displaystyle Br}{|}}}{\overset{\displaystyle H}{\underset{|}{R^1-C-SeC_6H_5}}}$ $\xrightarrow[\text{2) } R^2COR^3]{\text{1) } n\text{-BuLi}}$ $\underset{\underset{2}{\overset{\displaystyle C_6H_5Se\ \ OH}{}}}{\overset{\displaystyle H\ \ R^2}{R^1-C-C-R^3}}$

α,t-*Butyl carboxylic acid esters.* Esters of carboxylic acids (1) can be *t*-butylated in the α-position by conversion to the ketene ketal (2) followed by treatment with *t*-butyl chloride and zinc chloride. One example using adamantyl bromide as the alkylating agent has been reported.[5]

$\underset{R^2}{\overset{R^1}{}}\!\!\diagdown\!\!\underset{}{\text{CHCOOR}} \longrightarrow \underset{R^2}{\overset{R^1}{}}\!\!\diagdown\!\!\text{C}=\text{C}\!\!\overset{OR}{\underset{OSi(CH_3)_3}{\diagup}} \xrightarrow[\text{35-70\%}]{\overset{(CH_3)_3CCl,}{ZnCl_2,\ CH_2Cl_2}} \underset{\underset{3}{\overset{\displaystyle C(CH_3)_3}{|}}}{\overset{\displaystyle R^1}{R^2-\overset{|}{C}-COOR}}$

     1   2   3

[1] H.-D. Jakubke, Ch. Klessen, E. Berger, and K. Neubert, *Tetrahedron Letters*, 1497 (1978).

[2] H. W. Scheeren, R. W. M. Aben, P. H. J. Ooms, and R. J. F. Nivard, *J. Org.*, **42**, 3128 (1977).

[3] R. W. Aben and J. W. Scheeren, *Synthesis*, 400 (1978).

[4] W. Dumont and A. Krief, *Angew. Chem., Int. Ed.*, **16**, 540 (1977); W. Dumont, M. Sevrin, and A. Krief, *ibid.*, **16**, 541 (1977).

[5] M. T. Reetz and K. Schwellnus, *Tetrahedron Letters*, 1455 (1978).

INDEX OF REAGENTS
ACCORDING TO TYPES

thiomaleic anhydride. 1-Phenylseleno-2-trimethylsilyloxy-4-methoxy-1,3-butadiene. Phenyl vinyl sulfoxide. Silver oxide. 1-Trimethylsilyloxy-1,3-butadiene. 2-Trimethylsilyloxy-1,3-cyclohexadiene.

ELBS OXIDATION: Potassium persulfate.
ENAMINES: 2,5-Dimethylpyrrolidine.
ENE REACTIONS: Acetyl hexachloro-antimonate. Aluminum chloride. n-Butyllithium–Tetramethylethylene-diamine.
ENOL LACTONIZATION: Acetic anhy-dride.
EPOXIDATION: Benzeneperoxyseleninic acid. t-Butyl hydroperoxide. Diperoxo-oxohexamethylphosphoramidomolyb-denum. Hydrogen peroxide–Diiso-propylcarbodiimide. 2-Nitrobenzene-seleninic acid. Peracetic acid. Peroxy-acetimidic acid. Quinine. Sodium hypochlorite.
ESTERIFICATION: Cesium carbonate; cesium bicarbonate. Dicyclohexyl-carbodiimide-4-Dimethylaminopyri-dine. Dimethylchloromethylidene-ammonium chloride. Triethyloxonium tetrafluoroborate.

FINKELSTEIN REACTION: Nickel bromide–Zinc.
FISCHER INDOLE SYNTHESIS: Pyri-dinium chloride.
FLUORINATION: Diethylaminosulfur trifluoride. Potassium fluoride.
FORMYLATION: 2-Chloro-1,3-dithiane. Iron carbonyl. Vilsmeier reagent.
FRIES REARRANGEMENT: Aluminum chloride.

GABRIEL SYNTHESIS: t-Butyl methyl iminodicarboxylate potassium salt.
GROB FRAGMENTATION: Copper(I) trifluoromethanesulfonate.

HALLER-BAUER CLEAVAGE: Sodium amide.
HYDROALUMINATION: Lithium aluminum hydride.
HYDROBORATION: 9-Borabicyclo-

[3.3.1]nonane. Diborane. Dibromo-borane–Dimethyl sulfide. Dicyclo-hexylborane. Isopinocampheyl borane. Monobromoborane–Dimethyl sulfide.
HYDROBORATION-AMINATION: Hy-droxyl-O-sulfonic acid.
HYDROCYANATION: Potassium cyanide-Acetone cyanohydrin.
HYDRODEAMINATION: Hydroxylamine-O-sulfonic acid.
HYDROFORMYLATION: Rhodium(III) oxide.
HYDROGENATION, CATALYSTS: (1,5-Cyclooctadiene)bis(methyldiphenyl-phosphine)iridium(I) hexafluorophos-phate. Di-μ-chlorodichlorobis(penta-methylcyclopentadienyl)rhodium. Dichlorotris(triphenylphosphine)-ruthenium(II). Palladium catalysts. Palladium hydroxide. Rhodium.
HYDROXYLATION: Diperoxo-oxohexa-methylphosphoramidomolybdenum-(VI). Hypofluorous acid. Nitrobenzene.
HYDROZIRCONATION: Chlorobis(cyclo-pentadienyl)hydridozirconium.

IODOLACTONIZATION: Iodine.

JAPP-KLINGEMANN REACTION: Ben-zenediazonium tetrafluoroborate.

KNOEVENAGEL CONDENSATION: Ti-tanium(IV) chloride.

LACTAMIZATION: 2-Mesitylenesulfonyl chloride.
LACTONIZATION: Benzeneselenenyl chloride. 2,2′-Dipyridyl disulfide–Triphenylphosphine.

MALONIC ESTER SYNTHESIS: Diethyl malonate.
MANNICH REACTION: Dimethyl-(methylene)ammonium trifluoro-acetate. Formaldehyde.
MARSCHALK CYCLIZATION: Sodium dithionite.
MESYLATION: Methanesulfonyl chloride.
METHYLATION: Methyl fluorosulfonate. Trimethyloxosulfonium hydroxide.

METHYLENATION: μ-Chlorobis(η-cyclo-
pentadienyl)(dimethylaluminum)-μ-
methylenetitanium(I). Methylene
iodide–Zinc–Trimethylaluminum.

NEF REACTION: Sodium hydroxide.

OXIDATION, REAGENTS: 1,1'-(Azodicarbonyl)dipiperidine. Benzeneseleninic
acid. Benzeneseleninic anhydride.
Barium manganate. Bis[3-salicylidene-
aminopropyl]aminecobalt(II). Bis(tri-
n-butyltin) oxide. t-Butyl hydroperoxide. t-Butyl hydroperoxide–Selenium
dioxide. Caro's acid. Ceric ammonium
sulfate. Chromic acid–3,5-Dimethyl-
pyrazole. Chromic acid–Silica gel. 1,4-
Diazabicyclo[2.2.2]octane–Bromine.
2,3-Dichloro-5,6-dicyano-1,4-benzo-
quinone. Dimethyl selenoxide. Dimethyl sulfoxide–Oxalyl chloride.
3,5-Dinitrobenzoyl t-butyl nitroxyl.
Hydrogen peroxide. Hydrogen peroxide–Diisopropylcarbodiimide. Hydrogen peroxide–Sodium peroxide.
Iodine pentafluoride. Lead tetraacetate. Manganese dioxide. Mercuric
oxide. Nickel peroxide. Oxygen.
Ozone–Silica gel. Potassium dichromate. Potassium ferricyanide. Potassium permanganate. Potassium persulfate. Potassium superoxide. Pyridinium chlorochromate. Silver carbonate–
Celite. Tetra-n-butylammonium permanganate. Thallium(III) nitrate.
Thallium(III) perchlorate. Trimethylamine N-oxide. Trityl tetrafluoroborate. Vanadyl trifluoride.
OXIDATIVE BISDECARBOXYLATION:
Dicarbonylbis(triphenylphosphine)-
nickel.
OXIDATIVE CLEAVAGE:
ALKYL METHYL ETHERS: Nitronium
tetrafluoroborate.
N,N-DIMETHYLHYDRAZONES: Oxygen, singlet.
HYDRAZONES: Cobalt(III) fluoride.
OXIMES; TOSYLHYDRAZONES: Ben-
zeneseleninic anhydride.
OXIDATIVE COUPLING: Thallium(III) tri-
fluoroacetate. Vanadyl trichloride.

OXIDATIVE CYCLIZATION: Pyridinium
chlorochromate.
OXIDATIVE DECARBOXYLATION: N-
Chlorosuccinimide. Lead tetraacetate.
OXIDATIVE DEOXIMATION: Pyridinium
chlorochromate.
OXIDATIVE PHENOL-BENZYL COU-
PLING: Thallium(III) trifluoroacetate.
OXYAMINATION: Osmium tetroxide-
t-Butyl-N-argentocarbamate. Osmium
tetroxide-Chloramine-T. Trioxo(t-
butylnitride)osmium(VIII).
OXY-COPE REARRANGEMENT: Potas-
sium hydride.
OXYSELENATION: Acetoxymethyl
methyl selenide. Copper(II) chloride.

PERMETHYLATION:
KETONES: Potassium hydride–Methyl
iodide.
PFITZNER-MOFFATT OXIDATION: Dimethyl sulfoxide. Polymeric carbodiimide reagent.
PHOSPHORYLATION: Di-t-butyl phosphorobromidate. 4-Nitrophenyl phenyl
phosphorochloridate.
PROTECTION OF:
AMINO GROUPS: 9-Anthrylmethyl p-
nitrophenyl carbonate. 2-[(Chloroformyl)oxy]-ethyltriphenylphos-
phonium chloride. Diborane. 2-Nitro-
benzenesulfenyl chloride. 2,2,2-Tri-
chloro-t-butyloxycarbonyl chloride.
Vinyl chloroformate.
CARBOXYLIC GROUPS: Chloromethyl
methyl sulfide. 2-Hydroxymethyl-
anthraquinone. p-Nitrobenzyl bromide.
2-Trimethylsilylethanol.
β-DIKETONES: Dicyanodimethylsilane.
ENEDIOLS: Sulfur trioxide–Trimethylamine.
ENONES: Dimethylphenylsilyllithium.
HEMIACETAL GROUP: Pyridinium p-
toluenesulfonate.
HYDROXYL GROUPS: Crotonyl anhydride. Pivaloyl chloride. Pyridinium p-
toluenesulfonate. Vinyl chloroformate.
γ-LACTONES: m-Chloroperbenzoic acid.
PHENOLS: Chloromethyl methyl sulfide.
Pivaloyl chloride.

ACYL ISOTHIOCYANATES: Triphenylphosphine–Thiocyanogen.
ALCOHOLS: *m*-Chloroperbenzoic acid.
ALDEHYDES: 1,3-Benzodithiolylium
perchlorate. Bis(triphenylphosphine)-
copper tetrahydroborate. 5-Bromo-
1,4-diphenyl-3-methylthio-5-triazolium
bromide. Dimethyl sulfoxide. 2-
Ethoxyvinyllithium.
ALDEHYDES: Lithium methylthioformaldine. 2-(N-Methyl-N-formyl)-
aminopyridine. Methyl methylthio
sulfoxide. 2-Methylthio-1,4-diphenyl-
5-triazolium iodide. Sodium 4,6-
diphenyl-1-oxido-2-pyridone. Tetramethylammonium tetracarbonylhydridoferrate.
ALKENES: Calcium amalgam. N,N-
Dimethylformamide dimethyl acetal.
N,N-Methylphenylaminotriphenyl-
phosphonium iodide. N,N,N',N',-Tetra-
methyldiamidophosphorochloridate.
Titanium(O). *p*-Toluenesulfonylhydra-
zine. Tri-*n*-butyltin hydride. Triphenylphosphine–Diethyl azodicarboxylate.
Trityl tetrafluoroborate.
1-ALKENES: Dimethyl(methylene)-
ammonium salts. Methylene bromide–
Zinc–Titanium(IV) chloride. Methylene
bromide–Zinc–Trimethylaluminium.
ALKENYLARENES: Chlorobis(cyclopentadienyl)hydridozirconium.
vic-ALKOXYIODOALKANES: Iodine–
Copper(II) acetate.
ALKYL ARYL KETONES: 2-Trifluoro-
methanesulfonyloxypyridine.
ALKYL BROMIDES: *N*-Bromosuccini-
mide.
ALKYL CHLORIDES: 2-Chloro-3-
ethylbenzoxazolium tetrafluoroborate.
AKLYL HALIDES: 3-Ethyl-2-fluoro-
benzothiazolium tetrafluoroborate.
2-ALKYL-5-HYDROXYCYCLOPEN-
TENE-2-ONES: Acetic acid.
γ-ALKYLIDENEBUTENOLIDES: *t*-
Butoxyfurane.
ALKYL IODIDES: Iodotrimethylsilane.
ALKYNES: Diethyl phosphorochloridate. Lithium acetylides.

1-ALKYNES: Lithium aluminum hydride. Potassium 3-aminopropylamide.
1-ALKYNYL KETONES: Dichlorobis-
(triphenylphosphine)palladium(II).
ALLENES: α-Bromovinyltrimethylsilane. Lithium acetylide. Lithium
aluminum hydride. 1-(1-Naphthyl)-
ethyl isocyanate.
ALLENIC ALDEHYDES, KETONES: Di-
methylformanide.
ALLENIC AMIDES: Diethylformamide
diethyl acetal.
ALLYLIC ALCOHOLS: B-1-Alkenyl-9-
borabicyclo[3.3.1]nonanes. Benzeneselenenic acid. Borane–Dimethyl
sulfide.
ALLYLIC CHLORIDES: Triphenylphosphine–Carbon tetrachloride.
ALLYLIC HALIDES: Benzeneselenenyl
bromide(chloride).
π-ALLYLPALLADIUM COMPLEXES:
Disodium tetrachloropalladate.
AMIDES: Aluminum chloride–Ethanol.
Diethyl phosphorobromidate. Dimethylaluminum amides. 2-Fluoro-
1,3,5-trinitrobenzene. 4-(4'-Methyl-1'-
piperazinyl)-3-butyne-2-one.
AMINES: Iodotrimethylsilane. Lithium
aluminum hydride.
AMINO ACID ESTERS: Cesium
carbonate.
BOC-AMINO ACIDS: *t*-Butyl methyl
iminodicarboxylate potassium salt.
AMINOPHENOLS: Benzeneseleninic
anhydride.
ANTHRAQUINONES: 6-Methoxy-4-
methyl-2-pyrone.
ANTHRONES: *n*-Butyllithium–Tetra-
methylethylenediamine.
ARENE IMINES: Triphenylphosphine–
Carbon tetrachloride.
ARENE OXIDES: *m*-Chloroperbenzoic
acid. Hydrogen peroxide–Diisopropylcarbodiimide. Sodium hypochlorite.
AROYL CHLORIDES: Oxalyl choride–
Aluminum chloride.
ARYLACETYLENES: Chloro-
methylenetriphenylphosphonium
iodide.
α-ARYL CARBOXYLIC ACIDS: Di-
phenylphosphoroazidate.

α-PHENYLSELENO KETONES: Diphenyl diselenide–Bromine–Hexabutyldistannoxane.

PHENYLSELENOIMINES: Benzeneseleninic anhydride.

PHENYLSELENOLACTONES: Benzeneselenenyl chloride.

QUINONE KETALS: 2-Lithio-3,3,6,6-tetramethoxy-1,4-cyclohexadiene.

QUINONE METHIDE KETALS: N,N-Dimethyltrimethylsilylacetamide.

o-QUINONE METHIDES: Silver oxide.

RESORCINOLS: Methyl phenylsulfinylacetate.

RUBANES: Quinuclidine N-oxide.

SELENOACETALS: Tris(phenylseleno)borane.

SELENOCYANATES: Potassium selenocyanate.

SILYL ENOL ETHERS: Dodecamethylcyclohexasilane.

STYRENES: 4-Methyl-2,6-di-t-butylpyridine.

SULFINES: Sulfur dioxide.

SULFINIC ACIDS: N-Hydroxymethylphthalimide.

TETRAHYDROPYRANYL ETHERS: Pyridinium p-toluenesulfonate.

TETRALONES: Phthalide.

1,3-THIAZOLES: Tosylmethyl isocyanide.

THIOCYANATES: Triphenylphosphine-Thiocyanogen. 2,4,6-Triphenylpyrylium thiocyanate.

THIOL (SELENOL) ESTERS: N,N′-Carbodiimidazole. 4-Dimethylamino-3-butyne-2-one.

THIOESTERS: Tris(ethylthio)borane.

THIOETHERS: N-(Phenylthio)succinimide.

THIONITRITES: Dinitrogen tetroxide.

TRIMETHYLSILYLARENES: Trimethylsilypotassium.

TROPOLONES: Dimethyloxosulfonium methylide.

α,β-UNSATURATED ACIDS: Diethyl carboxymethanephosphonate. Diethyl trimethylsilyloxycarbonylmethanephosphonate.

α,β-UNSATURATED ALDEHYDES: Chloromethyl phenyl sulfoxide. N,N-Dimethylhydrazine. N,N-Dimethylthiocarbamoyl chloride. (Z)-2-Ethoxyvinyllithium. Lithium di-(E)-propenylcuprate. Oxygen, singlet. Methoxyallene. Methoxyallidenetriphenylphosphorane. 3-Methoxy-1-phenylthio-1-propene. Pyridinium chlorochromate.

α,β-UNSATURATED CYCLOHEXENONES: 2-Ethoxyallylidenetriphenylphosphorane. Oxygen, singlet.

α,β-UNSATURATED ESTERS: Oxygen, singlet. Phosphoryl chloride.

β,γ-UNSATURATED ESTERS: p-Toluenesulfonylhydrazine.

α,β-UNSATURATED KETONES: Benzeneselenenyl bromide. N,N-Dimethylthiocarbamoyl chloride. 2-Ethoxyvinyllithium. Palladium(II)acetate. Palladium(II) chloride. Trityl tetrafluoroborate.

β,γ-UNSATURATED KETONES: Acetyl hexachloroantimonate. Chloro-trimethylsilane.

γ,δ-UNSATURATED KETONES: Chlorobis(cyclopentadienyl)hydridozirconium.

α,β-UNSATURATED LACTONES: Phosphoryl chloride.

β,γ-UNSATURATED δ-LACTONES: Diethyl ketomalonate.

α,β-UNSATURATED NITRILES: Cyanomethyldiphenylphosphine oxide. Methylthioacetonitrile. Tetrakis(triphenylphosphine)palladium(O).

β,γ-UNSATURATED NITRILES: Methylthioacetonitrile.

α,β-UNSATURATED SULFIDES, SULFOXIDES, SULFONES: Benzeneselenenyl chloride.

γ,δ-UNSATURATED SULFONES: Palladium(II) chloride.

UREAS: Dicyclohexylcarbodiimide.

URETHANES: Methyl (carboxysulfamoyl)triethylammonium hydroxide inner salt.

VINYLACETYLENES: Chlorotris(triphenylphosphine)rhodium.

VINYL BROMIDES: N-Bromosuccinimide. Lithio-α-chloromethyltrimethylsilane.

AUTHOR INDEX

SUBJECT INDEX

Page numbers referring to reagents are indicated in **boldface**.